# Communications
# in Computer and Information Science 1059

*Commenced Publication in 2007*
Founding and Former Series Editors:
Phoebe Chen, Alfredo Cuzzocrea, Xiaoyong Du, Orhun Kara, Ting Liu,
Krishna M. Sivalingam, Dominik Ślęzak, Takashi Washio, and Xiaokang Yang

## Editorial Board Members

More information about this series at http://www.springer.com/series/7899

Rui Mao · Hongzhi Wang ·
Xiaolan Xie · Zeguang Lu (Eds.)

# Data Science

5th International Conference
of Pioneering Computer Scientists,
Engineers and Educators, ICPCSEE 2019
Guilin, China, September 20–23, 2019
Proceedings, Part II

 Springer

*Editors*
Rui Mao
College of Computer Science
and Software Engineering
Shenzhen University
Shenzhen, China

Xiaolan Xie
Guilin University of Technology
Guilin, China

Hongzhi Wang
Harbin Institute of Technology
Harbin, China

Zeguang Lu
National Academy of Guo Ding
Institute of Data Science
Harbin, China

ISSN 1865-0929           ISSN 1865-0937   (electronic)
Communications in Computer and Information Science
ISBN 978-981-15-0120-3      ISBN 978-981-15-0121-0   (eBook)
https://doi.org/10.1007/978-981-15-0121-0

This Springer imprint is published by the registered company Springer Nature Singapore Pte Ltd.
The registered company address is: 152 Beach Road, #21-01/04 Gateway East, Singapore 189721, Singapore

# Preface

As the program chairs of the 5th International Conference of Pioneer Computer Scientists, Engineers and Educators 2019 (ICPCSEE 2019, originally ICYCSEE), it is our great pleasure to welcome you to the proceedings of the conference, which was held in Guilin, China, September 20–23, 2019, hosted by Guilin University of Technology, Guilin University of Electronic Technology, and National Academy of Guo Ding Institute of Data Science. The goal of this conference was to provide a forum for computer scientists, engineers, and educators.

The call for papers of this year's conference attracted 395 paper submissions. After the hard work of the Program Committee, 104 papers were accepted to appear in the conference proceedings, with an acceptance rate of 26.4%. The major topic of this conference was data science. The accepted papers cover a wide range of areas related to Basic Theory and Techniques for Data Science including Mathematical Issues in Data Science, Computational Theory for Data Science, Big Data Management and Applications, Data Quality and Data Preparation, Evaluation and Measurement in Data Science, Data Visualization, Big Data Mining and Knowledge Management, Infrastructure for Data Science, Machine Learning for Data Science, Data Security and Privacy, Applications of Data Science, Case Study of Data Science, Multimedia Data Management and Analysis, Data-Driven Scientific Research, Data-Driven Bioinformatics, Data-Driven Healthcare, Data-Driven Management, Data-Driven eGovernment, Data-Driven Smart City/Planet, Data Marketing and Economics, Social Media and Recommendation Systems, Data-Driven Security, Data-Driven Business Model Innovation, and Social and/or Organizational Impacts of Data Science.

We would like to thank all the Program Committee members, 203 coming from 93 institutes, for their hard work in completing the review tasks. Their collective efforts made it possible to attain quality reviews for all the submissions within a few weeks. Their diverse expertise in each individual research area helped us to create an exciting program for the conference. Their comments and advice helped the authors to improve the quality of their papers and gain deeper insights.

Great thanks should also go to the authors and participants for their tremendous support in making the conference a success.

We thank Dr. Lanlan Chang and Jane Li from Springer, whose professional assistance was invaluable in the production of the proceedings.

Besides the technical program, this year ICPCSEE offered different experiences to the participants. We hope you enjoy the conference proceedings.

June 2019

Xiaohui Cheng
Rui Mao

# Organization

The 5th International Conference of Pioneering Computer Scientists, Engineers and Educators (ICPCSEE, originally ICYCSEE) 2019 (http://2019.icpcsee.org) was held in Guilin, China, during September 20–23 2019, hosted by Guilin university of Technology, Guilin University of Electronic Technology, and National Academy of Guo Ding Institute of Data Science.

## General Chair

Mei Wang                          Guilin University of Technology, China

## Program Chairs

Xiaohui Cheng                     Guilin University of Technology, China
Rui Mao                           Shenzhen University, China

## Program Co-chairs

XianXian Li                       Guangxi Normal University, China
Liang Chang                       Guilin University of Electronic Technology, China
Wei Li                            Central Queensland University, Australia
Goi Bok Min                       Universiti Tunku Abdul Rahman (UTAR), Malaysia
Xiaohua Ke                        Guangdong University of Foreign Studies, China

## Organization Chairs

Xiaolian Xie                      Guilin University of Technology, China
Yong Ding                         Guilin University of Electronic Technology, China
Zhenjun Tang                      Guangxi Normal University, China
Donghong Qin                      GuangXi University for Nationalities, China
Zeguang Lu                        National Academy of Guo Ding Institute
                                     of Data Science, China

## Organization Co-chairs

Chao Jing                         Guilin University of Technology, China
Qin Yang                          Guilin University of Technology, China
Yu Wang                           Guilin University of Technology, China
Hanying Liu                       Guilin University of Technology, China
Shouxue Chen                      Guilin University of Technology, China
Jili Chen                         Guilin University of Technology, China

| | |
|---|---|
| Wenpeng Chen | Guilin University of Technology, China |
| Pinle Qin | North University of China, China |
| Jianhou Gan | Yunnan Normal University, China |
| Mingrui Chen | Hainan University, China |

## Publication Chairs

| | |
|---|---|
| Hongzhi Wang | Harbin Institute of Technology, China |
| Guanglu Sun | Harbin University of Science and Technology, China |

## Publication Co-chairs

| | |
|---|---|
| Weipeng Jing | Northeast Forestry University, China |
| Xianhua Song | Harbin University of Science and Technology, China |
| Xie Wei | Harbin University of Science and Technology, China |
| Guoyong Cai | Guilin University of Electronic Technology, China |
| Minggang Dong | Guilin University of Technology, China |
| Canlong Zhang | Guangxi Normal University, China |

## Forum One Chairs

| | |
|---|---|
| Fudong Liu | Information Engineering University, China |
| Feng Wang | RoarPanda Network Technology Co., Ltd., China |

## Forum Two Chairs

| | |
|---|---|
| Pinle Qin | Zhongbei University, China |
| Haiwei Pan | Harbin Engineering University, China |

## Forum Three Chairs

| | |
|---|---|
| Jian Wang | RADI, CAS, China |
| Weipeng Jing | Northeast Forestry University, China |

## Forum Four Chairs

| | |
|---|---|
| Liehuang Zhu | Beijing Institute of Technology, China |
| Yong Ding | Guilin University of Electronic Science and Technology, China |

## Forum Five Chairs

| | |
|---|---|
| Junyu Lin | Information Institute of Chinese Academy of Sciences, China |
| Haofen Wang | Shanghai Leyan Technologies Co. Ltd., China |

## Forum Six Chair

Qiguang Miao                Xidian University, China

## Forum Seven Chair

Canlong Zhang              Guangxi Normal University, China

## Education Chairs

Xiaomei Tao                Guilin University of Technology, China
Hui Li                     Guangxi University of Technology, China

## Industrial Chairs

Li'e Wang                  Guangxi Normal University, China
Zheng Shan                 Information Engineering University, China

## Demo Chairs

Li Ma                      Guilin Medical University, China
Xia Liu                    Sanya Aviation and Tourism College, China

## Panel Chairs

Yun Deng                   Guilin University of Technology, China
Rifeng Wang                Guangxi University of Technology, China
Peng Liu                   Guangxi Normal University, China

## Post Chair

Panfeng Zhang              Guilin University of Technology, China

## Expo Chairs

Chaoquan Chen              Guilin University of Technology, China
Jingli Wu                  Guangxi Normal University, China
Jingwei Zhang              Guilin University of Electronic Technology, China

## Registration/Financial Chair

Chunyan Hu                 National Academy of Guo Ding Institute
                             of Data Science, China

# ICPCSEE Steering Committee

| | |
|---|---|
| Jiajun Bu | Zhejiang University, China |
| Wanxiang Che | Harbin Institute of Technology, China |
| Jian Chen | ParaTera, China |
| Wenguang Chen | Tsinghua University, China |
| Xuebin Chen | North China University of Science and Technology, China |
| Xiaoju Dong | Shanghai Jiao Tong University, China |
| Qilong Han | Harbin Engineering University, China |
| Yiliang Han | Engineering University of CAPF, China |
| Yinhe Han | Institute of Computing Technology, Chinese Academy of Sciences, China |
| Hai Jin | Huazhong University of Science and Technology, China |
| Weipeng Jing | Northeast Forestry University, China |
| Wei Li | Central Queensland University, Australia |
| Min Li | Central South University, China |
| Junyu Lin | Institute of Information Engineering, Chinese Academy of Sciences, China |
| Yunhao Liu | Michigan State University, China |
| Zeguang Lu | National Academy of Guo Ding Institute of Data Sciences, China |
| Rui Mao | Shenzhen University, China |
| Qi Guang Miao | Xidian University, China |
| Haiwei Pan | Harbin Engineering University, China |
| Pinle Qin | North University of China, China |
| Zhaowen Qiu | Northeast Forestry University, China |
| Zheng Shan | The PLA Information Engineering University, China |
| Guanglu Sun | Harbin University of Science and Technology, China |
| Jie Tang | Tsinghua University, China |
| Tian Feng | Institute of Software Chinese Academy of Sciences, China |
| Tao Wang | Peking University, China |
| Hongzhi Wang | Harbin Institute of Technology, China |
| Xiaohui Wei | Jilin University, China |
| lifang Wen | Beijing Huazhang Graphics and Information Co., Ltd., China |
| Liang Xiao | Nanjing University of Science and Technology, China |
| Yu Yao | Northeastern University, China |
| Xiaoru Yuan | Peking University, China |
| Yingtao Zhang | Harbin Institute of Technology, China |
| Yunquan Zhang | Institute of Computing Technology, Chinese Academy of Sciences, China |
| Baokang Zhao | National University of Defense Technology, China |

| | |
|---|---|
| Min Zhu | Sichuan University, China |
| Liehuang Zhu | Beijing Institute of Technology, China |

## ICPCSEE 2019 Program Committee Members

### Program Committee Area Chairs

| | |
|---|---|
| Wanxiang Che | Harbin Institute of Technology, China |
| Cheng Feng | Northeast Forestry University, China |
| Min Li | Central South University, China |
| Fudong Liu | State Key Laboratory of Mathematical Engineering Advanced Computing, China |
| Zeguang Lu | National Academy of Guo Ding Institute of Data Science, China |
| Rui Mao | Shenzhen University, China |
| Qiguang Miao | Xidian University, China |
| Haiwei Pan | Harbin Engineering University, China |
| Qin Pinle | North University of China, China |
| Zheng Shan | State Key Laboratory of Mathematical Engineering Advanced Computing, China |
| Guanglu Sun | Harbin University of Science and Technology, China |
| Hongzhi Wang | Harbin Institute of Technology, China |
| Yuzhuo Wang | Harbin Institute of Technology, China |
| Xiaolan Xie | Guilin University of Technology, China |
| Yingtao Zhang | Harbin Institute of Technology, China |

### Program Committee Members

| | |
|---|---|
| Chunyu Ai | University of South Carolina Upstate, USA |
| Zhipeng Cai | Georgia State University, USA |
| Richard Chbeir | LIUPPA Laboratory, France |
| Zhuang Chen | Guilin University of Electronic Technology, China |
| Vincenzo Deufemia | University of Salerno, Italy |
| Minggang Dong | Guilin University of Technology, China |
| Longxu Dou | Harbin Institute of Technology, China |
| Pufeng Du | Tianjin University, China |
| Zherui Fan | Xidian University, China |
| Yongkang Fu | Xidian University, China |
| Shuolin Gao | Harbin Institute of Technology, China |
| Daohui Ge | Xidian University, China |
| Yingkai Guo | National University of Singapore, Singapore |
| Meng Han | Georgia State University, USA |
| Meng Han | Kennesaw State University, USA |
| Qinglai He | Arizona State University, USA |
| Tieke He | Nanjing University, China |
| Zhixue He | North China Institute of Aerospace Engineering, China |

Tao He                  Harbin Institute of Technology, China
Yutai Hou               Harbin Institute of Technology, China
Xu Hu                   Xidian University, China
Kuan Huang              Utah State University, USA
Hekai Huang             Harbin Institute of Technology, China
Cun Ji                  Shandong Normal University, China
Xiaoyan Jiang           Shanghai University of Engineering Science, China
Wanchun Jiang           Central South University, China
Xin Jin                 Beijing Electronic Science and Technology Institute,
                            China
Chao Jing               Guilin University of Technology, China
Wei Lan                 Guangxi University, China
Mingzhao Li             RMIT University, Australia
Wei Li                  Georgia State University, USA
Yunan Li                Xidian University, China
Hongdong Li             Central South University, China
Xiangtao Li             Northeast Normal University, China
Xia Liu                 Sanya Aviation Tourism College, China
Yarong Liu              Guilin University of Technology, China
Shuaiqi Liu             Tianjin Normal University, China
Yan Liu                 Harbin Institute of Technology, China
Jin Liu                 Central South University, China
Yijia Liu               Harbin Institute of Technology, China
Zeming Liu              Harbin Institute of Technology, China
Mingming Lu             Central South University, China
Junwei Luo              Henan Polytechnic University, China
Zhiqiang Ma             Inner Mongolia University of Technology, China
Chenggang Mi            Northwestern Polytechnical University, China
Xiangda Qi              Xidian University, China
Libo Qin                Research Center for Social Computing and Information
                            Retrieval, China
Chang Ruan              Central South University, China
Yingshan Shen           South China Normal University, China
Meng Shen               Xidian University, China
Feng Shi                Central South University, China
Yuanyuan Shi            Xi'an University of Electronic Science
                            and Technology, China
Xiaoming Shi            Harbin Institute of Technology, China
Shoubao Su              Jinling Institute of Technology, China
Dechuan Teng            Harbin Institute of Technology, China
Vicenc Torra            Högskolan i Skövde, Sweden
Qingshan Wang           Hefei University of Technology, China
Wenfeng Wang            Chinese Academy of Sciences, China
Shaolei Wang            Harbin Institute of Technology, China
Yaqing Wang             Xidian University, China
Yuxuan Wang             Harbin Institute of Technology, China

| | |
|---|---|
| Wei Wei | Xi'an Jiaotong University, China |
| Haoyang Wen | Harbin Institute of Technology, China |
| Huaming Wu | Tianjin University, China |
| Bin Wu | Institute of Information Engineering, Chinese Academy of Sciences, China |
| Yue Wu | Xidian University, China |
| Min Xian | Utah State University, USA |
| Wentian Xin | Xidian University, China |
| Qingzheng Xu | National University of Defense Technology, China |
| Yang Xu | Harbin Institute of Technology, China |
| Yu Yan | Harbin Institution of Technology, China |
| Cheng Yan | Central South University, China |
| Shiqin Yang | Xidian University, China |
| Jinguo You | Kunming University of Science and Technology, China |
| Lei Yu | Georgia Institute of Technology, USA |
| Xiaoyi Yu | Peking University, China |
| Yue Yue | SUTD, Singapore |
| Boyu Zhang | Utah State University, USA |
| Wenjie Zhang | The University of New South Wales, Australia |
| Jin Zhang | Beijing Normal University, China |
| Dejun Zhang | China University of Geosciences, China |
| Zhifei Zhang | Tongji University, China |
| Shigeng Zhang | Central South University, China |
| Mengyi Zhang | Harbin Institute of Technology, China |
| Yongqing Zhang | Chengdu University of Information Technology, China |
| Xiangxi Zhang | Harbin Institute of Technology, China |
| Meiyang Zhang | Southwest University, China |
| Zhen Zhang | Xidian University, China |
| Peipei Zhao | Xidian University, China |
| Bo Zheng | Harbin Institute of Technology, China |
| Jiancheng Zhong | Central South University, China |
| Yungang Zhu | Jilin University, China |
| Bing Zhu | Central South University, China |

# Contents – Part II

## System

## Education

## Application

# Contents – Part I

**Network**

**Security**

## Machine Learning

# Bioinformatics

# Survey on Deep Learning for Human Action Recognition

Zirui Qiu[1], Jun Sun[1], Mingyue Guo[1], Mantao Wang[1(✉)], and Dejun Zhang[2]

[1] Sichuan Agricultural University, Yaan 625014, China
wangmantao@sicau.edu.cn
[2] China University of Geosciences, Wuhan 430074, China

**Abstract.** Human action recognition has gained popularity because of its worldwide applications such as video surveillance, video retrieval and human–computer interaction. This paper provides a comprehensive overview of notable advances made by deep neural networks in this field. Firstly, the basic conception of action recognition and its common applications were introduced. Secondly, action recognition was categorized as action classification and action detection according to its respective research goals. And various deep learning frameworks for recognition tasks were discussed in detail and the most challenging datasets and taxonomies were briefly reviewed. Finally, the limitations of the state-of-the-art and promising directions of the research were briefly outlined.

**Keywords:** Action recognition · Deep neural network · Action classification · Action detection

## 1 Introduction

Human action recognition is a multi-disciplinary research which involved computer vision, machine learning and artificial intelligence. Its achievements are commonly used in smart video surveillance, multimedia content understanding, virtual reality, human-computer interaction and so on. Therefore, researches with respect to the key technologies of human action recognition in video have significant academic importance and strong application value.

One of the simplest action recognition tasks is called action classification, which refers to classify and identify the actions being performed by humans in an unknown video with the help of machine learning and pattern recognition. Traditional action recognition algorithms generally represent the actions through hand-crafted features (e.g. iDT [56, 57]), and use classifiers such as support vector machines, decision trees or random forest to discriminate action representations. Besides, action detection is a relatively complex action recognition task. It not only needs to classify the actions appearing in the video, but also needs to estimate the time interval and spatial extent of the action in the video sequences.

With the remarkable progress in image recognition [24, 27, 44] and natural language processing [17, 18, 65] that achieved by deep learning in recent years, the deep

R. Mao et al. (Eds.): ICPCSEE 2019, CCIS 1059, pp. 3–21, 2019.
https://doi.org/10.1007/978-981-15-0121-0_1

learning based human action recognition methods have become the major research direction and gradually replaced the traditional approaches that extract the hand-crafted features. This paper will focus on deep learning and cover the related work of video action recognition from the two aspects of action classification and action detection. To have a glance, the main contents have been summarized in Fig. 1.

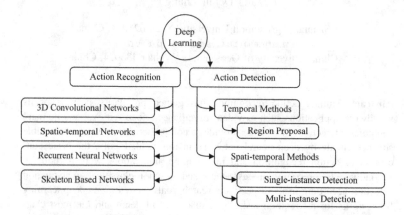

**Fig. 1.** Taxonomy of action recognition based on deep learning methods

## 2    Action Classification

Action classification based on video mainly solves two problems: (1) how to extract the appearance information of a moving human body and (2) how to simulate the temporal dynamics of actions.

Convolutional neural network (CNN/ConvNet), which has the characteristics of local perception, weight sharing and pooling, can capture complex spatial appearance information from images and has good performance in image classification tasks [24, 44, 45, 47, 64]. One of the direct way to apply convolutional network to a video recognition task is that separate the video sequence into single still images, and then use a convolutional network to perform feature extraction on each single video frame [23, 66]. However, roughly treating the video as static images will cause the loss of temporal context information in the video sequences, so the model cannot discriminate between highly similar actions such as "opening the door" and "closing the door".

Therefore, how to model the temporal context information of a video is the most important problem faced by the current action classification algorithms. On this basis, this paper divides the human action classification methods (i.e., the temporal context modeling methods) into four categories.

1. Action classification based on 3D convolutional networks.
2. Action classification based on spatio-temporal two-stream networks.
3. Action classification based on recurrent neural networks.
4. Action classification based on skeleton sequences.

## 2.1 Action Classification Based on 3D Convolutional Networks

In order to effectively utilize the temporal context information of the video sequence, the literature [1, 12, 21, 23, 52] introduced a 3D ConvNets model. That is to say, the 2D convolution operation can be extended to 3D by stacking consecutive video frames, and then use 3D convolution kernel to extract the temporal and spatial hierarchical feature representation of the video data. A comparison of 2D convolution and 3D convolution is shown in Fig. 2, in which the same type of connection represents the shared weight. Multiple 3D convolution kernels are stacked on the same convolutional layer to capture different feature representations in the same spatial location.

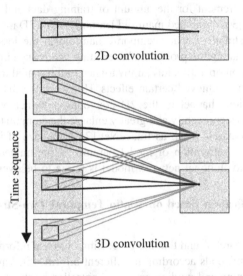

**Fig. 2.** Comparison between 2D convolution and 3D convolution [21]

Ji et al. [21] were the first people who proposed an action classification approach based on 3D ConvNets. Seven consecutive frames of images are randomly selected from the video sequence, and a hand-wired data processing method is designed to generated 33 feature channels (such as gray-scale and optical flow map) from the original image which become the input to a convolution network. Their algorithm uses an idea of group convolution to abstract the high-level feature expressions from different types of feature channels in stages. Finally, all features are fused and classified with the fully connection layer. The method is applicable to scenes such as noise, background interference, and occlusion. The experimental results show its effectiveness in real environments. In order to improve the hand-wired data processing method in literature [21], Baccouche et al. [1] presented a fully automatic feature learning framework.

The action classification method which based on 3D ConvNet was applied to large-scale video datasets by Karpathy et al. [23], and they found that the slow fusion outperforms other temporal domain fusion methods. Besides, the training time reduced to one quarter to one half by constructing a multiresolution 3D ConvNets model. Tran et al. [52] analyzed the influence of the 3D convolution kernel size on algorithm performance. It has been proved by numerous experiments that the best performance can be obtained by the $3 \times 3 \times 3$ convolution kernels, and then comes the design of C3D (Convolutional 3D) architectures. The model is simple to implement, easy to train, and can be directly applied to different visual tasks such as scene and object recognition and action similarity labeling.

Sun et al. [50] suggest that factorizing the original 3D convolution kernel to a 2D spatial convolution kernel and a 1D temporal kernels. The lower layers use the 2D kernels to extract the spatial features of the video, and the upper network uses 1D convolution to verify the feature fusion on the time domain. The factorization operation of the 3D convolution kernel not only reduces the time complexity of the model, but also reduces the requirement for the amount of training data and reduces the risk of over-fitting. Li et al. [30] combined manifold learning with C3D model [52], and added a regularization constraint of spatio-temporal manifold in the loss function of paper [52], effectively alleviating the problem of overfitting and reduce intra-class variations.

The 3D convolution operation has an invariance to the spatial transformation in the time sequence, and has achieved certain effects. However, the 3D convolution kernel adds a time dimension channel to the 2D convolution kernel, which increases the number of parameters significantly and greatly enlarge the computational cost. Besides, compared to image datasets (such as ImageNet [24]), the size of the video dataset is small. Thus, the direct migration of model parameters pre-trained on the 2D image dataset to the 3D convolutional network model is prone to be overfitting.

## 2.2 Action Classification Based on Spatio-Temporal Two-Stream Networks

Visual perception research found that the processing of visual information can take two different processing methods according to different information functions [52]. Visual information for adjusting and guiding action is controlled by the dorsal channel, while visual information for description and cognitive judgment is controlled by the ventral channel. The two are independent of one another, and the injury of one channel does not affect the normal operation of the other stream.

Simonyan et al. [43] applied the visual two-streams hypothesis to the research of video action classification, and constructed two independent convolutional architectures to extract the spatial and temporal features of the video. First, a frame of still image is randomly selected from the video sequence as an input of the spatial network, and the object's appearance feature of interest is extracted. Then, the optical flow information (the optical flow 10 frames before and after the image) associated with this frame of image is input to the temporal network, in order to extract motion features. The spatial network architecture is similar to the image recognition network architecture, so it can be pre-trained on the ImageNet dataset [24]. The two-stream convolutional framework [43] has a shallow network layer, combined with a multi-task learning strategy, can achieve better performance under the condition of small dataset.

Feichtenhofer et al. [12] extended the 2D two-stream convolution architecture [43] to 3D and designed the 3D fusion operation in the spatio-temporal domain. The experimental results show that 3D convolution fusion at the top of the last convolutional layer of the network can significantly improve the performance of the algorithm and shorten the training time of the parameters.

A trajectory-pooled deep convolution descriptors (TDD) is proposed by Wang et al. [58]. Firstly, the two-stream convolutional network is trained on a relatively large video dataset, and the inter-frame action trajectory of the pixel is detected using the iDT [56, 57] feature extraction method. Secondly, each trained convolutional network is used to extract each multiscale feature maps of the training samples. Thirdly, based on the convolutional feature and the iDT feature, the local convolution response of the image is appended to the center of the trajectory, and the resulting feature descriptor is TDD. Finally, the literature [58] utilizes the Fisher vector to encode the TDDs of the video into a global feature vector and achieve action recognition with a linear SVM classifier.

The above methods can only capture action changes within a pre-fixed sampling interval, whereas Wang et al. [59] proposed a temporal segment network (TSN), which is the first end-to-end temporal series modeling framework that based on complete video, as well as a method to prevent overfitting [60]. TSN implemented a two-stream convolutional network using VGG structure [44]. The long video sequence is equally divided into non-overlapping segments using a sparse temporal sampling strategy. Then, all video segments are trained separately using a two-stream convolution structure. Finally, the consensus functions are used to fuse feature across different sequence segments to get the video-level feature descriptor. The sparse temporal sampling strategy can effectively model the temporal context of video and has many applications in long-term video understanding tasks [6, 42, 63, 67].

Feichtenhofer et al. [11] implemented a two-stream residual ConvNets model (see Fig. 3 for an illustration). The method introduces a residual connection between equal-interval video frames, and connects the temporal network with the corresponding channel of the spatial network. The dimension of feature channel is upgraded and reduced by a $1 \times 1$ convolution kernel, and finally combined with the spatio-temporal network . The training mechanism improves the accuracy of action classification. Tu et al. [53] combined the different visual features and constructed a multi-stream ConvNets structure that choose the complete video frame, the region of interest and the motion saliency region as input. Their techniques acquired a high accuracy in the action recognition tasks.

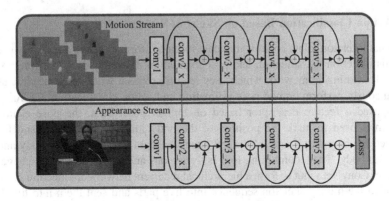

**Fig. 3.** Two-stream residual ConvNet model [11]

## 2.3   Action Classification Based on Recurrent Neural Networks

The action classification method for modeling complete video sequences is computationally expensive. The literatures [1, 7, 33, 39] use recurrent neural networks to model time series of the video sequences with lengths of 64–120 frames. This method is used for long-term video recognition and has achieved good performances. The sequential modeling method based on recurrent neural network mainly includes two solutions: (1) simultaneous training of spatiotemporal two-stream ConvNet, (2) using convolutional feature extracted by spatial network to train the temporal network.

For the former solution, Donahue et al. [7] proposed Long-term Recurrent Convolutional Networks (LRCNs), which extract spatial information of actions in video through deep ConvNet. LRCNs can extract long-term information of actions. Sharma et al. [39] applied a differentiable attention mechanism to the domain of action classification. The multi-layer recurrent neural network is constructed by stacking multiple LSTM units, and a translation and zooming mechanism of the attention region is added to the LSTM network aim for the learning of the key moving parts of the video. Li et al. [31] constructed a multi-channel ConvNet framework with multiple granularities from single frame to the entire video. Each network stream consists of a 2D or 3D ConvNest and dynamically models the output of the different channels through the LSTM layer. Srivastava et al. [48] combined the self-encoder with LSTM to encode the input video sequence into feature vectors of a specific length.

For the latter solution, Baccouche et al. [1] added an LSTM layer on top of the 3D convolutional network to capture the temporal order and long-term dependency information of the action by encoding object states at different times. Carreira et al. [3] used batch normalization [19] for the LSTM layer and got better performance. Lev et al. [28] performed sequences modeling in conjunction with RNN (recurrent neural networks) and Fisher vector representation [58]. First, the spatial features of the moving human body are extracted from the video frames one by one using the VGG framework [35], and the RNN is trained to predict the temporal order of the features. Afterwards, the gradient of the RNN is obtained using a back propagation algorithm in order to compute the Fisher vector. Escorcia et al. [10] proposed an automatic retrieval method for action segments, and put the LSTM unit on the top of the ConvNets to learn spatial features of the sequence data. This method can be used for action synthesis.

## 2.4   Action Classification Based on Skeleton Sequences

The human skeleton joints contain abundant motion information and are widely used in the fields of action classification and pose estimation. Gkioxari et al. [13] proposed a multi-task learning framework which uses region proposal with ConvNet structure to simultaneously perform human pose estimation and action classification. Chéron et al. [5] proposed a feature descriptor based on human pose. First, the pose estimation algorithm is used to mark the position of the body joints in the image, and further calculated the boundary frame of the left hand, the right hand, the upper body and the full human body. Then, combining four body regions and a complete video image as a two-steam ConvNet input to form a hierarchical recurrent network architecture. Du et al. [9] divided the human skeleton sequence into five parts and sent them into five RNN sub-network learning sequence representations. As the layer increases, the feature fusion results of the previous layer will be fed into the higher layer. The final fused skeleton sequence representations will be sent to a single layer perceptron for classification.

Zolfaghari et al. [69] first proposed an action classification method for automatic skeleton sequence extraction using deep neural networks. Firstly, the image segmentation convolution network is used to extract the joints of the moving human body. Then, the multi-stream convolution network is constructed by combining the joints motion optical flow, the inter-frame dense optical flow and the video RGB image.

The literature [8, 29] converts the temporal information of the skeleton sequence into the spatial structure information of the depth image, and uses the deep convolution network to extract the features of the spatiotemporal map. Du et al. [8] connected the joints of the extremities and the trunk as the characteristic representation of the complete video frame, and the (R, G, B) channels of the spatiotemporal map are represented by the (x, y, z) coordinates of the 3D joint points. Li et al. [29] proposed an improvement to the literature [8] to solve the translational and scale changes of joints. They designed a skeleton transforming method that automatically learns the optimal joint ordering directly from spatiotemporal images. Liu et al. [33] proposed a tree-shaped skeleton sequence modeling method based on Spatio-Temporal LSTM (ST-LSTM) via the relationship between 3D skeleton joints in time and space. In the 2D LSTM, the spatio-temporal context information of the joints is coded by the trust gating mechanism. The network structure is shown in Fig. 4.

Recently, Tang et al. [51] combine reinforcement learning with action recognition. Their intended to select informative frames from the input video and describes the selection as a Markov decision process. Therefore, they can leverage deep progressive reinforcement learning (DPRL) method to train a selecting network.

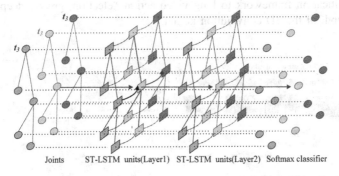

Joints    ST-LSTM units(Layer1)  ST-LSTM units(Layer2)  Softmax classifier

**Fig. 4.** Deep tree structured ST-LSTM network [33]

## 3  Action Detection

Video clips used for human action recognition have a short duration (usually a few seconds) and contain only a single explicit action. However, video under a real scene is often an untrimmed long video, and contains multiple motion instances and complex background content, and human action often occurs within a certain time and space of the video sequence. This led to a new research trend in the field of action recognition, i.e., action detection.

Action detection can be divided into temporal action detection and spatiotemporal action detection. Among them, the temporal action detection requires to locate the start and end time of the action in the video, and the datasets used only contain one action segment. Recently, the temporal detection method based on region proposal and has gained extensive attention. It only needs to distinguish whether the region proposals contain action or not, and does not need to classify the specific action classes. Spatiotemporal action detection not only requires locating the start and end time of the action in the video, but also requires further determination of the spatial extent of the action in the video frame, that is, constructing the space-time action tubes. According to the difficulty of the task, the spatiotemporal action detection is divided into single instance spatiotemporal action detection and multi-instance spatiotemporal action detection. Therefore, the datasets used contain one or more clips of action.

## 3.1 Temporal Action Detection

**Action Detection Based on Region Proposal.** Shou et al. [42] proposed a multi-stage 3D ConvNet (Segment-CNN, S-CNN), as shown in Fig. 5. Firstly, the multi-scale video clips are generated using the sliding window mechanism and then sent to the multi-stage feature extractor. S-CNN mainly includes 3 sub-networks: (1) The proposal network for judging the action score of each video segments; (2) The classification network for judging the category score of the current video clip and initializing the localization network; (3) The localization network, after optimization, the starting and ending boundaries of the action instance are located. S-CNN applies the C3D [52] action classification framework to long video action detection, greatly deepening the temporal depth of the 3D convolution kernel.

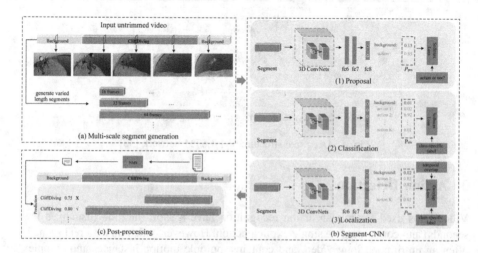

**Fig. 5.** Overall framework of S-CNN for temporal action recognition [42]

Dai et al. [6] introduced a temporal context network, first using region proposal networks [37] to generate proposals of multiple scales at equal intervals in a video. Then training classifiers for region proposals among the different time interval and ranking the scores.

Zhao et al. [67] built a structured segment networks, modeling the temporal structure of each action instance through a structured temporal pyramid. Constructing two classifiers at the top of the pyramid for action classification and completeness distinction. This architecture effectively separate positive samples from background or incomplete samples. In addition, a more efficient temporal action region proposal algorithm is designed to produce high quality candidate actions.

Shou et al. [41] proposed Convolutional-De-Convolutional Networks (CDC) that adding CDC filters on top of 3D convolution networks, and simultaneously performing upsample in time and downsample in space for each proposal to achieve frame-level accurate action prediction.

Motivated by the Faster R-CNN architecture [37], Chao et al. proposed an improved structure called Temporal Action Localization Network (TAL-Net) [4]. Because unreasonable receptive field size will cause issues of extracted feature with too much or too less information, TAL-net implements receptive field alignment through extra networks to address this problem. This framework achieves very competitive capability with other novel techniques of late fusion and temporal context extraction leverage 1D RoI pooling.

**Other Methods.** Temporal action detection methods based on region proposals [6, 42, 67] sort the category scores of video segments according to predefined action boundaries. This method can only process candidate segments with predefined action boundaries. Lin et al. [32] proposed a single shot temporal action detection framework based on the Ssd model [34]. First, training the classifier with multi-scale video segments through the temporal convolution network to calculate the action score. Then, the anchor mechanism is used to predict the temporal boundary of the action instance.

Yuan et al. [63] enabled the processing of arbitrary length video segments with a structured maximal sums algorithm, that is, sum the frame-wise detection results, and use this as the category score of the current video segment.

Sun et al. [49] assumed that the temporal action detection training set only annotates the category labels of the video, and based on this assumption, they propose a bidirectional transfer learning method between video images and web images. A deep convolution network pre-train on the web images or video images, and extract the convolutional features from the images at first. Then the frame-level action prediction is performed on the video according to the extracted features. Finally, a LSTM model combined with the action scores implements the category prediction. Singh et al. [45] constructed a temporal action detection architecture based on multi-stream ConvNets. Firstly, a tracking algorithm is used to generate a bounding box around the moving human body. Then, the dense trajectories of the bounding box across frames and the pixels is used as an input of the multi-steam spatiotemporal convolution network. Finally, a bi-directional LSTM layer is applied to accomplish the multi-stream information fusion.

Yuan et al. [62] proposed a method of extracting Pyramid of Score Distribution Feature (PSDF) based on iDT [56, 57], which uses LSTM unit to process the PSDF feature sequences, and the category prediction of the action segment is then obtained based on the action category confidence score of each frame of the image.

## 3.2   Spatio-Temporal Action Detection

**Single-Instance Action Detection.** Convolutional neural networks based on region proposal [68] has made a breakthrough in the field of image detection. Gkioxari et al. combined the regional proposal network with the visual two-stream theory and proposed a spatiotemporal two-stream action detection framework based on regional proposal [14]. At first, about 2000 region proposals were generated for each video frame using the selective search algorithm [54]. Then, the 2000 proposals were detected for saliency, and the regions with significant motion were selected. Afterwards, constructing class-independent SVM classifiers on time and space networks to classify and identify regions with significant motion. Finally, based on the overlap rate of the proposals of adjacent frame regions, via the Viterbi algorithm to find the optimal space-time tubes. Weinzaepfel et al. [61] combined the boundary bounding box region proposal algorithm [68] with the detection-based visual tracking framework to achieve separate localization in time and space. However, this method uses a multi-scale sliding window approach to generate space-time tubes for each tracking object, which is inefficient when processing long sequence video.

Gemert et al. [55] proposed a region proposal algorithm based on dense trajectories, and used k-nearest neighbor search to generate spatiotemporal candidate regions for each pixel trajectory. Kalogeiton et al. [22] introduced the temporal continuity of action, using continuous multi-frame images as input to the SSD object detection framework [34], and simultaneously output multiple space-time tubes.

Most of the spatiotemporal action detection methods mentioned above combine the unsupervised region proposal algorithm and the multi-stage joint training idea. The feature extraction module and the action classification module are respectively trained, resulting in the training output often being a non-optimal solution.

Recently, Peng et al. [36] used the supervised region proposal method [37] to generate frame-level action proposals, and then train the end-to-end convolutional network structure (see Fig. 6) for action classification and bounding box regression to convert multi-stage detection problems into single step detection problem. Finally, the two-step recursive dynamic programming algorithm is used to trim and annotate the space-time tubes. Zolfaghari et al. [69] constructed a three-stream 3D convolution network combining visual features including body joint locations, dense optical flow, and RGB image three. Since the targets of different streams are independent, a Markov chain is introduced to fuse the prediction results of the three streams.

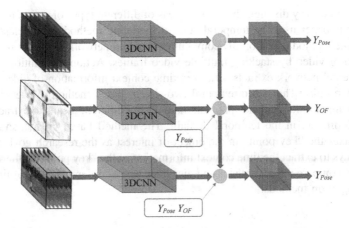

**Fig. 6.** Three-stream 3D ConvNets for spatio-temporal action detection [69]

**Multi-instance Action Detection.** Since there are often multiple instances of action in long video, Saha et al. [38] proposed a multi-instance spatiotemporal action detection framework. First, the space-time two-stream region proposal network [37] works on the RGB image and the optical flow image respectively. Then, the generated region proposals are sent to the object detection framework separately, and the bounding boxes' regression is performed for each video frame region proposal, while predicting whether the region proposal include a motion instance or not. During the fusion of spatial feature and temporal feature, combine the overlapping ratios of spatial network and temporal network prediction results to enhance the prediction results of the temporal network. Finally, use the two-step dynamic programming algorithm to perform optimization and cropping of the multi-instance action tubes.

To further improve the computational efficiency of the spatiotemporal detection algorithm [38], Singh et al. [46] proposed a real-time multi-instance spatiotemporal action detection framework. In the feature extraction phase, combined with single-stage real-time object detector [34] and real-time optical flow algorithm [25]. In the space-time tubes generation phase, an improved one-step dynamic programming algorithm is proposed to achieve incremental real-time multi-instance action detection. Figure 7 shows the basic structure of the multi-instance real-time detection framework [46].

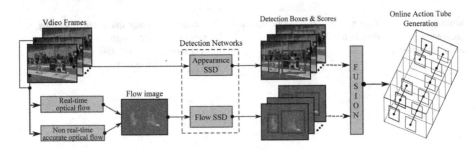

**Fig. 7.** Overall framework of online multi-instance action detection [46]

In order to clearly distinguish the core ideas of different types of action recognition methods, this paper uses the temporal context modeling method as the classification criterion. The method based on 3D convolutional neural network extracts time context information of video by stacking multiple video frames. Action recognition based on recurrent neural network extracts serialized time context information of video through multiple loop units. The spatio-temporal two-stream based method divide video into two parts: temporal domain and spatial domain. Using different means to extracted time context information in the temporal domain. The method based on human skeleton sequence takes the "key point" in the object of interest as the research goal, and uses different ways to extract the time context information of the "key point" in the video. In Table 1, we compare the advantages and disadvantages of different deep learning based action recognition methods.

**Table 1.** Comparison of deep learning based action recognition methods

| Method | Category | Applicable datasets | Advantages | Shortages |
|---|---|---|---|---|
| Action classification based on 3D CNNs | Action classification | Trimmed datasets | Simple to implement, easy to train | Large quantities of training data, high computational demand, unfit to transfer learning, not for real scenes |
| Action classification based on spatio-temporal two-stream networks | Action classification | Trimmed datasets | Simple to implement, easy to train, fit transfer learning | Simple to implement, easy to train, fit transfer learning |
| Action classification based on RNN | Action classification | Trimmed datasets | Simple to implement, easy to train | Large quantities of training data, high computational demand, not for real scenes |
| Action classification based on skeleton | Action classification | Trimmed datasets | Strong ability to model action information | Lack of training data, difficult to label, model's complexity, unfit to transfer learning, not for real scenes |
| Temporal detection based on region proposal | Action detection | Trimmed/untrimmed datasets | Simple to implement, fit transfer learning | Large quantities of training data, high computational demand, not for real scenes |
| Single-instance spatiotemporal action detection | Trimmed datasets | Trimmed datasets | Simple to implement, fit transfer learning | Large quantities of training data, high computational demand, not for real scenes |
| Multi-instance spatiotemporal action detection | Action detection | Trimmed/untrimmed datasets | Fit transfer learning, used for real scenes | Large quantities of training data, high computational demand, task complexity, low accuracy |

# 4  Human Action Datasets

Depending on the duration of the action in the video, we can divide the human action dataset into two broad categories: trimmed datasets and untrimmed datasets. As listed in Table 2, the trimmed datasets mainly include: UCF-101 [47], HMDB-51 [26] and JHMDB-21 [20]. And untrimmed datasets include: THUMOS Challenge [16] and ActivityNet Challenge [2].

**Table 2.** Human action datasets

| Dataset | Year | Clips | Condition | Action categories | Action classification | Action detection | State-of-the-art method |
|---|---|---|---|---|---|---|---|
| UCF-101 [47] | 2012 | 13,320 | Trimmed | 101 | ✓ | ✗ | Zolfaghari et al. [69] |
| HMDB-51 [26] | 2011 | 6,766 | Trimmed | 51 | ✓ | ✗ | Shi et al. [40] |
| JHMDB-21 [20] | 2013 | 928 | Trimmed | 21 | ✓ | ✓ | Zolfaghari et al. [69] |
| THUMOS13/UCF-101-24 | 2014 | 24 | Untrimmed | 25 | ✓ | ✗ | Singh et al. [46] |
| THUMOS14 [16] | 2015 | 15,906 | Untrimmed | 101 | ✓ | ✓ | Zhao et al. [67] |
| THUMOS15 [16] | 2016 | 18,420 | Untrimmed | 101 | ✓ | ✓ | Yuan et al. [62] |
| ActivityNet v1.2 [2] | 2015 | 9,682 | Untrimmed | 100 | ✓ | ✓ | Wang et al. [60] |
| ActivityNet v1.3 [2] | 2016 | 19,994 | Untrimmed | 200 | ✓ | ✓ | Zhao et al. [67] |

## 4.1  Trimmed Datasets

**UCF-101**[1]. There are 13,320 video clips in the UCF-101 dataset, covering 101 different outdoor action categories with the frame rate of 25 and the resolution of $320 \times 240$. Each action clip is randomly divided into 25 groups and each group containing 4 to 7 video clips. Among them, the average length of each video clip is about 7.21 s, for the total video length about 1,600 min. The shortest video clip is about 1.06 s, while the longest is about 71.04 s.

**HMDB-51**[2]. The HMDB-51 dataset is a trimmed human action dataset that extract the movies and videos from the YouTube, with a total of approximately 7,000 hand-labeled action segments covering 51 actions categories. The dataset is divided into 3 training sets and 3 test sets, and the action segments from the same video clips cannot exist in the training sets or test sets at the same time.

---

[1] http://crcv.ucf.edu/data/UCF101.php.

[2] http://www.hmdb.ca/.

**JHMDB-21.** http://jhmdb.is.tue.mpg.de/. The JHMDB-21 dataset is a subset of the HMDB51 dataset with 928 videos and 21 action categories. The dataset not only contains action tags, but also other annotation information, such as the location of key points of human bones, the mask of moving human body, and the dense optical flow of moving human body.

## 4.2    Untrimmed Datasets

In a trimmed dataset, each video clip contains only one action, and its motion time runs through the entire video clip. Obviously, this does not match the real scene. Therefore, the THUMOS Challenge and the ActivityNet Challenge presented two more challenging untrimmed datasets.

**THUMOS Challenge.** The THUMOS Challenge has gone through three phases: THUMOS13, THUMOS14, THUMOS15.

*THUMOS134.* The THUMOS134[3] used the dataset which consists of 24 subsets of 101 untrimmed videos in UCF-101. Each video contains one action category and multiple motion instances (average 1.5), where each instance's motion time occupies 70% of the total video length.

*THUMOS145.* The THUMOS145[4] dataset consists of four parts: the training set, the verification set, the test set, and the background set. Among them, the training set includes 13,320 video clips in UCF-101, the verification set and test set include 1,010 and 1,574 untrimmed videos, respectively, and the background set consists of 2,500 untrimmed long videos. This data provides the iDT characteristics of the complete dataset.

*THUMOS156.* The THUMOS156[5] has more than 430 h of video, about 45 million video frames, which is 1.7 times the size of THUMOS14. It has the same training set as THUMOS14. The verification set has more than 2,100 untrimmed long videos with temporal annotations. The test set includes 5,600 untrimmed long videos, and the background set has approximately 3,000 untrimmed long videos.

**ActivityNet Challenge.** ActivityNet is currently the largest dataset in video action recognition. Up to now, there are two major versions of ActivityNet datasets, namely ActivityNet v1.2 and ActivityNet v1.3. In version 1.2, there were 100 types of actions, including 4 819 training video (7,151 instances in total), 2,383 verification video (3,582 instances in total) and 2,480 test video (category label retained). Version 1.3 has a total of 200 actions, including 10,024 training video (with 15,410 instances), 4,926 verification video (with 7,654 instances), and 5,044 test video (with category labels retained).

---

[3] http://crcv.ucf.edu/ICCV13-Action-orkshop/download.html .

[4] http://crcv.ucf.edu/THUMOS14/ .

[5] http://www.thumos.info/home.html .

# 5  Summary

The researches of the human action recognition in videos, as the basis of human-computer interaction and smart video surveillance, has attracted more and more attention in recent years. This paper analyzes and reviews the current mainstream action recognition methods based on deep neural networks from the perspectives of action classification and action detection. Numerous investigations have found that although deep neural networks have had considerable success in image recognition, natural language processing and other fields, there are still many problems remain to be solved when deep learning models are applied to action recognition in video. On the one hand, video data contains rich temporal context information, which is different from the static image data. The current temporal modeling method that directly stacks multiple consecutive video frames to construct the temporal channels (such as C3D [52]) or takes it as the continuous input of LSTM, which failing to effectively utilize the temporal information of video. To design a more reasonable temporal modeling method becomes the current research difficulty. On the other hand, the existing action dataset with annotations is small in scale, and the combination with the deep neural network with a large number of parameters is faced with serious overfitting risk. It has become a hot issue that how to design a more reasonable model training skills. Moreover, compared with image-based visual tasks, video-based visual tasks is far more complicated, which require higher computer performance.

**Acknowledgments.** This paper was supported by the National Science Foundation of China (Grant No. 61702350).

# References

1. Baccouche, M., Mamalet, F., Wolf, C., Garcia, C., Baskurt, A.: Sequential deep learning for human action recognition. In: Salah, A.A., Lepri, B. (eds.) HBU 2011. LNCS, vol. 7065, pp. 29–39. Springer, Heidelberg (2011). https://doi.org/10.1007/978-3-642-25446-8_4
2. Caba Heilbron, F., Escorcia, V., Ghanem, B., Carlos Niebles, J.: ActivityNet: a large-scale video benchmark for human activity understanding. In: Proceedings of the IEEE Conference on Computer Vision and Pattern Recognition, pp. 961–970 (2015)
3. Carreira, J., Zisserman, A.: Quo vadis, action recognition? A new model and the kinetics dataset. In: Proceedings of the IEEE Conference on Computer Vision and Pattern Recognition, pp. 6299–6308 (2017)
4. Chao, Y.W., Vijayanarasimhan, S., Seybold, B., Ross, D.A., Deng, J., Sukthankar, R.: Rethinking the faster R-CNN architecture for temporal action localization. In: Proceedings of the IEEE Conference on Computer Vision and Pattern Recognition, pp. 1130–1139 (2018)
5. Chéron, G., Laptev, I., Schmid, C.: P-CNN: pose-based cnn features for action recognition. In: Proceedings of the IEEE international conference on computer vision. pp. 3218–3226 (2015)
6. Dai, X., Singh, B., Zhang, G., Davis, L.S., Qiu Chen, Y.: Temporal context network for activity localization in videos. In: Proceedings of the IEEE International Conference on Computer Vision, pp. 5793–5802 (2017)

7. Donahue, J., et al.: Long-term recurrent convolutional networks for visual recognition and description. In: Proceedings of the IEEE Conference on Computer Vision and Pattern Recognition, pp. 2625–2634 (2015)
8. Du, Y., Fu, Y., Wang, L.: Skeleton based action recognition with convolutional neural network. In: 2015 3rd IAPR Asian Conference on Pattern Recognition (ACPR), pp. 579–583. IEEE (2015)
9. Du, Y., Wang, W., Wang, L.: Hierarchical recurrent neural network for skeleton based action recognition. In: Proceedings of the IEEE Conference on Computer Vision and Pattern Recognition, pp. 1110–1118 (2015)
10. Escorcia, V., Caba, H.F., Niebles, J.C., Ghanem, B.: DAPs: deep action proposals for action understanding. In: Leibe, B., Matas, J., Sebe, N., Welling, M. (eds.) ECCV 2016. LNCS, vol. 9907, pp. 768–784. Springer, Cham (2016). https://doi.org/10.1007/978-3-319-46487-9_47
11. Feichtenhofer, C., Pinz, A., Wildes, R.P.: Spatiotemporal multiplier networks for video action recognition. In: Proceedings of the IEEE Conference on Computer Vision and Pattern Recognition, pp. 4768–4777 (2017)
12. Feichtenhofer, C., Pinz, A., Zisserman, A.: Convolutional two-stream network fusion for video action recognition. In: Proceedings of the IEEE Conference on Computer Vision and Pattern Recognition, pp. 1933–1941 (2016)
13. Gkioxari, G., Hariharan, B., Girshick, R., Malik, J.: R-CNNs for pose estimation and action detection. arXiv preprint arXiv:1406.5212 (2014)
14. Gkioxari, G., Malik, J.: Finding action tubes. In: Proceedings of the IEEE Conference on Computer Vision and Pattern Recognition, pp. 759–768 (2015)
15. Goodale, M.A., Milner, A.D.: Separate visual pathways for perception and action. Trends Neurosci. **15**(1), 20–25 (1992)
16. Gorban, A., et al.: Thumos challenge: action recognition with a large number of classes (2015)
17. Graves, A., Mohamed, A.R., Hinton, G.: Speech recognition with deep recurrent neural networks. In: 2013 IEEE International Conference on Acoustics, Speech and Signal Processing, pp. 6645–6649. IEEE (2013)
18. Hochreiter, S., Schmidhuber, J.: Long short-term memory. Neural Comput. **9**(8), 1735–1780 (1997)
19. Ioffe, S., Szegedy, C.: Batch normalization: accelerating deep network training by reducing internal covariate shift. arXiv preprint arXiv:1502.03167 (2015)
20. Jhuang, H., Gall, J., Zuffi, S., Schmid, C., Black, M.J.: Towards understanding action recognition. In: Proceedings of the IEEE International Conference on Computer Vision, pp. 3192–3199 (2013)
21. Ji, S., Xu, W., Yang, M., Yu, K.: 3D convolutional neural networks for human action recognition. IEEE Trans. Pattern Anal. Mach. Intell. **35**(1), 221–231 (2013)
22. Kalogeiton, V., Weinzaepfel, P., Ferrari, V., Schmid, C.: Action tubelet detector for spatio-temporal action localization. In: Proceedings of the IEEE International Conference on Computer Vision, pp. 4405–4413 (2017)
23. Karpathy, A., Toderici, G., Shetty, S., Leung, T., Sukthankar, R., Fei-Fei, L.: Large-scale video classification with convolutional neural networks. In: Proceedings of the IEEE Conference on Computer Vision and Pattern Recognition, pp. 1725–1732 (2014)
24. Krizhevsky, A., Sutskever, I., Hinton, G.E.: ImageNet classification with deep convolutional neural networks. In: Advances in Neural Information Processing Systems, pp. 1097–1105 (2012)

25. Kroeger, T., Timofte, R., Dai, D., Van Gool, L.: Fast optical flow using dense inverse search. In: Leibe, B., Matas, J., Sebe, N., Welling, M. (eds.) ECCV 2016. LNCS, vol. 9908, pp. 471–488. Springer, Cham (2016). https://doi.org/10.1007/978-3-319-46493-0_29

26. Kuehne, H., Jhuang, H., Garrote, E., Poggio, T., Serre, T.: HMDB: a large video database for human motion recognition. In: 2011 International Conference on Computer Vision, pp. 2556–2563. IEEE (2011)

27. LeCun, Y., Bengio, Y., Hinton, G.: Deep learning. Nature **521**(7553), 436 (2015)

28. Lev, G., Sadeh, G., Klein, B., Wolf, L.: RNN fisher vectors for action recognition and image annotation. In: Leibe, B., Matas, J., Sebe, N., Welling, M. (eds.) ECCV 2016. LNCS, vol. 9910, pp. 833–850. Springer, Cham (2016). https://doi.org/10.1007/978-3-319-46466-4_50

29. Li, B., Dai, Y., Cheng, X., Chen, H., Lin, Y., He, M.: Skeleton based action recognition using translation-scale invariant image mapping and multi-scale deep CNN. In: 2017 IEEE International Conference on Multimedia & Expo Workshops (ICMEW), pp. 601–604. IEEE (2017)

30. Li, C., Chen, C., Zhang, B., Ye, Q., Han, J., Ji, R.: Deep spatio-temporal manifold network for action recognition. arXiv preprint arXiv:1705.03148 (2017)

31. Li, Q., Qiu, Z., Yao, T., Mei, T., Rui, Y., Luo, J.: Action recognition by learning deep multi-granular spatio-temporal video representation. In: Proceedings of the 2016 ACM on International Conference on Multimedia Retrieval, pp. 159–166. ACM (2016)

32. Lin, T., Zhao, X., Shou, Z.: Single shot temporal action detection. In: Proceedings of the 25th ACM International Conference on Multimedia, pp. 988–996. ACM (2017)

33. Liu, J., Shahroudy, A., Xu, D., Wang, G.: Spatio-temporal LSTM with trust gates for 3D human action recognition. In: Leibe, B., Matas, J., Sebe, N., Welling, M. (eds.) ECCV 2016, LNCS, vol. 9907, pp. 816–833. Springer, Cham (2016). https://doi.org/10.1007/978-3-319-46487-9_50

34. Liu, W., et al.: SSD: single shot multibox detector. In: Leibe, B., Matas, J., Sebe, N., Welling, M. (eds.) ECCV 2016, LNCS, vol. 9905, pp. 21–37. Springer, Cham (2016). https://doi.org/10.1007/978-3-319-46448-0_2

35. Ngiam, J., Chen, Z., Koh, P.W., Ng, A.Y.: Learning deep energy models. In: Proceedings of the 28th International Conference on Machine Learning, ICML 2011, pp. 1105–1112 (2011)

36. Peng, X., Schmid, C.: Multi-region two-stream R-CNN for action detection. In: Leibe, B., Matas, J., Sebe, N., Welling, M. (eds.) ECCV 2016, LNCS, vol. 9908, pp. 744–759. Springer, Cham (2016). https://doi.org/10.1007/978-3-319-46493-0_45

37. Ren, S., He, K., Girshick, R., Sun, J.: Faster R-CNN: towards real-time object detection with region proposal networks. In: Advances in Neural Information Processing Systems, pp. 91–99 (2015)

38. Saha, S., Singh, G., Sapienza, M., Torr, P.H., Cuzzolin, F.: Deep learning for detecting multiple space-time action tubes in videos. arXiv preprint arXiv:1608.01529 (2016)

39. Sharma, S., Kiros, R., Salakhutdinov, R.: Action recognition using visual attention. arXiv preprint arXiv:1511.04119 (2015)

40. Shi, Y., Tian, Y., Wang, Y., Zeng, W., Huang, T.: Learning long-term dependencies for action recognition with a biologically-inspired deep network. In: Proceedings of the IEEE International Conference on Computer Vision, pp. 716–725 (2017)

41. Shou, Z., Chan, J., Zareian, A., Miyazawa, K., Chang, S.F.: CDC: convolutional-de-convolutional networks for precise temporal action localization in untrimmed videos. In: Proceedings of the IEEE Conference on Computer Vision and Pattern Recognition, pp. 5734–5743 (2017)

42. Shou, Z., Wang, D., Chang, S.F.: Temporal action localization in untrimmed videos via multi-stage CNNs. In: Proceedings of the IEEE Conference on Computer Vision and Pattern Recognition, pp. 1049–1058 (2016)

43. Simonyan, K., Zisserman, A.: Two-stream convolutional networks for action recognition in videos. In: Advances in Neural Information Processing Systems, pp. 568–576 (2014)

44. Simonyan, K., Zisserman, A.: Very deep convolutional networks for largescale image recognition. arXiv preprint arXiv:1409.1556 (2014)

45. Singh, B., Marks, T.K., Jones, M., Tuzel, O., Shao, M.: A multi-stream bi-directional recurrent neural network for fine-grained action detection. In: Proceedings of the IEEE Conference on Computer Vision and Pattern Recognition, pp. 1961–1970 (2016)

46. Singh, G., Saha, S., Sapienza, M., Torr, P.H., Cuzzolin, F.: Online real-time multiple spatiotemporal action localisation and prediction. In: Proceedings of the IEEE International Conference on Computer Vision, pp. 3637–3646 (2017)

47. Soomro, K., Zamir, A.R., Shah, M.: UCF101: a dataset of 101 human actions classes from videos in the wild. arXiv preprint arXiv:1212.0402 (2012)

48. Srivastava, N., Mansimov, E., Salakhudinov, R.: Unsupervised learning of video representations using LSTMs. In: International Conference on Machine Learning, pp. 843–852 (2015)

49. Sun, C., Shetty, S., Sukthankar, R., Nevatia, R.: Temporal localization of fine-grained actions in videos by domain transfer from web images. In: Proceedings of the 23rd ACM International Conference on Multimedia, pp. 371–380. ACM (2015)

50. Sun, L., Jia, K., Yeung, D.Y., Shi, B.E.: Human action recognition using factorized spatio-temporal convolutional networks. In: Proceedings of the IEEE International Conference on Computer Vision, pp. 4597–4605 (2015)

51. Tang, Y., Tian, Y., Lu, J., Li, P., Zhou, J.: Deep progressive reinforcement learning for skeleton-based action recognition. In: Proceedings of the IEEE Conference on Computer Vision and Pattern Recognition, pp. 5323–5332 (2018)

52. Tran, D., Bourdev, L., Fergus, R., Torresani, L., Paluri, M.: Learning spatiotemporal features with 3D convolutional networks. In: Proceedings of the IEEE International Conference on Computer Vision, pp. 4489–4497 (2015)

53. Tu, Z., Xie, W., Qin, Q., Poppe, R., Veltkamp, R.C., Li, B., Yuan, J.: Multistream CNN: Learning representations based on human-related regions for action recognition. Pattern Recogn. **79**, 32–43 (2018)

54. Uijlings, J.R., Van De Sande, K.E., Gevers, T., Smeulders, A.W.: Selective search for object recognition. Int. J. Comput. Vis. **104**(2), 154–171 (2013)

55. Van Gemert, J.C., Jain, M., Gati, E., Snoek, C.G., et al.: APT: action localization proposals from dense trajectories. In: BMVC, vol. 2, p. 4 (2015)

56. Wang, H., Kläser, A., Schmid, C., Liu, C.L.: Dense trajectories and motion boundary descriptors for action recognition. Int. J. Comput. Vis. **103**(1), 60–79 (2013)

57. Wang, H., Schmid, C.: Action recognition with improved trajectories. In: Proceedings of the IEEE International Conference on Computer Vision, pp. 3551–3558 (2013)

58. Wang, L., Qiao, Y., Tang, X.: Action recognition with trajectory-pooled deep-convolutional descriptors. In: Proceedings of the IEEE Conference on Computer Vision and Pattern Recognition, pp. 4305–4314 (2015)

59. Wang, L., Xiong, Y., Wang, Z., Qiao, Y.: Towards good practices for very deep two-stream convnets. arXiv preprint arXiv:1507.02159 (2015)

60. Wang, L., et al.: Temporal segment networks: towards good practices for deep action recognition. In: Leibe, B., Matas, J., Sebe, N., Welling, M. (eds.) ECCV 2016. LNCS, vol. 9912, pp. 20–36. Springer, Cham (2016). https://doi.org/10.1007/978-3-319-46484-8_2

61. Weinzaepfel, P., Harchaoui, Z., Schmid, C.: Learning to track for spatiotemporal action localization. In: Proceedings of the IEEE International Conference on Computer Vision, pp. 3164–3172 (2015)

62. Yuan, J., Ni, B., Yang, X., Kassim, A.A.: Temporal action localization with pyramid of score distribution features. In: Proceedings of the IEEE Conference on Computer Vision and Pattern Recognition, pp. 3093–3102 (2016)
63. Yuan, Z., Stroud, J.C., Lu, T., Deng, J.: Temporal action localization by structured maximal sums. In: Proceedings of the IEEE Conference on Computer Vision and Pattern Recognition, pp. 3684–3692 (2017)
64. Zhang, D., He, F., Han, S., Zou, L., Wu, Y., Chen, Y.: An efficient approach to directly compute the exact Hausdorff distance for 3D point sets. Integr. Comput-Aided Eng. **24**(3), 261–277 (2017)
65. Zhang, D., Tian, L., Hong, M., Han, F., Ren, Y., Chen, Y.: Combining convolution neural network and bidirectional gated recurrent unit for sentence semantic classification. IEEE Access **6**, 73750–73759 (2018)
66. Zhang, D., et al.: Part-based visual tracking with spatially regularized correlation filters. Vis. Comput. 1–19 (2019)
67. Zhao, Y., Xiong, Y., Wang, L., Wu, Z., Tang, X., Lin, D.: Temporal action detection with structured segment networks. In: Proceedings of the IEEE International Conference on Computer Vision, pp. 2914–2923 (2017)
68. Zitnick, C.L., Dollár, P.: Edge boxes: locating object proposals from edges. In: Fleet, D., Pajdla, T., Schiele, B., Tuytelaars, T. (eds.) ECCV 2014. LNCS, vol. 8693, pp. 391–405. Springer, Cham (2014). https://doi.org/10.1007/978-3-319-10602-1_26
69. Zolfaghari, M., Oliveira, G.L., Sedaghat, N., Brox, T.: Chained multi-stream networks exploiting pose, motion, and appearance for action classification and detection. In: Proceedings of the IEEE International Conference on Computer Vision, pp. 2904–2913 (2017)

# Natural Language Processing

# Superimposed Attention Mechanism-Based CNN Network for Reading Comprehension and Question Answering

Mingqi Li[1,2], Xuefei Hou[1,2], Jiaoe Li[1,2], and Kai Gao[1,2(✉)]

[1] Princeton University, Princeton, NJ 08544, USA
[2] School of Information Science and Engineering,
Hebei University of Science and Technology, Shijiazhuang 050700, China
gaokai@hebust.edu.cn

**Abstract.** In recent years, end-to-end models have been widely used in the fields of machine comprehension (MC) and question answering (QA). Recurrent neural network (RNN) or convolutional neural network (CNN) is combined with attention mechanism to construct models to improve their accuracy. However, a single attention mechanism does not fully express the meaning of the text. In this paper, recurrent neural network is replaced with the convolutional neural network to process the text, and a superimposed attention mechanism is proposed. The model was constructed by combining a convolutional neural network with a superimposed attention mechanism. It shows that good results are achieved on the Stanford question answering dataset (SQuAD).

**Keywords:** Convolutional neural network · Attention mechanism · Machine comprehension · Question answering

## 1 Introduction

Since the birth of the computer, people's research on natural language has never stopped. In 1950, Alan Turing proposed the "Turing test" to test whether the machine have "intelligence" [1]. In recent years, the exploration of machine comprehension and question answering has been deepened, and researchers have conducted extensive researches on the SQuAD dataset [2] and made great progresses.

In previous model, the end-to-end model [3] commonly used two neural networks, namely the convolutional neural network [4] and the recurrent neural network [5]. The recurrent neural network model is good at modeling natural language. It converts sentences of arbitrary length into floating-point numbers of specific dimensions, thus realizing the "memory" of sentences. RNN is usually used in the field of text. Although it is superior to other neural networks in its own characteristics, it takes a long time to train the model. The processing of the convolutional neural network model is to simulate the visual perception mechanism of the organism, then performs module convolution on the image and extracts the features from the different positions, so as to express the data information effectively. Therefore, CNN is usually used in the image field. RNN is the first choice in language models, but the CNN can often produce

© Springer Nature Singapore Pte Ltd. 2019
R. Mao et al. (Eds.): ICPCSEE 2019, CCIS 1059, pp. 25–37, 2019.
https://doi.org/10.1007/978-981-15-0121-0_2

unexpected effects. For example, the paper [6] applies CNN to text classification, and the text translation constructed by study [7] is several times faster than before, and achieved good effect.

In the work of Bi-Directional Attention Flow (BiDAF), a bidirectional attention network is proposed for modeling the representation of context fragments at different intensity levels by constructing a hierarchical structure comprising character level and word level. Embedding the hierarchical structure and the context is to query the bidirectional attention of the context.

In the process of sentence modeling based on CNN, considering the interaction between sentences, it is necessary to extract the linguistic features of the sentences in the context. On this basis, it can be applied to various tasks of sentence modeling and the influencing factors of sentences entering CNN, which means that some inter dependent sentences are more powerful than individual sentences. The Reading Wikipedia work is to combine bigram hash and TF-IDF search components with a multi-layer recurrent neural network model based on Wikipedia, and to search through large-scale text reading. In the paper [8], the convolutional neural network was used to model machine comprehension and question answering, which greatly improved the training speed and reasoning speed, and the corpus translation of the neural translation model was proposed. Combining the data generated by the inverse translation of the neural machine translation model increases the amount of data and greatly shortens the training time. Some models consider that problem understanding is important to solve problems [9]. The model [10] can help the problems in the end-to-end neural frame-work, modeling different types of problems and sharing information among the problems and building adaptive models.

With the development of neural networks, the paper [11] proposed two global and local Attention mechanisms, which showed the expansion of attention in RNN and did research work to reduce computation time. Attention has been widely used in images [12] and texts [13], and the traditional algorithms and effects have been greatly improved by combining with RNN or CNN. BiDAF [14] successfully combined attention mechanism and applied them to neural network with machine question answering. In the previous works, the Attention has been applied to encoders or decoders to make it easier to correct and improve.

In this paper, we focus on a mechanism for parallel enhancement of attention and use the attention to superimpose the vectors to achieve the effect of attention enhancement. By comparing different models, the intensive attention mechanism sig-nificantly improves the effect of question and answer. This paper focuses on optimizing the Attention layer. On the SQuAD public dataset, this approach significantly improves the accuracy and matching rate.

The rest of our paper is structured as follows: Sect. 2 introduces the key back-ground. Section 3 describes the architecture of the model. Section 4 introduces our experimental results. Section 5 reviews related work.

# 2  Related Work

This section introduces the models related to this paper. We will cover some of the work in our model and analyze its main architecture.

## 2.1  Convolutional Neural Network

The traditional convolutional neural network (CNN) [4] consists of four parts: the input layer, the activation layer, the pooling layer, and the output layer. When constructing models, some parts can be combined to obtain different neural models. In the LeNet [15] model, three convolutional layers are used, each of which includes convolution, pooling, and activation functions (sigmoid), two sampling layers, and a fully connected layer. The LeNet model is considered to be the earliest convolutional neural network model. Handwritten characters are identified by convolution extraction of spatial features. AlexNet [16] uses an 8-layer structure with five convolution layers and three fully-connected layers. Unlike LeNet, AlexNet uses a dropout layer in the fully connected layer. By introducing the Dropout Layer, each layer of neurons can be removed randomly according to a certain probability P, thereby alleviating the problem of overfitting in the CNN. The MHAM [17] captures multiple semantic aspects from the user utterances to solve different problems of semantic aspects, and the RASG [18] proposes a conventional sequence-to-sequence method for abstractive text summarization.

In 2014, VggNet [19] won the first place in the ImageNet competition. Except for the pooling layer and softmax layers, the model has 16 to 19 layers. In the model, 2–3 convolutional layers are combined into one convolutional block to reduce network parameters, and more linear conversions are performed by using the ReLu activation function. However, the ResNet [20] model uses a deeper 152-layer network structure to avoid the phenomenon of accuracy saturation, as shown in the Fig. 1 shows, This figure shows the composition of the block, left: a building block (on 5656 feature maps) for ResNet34, right: a bottleneck building block for ResNet-50/101/152.

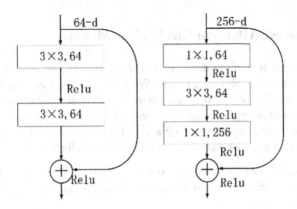

**Fig. 1.** ResNet network structure

With the development of deep learning, the number and layers of neural networks are increasing. In the subsequent experiments of ResNet, the number of layers has been added to 1000 layers, the learning objectives and difficulty have been simplified, and better experimental results have been achieved.

## 2.2   Attention Structure

The attention mechanism first appeared in the image field. In the work of [12], RNN + Attention was used to classify the images. The current state will be learned according to the previous state and the current position. The first application of the attention mechanism in natural language processing is the machine translation model [13], in which a perceptron formula is used to connect the target language with the source language.

A probability matrix is formed by the normalization of the softmax function. Just like the alignment function in statistical machine translation, each word takes the context into consideration with is more advantageous in dealing with long sentences. In the first part, we mentioned the extension of the attention mechanism, one is the global attention mechanism, and the other is the local attention mechanism, which is expressed in the author's experiment by using the local attention mechanism. It works better than using the global attention mechanism. ABCNN [21] is a successful model for early exploration of CNN + Attention. The model proposes three methods. The first is to use the attention mechanism before convolution and input the volume through the attention features and the original features. The second is to add attention mechanism into the pooling layer, by weighting the convolution layer, and then performing the pooling layer operation. The third method is to integrate the first two methods into the convolution layer and to add attention mechanism to the convolution layer and the pooling layer, The effect is better than that of a single method.

In this paper, we propose a framework for parallel superimposition enhancement of Self-Attention. Through the superposition of these two Self-Attentions, we can strengthen the part and even the whole of the text, thereby improving the correlation between the question and the text.

## 3   Superimposed Attention Mechanism Model

In this paper, the structure of our model is similar to most existing models. It consists of the following six parts as the Fig. 2 shows. We encode the text and the problem, then input them into the convolutional neural network, and then parallel superimpose two Attentions to achieve the effect of enhancement. The effect can be optimized for local and even the whole, and then the starting and ending position can be modeling by the convolutional neural network output to achieve the final effect.

(a)  Character and Word Embedding Layer Similar to BiDAF [14], characters and words are mapped into vector space by character-level coding and word-level coding, respectively;

(b)  Encoder Layer The vector is processed by using a convolutional neural network;

(c) Attention Layer The attention layer generates feature vectors and performs feature operations by integrating context information;

(d) Superimposed Attention Layer The main innovation layer. It achieves the superposition of feature vectors by integrating two self-attention layers;

(e) Modeling Layer Using convolutional neural networks to scan the context;

(f) Output Layer Output the answer that is most likely to correspond to the question.

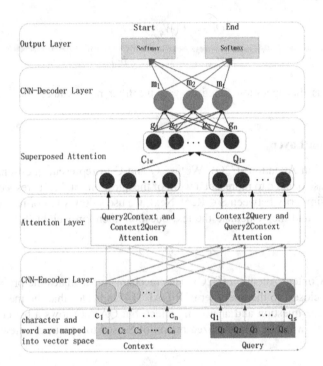

**Fig. 2.** Superimposed attention mechanism-based CNN network model

### 3.1 Character and Word Embedding Layer

Assuming that the problem sequence is represented as Q = {Q1, Q2, Q3 ... Qs}, while the context sequence is expressed as C = {C1, C2, C3 ... Cn}, and the word vector training is performed by using Glove [22]. Each word is mapped into a vector space, and at the same time, a high-dimensional vector is mapped to the character set by using CNN [6], which obtain the same vector as CNN dimension by inputting a one-dimensional vector [23]. In the Fig. 2, c = {c1, c2, c3 ... cn} and q = {q1, q2, q3 ... qs} represent the vector representation of the content and the problem, respectively.

### 3.2 Encoder Layer

At the coding level, we use CNN as the main algorithm for calculation. In the process of constructing, we borrows the coding layer of ABCNN [21]. In recent years, CNN has been used to deal with linguistic sequences, and often achieves extraordinary results. QANet [8] has proved that the applications of CNN in the language sequences have achieved very good results. By convolving{c1, c2, c3 ...cn} and {q1, q2, q3 ... qs}, we set j and i respectively as the position of the question and content, where $0 < j < s + 1$ and $0 < i < n + 1$, the range of i is like j, we can get the input block of the problems, Wq, and the input block of text, Wc. Then the nth node in the matrix takes the value:

$$Q_{(s)} = F(W_q \cdot y_i + b^j) \tag{1}$$

$$P_{(n)} = F(W_c \cdot y_i + b^j) \tag{2}$$

Where F is the operation function, y is the filter node value, and b is the bias parameter.

## 3.3   Attention Layer

Content-Question Attention (C2Q): We use Q and C to represent questions and texts respectively, use q and c to represent a word of the questions and texts respectively, and calculate the similarity between $S \in Rc*t$. Next, we use softmax to normalize S to get a new matrix S', and then calculate the text-question attention weight [8]:

$$Z = S' \cdot Q^T \in R^{c*d} \tag{3}$$

Question-Content Attention (Q2C): This mechanism calculates the probability that the context is closest to the query term. BiDAF [14] indicates that the mechanism has an impact on the question and answering effect, but QANet shows a better DCN [27] attention model, which is a normalized matrix of S used to construct a new question - text attention:

$$Z' = S \cdot S'^T \cdot C^T \tag{4}$$

Our calculation of text-question attention and question-text attention can maximize the similarity between question and text, so as to ensure that we can achieve the greatest correlation.

## 3.4   Superimposed Attention Layer

This layer is the main innovation layer of the thesis. The structure usually has five layers. In this paper, the intensive attention layer is added in the middle, and the reinforcement effect is achieved by superimposing the two attentions.

We set two attention layers as C1w and Q1w, and then calculate the weights of the text and the problem respectively to get $C_1', Q_1'$. We sum the two weight matrices and calculate the mean value to get the value of the position of each matrix. The formula is as follows:

$$S = \frac{1}{n}\sum_{i}^{n}[C'_{1ij} + Q'_{1ij}] \tag{5}$$

### 3.5 Modeling Layer

At the modeling layer, we still use convolutional neural networks for coding, but in principle it is consistent with the work of BiDAF, which uses recurrent neural network (RNN). The modeling layer is related to the relationship between the contexts of the query conditions by convolving the attention matrix separately and performing weight sharing.

### 3.6 Output Layer

The output layer is not improved in this paper. Similar to the work done by BiDAF, we model the probabilities of the starting and ending positions, P1 and P2 as follows:

$$P_1 = Softmax(W_1[G : M]) \tag{6}$$

$$P_2 = Softmax(W_2[G : M]) \tag{7}$$

$W_1$ and $W_2$ are two trainable weight vectors, and G, M, and $M'$ are the outputs of the three models.

Then the objective function is defined by the starting positions $P'_1$ and $P'_2$. By summing up the logarithms of all positions, the objective function is calculated by taking the average value. The formula is as follows:

$$L(\theta) = -\frac{1}{N}\sum_{i}^{N}[log(P'_{2i})] \tag{8}$$

Where $\theta$ is all trainable parameters.

## 4 Experiment Result

### 4.1 SQuAD Dataset

With the release of the dataset, machine comprehension and question answering are constantly updating models and effects. The main source of data for SQuAD is Wikipedia [2]. Stanford University released the SQuAD 2.0 version [24] in 2018. The paper [25] released the NewsQA dataset, and the data published in the paper [26] came mainly from the Internet. The release of these data greatly promoted the efficiency of the question and answer model.

Since Stanford University released the open source dataset SQuAD, many scholars have used this as a standard for question answering experiments. Through dynamic attention mechanism, the context information is queried. By modeling the problem, an adaptive model is proposed to solve the problem that different questions correspond to the same answer in the question and answer. The local error answer recovery problem [27] and the contextual information were also studied in depth and predicted by answering the text span in the paragraph [28]. The work of QANet [8] surpassed the predictions of human experts and achieved remarkable results on the SQuAD dataset.

## 4.2  Experimental Settings

We use the SQuAD dataset for parameter adjustment. We consider on the Stanford dataset SQuAD and calculate the exact matching (EM) and F1 values, respectively. The EM is the score that exactly matches the predicted answer with the original answer, and the F1 score is the weighted average of the precision and recall rate at the character level.

**Table 1.**  The parameters for our experiments

| Parameters | Values |
| --- | --- |
| Batch_size | 32 |
| Hidden layer size | 128 |
| Content_limit | 300 |
| Question_limit | 50 |
| Dropout | 0.1 |

We use Glove to pre-train on the 840B corpus, and set the maximum text input length to 300, the question length to 50, and the hidden layer to 128. We spend about 7.2 h on a GTX1080ti GPU. Compared to other models, the speed is much faster, and the EM value of 71.2 and the F1 value of 80.2 are achieved, which is much better than the baseline model. The hyperparameters of this paper are shown in Table 1.

## 4.3  Results and Discussion

This section mainly analyses the experimental results, Through the comparison of the architecture itself and comparison with other models, the advantages of the model are demonstrated, and the shortcomings of the model are analysed.

First, we analyze by comparing our own architecture. Figure 3 shows the comparison in the absence of superimposed attention. The ordinate indicates the scores of EM and F1. The blue part of the figure is the result of the work without superimposed attention, and the effect is that the EM value is 70.4 and the F1 value is 79.7. The yellow part of the figure is the result of our work with intensive attention, with the EM value of 71.2 and the F1 value of 80.2, this scores are within the error range ($\pm$ 0.5). The experiments prove that our EM and F1 values are increased by 0.8 and 0.5, respectively, in the case of parallel superposition and enhancement of Attention.

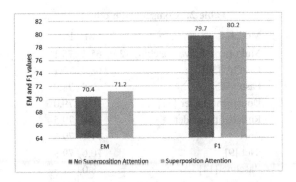

**Fig. 3.** The comparison of two architectures (Color figure online)

The experimental results show that it is helpful to superimpose the text and the problem. By superimposing two times, we can see that two features are extracted, and different features link the text to the problem more fully, as shown in the Fig. 4. We divide the texts and the problems into two parts. The lower part of the figure is the first attention mechanism, the upper part is the second attention mechanism, and the red line represents the association between the texts and the questions. The mechanism of superposition and attention enhancement divide the texts into "Its" and "very good", and devide are divided into "What do you think of" and "this book". By associating different parts, we can establish a connection between the two subjects. For example, using "very good" as the subject of the text and "this book" as the subject of the question, and making "this book is very good" as a pre-selected answer, the purpose of the final attention superposition is achieved by performing a similarity calculation on the pre-selected answer and the original answer. By calculating the similarity between the pre-selected answers and the original answers, the purpose of the final attention superposition is achieved.

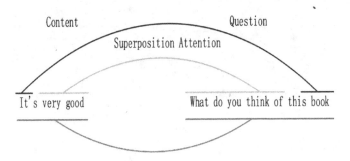

**Fig. 4.** Example for superimposed attention (Color figure online)

Next, we compare the results of other models with those of our model. Table 2 shows a comparison with Baseline and other models.

**Table 2.** Compare with others

| Model | EM | F1 |
|---|---|---|
| Multi-perspective matching [29] | 65.5 | 75.1 |
| FastQA [30] | 68.4 | 77.1 |
| FastQAExt | 70.8 | 78.9 |
| Document reader [31] | 70.0 | 79.0 |
| Ruminating reader | 70.6 | 79.5 |
| Rnet | 71.1 | 79.5 |
| jNet [9] | 70.6 | 79.8 |
| Baseline [14] | 68.0 | 77.3 |
| Attention superposition (Ours) | 71.2 | 80.2 |
| Human [2] | 81.4 | 91.0 |

Based on the above comparison, our model is 3.2 and 2.9 higher than Baseline's EM and F1 values, respectively, and the effect has been significantly improved. Compared with Rnet, the EM value is flat, and the F1 value is slightly higher. This may be due to the fact that the overall matching has little effect on the predicted answers, but it is better than most models. Therefore, the model is effective for the enhanced question answering of Self-Attention.

We trained on GTX1080 for about 7 h, and by comparison, after about 2.5 h of training, the F1 value of our model reached the effect of the baseline model.

As shown in Fig. 5, We replace the number of steps with hours, during the 0–3 h period, the performance of the model was improved by a large margin, and then began to slow down gradually. We extended our training time and got an improvement in F1 value.

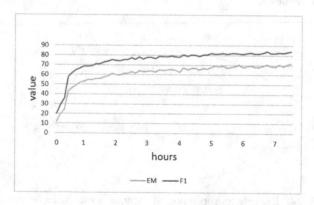

**Fig. 5.** F1 and EM values in the model

Figure 6 shows the change of the loss function during the training process, which is the same as the change of F1 and EM values. During the first three hours, the loss function decreases rapidly, then gradually slows down, and the loss function ends up around 2.0. We can try to minimize the loss function of the model to improve the effect of the model. After we extend the training step, the loss function is reduced to below 2.0, and the performance of the model is improved.

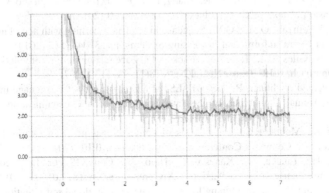

**Fig. 6.** Loss function model

In this experiment, we pay attention to the enhancement of different texts and problems. In the process of experiment, by adjusting the experimental parameters and optimizing the model, some progress has been made to some extent. But there are still some shortcomings that need to be improved in the next experiment.

## 5   Experiment Result

In this paper, a model of the enhanced attention mechanism for the end-to-end model is proposed. By using convolutional neural network instead of traditional recurrent neural network as the basic model algorithm, the training speed of the model can be accelerated to only about 7 h on the GTX1080ti. It can be proved that the convolutional neural network is more efficient for text processing than the recurrent neural network. In the model, a mechanism for strengthening Self-Attention is proposed, The main idea of this paper is to superimpose the weights through two attention mechanisms, and to calculate the similarity between the text and problem based on the weights after the superposition. By enhancing the attention mechanism, the correlation between text-question and question-text can be enhanced, and the effect on the SQuAD dataset is obviously improved.

**Acknowledgement.** This paper is sponsored by National Science Foundation of China (61772075) and National Science Foundation of Hebei Province (F2017208012) and Humanities and social sciences research projects of the Ministry of Education (17JDGC022).

# References

1. Turing, A.M.: Computing machinery and intelligence. In: Epstein, R., Roberts, G., Beber, G. (eds.) Parsing the Turing Test, pp. 23–65. Springer, Dordrecht (2009). https://doi.org/10.1007/978-1-4020-6710-5_3
2. Rajpurkar, P., Zhang, J., Lopyrev, K., Liang, P.: SQuAD: 100,000 + questions for machine comprehension of text. arXiv preprint arXiv:1606.05250 (2016)
3. Sutskever, I., Vinyals, O., Le, Q.V.: Sequence to sequence learning with neural networks. In: Advances in Neural Information Processing Systems, pp. 3104–3112 (2014)
4. Lawrence, S., Giles, C.L., Tsoi, A.C., Back, A.D.: Face recognition: a convolutional neural-network approach. IEEE Trans. Neural Netw. 8(1), 98–113 (1997)
5. Sundermeyer, M., Schlüter, R., Ney, H.: LSTM neural networks for language modeling. In: Thirteenth Annual Conference of the International Speech Communication Association (2012)
6. Vieira, J.P.A., Moura, R.S.: An analysis of convolutional neural networks for sentence classification. In: Computer Conference (CLEI), pp. 1–5. IEEE (2017)
7. Kameoka, H., Tanaka, K., Kaneko, T., Hojo, N.: ConvS2S-VC: fully convolutional sequence-to-sequence voice conversion. arXiv preprint arXiv:1811.01609 (2018)
8. Yu, A.W., et al.: QANet: combining local convolution with global self-attention for reading comprehension. arXiv preprint arXiv:1804.09541 (2018)
9. Zhang, J., Zhu, X., Chen, Q., Dai, L., Wei, S., Jiang, H.: Exploring question understanding and adaptation in neural-network-based question answering. arXiv preprint arXiv:1703.04617 (2017)
10. Gao, S., Zhao, Y., Zhao, D., Yin, D., Yan, R.: Product-Answer Generation in ECommerce Question-Answering (2019)
11. Luong, M.-T., Pham, H., Manning, C.D.: "Effective approaches to attention-based neural machine translation." arXiv preprint arXiv:1508.04025 (2015)
12. Mnih, V., Heess, N., Graves, A.: Recurrent models of visual attention. In: Advances in Neural Information Processing Systems, pp. 2204–2212 (2014)
13. Bahdanau, D., Cho, K., Bengio, Y.: Neural machine translation by jointly learning to align and translate. arXiv preprint arXiv:1409.0473 (2014)
14. Seo, M., Kembhavi, A., Farhadi, A., Hajishirzi, H.: Bidirectional attention flow for machine comprehension. arXiv preprint arXiv:1611.01603 (2016)
15. LeCun, Y., Bottou, L., Bengio, Y., Haffner, P.: Gradient-based learning applied to document recognition. Proc. IEEE 86(11), 2278–2324 (1998)
16. Krizhevsky, A., Sutskever, I., Hinton, G.E.: Imagenet classification with deep convolutional neural networks. In: Advances in Neural Information Processing Systems, pp. 1097–1105 (2012)
17. Tao, C., Gao, S., Shang, M., Wu, W., Zhao, D., Yan, R.: Get the point of my utterance! learning towards effective responses with multi-head attention mechanism. In: IJCAI, pp. 4418–4424 (2018)
18. Gao, S., Chen, X., Ren, Z., Bing, L., Zhao, D., Yan, R.: Abstractive Text Summarization by Incorporating Reader Comments (2019)
19. Simonyan, K., Zisserman, A.: Very deep convolutional networks for large-scale image recognition. arXiv preprint arXiv:1409.1556 (2014)
20. He, K., Zhang, X., Ren, S., Sun, J.: Deep residual learning for image recognition. In: Proceedings of the IEEE Conference on Computer Vision and Pattern Recognition, pp. 770–778 (2016)

21. Yin, W., Schütze, H., Xiang, B., Zhou, B.: ABCNN: attention-based convolutional neural network for modeling sentence pairs. Trans. Assoc. Comput. Linguist. **4**, 259–272 (2016)
22. Pennington, J., Socher, R., Manning, C.: Glove: Global vectors for word representation. In: Proceedings of the 2014 Conference on Empirical Methods in Natural Language Processing (EMNLP), pp. 1532–1543 (2014)
23. Srivastava, R.K., Greff, K., Schmidhuber, J.: Highway networks. arXiv preprint arXiv:1505. 00387 (2015)
24. Rajpurkar, P., Jia, R., Liang, P.: Know what you don't know: unanswerable questions for SQuAD. arXiv preprint arXiv:1806.03822 (2018)
25. Trischler, A., et al.: NewsQA: a machine comprehension dataset. arXiv preprint arXiv:1611. 09830 (2016)
26. Nguyen, T., et al.: MS MARCO: a human generated machine reading comprehension dataset. arXiv preprint arXiv:1611.09268 (2016)
27. Xiong, C., Zhong, V., Socher, R.: Dynamic coattention networks for question answering. arXiv preprint arXiv:1611.01604 (2016)
28. Lee, K., Salant, S., Kwiatkowski, T., Parikh, A., Das, D., Berant, J.: Learning recurrent span representations for extractive question answering. arXiv preprint arXiv:1611.01436 (2016)
29. Wang, Z., Mi, H., Hamza, W., Florian, R.: Multi-perspective context matching for machine comprehension. arXiv preprint arXiv:1612.04211 (2016)
30. Weissenborn, D., Wiese, G., Seiffe, L.: Making neural QA as simple as possible but not simpler. arXiv preprint arXiv:1703.04816 (2017)
31. Gong, Y., Bowman, S.R.: Ruminating reader: reasoning with gated multi-hop attention. arXiv preprint arXiv:1704.07415 (2017)

# Software Engineering

# Empirical Study of Hybrid Optimization Strategy for Evolutionary Testing

Chunling Hu[1(✉)], Bixin Li[2], Xiaofeng Wang[1], and Gang Lv[1]

[1] Hefei University, No. 99 Jinxiu Avenue, Hefei, Anhui, China
huchunling@hfuu.edu.cn
[2] Southeast University,
No. 2 Southeast University Road, Nanjing, Jiangsu, China

**Abstract.** Evolutionary testing (ET) is an effective test case generation technique which uses some meta-heuristic search algorithm, especially genetic algorithm, to generate test case automatically. However, the population prematurity problem may decrease the performance of ET. In this paper, a hybrid optimization strategy is proposed based on extended cataclysm which integrates both static configuration strategies and dynamic optimization strategy. Dynamic optimization strategy included the optimization of initial population and the dynamic population optimization based on extended cataclysm, where the diversity of population was monitored during the evolution process of ET, and once the population prematurity was detected, extended cataclysm operation was used to renew the diversity of the population. Experimental results show that the hybrid optimization strategy can improve the performance of ET.

**Keywords:** Evolutionary testing · Population prematurity ·
Extended cataclysm · Diversity measurement

## 1 Introduction

Evolutionary Testing (ET) is a promising automatic testing technique, which converts the problem of generating test case into an optimization problem. ET uses a kind of meta-heuristic search method (for example, Genetic Algorithm (GA) to search the test case for fulfilling a test objective [1–7]. In general, the search space consists of all possible inputs of test objects, and the global optimal solution of the search is the test case that can fulfill the test objective. Compared with random testing, ET can work better in many aspects including the effectiveness and efficiency of automated test case generation.

If the population of ET is premature, ET can only find a local optimal solution that is undesirable, and the performance of ET is greatly reduced. To overcome the problem of population prematurity of ET, some static configuration strategies have been proposed [8, 9]. In fact, some of them can effectively prevent the occurrence of the population prematurity. However, even if we have the best static configuration strategy, there is no guarantee that the global optimal solution for ET can be generated. Researchers have proposed some dynamic optimization techniques to deal with the population prematurity problem [14, 16].

R. Mao et al. (Eds.): ICPCSEE 2019, CCIS 1059, pp. 41–53, 2019.
https://doi.org/10.1007/978-981-15-0121-0_3

During the past few years, more and more researchers have studied ET and apply it in functional testing, structural testing, safety testing, object testing, with automated generation and selection of test data [2, 3, 8–13]. Related works of ET can be divided into two classes. The first class of works aim at preventing the occurrence of population prematurity [5, 8, 9, 15, 17, 19, 21–23]. The second class of works are proposed to solve the population prematurity problem when it does happen [14, 16, 24, 25].

For the first class, Harman et al. suggested Testability Transformation for the flag problem in ET [5, 19]. Testability Transformation focuses on the transformation of the source code. It seeks to transform a program in order to make it easier to generate test case. McMinn et al. suggested the extended chain approach for the state problem in ET [15, 17]. Arcuri et al. discussed the configuration strategies for evolutionary testing, where annealing genetic algorithm and restricted genetic algorithm are two strategies used to help a globally optimal solution in fewer generations [8]. Annealing genetic algorithm combines simulated annealing and genetic algorithm to deal with the population prematurity. Restricted genetic algorithm adds some restriction, a punishment factor, to the fitness function. Arcuri et al. carried out a series of experiments to examine the relative performances of different configuration strategies [9]. The experiments focus on structural evolutionary testing that generates test case to cover a statement in a function without loop conditions. Their results highlight several differences between the configuration strategies and provide some advices that are helpful to select a configuration strategy in practice.

For the second class, Xie et al. offered DOMP (Dynamic Optimization of Mutation Possibility) [16]. The key idea of DOMP is to adjust the mutation possibility Pm according to the current population diversity. During the normal evolution of ET, Pm is quite small. But once the population prematurity occurs, Pm is enlarged to a predefined value. In the following population evolution, if the population diversity is recovered, Pm is restored. After a lot of experiments, it can be found that DOMP can deal with the population prematurity. However, in certain cases, DOMP does not work well. Jin et al. introduced cataclysm-genetic algorithm to design the DNA sequences satisfying H distance and hamming distance constraints [24]. Eichie et al. introduced another cataclysm operation and used selected and easily measurable atmospheric parameters as input parameters in developing two new models for computing the Rxlevel of GSM signal through a three-step approach [25].

In this paper, we integrate the advantages of the above two class works and propose a novel hybrid optimization strategy. Our main contributions include: (1) static configuration strategies are explored by experiments for determining which one is better for specific test objects. (2) dynamic optimization strategies based on extended cataclysm are proposed for structural evolutionary testing. In our research, static configuration strategies are applied to prevent population prematurity, and dynamic optimization strategies are used to deal with the population prematurity when it does happen. The dynamic optimization strategy works as follows: it monitors the diversity of the population during the evolution process, detects the population prematurity by calculating the diversity and comparing it with a threshold value, and uses the extended cataclysm operation to recover the diversity of the population when population prematurity happens.

The rest of the paper is organized as follows: Sect. 2 briefly introduces ET, fitness function, population diversity and the population prematurity problem; Sect. 3 introduces our hybrid optimization strategy in detail; Sect. 4 presents experimental results; Sect. 5 concludes this paper.

## 2 Background

We will first give an overview of ET, then define several common fitness functions, and population diversity. Finally, we explain the population prematurity problem.

### 2.1 Evolutionary Testing

ET refers to the use of a kind of meta-heuristic search method, especially Genetic Algorithm (GA), for the test case generation. ET borrows the principle of biological evolution theory to optimize test case generation, where a fitness function is used to evaluate an individual (a test case) in an initial population (a set of test cases), and some genetic operations are used to generate offspring population. After several iterative generations, we hope to obtain a global optimal solution. The fitness is an important control factor for improving the performance of ET.

In ET, the input domain of the test object forms the search space in which ET searches for a test case that fulfills the test objective [1, 2]. Figure 1 shows a typical process of ET, where the initial population is generated randomly. The evaluation block shows the evaluation process of population, which consists of the following tasks: extraction of individuals, decode each individual into a test case, execute each test case, monitor the test execution and result, and calculate the fitness of each individual iteratively. The evolution block shows an evolution process of the population based on GA, which performs operations of Selection, Crossover, Mutation and Survival.

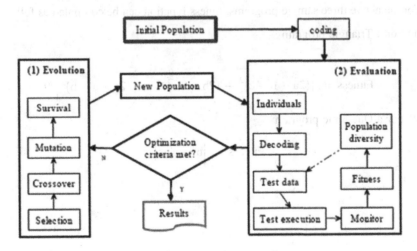

**Fig. 1.** A traditional evolutionary testing process.

The final goal of ET is to find a test case that realizes the specified test objective, which shows that the test case is in fact the global optimal solution of GA.

## 2.2  Fitness Function

A fitness function is a particular type of objective function that prescribes the optimality of a solution. In ET, an individual (corresponding to a chromosome in GA) of the population represents a test case, and the fitness function is designed according to the specific test objective, which gives higher values to better individuals [16, 17]. The goal of ET is to find a test case to realize test objective, which is in fact the global optimal solution of a Generic Algorithm.

In structural testing, the test objective is usually based on some coverage criteria such as branch coverage, condition coverage, combination coverage, path coverage, etc. For three simple programs used as the test objects shown in Table 1 [16, 20], the test criterion of experiments is to cover a particular branch of the test objects (Branch Coverage).

**Table 1.** Different functions in experiments.

| Function name | Input | Condition to satisfy (testing objective) | Corresponding result |
|---|---|---|---|
| Triangle | a, b, c | (a==b) && (b==c) | The triangle is an equilateral one |
| Quadratic | a, b, c | b²−4*a*c ==0 | The equation has two equal real solutions |
| LineCover | 11x, 12x<br>11y, 12y<br>rx, ry<br>w, h | (rx ==min(11x, 12x))<br>&& (ry ==min(11y, 12y))<br>&& (rx + w== max(11x, 12x))<br>&& (ry + h== max(11y, 12y)) | (w, h) are the width and height of a rectangle. The line segment is the diagonal of the rectangle |

For the above three sample programs, fitness function can be computed as follows.

(1) For a Triangle program

$$\text{Fitness} = \left[(2a - b - c)^2 + (2b - a - c)^2 + (2c - a - b)^2\right]/9$$

(2) For a Quadratic program

$$\text{Fitness} = \left|b^2 - 4ac\right|$$

(3) For a LineCover program

$$Fitness = (rx - \frac{|11x+12x|-|11x-12x|}{2})^2 + (ry - \frac{|11y+12y|-|11y-12y|}{2})^2$$
$$+ (rx+w - \frac{|11x+12x|-|11x-12x|}{2})^2 + (ry+h - \frac{|11y+12y|-|11y-12y|}{2})^2$$

## 2.3  Population Diversity and Prematurity

In structural evolutionary testing, a population is a set of test cases, which are generated from the input domain of test object. The size of a population can be controlled according to the test objective or test requirement. Population diversity denotes the distribution scope of test cases in input domain, which directly affect the performance of ET. Whether the diversity is too high or too low, the performance of ET may decrease a lot.

For our study, we use Entropy Diversity Measurement $D_E(P) = -\sum p_k \log p_k$ to measure Diversity of Current Population (DCP) [18], where $p_k$ denote the proportion of the k class fitness in the current population, i.e., DCP = $D_E(P)$ is the diversity of the population. The higher DCP is, the higher the population diversity is. We take DCP as the way to measure the population diversity, which can reflect the state of the population and cost less time. And the normalized range of the population diversity is in [0, 1].

During the population evolution of GA, population prematurity may occur. The causes of population prematurity are: (1) most individuals of the population are similar; (2) the population diversities of several neighbor generations are similar. When the population prematurity happens, the population is only consisting of a few kinds of individuals, and the population will evolve with these individuals until the iteration times reach the maximum.

## 3  Hybrid Optimization Strategy

In our research, the optimization of ET is done based on two key processes: (1) the selection of static configuration strategies; (2) dynamic optimization process, which includes the optimization of initial population and the dynamic population optimization based on extended cataclysm. In this section, these aspects are explored and discussed in detail. Since our strategy is a combination of static configurations and dynamic optimizations, we call it a hybrid optimization strategy.

### 3.1  Static Configurations Selection

ET use GA to search the test case space, different test objects and test objectives have different test case spaces, it is the diversity of test case spaces that require testers to configure reasonably and correctly all parameters of ET according to the features of real applications, so as to generate test case satisfying test objectives effectively and efficiently, which is in fact a process of static configuration. Good static configuration strategies can prevent population prematurity [1, 2, 8, 9]. In this section, we discuss some factors which affect the selection of a good static configuration strategy. As an example, we show 24 static configuration strategies in Table 2, built from different configurations of parameters of Code, Selection, Crossover, and Survival described as following. B (Binary Code), G (Gray Code), R (Random Selection), S1 (Spin Selection), S2 (Single Crossover), U1 (Uniform Crossover), S3 (Spin Survival), F (Fitness Survival), U2 (Unrepeatable Survival) in [16]. Obviously, code, selection, crossover and survival etc. are main factors for a good static configuration strategy.

**Table 2.** Configuration strategies of ET.

| | Code | | Selection | | Crossover | | Survival | | | | Code | | Selection | | Crossover | | Survival | | |
|---|---|---|---|---|---|---|---|---|---|---|---|---|---|---|---|---|---|---|---|
| | B | G | R | S1 | S2 | U1 | S3 | F | U2 | | B | G | R | S1 | S2 | U1 | S3 | F | U2 |
| C1 | * | | * | | * | | * | | | C13 | * | | | * | * | | * | | |
| C2 | * | | * | | * | | | * | | C14 | * | | | * | * | | | * | |
| C3 | * | | * | | * | | | | * | C15 | * | | | * | * | | | | * |
| C4 | | * | * | | * | | * | | | C16 | | * | | * | * | | * | | |
| C5 | | * | * | | * | | | * | | C17 | | * | | * | * | | | * | |
| C6 | | * | * | | * | | | | * | C18 | | * | | * | * | | | | * |
| C7 | * | | * | | | * | * | | | C19 | * | | | * | | * | * | | |
| C8 | * | | * | | | * | | * | | C20 | * | | | * | | * | | * | |
| C9 | * | | * | | | * | | | * | C21 | * | | | * | | * | | | * |
| C10 | | * | * | | | * | * | | | C22 | | * | | * | | * | * | | |
| C11 | | * | * | | | * | | * | | C23 | | * | | * | | * | | * | |
| C12 | | * | * | | | * | | | * | C24 | | * | | * | | * | | | * |

## 3.2 Dynamic Optimization

If the population diversity is too high, the individuals of the population are too decentralized to converge, which makes ET become random testing; On the contrary, if the population diversity is too low, the evolution of ET may be gotten trapped into a local optimal solution. In both cases, the performance of ET decreases greatly. Dynamic optimization method proposed in this paper uses the following two optimization processes to improve the performance of ET.

(1) *Initial Population Optimization:* If the population diversity is higher than $T_{high}$, we keep replacing the individuals that have low *fitness* with the individuals that have high *fitness* until the diversity is less than $T_{high}$.

(2) *Dynamic Optimization with Cataclysm:* If the current population diversity is lower than $T_{low}$, we perform extended cataclysm operation on the population to avoid the occurrence of population prematurity.

$T_{high}$ and $T_{low}$ are two predefined thresholds which offer the judgment standards for determining which optimization strategy should be adopted. DCP reflects the diversity of the current population. If DCP is higher than $T_{high}$, the individuals of the current population are too decentralized to converge and it is necessary to generate a new population. If DCP is less than $T_{low}$ and ET has not found the global optimal solution, the population prematurity occurs and the extended cataclysm operation need be taken. Obviously, both $T_{high}$ and $T_{low}$ are critical parameters of ETPc (evolutionary testing process with extended cataclysm). If $T_{high}$ and $T_{low}$ are not properly assigned, the performance of ET may greatly decrease.

## 3.3   Initial Population Optimization

Before carrying out the genetic operations, ET first generates the initial population randomly. The performance of ET varies with the distribution of the individuals of the initial population. On the one hand, if the diversity of the initial population is too high, it will cost more time searching for the global optimal solution. In the worst case, ET cannot find the global optimal solution when the iteration of ET reaches the maximum iteration. On the other hand, if the diversity of the initial population is too low, which means the individuals are distributed close to each other, the search process will converge much faster and the possibility of the occurrence of the population prematurity increases. In both cases, the performance of ET decreases. Therefore, we need to make sure that the diversity of the initial population must belong to some interval when we generate the initial population, and we need take some operations to help the individuals distribute well.

Although some researchers have concentrated on improving the diversity of the initial population, they have not extensively considered the problem caused by the high diversity of the initial population. In this paper, we proposed an optimization algorithm (IPOptimization) to solve the problem with high diversity of the initial population. The proposed Algorithm IPOptimization includes the following operation. If the DCP of initial population generated randomly is higher than $T_{high}$, algorithm IPOptimization keeps replacing the individuals that have low fitness with the individuals that have high fitness and are marked as unselected until the diversity DCP is less than $T_{high}$. With the optimization of the initial population, the individuals of the initial population do not distribute widely which helps to improve the search speed compared to using the traditional initial population. However, this optimization cannot solve the population prematurity problem.

## 3.4   Dynamic Optimization with Extended Cataclysm

Even with the best static configuration strategy and the optimization of the initial population, ET cannot completely avoid the population prematurity problem. In ET, when the diversity of a population is too low, we can use cataclysm operation to produce a new population with higher diversity, which can avoid the population prematurity problem effectively. The traditional cataclysm operation keeps the best individual and kill the rest and this cataclysm operation involves the following steps: once the population prematurity happens, keep the best individual, kill all the other individuals, generate new individuals randomly and merge the best individual with the randomly-generated individuals to form a new population. The new population becomes the input of the next genetic operation.

This kind of traditional cataclysm operation can recover the population diversity and avoid the population prematurity problem. However, there are some obvious drawbacks. The traditional cataclysm operation only keeps the best individual and generates other individuals randomly, which almost like starting a new search. And the individuals of the new population may be widely distributed, with some individuals having a low fitness that may slow down the searching speed.

In this paper, we proposed an extended cataclysm algorithm $ETP_c$ (evolutionary testing process with extended cataclysm) to solve the population prematurity problem. The extended cataclysm operation includes four steps: (1) keep the best individual of the premature population; (2) kill all the other individuals and generate new individuals randomly; (3) merge the best individual and the randomly-generated individuals; (4) replace the individuals that have the lower fitness of the new population with the individuals in the set $V$, which stores the individual that has the highest fitness of each evolved population. Finally, a new population is generated.

The process of adding the good individual into $V$ is as follows: (1) fetch the best individual of the current population; (2) if $V$ is not full and the best individual is not in $V$, add the best individual into $V$. If $V$ is full and the best individual is better than the worst individual of $V$, replace the worst individual of $V$ with the best individual; otherwise, no action is needed.

The extended cataclysm algorithm by integrating the initial population optimization, which is shown in Algorithm 1 (ETPc algorithm), where the line 15–31 are the description for extended cataclysm operations. The extended cataclysm operation can improve the performance of ET, because after the extended cataclysm operation has been performed, the individuals do not distribute too scatted any more in the searching space and the "bad" individuals of the population are also replaced by better ones, so that the searching speed is improved. Beside $T_{high}$ and $T_{low}$, two other parameters OFO (Observation Frequency of ETPc) and size of the set $V$ can be adjusted to help to improve the performance.

---

**Algorithm 1 ETPc algorithm**

---

1. Input: Initial Population $P=\{p_1, p_2,..., p_n\}$ which is generated randomly

2. Input: Predefined Thresholds $T_{high}$ and $T_{low}$

3. Output: Generated Test Case $p_t^{'}$

4. // The start of the algorithm

5. ETPc $(P, T_{high} , T_{low})$

6. Carry out initial optimization, get the optimized population $P'=$IPOptimization$(P, T_{high})$ and its fitness value $F(P')$

7. Initial the vector $V$

8. while (there is no individual $p_t^{'}$ that satisfies the terminal condition) do

9.    while $(DCP(P') < T_{low})$ do

10.    //$P'$has become the premature population

11.    Select the individual $p_h^{'}$ with the highest fitness values in $P'$

12.    Generate $n$-1 individuals randomly, named as $P_{gen}^{'}$

13.    Carry out cataclysm by merging $p_h^{'}$ and $P_{gen}^{'}$ as a cataclysmic population $P_c^{'}$, e.g. $P_c^{'} = \{P_h^{'}\} \cup P_{gen}^{'}$

14.    Calculate fitness values of $P_c^{'}$ and get $F(P_c^{'})$

15.    if $V$ is not empty then

16.    Suppose the number of individuals in $V$ is $m$, delete $m$ individuals in $P'$ with lower fitness value than other $n$-$m$ ones

17.    Merge $P_c^{'}$ and $V$ as an optimized cataclysmic population $P_{oc}^{'}$, e.g. $P_{oc}^{'} = P_c^{'} \cup V$

18.    end if

19.    end while

20.    $P' = P_{oc}^{'}$

21.    Carry out the selection, crossover and mutation operations on $P'$, and generate the offspring population $P_o^{'}$

22.    Carry out the survival operation on $P'$ and $P_o^{'}$, and generate the offspring population $P_s^{'}$

23.    Calculate fitness values $P_s^{'}$ and get $F(P_s^{'})$

24.    Select the individual $p_h^{'}$ in $P_s^{'}$ with the highest fitness value $f(p_h^{'})$

25.    if ($V$ is not full) then

26.    Add $p_h^{'}$ into $V$, e.g. $V = V \cup \{p_h^{'}\}$;

27.    else

28.    Select the individual $v_l$ in $V$ with the lowest fitness value $f(v_l)$

29.    if $( f(p_h^{'}) > f(v_l) )$then

---

30.     Replace $v_l$ with $p_h'$, e.g. $V = V \setminus \{v_l\} \cup \{p_h'\}$;
31.     end if
32.     end if
33.     $p_t' = P_s'$
34.   end while
35.   Return $p_t'$

---

# 4   Experimental Study

In this section, we will discuss how to evaluate some static configuration strategies for selection, how to improve the performance of ETPc with $T_{high}$ and $T_{low}$, and what about the result of comparing ETPc with DOMP proposed in [16] and traditional ET without optimization by performing a lot of experiments.

For some large and complex programs, it is difficult to trigger the occurrence of the population prematurity. Therefore, in our experiments, we use LineCover simple programs as the test objects (shown in Table 1), and because of the same examples as DOMP and traditional ET, we can do comparison easily. The test criterion of experiments is to cover a particular branch of the test objects (Branch Coverage).

## 4.1   Parameter Settings

For each configuration strategy of ET, we let the total number of experiments be $T_0$, the size of the population be $P_{size}$, and the maximum iteration of an experiment be *IterMax*. In addition, in order to compare the performance of $ETP_c$ with DOMP, we adopt the performance parameters $T_h$ and HAG used in [16]:

$$(1) \ \text{Hit Percent} (P_h) : P_h = T_h/T_0$$

where $T_h$ is the number of experiments that fulfill the testing objective. Obviously, $P_h$ is an essential index to evaluate the effectiveness of ET. The higher $P_h$ is, the more effective ET is.

Another performance indicator is HAG:

$$(2) \ \text{Hit Average Generation (HAG): HAG} = \left( \sum\nolimits_{i=1}^{T_0} H_i \right)/T_0$$

where $H_i$ denotes the iteration number of $i^{th}$ experiment. HAG is the average iteration number of an experiment, which reflects the converge speed of ET. The smaller HAG is, the faster ET is.

In our experiments, other parameters of ETPc are set as follows: the crossover possibility $P_c = 0.9$, the mutation possibility $P_m = 0.1$, the survive possibility $P_s = 0.5$, the size of the population $P_{size} = 100$, predefined thresholds $T_{high} = 0.85$, $T_{low} = 0.4$, the maximum iteration IterMax = 250 and the total number of experiments $T_0 = 300$. In addition, the value range of the input variable in LineCover is [1, 50]. The normalized range of the population diversity is [0, 1].

## 4.2  Performance Improvement of ETPc

Table 3 shows the comparison among ET without optimization(Un_O), DOMP and ETPc. As shown in Table 3, for LineCover in Table 1, $ETP_c$ improves better than DOMP. Firstly, for $P_h$, compared with ET without optimization, except C8 and C9, ETPc performs better for all the other configurations. Compared with DOMP, except C8, C9, C15 and C16, ETPc performs better for all the other configurations. It can be found that Ph is improved on average. Secondly, for HAG, compared with ET without optimization, except C8 and C9, ETPc performs better for all the other configurations. Compared with DOMP, except C8, C9 and C12, ETPc performs better for all the other configurations. It can be found that HAG is improved on average.

**Table 3.** Performance comparison between Un_O, DOMP and ETPc in LineCover.

|     | Ph | | | HAG | | |     | Ph | | | HAG | | |
| --- | --- | --- | --- | --- | --- | --- | --- | --- | --- | --- | --- | --- | --- |
|     | U | D | E | U | D | E |     | U | D | E | U | D | E |
| C1 | 0.26 | 0.42 | 0.54 | 221 | 195 | 175 | C13 | 0.28 | 0.34 | 0.54 | 210 | 201 | 194 |
| C2 | 0.02 | 0.0 | 0.04 | 249 | 250 | 243 | C14 | 0.62 | 0.62 | 0.04 | 155 | 155 | 101 |
| C3 | 0.02 | 0.02 | 0.1 | 250 | 250 | 235 | C15 | 0.08 | 0.04 | 0.23 | 231 | 246 | 214 |
| C4 | 0.68 | 0.78 | 0.9 | 156 | 144 | 92 | C16 | 0.62 | 0.72 | 0.7 | 160 | 150 | 97 |
| C5 | 0.02 | 0.0 | 0.1 | 248 | 250 | 239 | C17 | 1.0 | 1.0 | 1.0 | 49 | 49 | 47 |
| C6 | 0.04 | 0.04 | 0.5 | 244 | 244 | 135 | C18 | 0.04 | 0.12 | 0.6 | 246 | 235 | 116 |
| C7 | 0.34 | 0.4 | 0.56 | 215 | 215 | 131 | C19 | 0.44 | 0.26 | 0.5 | 205 | 231 | 129 |
| C8 | 0.08 | 0.06 | 0.04 | 244 | 244 | 246 | C20 | 0.56 | 0.58 | 0.75 | 156 | 155 | 92 |
| C9 | 0.02 | 0.02 | 0.02 | 248 | 248 | 249 | C21 | 0.04 | 0.02 | 0.14 | 247 | 249 | 238 |
| C10 | 0.74 | 0.5 | 0.8 | 136 | 136 | 86 | C22 | 0.66 | 0.74 | 0.74 | 152 | 121 | 120 |
| C11 | 0.04 | 0.06 | 0.08 | 248 | 248 | 241 | C23 | 1.00 | 1.00 | 1.00 | 49 | 49 | 49 |
| C12 | 0.06 | 0.04 | 0.1 | 136 | 240 | 241 | C24 | 0.02 | 0.06 | 0.38 | 248 | 246 | 165 |

U: ET without optimization. D: DOMP. E: ETPc

During the population evolution of ETPc, once the population prematurity occurs, the extended cataclysm operation starts working. It can be considered that the search space is enlarged. ETPc throws away the old search space and keep the best individual in the new search space. The best individual can help speed up the population evolution process. So the performance of ETPc is improved a lot. And the individuals of the vector $V$ can also help the population evolution.

## 5  Conclusion and Future Work

In this paper, we propose a hybrid optimization method to optimize evolutionary testing based on following several aspects: (1) select a suitable static configuration strategy; (2) optimization of initial population; (3) generate new population based on extended cataclysm. Where fitness and population diversity are two important factors used to guide optimization process, thresholds $T_{high}$ and $T_{low}$ are two important boundaries used to control the selection of population individuals for preventing population distribute too decentralized or two centralized. Empirical results show that our hybrid optimization strategy can greatly improve the performance of ET. And in most cases, $ETP_c$ works better than DOMP and traditional ET without optimization.

There are some situations that $ETP_c$ does not work well. If the branch condition in a program is very complicate, the performance of $ETP_c$ is not so good. And for some configuration strategies, the time cost of $ETP_c$ is very high. In future works, we will continue the research on the improvement of the performance of $ETP_c$.

**Acknowledgements.** This work was supported in part by National Natural Science Foundation of China under Grant No. 61806068, 61672204, by Visiting Scholar at Home and Aboard Funded Project of Universities of Anhui Province under Grant gxfxZD2016209, by Key Technologies R&D Program of Anhui Province under Grant 1804a09020058, by the Major Program for Scientific and Technological of Anhui Province under Grant 17030901026, by Talent Research Foundation Project of Hefei University under Grant 16-17RC23, by Humanities and Social Science Research Project of Universities of Anhui Province under Grant SK2018A0605.

# References

1. Pargas, R.P., Harrold, M.J., Peck, R.R.: Test - data generation using genetic algorithms. Softw. Test. Verif. Reliab. **9**(4), 263–282 (1999)
2. Pachauri, A., Srivastava, G.: Automated test data generation for branch testing using genetic algorithm: an improved approach using branch ordering, memory and elitism. J. Syst. Softw. **86**(5), 1191–1208 (2013)
3. Bauersfeld, S., Wappler, S., Wegener, J.: A metaheuristic approach to test sequence generation for applications with a GUI. In: Cohen, M.B., Ó Cinnéide, M. (eds.) SSBSE 2011. LNCS, vol. 6956, pp. 173–187. Springer, Heidelberg (2011). https://doi.org/10.1007/978-3-642-23716-4_17
4. Fraser, G., Arcuri, A., McMinn, P.: A memetic algorithm for whole test suite generation. J. Syst. Softw. **103**, 311–327 (2015)
5. Jia, Y., Harman, M.: An analysis and survey of the development of mutation testing. IEEE Trans. Softw. Eng. **37**(5), 649–678 (2011)
6. Vos, T.E.J., Baars, A.I., Lindlar, F.F., et al.: Industrial case studies for evaluating search based structural testing. Int. J. Softw. Eng. Knowl. Eng. **22**(08), 1123–1149 (2012)
7. Anand, S., Burke, E.K., Chen, T.Y., et al.: An orchestrated survey of methodologies for automated software test case generation. J. Syst. Softw. **86**(8), 1978–2001 (2013)
8. Arcuri, A., Fraser, G.: On parameter tuning in search based software engineering. In: Cohen, M.B., Ó Cinnéide, M. (eds.) SSBSE 2011. LNCS, vol. 6956, pp. 33–47. Springer, Heidelberg (2011). https://doi.org/10.1007/978-3-642-23716-4_6
9. Arcuri, A., Fraser, G.: Parameter tuning or default values? An empirical investigation in search-based software engineering. Empir. Softw. Eng. **18**(3), 594–623 (2013)
10. Dimitar, M., Dimitrov, I.M., Spasov, I.: Evotest-framework for customizable implementation of evolutionary testing. In: International Workshop on Software and Services (2008)
11. Inkumsah, K., Xie, T.: Improving structural testing of object-oriented programs via integrating evolutionary testing and symbolic execution. In: Proceedings of the 2008 23rd IEEE/ACM International Conference on Automated Software Engineering, pp. 297–306. IEEE Computer Society (2008)

12. McMinn, P., Harman, M., Lakhotia, K., et al.: Input domain reduction through irrelevant variable removal and its effect on local, global, and hybrid search-based structural test data generation. IEEE Trans. Softw. Eng. **38**(2), 453–477 (2012)
13. Sofokleous, A.A., Andreou, A.S.: Automatic, evolutionary test data generation for dynamic software testing. J. Syst. Softw. **81**(11), 1883–1898 (2008)
14. McMinn, P., Binkley, D., Harman, M.: Empirical evaluation of a nesting testability transformation for evolutionary testing. ACM Trans. Softw. Eng. Methodol. (TOSEM) **18** (3), 11 (2009)
15. McMinn, P., Holcombe, M.: Evolutionary testing using an extended chaining approach. Evol. Comput. **14**(1), 41–64 (2006)
16. Xie, X., Xu, B., Shi, L., et al.: A dynamic optimization strategy for evolutionary testing. In: 12th Asia-Pacific Software Engineering Conference, APSEC 2005, 8 pp. IEEE (2005)
17. McMinn, P.: Search-based software testing: past, present and future. In: 2011 IEEE Fourth International Conference on Software Testing, Verification and Validation Workshops (ICSTW), pp. 153–163. IEEE (2011)
18. Wu, X., Zhu, Z.: Research on diversity measure of genetic algorithms. Inf. Control Shenyang **34**(4), 416–422 (2005)
19. Harman, M.: Testability transformation for search-based testing. In: Keynote of the 1st International Workshop on Search-Based Software Testing (SBST) in Conjunction with ICST 2008 (2008)
20. Arcuri, A.: A theoretical and empirical analysis of the role of test sequence length in software testing for structural coverage. IEEE Trans. Softw. Eng. **38**(3), 497–519 (2012)
21. Harman, M.: Automated test data generation using search based software engineering. In: Proceedings of the Second International Workshop on Automation of Software Test, p. 2. IEEE Computer Society (2007)
22. Harman, M, McMinn, P.: A theoretical & empirical analysis of evolutionary testing and hill climbing for structural test data generation. In: Proceedings of the 2007 International Symposium on Software Testing and Analysis, pp. 73–83. ACM (2007)
23. Lakhotia, K., Harman, M., McMinn, P.: A multi-objective approach to search-based test data generation. In: Proceedings of the 9th Annual Conference on Genetic and Evolutionary Computation, pp. 1098–1105. ACM (2007)
24. Zhao, Y., Zhang, Q., Wang, B.: Improving the lower bounds of DNA encoding with combinational constraints. J. Comput. Theor. Nanosci. **9**(1), 50–54 (2012)
25. Eichie, J.O., Oyedum, O.D., Ajewole, M.O., et al.: Artificial Neural Network model for the determination of GSM Rxlevel from atmospheric parameters. Eng. Sci. Technol. Int. J. **20** (2), 795–804 (2017)

# Graphic Images

# Android Oriented Image Visualization Exploratory Search

Jianquan Ouyang[1,2]([✉]), Hao He[1,2], Minnan Chu[1,2], Dong Chen[1,2], and Huanrong Tang[1,2]

[1] College of Information Engineering, Xiangtan University,
Xiangtan 411105, China
oyjq@xtu.edu.cn
[2] Key Laboratory of Intelligent Computing and Information Processing,
Ministry of Education, College of Information Engineering, Xiangtan University,
Xiangtan 411105, China

**Abstract.** When using traditional image search engines, smartphone users often complain about their poor user interface including poor user experience, and weak interaction. Moreover, users are unable to find a desired picture partly due to the unclear key words. This paper proposes the word-bag co-occurrence scheme by defining the correlation between images. Through exploratory search, the search range can be expanded and help users refine retrieval of the expected images. Firstly, the proposed scheme applied the bag of visual words (BoVW) vector by processing images on Hadoop. Secondly, similarity matrix was constructed to organize the image data. Finally, the images in which users were interested was visually displayed on the android mobile phone via exploratory search. Comparing the proposed method to current methods by testing with image data sets on ImageNet, the experimental results show that the former is superior to the latter on visual representation, and the proposed scheme can provide a better user experience.

**Keywords:** Android · Hadoop · Visualization exploratory search · BoVW

## 1 Introduction

As the use of mobile devices grow, such as smart phones, tablet computers, and digital cameras, users can upload and browse images at anytime. In 2010, Beaver et al. [1] pointed out that Facebook, whose users uploaded about 1 million pictures each week, had stored around 260 billion images on its servers. According to the statistics of Intel [2], there were nearly 20 million pictures being viewed on Flickr and 3000 pictures being uploaded per minute. As of August 2011, there were 6 billion images on Flickr. It is important to retrieve a particularly one from massive images.

When searching for pictures, users are accustomed to use traditional image search engines (Such as Google, Yahoo, Bing, etc.). The typical work flow of these search engines starts by the user inputting keywords and then the system returns a sorted list of images. The returned results are uncorrelated images with mixed structure, and it will waste user's time [3]. Moreover, it is commonly the case that an appropriate image is

© Springer Nature Singapore Pte Ltd. 2019
R. Mao et al. (Eds.): ICPCSEE 2019, CCIS 1059, pp. 57–73, 2019.
https://doi.org/10.1007/978-981-15-0121-0_4

not returned and, in this case, the user will need to modify the query keywords or select another image for the next query. Thus, the current search engines seem to be more suitable for a "one-time hit of inquiry" than a content-oriented interaction. If users only have a vague concept of the query target or they just want to explore additional information associated with this target, current search engines are not adequate. We believe that in the process of retrieving images, relevant messages that guide users to find their required information should be returned, and it is also quite meaningful that relevant information should be given [4].

In recent years, the method of exploratory search has received wide attention and lots of studies have focused on it [4]. This method can combine querying with browsing strategies for users to explore, learn and discover new knowledge. Exploratory search originated from semantic exploration, and has gradually been applied in the field of image search. Cai [3] presents a hierarchical clustering method that separate the image search results into different semantic clusters. Jing et al. [5] show an image browsing system, which uses the image content to organize search results. It is a dynamic web-based user interface that enables users to browse images quickly and interactively. Yee [6] uses the query mode of hierarchical metadata for massive images. Liu [7] present an interactive visual exploratory tool for selecting billboard locations.

However, the existing exploratory search systems are not optimized for smart phones, because using the images retrieved by the scroll list as a mobile page is not user-friendly [8]. Exploratory search requires user to conduct multiple interactions before completing a search process. User interaction with smartphones requires using a finger to select, drag, and dig the information in the process of touching and clicking the screen. With their portability and multi-touch, mobile phones provide good conditions of human-computer interaction for the exploratory search. Several scholars [9, 10] have researched into exploratory search on mobile devices, but recent studies still have various limitations. According to a survey done in 2003 [9]. the limitation of the screen size of mobile devices and the inconvenience of soft keyboard, which provides input, make it uncomfortable for the users to do more search. Westlund [10] found that the most smooth and convenient way for users to interact with a system are touching, double-clicking, sliding, dragging, dropping, extruding with two or more fingers, etc.

In order to solve these problems, we have done following work:

(1) To solve the existed stand-alone image processing problems, such as insufficient memory and long processing time, we use the Hadoop platform to process massive images, and then use the spectral clustering methods to get the "bag of visual words" (BOVW), making the abstract image information to be "visual vocabulary".
(2) We first propose a method using the analysis of co-occurrence technology to define the correlation between images, which can make it easier to understand the relation between images.
(3) We implement a visualization exploratory search system of massive images based on correlation, which can solve the problem of un-convenient input to android smart-phone users, making it friendlier to users.

Compared with the existed search methods, the proposed method in this paper has three advantages as followed:

(1) The search results are displayed to the user with a graphical interface. User experience can be improved.

(2) User enable to distinguish the relationship between the query-image and result list image more clearly.

(3) The search system can guide the user to get richer and more useful information. User interaction can be improved.

Experimental result shows that the proposed method can effectively improve the ability of users to access relevant images and get the relationship between images.

## 2  Exploratory Search

The research area of exploratory search is nested in the field of human-computer interaction. It started in 2005 with the symposium of exploratory search interface held at the University of Maryland in United States. The goal is to advance the next generation search interface [11]. Even if the user's search goal is fuzzy, exploratory search supports the user to search for more information and learn from information to find the solution.

In 2006, Marchionini [4] made a detailed division and explanation on the exploratory search from the discovery to the understanding of information. In 2009, White and Roth [12] made a conclusion for exploratory search. They thought "exploratory search can be used to describe a scene of searching question based on information, which is unlimited, diversified, and the searching process needs many iterations and the combination of multiple strategies".

Exploratory search helps users to explore and discover new knowledge through a combination of strategies like query, browse, interaction and hints. Existing Web search engines pay much attention to the precision, especially the high correlation between the initial page search results and the final search target. Unlike these search engines, exploratory search returns the result ranked through numerous iterations, so it focuses more on the completeness and ensures that the results ranked back can also meet the needs of users. In order to reduce the density of information space, exploratory search usually returns results to users in way of media type like figures, trees, network or documents, and requires users to browse, compare and make a qualitative or quantitative evaluation [4]. Exploratory search requires users to participate in the search process, so that the two can interact, and it may require many interactions to complete a search process. Klouche [13] designed Exploration Wall, that allows incremental touch based exploration and sense-making of large information spaces by combining entity search and spatial configuration.

I find the introduction a bit misleading as it describes the task of image exploration. However the image exploration itself is performed in a simple relevance feedback manner - in each iteration the user selects a single image to get a list of its nearest neighbors in the dataset. There are more sophisticated methods for image exploration which are based on nearest neighbors queries (see missing references below).

A similar topic is relevance feedback. Since Rui [14] tried to bridge the semantics gap between the input image and low level visual features, many efforts of online iterative or learning were proposed to retrieve most relevant images in massive images [15]. Mohanan [16] gave an overview of subspace learning based relevance feedback algorithm for retrieving images. However, there is a little bit of difference between relevance feedback and exploratory search. Namely, the exploratory search focuses on correlation instead of similarity.

# 3  Image Feature Vectors

This paper uses the Bag of Visual Word (BoVW) model [17] as the image characteristics to establish the "Visual Dictionary". Since the features in the image and the visual vocabulary are not directly visible in the document, we need to extract the features and cluster images to obtain visual dictionary. In this paper, image feature vector generation process proceeds as follows (as shown in Fig. 1):

Step 1: Use SIFT (Scale Invariant Feature Transform) [18] to extract the SIFT features in all images by hadoop;

Step 2: Collect all SIFT features together; Calculate the similarity of sift feature points to generate a feature matrix; The similarity is adjusted with 7-neighbor.

Step 3: Apply spectral clustering NJW [19] algorithm to cluster SIFT features, and cluster similar points into "Visual Dictionary"; Insert each SIFT features extracted from each image into the closest object of image visual dictionary; so the image is mapped to an N-dimension BoVW vector.

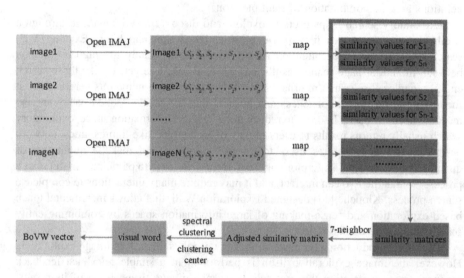

**Fig. 1.** Image feature vector generation process

## 3.1 Extract the Image Sift Feature by Hadoop

Firstly, we store the image in the HDFS file system, using the open source library Open IMAJ [20] to extract the image sift feature, and then input each feature point into the HDFS file system as <image name, sift feature vector>, so we can get all the sift feature points for each image.

---

**Algorithm 1:**

**Input:** <Text, KUImage>. Text is the image name, as the key; KUImage is image object of byte stream, as the value.

**Output:** <Text, Text>. The first Text is the image name. The second Text is a string of 128-dimensional sift feature points.

---

**Step1:** Read the image to KUImage object, Stored in HDFS;

**Step2:** Converts an Image object to an image type that can be processed in Open IMAJ ;

**Step3:** The image is extracted and saved in the feature list;

**Step4:** Remove the sift feature point from the feature list and write it to HDFS as <Text, Text>.

---

## 3.2 Generating Similarity Matrices

First, the sift feature points $(S_1, S_2, S_3, \ldots, S_i, \ldots, S_n,)$ for all the pictures are stored in the HDFS. Since the undirected complete graph with n vertices have $n \times (n - 1)/2$ edges, and the distance from $S_i$ to $S_j$ is equal to the distance from $S_j$ to $S_i$. Therefore, when calculating the similarity between vertices, we only need to calculate $n \times (n - 1)/2$ times instead of $n \times (n - 1)$ times. In calculating the similarity of the data points labeled i, we should calculate with $n - i + 1$ data points—calculate $\{(S_i, S_i), (S_i, S_{i+1}), \ldots, (S_i, S_n)\}\}$ Similarity. The similarity of $(S_i, S_j)$ is represented by $\delta_{ij}$.

Second, because each sift feature point is 128-dimensional, the matrix composed of feature points is quite large. We use HDFS and HBase to achieve the construction of similar matrices. We put the $S_1$ data list and $S_n$ data list together, calculate their respective similarity. The data list with $S_2$ are stored together with the data list with $S_{n-1}$ to calculate the similarity and in this order back analogy. It can achieve the purpose of load balancing. Then, we can get <imageName, i, j, $\delta_{ij}$>.

In order to improve the performance of spectral clustering algorithm. We adjust the value of $\delta_{ij}$ using the 7-neighbor approach proposed by Zelnik-Manor and Perona [21]. Experiments show the method—adjusting the value of $\delta_{ij}$ with the 7-neighbor approach proposed—is superior to the one clustering high-dimensional data.

### 3.3    Generating the Bag of Visual Word (BoVW) Vector

After obtaining a similar matrix, we apply spectral clustering NJW [19] algorithm to cluster similar matrix. We extract the data from HBase to the mahout spectral clustering algorithm in the form of <i, j, $\delta_{ij}$> (i: row number; j: column number; $\delta_{ij}$: the value of similarity). The result of cluster center is the visual word. We can get <imageName, i, j, $\delta_{ij}$, category number>. For each picture we can get <imageName, category list>. After sorting the category list, the BoVW vector can be obtained.

## 4    Image Correlation

### 4.1    Definition of Image Correlation

Inspired by vocabulary symbiotic methodology in text processing domain [22], we build the correlation network among images by studying the visual vocabularies' co-occurrence relationship among images. In this paper, we define the image correlation as follows:

1. Multiple visual vocabularies in the same image have a symbiotic relationship.
2. Within two different kind of images, if there exist some visual vocabularies in the same class, and then the remained visual vocabularies from the two images can be established a symbiotic relationship using the common visual vocabularies as bridge.

The degree of correlation is defined as follows:

1. The degree of correlation of any two visual vocabularies is equal to the product of their weights in the figure. The equation is shown as Eq. 1.

$$W_k(d_i, d_j) = \left(\text{freq}(d_i) \times \text{freq}(d_j)\right)/\text{size}(k) \tag{1}$$

In Eq. 1, Wk($d_i$, $d_j$) represents the weight value of visual vocabulary ($d_i$, $d_j$) in $k^{th}$ image. $\left(\text{freq}(d_i) \times \text{freq}(d_j)\right)$ denotes the product of numbers of visual vocabulary $d_i$ and $d_j$ in $k^{th}$ image. The greater value is, the stronger correlation of these two visual vocabularies in the image should be.

Size(k) denotes the number of visual vocabularies contained in $k^{th}$ image.

2. The degree of correlation between two different images is defined as half the sum of the weights of the common visual vocabularies shared by the two images, which is shown as the Eq. 2.

$$W(I_i, I_j) = \sum_{k=1}^{m} \text{freq}(v_k)_{V_k \in I_i}/\text{size}(I_i) + \text{freq}(v_k)_{V_k \in I_j}/\text{size}(I_j) \tag{2}$$

In Eq. 2, $W(I_i, I_j)$ infers to the correlation between image $I_i$ and $I_j$; m refers to the common visual vocabularies contained between two images. Among them, the symbiotic relationship of multiple visual vocabularies in the same image means that there is a correlation between any two visual vocabularies in the image, because it is an undirected complete graph, images that contain n visual vocabularies will have correlation of set $C_n^2 = n \times (n-1)$.

## 4.2    Establishment of *the* Image Correlation Diagram

To illustrate easily, we assume that there is an image library that consists of four images and the size of the Bag of Visual Words (BoVW) is 10. The data are shown in Table 1.

**Table 1.** The vector of BoVW in four images

| Image name | Visual words | | | | | | | | | |
|------------|---|---|---|---|---|---|---|---|---|---|
|  | A | B | C | D | E | F | G | H | I | K |
| Image 1 | 3 | 1 | 0 | 0 | 2 | 4 | 0 | 0 | 0 | 0 |
| Image 2 | 1 | 0 | 6 | 1 | 3 | 0 | 1 | 0 | 0 | 0 |
| Image 3 | 0 | 0 | 1 | 2 | 0 | 1 | 0 | 0 | 0 | 0 |
| Image 4 | 0 | 0 | 0 | 0 | 0 | 0 | 1 | 1 | 1 | 1 |

Rows in Table 1 represent the vector of BoVW of the corresponding image, columns represent the visual vocabulary of BoVW, and value indicates the number of occurrences of the visual words in the image.

---

**Algorithm 2 :**

**Input:** image name and the data of the corresponding image's BoVW vector.

**Output:** image information.

---

**Step1:** Read a row of data from the input data, and use a regular expression to match the row of data to get the image name and the BoVW vector of images.

**Step2:** Convert the BoVW vector into a one-dimensional array including m (m is the size of BoVW) Integer objects;

**Step3:** Calculate the frequency of non-zero values in the array and save it in the array weight. Note the corresponding subscript index and store it in the variable array visual_id. The value of frequency is the number of times the visual vocabulary appears in the image;

**Step4:** Calculate all permutations and combinations in array visual_id and list them in the order of format (visword_i, visword_j), which has a total of n*(n-1)/2 Group, and calculates the correlation between weight_ij according to the formula (1);

**Step5:** Store the data in a string image information, output image information and save it to disk;

**Step6:** Repeat Step 1 to 5 until the end of the file is encountered.

---

In accordance with the Algorithm 1: we first get the non-zero visual vocabularies of images and their corresponding weights. The data are as follows:

Images 1: (A, 3), (B, 1), (E, 2), (F, 4)
Images 2: (A, 1), (C, 6), (D, 1), (E, 3), (G, 3)
Images 3: (C, 1), (D, 2), (F, 1)
Images 4: (G, 1), (H, 1), (I, 1), (K, 1)

Then, the correlation of all the visual vocabularies in every picture is established. At this time, each image is represented by the edge correlation (as shown in Fig. 2), whose permutations and combinations are as follows:

Images 1: (A, B), (A, E), (A, F), (B, E), (B, F), (E, F)
Images 2: (A, C), (A, D), (A, E), (A, G), (C, D), (C, E), (C, G), (D, E), (D, G), (E, G)
Images 3: (C, D), (C, F), (D, F)
Images 4: (G, H), (G, I), (G, K), (H, I), (H, K), (I, K)

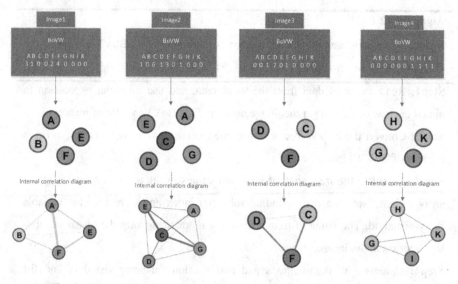

**Fig. 2.** The establishment process of the internal correlation among images

After building the internal correlation diagram among images, we also need to save the image details, including image name, weight of each visual vocabulary and all relevant edges of the image and their corresponding weights, which is used to provide data source for later correlation retrieval.

Next, according to the Algorithm 3, input the data which are from Algorithm 2 to create relationship diagrams among images.

---

**Algorithm 3:**
**Input:** Image information, which is the specific information about images in algorithm 2;
**Output:** Some visual word and all related_words associated with the visual word.

---

**Step1:** Use regular expressions to match image information, extract the information (visword_i, visword_j, weight_ij) and save it into the edge content;
**Step2:** Extract all related (visword_i, visword_j) of edge content and store it in the arrays sub edge;
**Step3:** Treat visword_I as key and visword_j as value and write them to disk;
**Step4:** Treat visword_i as key and visword_j as value and write them to disk again;
**Step5:** Process Iterator<rel_words_list> in the intermediate result <visword, Iterator<rel_words_list>> from the two steps (Step 3, Step 4) with the relationship duplicate removal function;
**Step6:** Sort the linear list vis words in ascending order;
**Step7:** Remove elements from vis_words and store them in string related_words
;
**Step8:** Output visual_word related_words.

---

After duplicated removal and sorting, we can get each visual vocabulary as well as lists of other words that has relationship with this one. In this way, we can construct the correlation diagram among images by lining up the related visual words with this visual vocabulary. In addition, it is important to note that because it is an undirected graph, there will be repeated edge. At this time, we need to locate the first number which is greater than its own visual vocabulary from the relevant visual vocabulary list (such as in {B, (A, E, F)}, we need to find the first visual vocabulary E larger than B, then simply B and E and E after the visual vocabulary of connection), then line up visual vocabularies of this number and after the number, as shown in Fig. 3.

### 4.3 Search Based on Correlation Diagram Among Images

After image correlation diagram has been built, we are now able to perform image retrieval. We proceed as follows:

Step 1. Upload the inquiry image to the Web Server, and Extract the image SIFT feature points and calculate with the center point of the dictionary database to get the query image's visual vocabulary vector.
Step 2. Sort the obtained visual vocabulary with Quicksort algorithm by their weights in descending order, take Top-K visual vocabularies, and map it into the correlation network diagram. The query images Top-K visual vocabularies are B, C, and E as shown in Fig. 4 (K = 3).

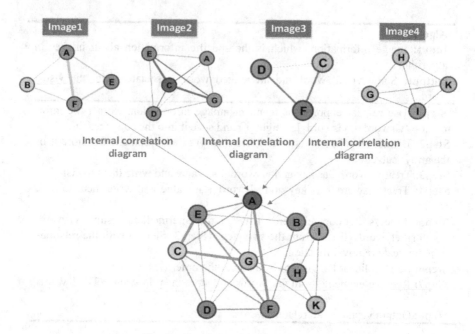

**Fig. 3.** Correlation diagram generation

Step 3. After mapping, choose the Top-K visual vocabularies in the form which stores these words and returns the one with the largest weighting values.

Step 4. Users select one image returned from one of the Top-K visual vocabularies, and then diffuse based on the image (this paper has spread only 1 layer), so we can obtain a set of edges constituted by the image and other visual vocabularies. As shown in Fig. 4, diffusing from point B, yields a set of {(B, A), (B, E), (B, F)}; diffusing from point E, a set of {(E, A), (E, B), (E, D), (E, G), (E, F), (E, C)} is obtained.

Step 5. Compute the union of the edge sets obtained from Step4, and for the recurring edges, only keep a record, such as computing the union of diffused edges of B, E, a set of {(B, A), (B, E), (B, F), (E, A), (E, D), (E, G), (E, F)} generates.

Step 6. For all images, identify whether there exists edges in each image; if so, then sum the score of correlation of these edges. Furthermore, sort the scores, and choose the most relevant results from the sorted Top-K images.

According to the results of above example, the chosen images are image 1 and image 2 as shown in Fig. 2.

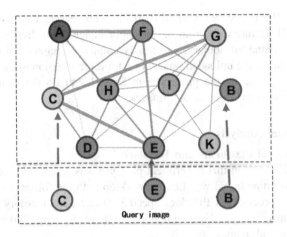

**Fig. 4.** Schematic diagram of maps and diffusion in image relevance retrieval

# 5  Realization of Image's Visual Exploratory Search

## 5.1  Visual Exploratory Search Framework

First of all, we get the vector representation of the image by spectral clustering in the platform of Hadoop. Then, we construct the image correlation diagram with the vector data. Finally, we store the data on the web Server. Through all of these steps, we can conduct the process of the visual exploratory search based on correlated numerous images. The search frame is as shown in Fig. 5.

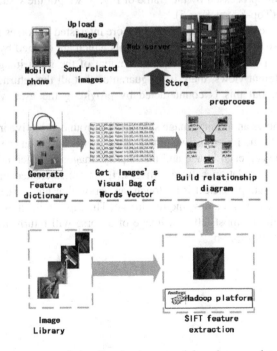

**Fig. 5.** The system prototype of visual exploratory search based on correlated numerous images

After getting the image, we upload it to the web server. The server first of all conducts the SIFT feature extraction on the input image, then divides these features into their nearest visual vocabulary, and finally counts the number of occurrences of each visual word to get the uploaded image's vector representation of the BoVW. In the Web server, we can follow the procedure in Sect. 4.3 to do the visual exploratory search based on the correlation among images.

## 5.2    Experimental Analysis

### 5.2.1    Experimental Platforms and Data

The main hardware used in this experiment platform was a single PC and an Android phone. The PC machine brand was Lenovo desktop with the following specifications: Intel Core i3-2130 processor, CPU clock speed 3.40 GHz, and memory (RAM) of 4G. The Android smart phone was Huawei honor 6, with a Hass Kirin 920 processor, CPU clock speed 1.3G, and memory of 3G.

Experimental data set chosen was a subset of ILSVRC2014 of ImageNet [22], in which a total of 1.2 million pictures1000 classes were stored.

### 5.2.2    Experimental Results

In the experiment, we conducted the feature extraction on each image using Hadoop platform. A total of 1,546,900,589 SIFT features were extracted in a period of 25,230 s. We used the spectral clustering to cluster the features into 53,620,387 classes; representing each image as a 53,620,387-dimensional vector. Finally, we stored, into the database, the correlated diagram built according to the method described in Sect. 4. While adding the image of "rider" by mobile phone randomly (Fig. 6a), after being processed by the frame of Fig. 5, we got the visualized results as shown in Fig. 6a (Top-K, set K = 5).

In the Fig. 6a, we can see the user may be more interested in horses and people, and at the same time we also get the trees and animals which are defined by the systems as "relevant" images. Due to the rich semantic content of the input image and different people having different views, there is no guarantee that all the returned images meet user's satisfaction, but this is counteracted by providing the user with other more relevant information.

The Fig. 6a, shows an inquiry image submitted surrounded by five images returned. These correspond to the five most weighted visual words based on the inquiry image. There is a line between each image and the inquiry image; the thickness of the line is proportional to the correlation between images. If we are interested in one of related images, we perhaps simply select it by touch, an action that has a cascading effect of turning the select image into a new inquiry image. The Fig. 6b illustrates this effect; In Fig. 6a, touching the topmost-leftmost image of a horse will return images related to the horse as shown in Fig. 6b.

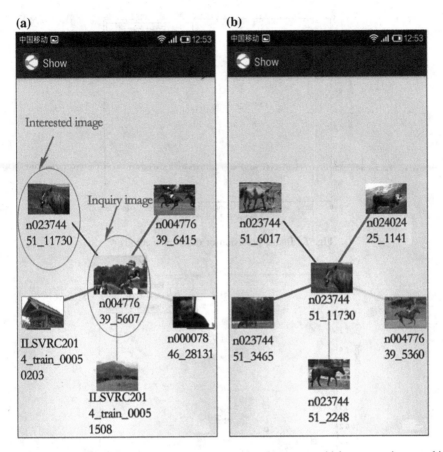

**Fig. 6.** (a): Correlated visual exploratory search results. (b) Images which users are interested in

### 5.2.3   Objective Evaluation of Experimental Algorithm

A fair image retrieval algorithm not only requires a satisfactory accuracy, but also an excellent performance of recall index [23]. Because of the limitation of the number of images that a phone can show in a single screen, our query algorithm was evaluated on a computer. Evaluation indices were mainly recall and precision. We chose 5 categories and 20 pictures in each category giving a total of 100 images to query. Each image was labelled according to its category. By changing the number of return image, we calculated the average recall and precision, and compared it with the experiments of Huang [24] and Lin et al. [25]. Experimental data indicates that the proposed algorithm has a better average recall and precision than the two algorithms on ILSVRC2014 datasets, as shown in Figs. 7 and 8.

The accuracy of image search could be greatly improved via several (usually 2–3) user interaction on the phone. As the number of different image categories and the properties of images are different, the search results may be varied (Fig. 9). After iteration 2–3 times, the search accuracy began to rise, and users could find the images that they were interested in. On the other hand, users often could not be patient to

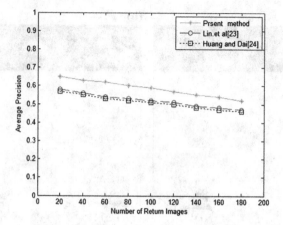

**Fig. 7.** Image precision of different methods

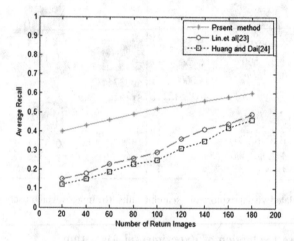

**Fig. 8.** Image recall of different methods

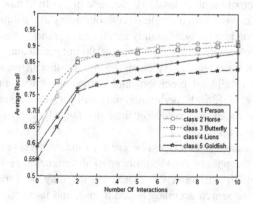

**Fig. 9.** Performance comparison of different types of image retrieval

interact with the retrieval system too much times. This is also in line with the psychology principle when users conduct the exploratory interaction:

### 5.2.4 Subjective Evaluation of the Users' Experience

In addition to objective evaluation, we designed a user survey to evaluate the system subjectively. The usability feature, of finding the most relevant image, and users' experience were used as the two major evaluation metrics. The evaluation scales are classified as three levels. Each item is scored using a single number in the range 0 to 2 inclusive; the scoring rules are shown in Table 2. We invited 20 users to participate in the survey, in which each user was assigned to submit 10 pictures from the phone. Participants were required to score each picture by scoring rules in Table 2. The evaluation criterion is based on the search results of interest (related) and users' experience. The rule of interaction profiling which includes Well-informed, Well-designed, New fashion, Response latence, Fast-conveyed can depict the ease of using visualization exploratory system.

**Table 2.** Scoring rules of evaluation

| Evaluation criterion | | Score | | |
| --- | --- | --- | --- | --- |
| | | 0 | 1 | 2 |
| Interesting (related) | | Unable to find the relevant image information | Find the relevant part of image information | Find all image information |
| User experience and interaction profiling | Well-informed | Poor | General | Good |
| | Well-designed | Poor | General | Good |
| | New fashion | Poor | General | Good |
| | Response latence | Slow | Medium | Fast |
| | Fast-conveyed | Poor | General | Good |

After the tests on mobile phone, participants were asked to rate the system against the image search engine of Google, Yahoo, and Bing. The average scores of the survey results from 20 participants (males 10 and females 10; Among them, three were teachers, three workers and 14 students) were calculated as shown in Table 3.

**Table 3.** Test score results

| Score | Search engine | | | |
| --- | --- | --- | --- | --- |
| | Proposed | Google | Yahoo | Bing |
| Interesting (related) | 1.8 | 1.5 | 1.35 | 1.55 |
| User experience and interaction profiling | 1.5 | 1.35 | 1.4 | 1.35 |
| Integrated | 1.65 | 1.425 | 1.325 | 1.45 |

Experimental results show that the performance of the proposed system is better than popular business search engines in retrieving related pictures on the mobile phone. It seems also attain better user experience and interaction on smart phone by the score of user evaluation in Table 3. Moreover, it demonstrates that the prototype system on Android platform is easier for users to interact than that in traditional keyword-based image search. However, relevant results need to be further improved and enhanced.

# 6 Conclusion

This paper investigates the image visual exploratory search scheme based on the platform of Android smart phone. Basing on Hadoop platform, it extracts the image features, gets the visual word bags by clustering features in the way of Mahout spectral clustering algorithm, and designs an image search based on the correlation. We build a correlation-based image visual exploratory search prototype system based on Android Smartphone. Experimental results show that our system provides more useful information to users than other popular systems in terms of finding the related images. The paper puts forward ideas on image exploratory search based on analysing the latest researches on exploratory search at home and abroad. However, our work on visual exploratory search for mobile platforms is currently still on the way.

**Acknowledgements.** We would thank Xiangtan University with the construction of key disciplines in Hunan program. This research has been supported by NSFC (61672495), Scientific Research Fund of Hunan Provincial Education Department (16A208), and the Open Project Program of The State Key Lab of Digital Technology And Application of Settlement Cultural Heritage, Hengyang Normal University.

# References

1. Beaver, D., Kumar, S., Li, H.C., et al.: Finding a needle in Haystack: Facebook's photo storage. In: Proceedings of the 9th USENIX Conference on Operating Systems Design and Implementation, OSDI 2010, pp. 47–60. USENIX Association, Berkeley (2010)
2. Temple, K.: What Happens In An Internet Minute? [EB/OL]. http://scoop.intel.com/what-happens-in-an-internet-minute. Accessed 13 Mar 2012/1 May 2014
3. Cai, D., He, X., Li, Z., et al.: Hierarchical clustering of WWW image search results using visual, textual and link information. In: Proceedings of the 12th Annual ACM International Conference on Multimedia, pp. 952–959. ACM (2004)
4. Marchionini, G.: Exploratory search: from finding to understanding. Commun. ACM **49**(4), 41–46 (2006)
5. Jing, Y., Rowley, H., Wang, J., et al.: Google image swirl: a large-scale content-based image visualization system. In: Proceedings of the 21st International Conference Companion on World Wide Web, pp. 539–540. ACM, New York (2012)
6. Yee, K.P., Swearingen, K., Li, K., et al.: Faceted metadata for image search and browsing. In: Proceedings of the SIGCHI Conference on Human Factors in Computing Systems, pp. 401–408. ACM, New York (2003)
7. Liu, D., et al.: SmartAdP: visual analytics of large-scale taxi trajectories for selecting billboard locations. IEEE Trans. Vis. Comput. Graph. **23**(1), 1–10 (2017)

8. Mukherjea, S., Hirata, K., Hara, Y.: Using clustering and visualization for refining the results of a WWW image search engine. In: Proceedings of the 1998 Workshop on New Paradigms in Information Visualization and Manipulation, pp. 29–35. ACM, New York (1998)

9. Pujol, J.M., Sangüesa, R., Bermúdez, J.: Porqpine: a distributed and collaborative search engine. In: Poster at the 12th International World Wide Web Conference WWW 2003, Budapest, Hungary (2003)

10. Westlund, O., Gomez-Barroso, J.L., Compañó, R., et al.: Exploring the logic of mobile search. Behav. Inf. Technol. **30**(5), 691–703 (2011)

11. White, R.W., Kules, B., Bederson, B.: Exploratory search interfaces: categorization, clustering and beyond: report on the XSI 2005 workshop at the Human-Computer Interaction Laboratory, University of Maryland. In: ACM SIGIR Forum, vol. 39, no. 2, pp. 52–56. ACM (2005)

12. White, R.W., Roth, R.A.: Exploratory search: beyond the query-response paradigm. Synth. Lect. Inf. Concepts Retr. Serv. **1**(1), 1–98 (2009)

13. Klouche, K., Ruotsalo, T., Cabral, D., Andolina, S., Bellucci, A., Jacucci, G.: Designing for exploratory search on touch devices. In: Proceedings of the 33rd Annual ACM Conference on Human Factors in Computing Systems, pp. 4189–4198, April 2015

14. Rui, Y., Huang, T.S., Ortega, M., et al.: Relevance feedback: a powerful tool for interactive content-based image retrieval. IEEE Trans. Circuits Syst. Video Technol. **8**(5), 644–655 (1998)

15. Suditu, N., Fleuret, F.: HEAT: iterative relevance feedback with one million images. In: International Conference on Computer Vision, pp. 2118–2125. IEEE Computer Society (2011)

16. Mohanan, A., Raju, S.: A Survey on different relevance feedback techniques in content based image retrieval. Int. Res. J. Eng. Technol. (IRJET) **04**(02), 582–585 (2017)

17. Fei-Fei, L., Perona, P.: A Bayesian hierarchical model for learning natural scene categories. In: IEEE Computer Society Conference on Computer Vision and Pattern Recognition, CVPR 2005, pp. 524–531. IEEE (2005)

18. Lowe, D.G.: Distinctive image features from scale-invariant keypoints. Int. J. Comput. Vis. **60**(2), 91–110 (2004)

19. Ng, A.Y., Jordan, M.I., Weiss, Y.: On spectral clustering analysis and an algorithm. In: Proceedings of Advances in Neural Information Processing Systems, vol. 14, pp. 849–856. MIT Press, Cambridge (2001)

20. Hare, J.S., Samangooei, S., Dupplaw, D.P.: OpenIMAJ and ImageTerrier: Java libraries and tools for scalable multimedia analysis and indexing of images. ACM (2011)

21. Zelnik-Manor, L., Perona, P.: Self-tuning spectral clustering. In: Advances in Neural Information Processing Systems (NIPS). MIT Press, Cambridge (2004)

22. Deng, J., Dong, W., Socher, R., et al.: ImageNet: a large-scale hierarchical image database. In: IEEE Conference on Computer Vision and Pattern Recognition, CVPR 2009, pp. 248–255. IEEE (2009)

23. Smith, J.R.: Image retrieval evaluation. In: IEEE Workshop on Content-based Access of Image and Video Libraries, pp. 112–113 (1998). http://www.ee.columbia.edu/~jrsmith/html/pubs/cbaivl98p.pdf

24. Huang, P.W., Dai, S.K.: Image retrieval by texture similarity. Pattern Recogn. **36**(3), 665–679 (2003)

25. Lin, C.-H., Chen, R.-T., Chan, Y.-K.: A smart content-based image retrieval system based on color and texture feature. Image Vis. Comput. **27**(6), 658–665 (2009)

# Blind Motion Deblurring for Online Defect Visual Inspection

Guixiong Liu[1(✉)], Bodi Wang[1], and Junfang Wu[2]

[1] School of Mechanical and Automotive Engineering,
South China University of Technology, Guangzhou 510640, China
megxliu@scut.edu.cn
[2] School of Physics, South China University of Technology,
Guangzhou 510640, China

**Abstract.** Online defect visual inspection (ODVI) works while the object has to be static, otherwise the relative motion between camera and object will create motion blur in images. In order to implement ODVI in dynamic scene, it developes one blind motion deblurring method whose objective is to estimate blur kernel parameters precisely. In the proposed method, Radon transform on superpixels determined the blur angle, and the autocorrelation function based on magnitude (AFM) of the preprocessed blurred image was utilized to identify the blur length. With the projection relationship discussed in this study, it will be unnecessary to rotate the blurred image or the axis. The proposed method is of high accuracy and robustness to noise, and it can somehow handle saturated pixels. To validate the proposed method, experiments have been carried out on synthetic images both in noise free and noisy situations. The results show that the method outperforms existing approaches. With the modified Richardson–Lucy deconvolution, it demonstrates that the proposed method is effective for ODVI in terms of subjective visual quality.

**Keywords:** Blind motion deblurring · Blur kernel estimation ·
Radon transform · Autocorrelation function · Saturated pixels

## 1 Introduction

Blurred image resulted from the relative motion between camera and object during the exposure time is commonly seen. It leads to poor image quality and poses obstacles against the information acquisition in various areas [1–4]. The sharp edge of an image can be deteriorated at the mercy of motion blur kernel, which characterizes the process how the latent image is blurred. Motion deblurring is involved with deconvolution where the degradation process is inversed. The early approaches, Richardson–Lucy [5] and Wiener deconvolution [6], are still popular thanks to their efficiency and high computational speed, but they can produce significant visual artifacts [7]. In recent years, deblurring performance is further improved in [8–10]. What's more, convolutional neural network based approaches [11, 12] have also been gaining attention. However, these methods are based on known blur kernel, a.k.a., non-blind deblurring, which is not the case in practice as blur kernel is frequently unknown. It's much more challenging for blind motion deblurring than non-blind deblurring.

© Springer Nature Singapore Pte Ltd. 2019
R. Mao et al. (Eds.): ICPCSEE 2019, CCIS 1059, pp. 74–89, 2019.
https://doi.org/10.1007/978-981-15-0121-0_5

With the prevalence of computer vision, industrial automatic inspection is widely utilized in defect detection [13]. Often, images are acquired by fixing the objects and using special light sources to improve the quality of image. This hinders the on-line visual inspection of defects, in which case the latent image will be blurred due to the motion of object moving on lines. Provided with the blur kernel, it will be possible to restore the latent image from the blurred version. Thus, good blur kernel estimation remains critical.

Numerous approaches for blur kernel estimation have been proposed over the past years, including parameterized methods and non-parameterized ones. The non-parameterized methods perform in an iterative coarse-to-fine kernel estimation process [14–16], which apply to complicated motion blurred image. Parameterized methods assume that the object moves along a straight line and the blur kernel can be modeled as two parameters, namely blur angle and blur length. Simplicity is highlighted here as well as low computational cost. By means of spatial, frequency or cepstrum domain, the orientation of the relative linear motion and its scale can be addressed. Lokhande et al. identify the blur kernel parameters from the log spectrum of the blurred images via Hough transform and periodic zeroes [17]. Moghaddam et al. introduce Radon transform and bispectrum modeling to find the blur angle and its length respectively [18], and their related work using fuzzy set concepts can be referred to in [19]. Besides, Dash et al. utilize pattern of the Gabor response of the blurred image to find the blur angle, and fit the relationship between the magnitude of Fourier coefficients and blur length via radial basis function neural network [20]. Researchers in [21] have indicated that the fourth bit plane of the modified cepstrum plays an important role for blur angel estimation. Lu et al. have observed that restored image has the most sparse representation once the kernel angle corresponds to the true blur angle, and measure blur length in Fourier domain [22]. Kumar et al. remove blur based on Histogram of oriented gradients and statistical properties as blur kernel rotates the gradient vector to a new direction and histogram of sharp image gradients exhibits higher Kurtosis than that of blurred image gradients [23].

Nevertheless, those approaches tend to rotate the image in order to align the motion direction with the horizon, in which case interpolation and out of range pixels occur, impeding blur length estimation. [18] and [19] have suggested rotating the coordinate system, but it's highly operational and doesn't actually settle the matter. In addition, not only motion blur but also non-Gaussian noise, saturated pixels and nonlinear camera response function can further degrade an image in practice [24], and blur kernel estimation from blurred image with these outliers cannot be effectively addressed via existing approaches. Different from [25–27] which are tailored for handling outliers in complicated blur kernel estimation, our work falls in linear blur kernel estimation for it's the main degradation factor in on-line defect visual inspection (ODVI).

Here, Radon transform and the autocorrelation function based on magnitude (AFM) have been introduced to estimate blur angle and blur length separately. Besides, superpixels and sharp edge selection are exploited respectively in Radon transform and AFM to boost their performance. Given the estimated blur kernel and Richardson–Lucy (RL) deconvolution modified for partially saturated image deblurring, the restoration can be carried out. Experiments indicate that the proposed method can survive noise and stay robust for saturated pixels. Note that there is no need to rotate the image or the axis to obtain the equivalent horizontal blurred version. Therefore, the nature of blurred image can remain intact with no valid data lost. The proposed method fruits favorably for synthetically as well as naturally blurred and noisy images containing saturated pixels, which makes it applicable in blind motion deblurring for ODVI.

The outline of this paper is as follows. In Sect. 2, the motion blur model for ODVI is explained. Blur kernel estimation exploiting Radon transform on superpixels and the proposed AFM method is presented in Sect. 3. Section 4 shows the difference between existing approaches and ours, and illustrates the effectiveness of the proposed method to handle saturated pixels. Section 5 deals with blur kernel estimation for synthetic and natural blur in ODVI respectively as well as comparison with other existing approaches. The conclusions are summarized in Sect. 6.

(a)    (b)    (c)    (d)    (e)

**Fig. 1.** (a) Original *boat* image, (b) Blurred *boat* image with $\theta = 60°$ and $L = 30$ *PX*, (c) Fourier spectrum of (b), (d) Blurred and noisy *boat* image with $\theta = 60°$, $L = 30$ *PX* and $\sigma^2 = 1e - 3$, (e) Fourier spectrum of (d).

## 2    Motion Blur Model

The relationship between the spatially invariant blurred image $g(x, y)$ and the latent image $f(x, y)$ can be generally expressed as

$$g(x, y) = f(x, y) \otimes h(x, y) + \eta(x, y), \tag{1}$$

where $h(x, y)$ denotes the blur kernel, $\otimes$ means the convolution operator, and $\eta(x, y)$ is the additive noise which is supposed to be zero under noise-free condition.

Blind deblurring is to recover the latent image from observation unknowing the blur kernel, and the restored image is not mathematically unique even with slightly different blur kernels. Therefore it's demanding to maximize the likelihood of the restored image being an instance of the original image, while blur kernel estimation holds the key apparently.

Linear motion blur kernel between object and camera in ODVI can be modeled as:

$$h(x, y) = \begin{cases} \frac{1}{L} & \text{if } \sqrt{x^2 + y^2} \leq \frac{L}{2}, \frac{y}{x} = \tan\theta. \\ 0 & \text{otherwise} \end{cases} \tag{2}$$

As is mentioned above, linear motion blur kernel is described as blur angle $\theta$ and blur length $L$, and the task is targeted on parameters estimation. Once the nature of the degradation is obtained, the ill-posed blind deblurring can turn into the non-blind deblurring.

# 3   The Proposed Method

In this section, blur kernel estimation is covered in detail. Identification of blur angle via Radon transform is organized in Sect. 3.1, while the AFM method is proposed to determine the blur length in pixels (PX) in Sect. 3.2.

## 3.1   The Blur Angle Estimation

The blurred version ($\theta = 60°$ and $L = 30$ PX) of one standard test image named *boat* Fig. 1(a) is taken as an example in Fig. 1(b). Besides the linear motion blur, Fig. 1(d) also involves itself with an additive noise ($\sigma^2 = 1e - 3$). It's shown from Fig. 1(c) and (e) that in the Fourier spectrum of the degraded image, there exist parallel dark and bright stripes reflecting the characteristic of blur kernel even in the presence of noise. The blur angle is the very angle between any of these parallel stripes and vertical axis. Amongst various line detection approaches, Radon transform is exploited here on account of its explicit calculation about the location of stripes.

Formally, the Radon transform $Rf$ of a real-valued function $\phi(x, y)$, at angle $\alpha$ and distance $\rho$ from the origin, is given by

$$Rf(\phi, \rho, \alpha) = \int_{-\infty}^{\infty} \int_{-\infty}^{\infty} \phi(x, y)\delta(\rho - x \cos \alpha - y \sin \alpha)dxdy, \qquad (3)$$

where $\delta$ denotes the dirac delta function. The results of a Radon transformed image from the boundary to its center at the range of 0 to 179° exist as one matrix, and every maximum in it corresponds to a straight line in the original image. Location of the maximum throws light on the direction of the straight line, making it possible for Radon transform to estimate the blur angle.

From above, it can be inferred that feature extracted from the stripes in Fourier spectrum helps blur angle estimation. Instead of directly Radon transforming the Fourier spectrum of blurred image [18, 19], we exploit the recently proposed super-pixels detection method [28, 29] to better extract the parallel stripes and wipe off other unwanted features in the Fourier spectrum. Let $\Delta G(\mu, v)$ be the superpixels detected result, then the estimation of blur angle arrives as follows:

$$\theta = \arg \max_{\alpha}\{Rf(\Delta G(\mu, v), \rho, \alpha)\}. \qquad (4)$$

The steps of blur angle estimation are sketched in Algorithm 1.

## Algorithm 1

Input: The blurred image.
Step1: Calculate the Fourier spectrum of the blurred image.
Step2: Detect superpixels in the Fourier spectrum.
Step3: Compute the Radon transform.
Step4: Locate the maximum.

### 3.2    The Blur Length Estimation

Here, the autocorrelation function based on magnitude i.e. AFM is proposed to tackle the problem of blur length estimation. In this method one doesn't have to rotate the blurred image or the axis to align it with the motion direction, ensuring that the entire valid data are available and the blur length estimation works unaffectedly, and it's demonstrated that the proposed AFM method is robust to saturated pixels.

Blind deblurring relies on sharp edge selection for blur kernel estimation [28]. In order to enhance sharp edge for blur length estimation and compromise between performance and computational expense, blurred image is filtered via edge detector, which is referred to as preprocessing in our method. The energy spectrum and the autocorrelation function of an image have formed one Fourier transformation pair. Let $o(x, y)$ and $O(\mu, v)$ be an $M \times N$ image and its Fourier transform respectively, and the following formulas (5), (6), and (7) give the derivation of energy spectrum $E(\mu, v)$ for an image:

$$O(\mu, v) = \frac{1}{MN} \sum_{x=0}^{M-1} \sum_{y=0}^{N-1} o(x, y) e^{-j2\pi\left(\frac{\mu x}{M} + \frac{vy}{N}\right)}, \tag{5}$$

$$E(\mu, v) = |O(\mu, v)|^2, \tag{6}$$

$$\psi(x, y) = \sum_{\mu=0}^{M-1} \sum_{v=0}^{N-1} E(\mu, v) e^{j2\pi\left(\frac{\mu x}{M} + \frac{vy}{N}\right)}. \tag{7}$$

In Eq. (7) the inverse Fourier transform of $E(\mu, v)$ yields the autocorrelation function of $o(x, y)$ denoted by $\psi(x, y)$. Plainly, $E(\mu, v)$ and $\psi(x, y)$ make up one Fourier transformation pair.

Autocorrelation function based on magnitude (AFM) is represented by $\Phi_{AFM}(x, y)$ as below:

$$\Phi_{AFM}(x, y) = \sum_{\mu=0}^{M-1} \sum_{v=0}^{N-1} |O(\mu, v)| e^{j2\pi\left(\frac{\mu x}{M} + \frac{vy}{N}\right)}. \tag{8}$$

Plotting the $\Phi_{AFM}$ matrix of the blurred *boat* image ($\theta = 30°$ and $L = 30$ *PX*) with its each column as a separate line in 2-D plane generates a cluster of curves in Fig. 2(a). Figure 2(b) shows another cluster by the AFM of blurred and noisy *boat* image ($\theta = 30°$, $L = 30$ *PX* and $\sigma^2 = 1e - 4$). As is illustrated in Fig. 2, two symmetric valleys can be easily found in the absence and presence of noise, ruling out the centered ones. Since linear motion blur kernel exhibits sinc-like behavior [22], the interval between them is related to the desired blur length in pixels.

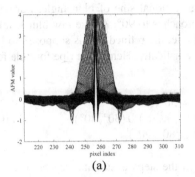 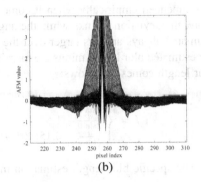

                        (a)                                          (b)

**Fig. 2.** (a) Cluster of curves plotted by AFM of blurred *boat* image with $\theta = 30°$ and $L = 30\ PX$, (b) Cluster of curves plotted by AFM of blurred and noisy *boat* image $\theta = 30°$, $L = 30\ PX$ and $\sigma^2 = 1e - 4$.

An example of relationship between the desired blur length and its projection, i.e., the interval mentioned above is displayed in Fig. 3. It reveals that half the interval is exactly the casting of blur length's shadow onto the axis, with the estimated blur angle taking its part in. On the basis of the projection, it dawn on us that the blur length can be calculated with the following equation:

$$L = \frac{d}{2\sin\theta},\tag{9}$$

where $d$ represents the interval between the two symmetric valleys, and $\theta$ is nothing but the estimated blur angle.

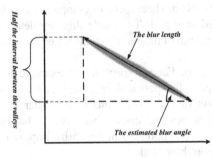

**Fig. 3.** An example of blur length projecting onto the axis.

Note that in this case, we can escape the trouble that a rotated image or axis might bring. Nevertheless, under the circumstances of the sine of blur angle being an irrational number, i.e., the estimated blur angle doesn't seem to be 30° or 90°. Technically the projection of blur length will be in sub-pixels, giving rise to an approximate 2-*PX* error deviation from the actual value to the estimated one which can be even worsened by a strong noise. In order to eliminate the deviation, a margin of 1.5-*PX* is recommended empirically when it comes to an irrational sine of blur angle. Furthermore, the deviation shrinks while the angle approaches to 90°. As the raw blur length estimated always appears larger over the actual one, the refined one is supposed to be the estimated blur length minus the margin mathematically. Hence, the ripe formula for blur length comes as follows:

$$L = \begin{cases} \frac{d}{2\sin\theta} & \theta = 30°, 90° \\ \frac{d}{2\sin\theta} - 1.5 & \theta \in (0, 30°) \cup (30°, 80°) \, . \\ \frac{d}{2\sin\theta} - 0.15(90° - \theta) & \theta \in [80°, 90°) \end{cases} \tag{10}$$

The specific blur length estimation method in the steps is summarized below.

**Algorithm 2**

Input: the blurred image.
Step1: Select the sharp edge of blurred image with edge detector.
Step2: Calculate the AFM and plot it in a cluster of curves.
Step3: Detect the twin valleys and measure the interval in pixels.
Step4: Compute the blur length using Eq. (10).

Given the estimated blur angle and blur length, the linear motion blur kernel is acquired. What's left for restoration is to deconvolve the blurred image.

## 4  Difference from Existing Approaches

Existing approaches using Radon/Hough transform often directly detect parallel stripes based on Fourier spectrum of blurred image [17–19]. Naturally blurred image tends to contain outliers which might cover up the stripe features, and in our case outliers are assumed to be saturated pixels as sheet metal of computer cases in ODVI often reflects too much light, illustrated in Fig. 4. To handle this, we detect sticky edge adhesive superpixels in Fourier spectrum of blurred image to better extract features for blur angle estimation.

As is shown in Fig. 5(a), Fourier spectrum of example 3 (see Sect. 5.2.) from ODVI is presented with periodic spots and irregular parallel lines, probably caused by outliers. Superpixels group pixels into regions with similar values, reduce the complexity of subsequent operations and discard bad pixels. Thanks to superpixels operation, the undesired pattern is suppressed in Fig. 5(b), improving the performance of Radon transform to estimate blur angle.

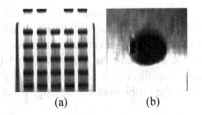

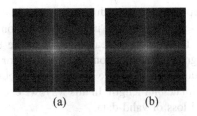

**Fig. 4.** Examples of sheet metal of computer cases in OVDI reflecting too much light

**Fig. 5.** (a) Fourier spectrum of example 3 from ODVI, (b) (a) in superpixels.

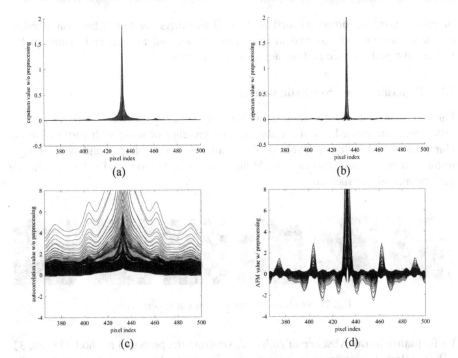

**Fig. 6.** (a) Surface plotted by cepstrum of unpreprocessed example 3 from ODVI, (b) Surface plotted by cepstrum of preprocessed example 3 from ODVI, (c) Cluster of curves plotted by autocorrelation of unpreprocessed example 3 from ODVI, (d) Cluster of curves plotted by AFM of preprocessed example 3 from ODVI.

Now that there is no log in our blur length estimation, the proposed method is different from the cepstrum based ones [18, 19, 31]. Suffering from saturated pixels, cepstrum of blurred image from ODVI doesn't exhibit sinc-like behavior as no salient symmetry valleys occur (see Fig. 6(a)) even when sharp edge is selected via our preprocessing procedure (see Fig. 6(b)). Likewise, as Fig. 6(c) illustrates, conventional autocorrelation based method [17] which is similar to our AFM method cannot preserve well the sinc-like behavior of partially saturated linear motion blur, leading to failure in blur length estimation for ODVI. Fueled by the sharp edge selected via the preprocessing procedure, our method is able to estimate the blur length with symmetry valleys standing out in the visualization of AFM as cluster of curves (see Fig. 6(d)).

In addition, instead of rotating the blurred image or its derivative, or the axis to align motion direction with horizon for the sake of collapsing the 2-D data into 1-D data [17, 20, 21], we apply the projection relationship discussed in Sect. 3.2 to fully estimate blur length. In this way, we can circumvent the problem of out of range pixels, and loss of valid data.

# 5   Experiments Results

In this section, the proposed method as well as others are tested both on standard synthetic and naturally blurred images from ODVI, and metrics and visual quality validate the performance and advantage of our method.

## 5.1   Experiments on Synthetic Images

Figure 7 shows the synthetic images used in our experiments. For these images, experiments are carried out in the absence and presence of noise with varying motion blur kernels and different levels of Gaussian noise. For blurred and noisy image, restoration via RL using the estimated blur kernel is shown to visually demonstrate the performance of our method.

**Fig. 7.** Synthetic images used in our experiments.

**Performance in the Absence of Noise.** Apart from the proposed method, [18, 20, 32, 33] are selected to estimate the parameters of synthetic blur kernel for a comparative analysis. In the absence of noise, Tables 1 and 2 summary the performance of those methods where standard images like *Lena, Cameraman, Tree* are subjected to different blur kernels.

Table 1 shows different blur angle estimations via the proposed method compared with others while the blur length is set to be 20 *PX*, and it has suggested that the proposed method has managed to determine the blur angle. Moreover, accuracy of the proposed method appears higher than existing ones except the blurred image *Checkboard128* where a poor estimation of a 2° error takes place due to the lack of textures itself. Blur length estimations via the proposed method compared with others while the blur angle is set to be 30°are reported in Table 2, and it has unveiled that the proposed AFM method has better access to blur length than existing ones in all the cases shown there.

**Table 1.** Blur angle estimations in the absence of noise.

| Image | The actual angle/° | [32] | [18] | [20] | Ours |
|---|---|---|---|---|---|
| Lena256 | 30 | 28 | 29 | 29 | **30** |
| C.man256 | 40 | 43 | 41 | 39 | **40** |
| Tree256 | 45 | 48 | 40 | 46 | **45** |
| Checkboard128 | 40 | 38 | 37 | **39** | 42 |
| Stick64 | 35 | 33 | 32 | **36** | **36** |

**Table 2.** Blur length estimations in the absence of noise.

| Image | The actual length/PX | [32] | [33] | [20] | Ours |
|---|---|---|---|---|---|
| Lena256 | 15 | 19 | 18 | **14** | **14** |
| C.man256 | 30 | 32 | 27 | 32 | **30** |
| Tree256 | 20 | 18 | 17 | 18 | **20** |
| Checkboard128 | 30 | 32 | 27 | 32 | **30** |
| Stick64 | 30 | 32 | 27 | 32 | **30** |

**Performance in the Presence of Noise.** Here, Gaussian noise with different *SNR* defined by Eq. (11) is penetrated into the blurred image for degradation always involves itself with noise in practice:

$$SNR = 10\log_{10}\left(\frac{\sigma_f^2}{\sigma_n^2}\right), \tag{11}$$

where $\sigma_f^2$ denotes the original image variance while $\sigma_n^2$ is the noise variance.

Displayed in Table 3 are experiments where our method together with [20] are conducted on different synthetic blurred and noisy images, and their *SNR* is set to be 25 dB and 30 dB separately. The comparison indicates that the proposed method works satisfactorily even up against a noise of 25 dB, and harbors more precise and acceptable estimations than [20].

Since the parameters of blur kernel are estimated via the proposed method along with others, Figs. 8 and 9 depict the comparative restorations with [20] exploiting RL deconvolution to further validate our method. It can be seen that despite of the noise amplified by RL, the proposed method outperforms the existing approaches as its restoration contains fewer ringing artifacts especially in Fig. 8.

Further experiments under lower *SNR* are collected in Table 4 with visual validation in Figs. 10 and 11, and it demonstrates that our method still produces favorable results confronted with *SNR* = 20 dB. The specified valleys are to be overwhelmed as the SNR decreases, resulting in poor estimation, and any error to blur angle will even make it worse. Nonetheless, the impact of noise can be alleviated once the motion direction is vertical even confronted with a lower *SNR* down to 15 dB as our preprocessing procedure mainly enhances sharp edge in vertical direction.

**Table 3.** Blur kernel estimations in the presence of a noise of 25 dB and 30 dB.

| Image | | | Lena | Tree | C.man | Vase |
|---|---|---|---|---|---|---|
| The actual angle $\theta/^\circ$ | | | 30 | 45 | 40 | 60 |
| The actual length $L/PX$ | | | 15 | 20 | 30 | 40 |
| [20] | $\theta_{est}/^\circ$ | $SNR = 25$ dB | 31 | 46 | 39 | 59 |
| | $L_{est}/PX$ | | | 13.3 | 18.4 | 28.5 | 39 |
| | $\theta_{est}/^\circ$ | $SNR = 30$ dB | 31 | **46** | 39 | 59 |
| | $L_{est}/PX$ | | 13.6 | 18.7 | 28.3 | 38.4 |
| Ours | $\theta_{est}/^\circ$ | $SNR = 25$ dB | **30** | **45** | **40** | **60** |
| | $L_{est}/PX$ | | **16** | **20.1** | **30.8** | **40.5** |
| | $\theta_{est}/^\circ$ | $SNR = 30$ dB | **30** | **44** | **40** | **60** |
| | $L_{est}/PX$ | | **16** | **20.5** | **30.8** | **40.5** |

**Table 4.** Blur kernel estimations in the presence of a noise of 20 dB and 22 dB.

| Image | | | Lena | Tree | C.man | Vase |
|---|---|---|---|---|---|---|
| The actual angle $\theta/^\circ$ | | | 30 | 45 | 40 | 60 |
| The actual length $L/PX$ | | | 15 | 20 | 30 | 40 |
| Ours | $\theta_{est}/^\circ$ | $SNR = 20$ dB | 31 | 44 | 39 | 61 |
| | $L_{est}/PX$ | | 15.6 | 20.5 | 29.8 | 40.1 |
| | $\theta_{est}/^\circ$ | $SNR = 22$ dB | 30 | 44 | 40 | 60 |
| | $L_{est}/PX$ | | 16 | 20.5 | 30.8 | 41.8 |

(a)                                          (b)

**Fig. 8.** Comparison of restorations on the degraded *Tree* image ($\theta = 45°$, $L = 20$ PX) using RL with the blur kernel estimated via the proposed method along with others. (a) From left to right: blurred and noisy image ($SNR = 25$ dB), restored image via [20], restored image via the proposed method; (b) From left to right: blurred and noisy image ($SNR = 30$ dB), restored image via [20], restored image via the proposed method.

(a)                                          (b)

**Fig. 9.** Comparison of restorations on the degraded *Vase* image ($\theta = 60°$, $L = 40$ PX) using RL with the blur kernel estimated via the proposed method along with others. (a) From left to right: blurred and noisy image ($SNR = 25$ dB), restored image via [20], restored image via the proposed method; (a) From left to right: blurred and noisy image ($SNR = 30$ dB), restored image via [20], restored image via the proposed method.

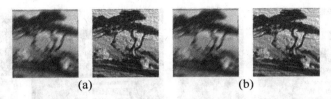

**Fig. 10.** Restorations on the degraded *Tree* image ($\theta = 45°$, $L = 20$ *PX*) using RL with the blur kernel estimated via the proposed method. (a) From left to right: blurred and noisy image (*SNR* = 20 dB), restored image via the proposed method; (b) From left to right: blurred and noisy image (*SNR* = 22 dB), restored image via the proposed method.

**Fig. 11.** Restorations on the degraded *Vase* image ($\theta = 60°$, $L = 40$ *PX*) using RL with the blur kernel estimated via the proposed method. (a) From left to right: blurred and noisy image (*SNR* = 20 dB), restored image via the proposed method; (b) From left to right: blurred and noisy image (*SNR* = 22 dB), restored image via the proposed method.

## 5.2   Experiments on Natural Images

It's been demonstrated that the proposed method is robust to a noise down to 20 dB, ensuring its functionality in practice. Experiments has been conducted on naturally blurred images of computer cases to verify the method's feasibility in ODVI. Due to the fact that the ground truth of blur kernel remains unknown, and naturally blurred image from ODVI contains saturated pixels shown in Sect. 4, which not only hinder blur kernel estimation but also produce ringing artifacts in deconvlution, the RL modified for partially saturated image deblurring [34] is adopted to assess the performance of blur kernel estimation.

Figures 12, 13, 14 and 15 depict the restoration examples on computer cases from ODVI compared with the blurred versions and their close-up views, and textures and details once deteriorated and submerged by motion blur just reflourish. Specifically, smears that relative motion has caused to the labels, words and geometric features have been removed almost completely, generating recognizable and plausible results. In blind motion deblurring for ODVI, metrics like *PSNR* do not work with no reference available. From the recovered visual quality, it can be inferred that the proposed method functions well for ODVI even in the presence of outliers.

Table 5 lists the average run times (in seconds) of our method on different images displayed in this paper (synthetic and natural, w/ and w/o noise) for blur kernel estimation, and the experiments are conducted on an Intel Core i7-4790 K CPU. One can see that our method has a fast testing speed for different images, and it still has a satisfactory performance as is shown in this section.

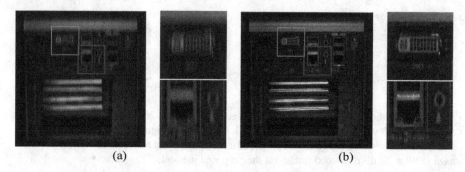

(a)                                        (b)

**Fig. 12.** (a) Example 1 on the naturally blurred images of computer cases from ODVI and its close-up views, (b) Restoration of (a) via modified RL with $\theta = 90°$ and $L = 28\ PX$ estimated using the proposed method and its close-up views.

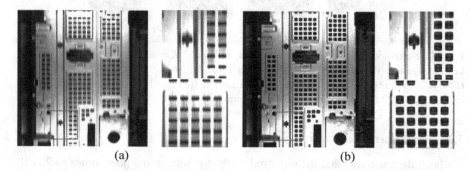

(a)                                        (b)

**Fig. 13.** (a) Example 2 on the naturally blurred images of computer cases from ODVI and its close-up views, (b) Restoration of (a) via modified RL with $\theta = 90°$ and $L = 20\ PX$ estimated using the proposed method and its close-up views.

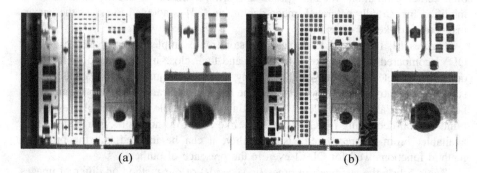

(a)                                        (b)

**Fig. 14.** (a) Example 3 on the naturally blurred images of computer cases from ODVI and its close-up views, (b) Restoration of (a) via modified RL with $\theta = 90°$ and $L = 22\ PX$ estimated using the proposed method and its close-up views.

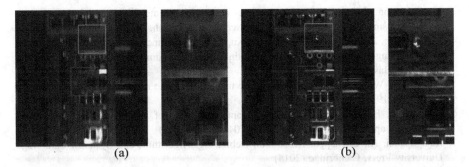

(a)                                                          (b)

**Fig. 15.** (a) Example 4 on the naturally blurred images of computer cases from ODVI and its close-up views, (b) Restoration of (a) via modified RL with $\theta = 90°$ and $L = 27$ PX estimated using the proposed method and its close-up views.

**Table 5.** Average run time (in seconds) of our method for blur kernel estimation.

| Image | Synthetic64 | Synthetic128 | Synthetic256 | Natural856 |
|---|---|---|---|---|
| Run time | 0.016 | 0.017 | 0.060 | 0.585 |

## 6  Conclusions

In this study, we have developed one method on blind motion deblurring for ODVI which focuses on linear motion blur kernel estimation. The proposed method finds blur angle via the Radon transform fueled by superpixels, and applies the AFM method on selected sharp edge to measure blur length. Performance of our method has been demonstrated on synthetic images both in the absence and presence of noise. The accuracy and robustness of our method prove superior as compared to existing approaches. Note that in the proposed method side effects brought by rotating the image or the coordinate system can be bypassed. Further experiments also show that the proposed method can even withstand a noise down to 20 dB. Another advantage of our method is that it can handle outliers, typically saturated pixels in blur kernel estimation. With naturally blurred images of computer cases restored via modified RL using the estimated blur kernel, feasibility of the proposed method for ODVI is verified.

**Acknowledgments.** Guangzhou Science and Technology Plan Project (201802030006), the Open Project Program of Guangdong Key Laboratory of Modern Geometry and Mechanics Metrology Technology (SCMKF201801).

## References

1. Zhang, W., Quan, W., Guo, L.: Blurred star image processing for star sensors under dynamic conditions. Sensors **12**, 6712–6726 (2012)
2. Kang, J.S., Kim, C.S., Lee, Y.W., Cho, S.W., Park, K.R.: Age estimation robust to optical and motion blurring by deep Residual CNN. Symmetry-Basel **10**, 108 (2018)

3. Li, Q., Liang, S.Y.: Microstructure images restoration of metallic materials based upon KSVD and smoothing penalty sparse representation approach. Materials **11**, 637 (2018)

4. Si, L., Wang, Z., Xu, R., Tan, C., Liu, X., Xu, J.: Image enhancement for surveillance video of coal mining face based on single-scale retinex algorithm combined with bilateral filtering. Symmetry-Basel **9**, 93 (2017)

5. Richardson, W.H.: Bayesian-based iterative method of image restoration. J. Opt. Soc. Am. **62**, 55–59 (1972)

6. Wiener, N.: Extrapolation, interpolation, and smoothing of stationary time series: with engineering applications. MIT Press **113**, 1043–1054 (1964)

7. Rajagopalan, A., Chellappa, R.: Motion Deblurring: Algorithms and Systems. Cambridge University Press, Cambridge (2014)

8. Krishnan, D., Fergus, R.: Fast Image deconvolution using hyper-Laplacian priors. In: Proceedings NIPS, pp. 1033–1041 (2009)

9. Danielyan, A., Katkovnik, V., Egiazarian, K.: BM3D frames and variational image deblurring. IEEE Trans. Image Process. **21**, 1715–1728 (2012)

10. Yair, N., Michaeli, T.: Multi-scale weighted nuclear norm image restoration. In: Proceedings IEEE CVPR, pp. 3165–3174 (2018)

11. Xu, L., Ren, J.S.J., Liu, C., Jia, J.: Deep convolutional neural network for image deconvolution. In: Proceedings NIPS, pp. 1790–1798 (2014)

12. Zhang, K., Zuo, W., Gu, S., Zhang, L.: Learning Deep CNN denoiser prior for image restoration. In: Proceedings IEEE CVPR, pp. 2808–2817 (2017)

13. Peng, Y., Wu, T., Wang, S., Kwok, N., Peng, Z.: Motion-blurred particle image restoration for on-line wear monitoring. Sensors **15**, 8173–8191 (2015)

14. Pan, J., Hu, Z., Su, Z., Yang, M.-H.: Deblurring text images via $L_0$-regularized intensity and gradient prior. In: Proceedings IEEE CVPR, pp. 2901–2908 (2014)

15. Michaeli, T., Irani, M.: Blind deblurring using internal patch recurrence. In: Fleet, D., Pajdla, T., Schiele, B., Tuytelaars, T. (eds.) ECCV 2014. LNCS, vol. 8691, pp. 783–798. Springer, Cham (2014). https://doi.org/10.1007/978-3-319-10578-9_51

16. Zuo, W., Ren, D., Gu, S., Lin, L., Zhang, L.: Discriminative learning of iteration-wise priors for blind deconvolution. In: Proceedings IEEE CVPR, pp. 3232–3240 (2015)

17. Lokhande, R., Arya, K.V., Gupta, P.: Identification of parameters and restoration of motion blurred images. In: Proceedings ACM Symposium on Applied Computing, pp. 301–305 (2006)

18. Moghaddam, M.E., Jamzad, M.: Motion blur identification in noisy images using mathematical models and statistical measures. Pattern Recogn. **40**, 1946–1957 (2007)

19. Moghaddam, M.E., Jamzad, M.: Linear motion blur parameter estimation in noisy images using fuzzy Sets and power spectrum. EURASIP J. Adv. Sig. Process. **1**, 068985 (2007)

20. Dash, R., Majhi, B.: Motion blur parameters estimation for image restoration. Optik **125**, 1634–1640 (2014)

21. Deshpande, A.M., Patnaik, S.: A novel modified cepstral based technique for blind estimation of motion blur. Optik **125**, 606–615 (2014)

22. Lu, Q., Zhou, W., Fang, L., Li, H.: Robust blur kernel estimation for license plate images from fast moving vehicles. IEEE Trans. Image Process. **25**, 2311–2323 (2016)

23. Kumar, A.: Deblurring of motion blurred images using histogram of oriented gradients and geometric moments. Sig. Process. Image Commun. **55**, 55–65 (2017)

24. Lai, W.-S., Huang, J.-B., Hu, Z., Ahuja, N., Yang, M.-H.: A comparative study for single image blind deblurring. In: Proceedings IEEE CVPR, pp. 1701–1709 (2016)

25. Hu, Z., Cho, S., Wang, J., Yang, M.-H.: Deblurring low-light images with light streaks. In: Proceedings IEEE CVPR, pp. 3382–3389 (2014)

26. Pan, J., Lin, Z., Su, Z., Yang, M.-H.: Robust kernel estimation with outliers handling for image deblurring. In: Proceedings IEEE CVPR, pp. 2800–2808 (2016)

27. Dong, J., Pan, J., Su, Z., Yang, M.-H.: Blind image deblurring with outlier handling. In: Proceedings IEEE ICCV, pp. 2497–2505 (2017)

28. Dollar, P., Zitnick, C.L.: Structured forests for fast edge detection. In: Proceedings IEEE ICCV, pp. 1841–1848 (2013)

29. Dollar, P., Zitnick, C.L.: Fast edge detection using structured forests. IEEE Trans. Pattern Anal. Mach. Intell. **37**, 1558–1570 (2015)

30. Xu, X., Pan, J., Zhang, Y.-J., Yang, M.-H.: Motion blur kernel estimation via deep learning. IEEE Trans. Image Process. **27**, 194–205 (2018)

31. Wu, S., Lu, Z., Ong, E.P., Lin, W.: Blind image blur identification in cepstrum domain. In: Proceedings IEEE International Conference on Computer Communications and Networks, pp. 1166–1171 (2007)

32. Rekleitis, I.M.: Optical Flow recognition from the power spectrum of a single blurred image. In: Proceedings IEEE International Conference on Image Processing, pp. 791–794 (1996)

33. Rekleitis, I.M.: Steerable filters and cepstral analysis for optical flow calculation from a single blurred image. In: Proceedings Vision Interface, pp. 159–166 (1996)

34. Whyte, O., Sivic, J., Zisserman, A.: Deblurring shaken and partially saturated images. Int. J. Comput. Vis. **110**, 185–201 (2014)

# Fast Nonlocal Diffusion by Stacking Local Filters for Image Denoising

Peng Qiao[1(✉)], Weichu Sun[2], Yong Dou[1], and Rongchun Li[1]

[1] Science and Technology on Parallel and Distributed Laboratory,
National University of Defense Technology, Changsha 410073, Hunan, China
`pengqiao@nudt.edu.cn`
[2] University of South China, Hengyang 421000, Hunan, China

**Abstract.** Many image denoising methods achieve state-of-the-art performance with little consideration of computation efficiency. With the popularity of high-definition imaging devices, these denoising methods do not scale well to the high-definition images. Therefore, a fast image denoising method is proposed to meet the demand of processing high-definition images. Based on the analysis of the distribution of the distance $dist_{rs}$ between the similar patches and their reference patches from a semantic aspect, the large $dist_{rs}$s was found to occur while their contribution to the overall distribution was small. Therefore, the nonlocal filters was replaced in trainable non-local reaction diffusion (TNLRD) with local filters. For image with $4096 \times 4096$ resolution, the proposed method runs about 6 times faster than TNLRD via a single-thread CPU implementation. And the GPU implementation of the proposed method is about 10 times faster than the CPU implementation. Furthermore, the proposed model achieves competing denoising performance compared with TNLRD in terms of PSNR and SSIM.

**Keywords:** Gaussian image denoising · Nonlocal self-similarity · Diffusion process · Discriminative learning

## 1 Introduction

Image denoising is a widely studied problem and is still an active research topic in image processing and low level vision. Aiming at recovering the latent image $u$ from its noisy observation $f$, which is formulated as

$$f = u + n, \tag{1}$$

where $n$ is the additive noise. In this work, we focus on Gaussian image denoising, which means $n$ is assumed to be additive Gaussian noise. During the past

---

P. Qiao and W. Sun—Equally contributed.

© Springer Nature Singapore Pte Ltd. 2019
R. Mao et al. (Eds.): ICPCSEE 2019, CCIS 1059, pp. 90–101, 2019.
https://doi.org/10.1007/978-981-15-0121-0_6

decades, many image denoising methods are proposed, e.g., [1,5,6,9,13,17,23–25,27]. It is well-known that image denoising is an ill-posed problem. To solve this ill-posed problem, variational method is one of the feasible approaches. With the help of machine learning, many denoising methods that exploit the variational disciplines are proposed, e.g., Cascade Shrinkage Fields (CSF, [17]), Trainable Non-linear Reaction Diffusion (TNRD, [5]) and Trainable NonLocal Reaction Diffusion (TNLRD, [15]). While TNLRD achieves superior denoising performance, it costs too much time to denoise an image.

Therefore, TNLRD is far from the demands for high-definition image denoising in terms of processing speed.

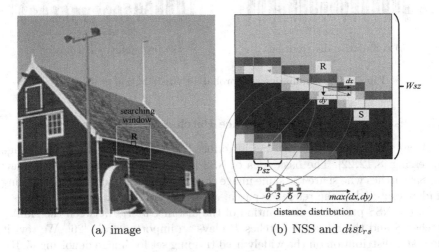

(a) image                            (b) NSS and $dist_{rs}$

**Fig. 1.** The distance distribution calculation diagram. (Color figure online)

To this end, we are highly motivated to propose a fast denoising method to meet the growing demand for high-definition image processing. From the $dist_{rs}$ distribution, we found that about 77% of the distance $dist_{rs}$ in training set [5] distributes within a $9 \times 9$ spatial range. We then analyzed the $dist_{rs}$ distribution from a semantic aspect, and found that in artifact and natural object images the large $dist_{rs}$s occur while their contribution to the overall distribution is small. Based on this observation, we proposed to replace the nonlocal filter in TNLRD with local filter, coined as $2^{nd}$layer filter. The original local filter in TNLRD is coined as $1^{st}$layer filter. The proposed model is coined as 2layerTNRD. Experimental results show that 2layerTNRD is more computationally efficient than TNLRD. Meanwhile, 2layerTNRD achieves competing denoising performance compared with TNLRD in terms of PSNR and SSIM.

## 2   Nonlocal vs. Local

In this section, we introduce the $dist_{rs}$ distribution and give a semantic explanation for the $dist_{rs}$ distribution.

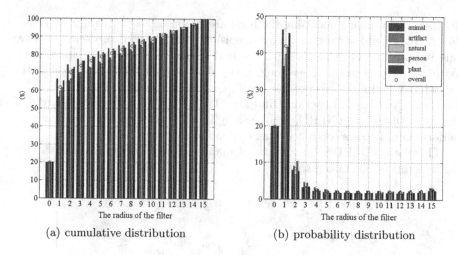

(a) cumulative distribution     (b) probability distribution

**Fig. 2.** The cumulative and probability distribution of $dist_{rt}$.

## 2.1 Distance Between a Reference Patch and Its Similar Patches

Nonlocal self-similarity (NSS) was introduced in [3] and widely used in image processing [8,12,22]. The basic idea of NSS is that a reference patch has a few similar patches who share the same image context and locate in a larger searching window centered at the reference patch, as shown in Fig. 1.

In the NSS prior, the distribution of the distance $dist_{rs}$ between the similar patches $S$ and the reference patches $R$ plays an important role [26]. We revisit the $dist_{rs}$ distribution on the widely used training set for image denoising [5,15–17,23]. The training set is consist of 400 images with 180 × 180 resolution. As shown in Fig. 1(a) and (b), given an image $I$, a reference patch $R$ and a searching window with size $Wsz$, the similar patches w.r.t. the reference patch $R$ in the searching window are selected using k-NN algorithms [15,19], illustrated via color bounding boxes in Fig. 1(b).

For the reference patch $R$ and its similar patches $S$ in Fig. 1(b), the spatial distance between them in x-axis and y-axis is dx and dy respectively. We measure the distance between $R$ and $S$ using the maximum value between dx and dy, i.e., the $\ell_\infty$ norm. This distance measure is a reasonable alternative to the $\ell_1$ or $\ell_2$ distance measure, and is more suitable to determine the filter size. Note that, similar patches with $dist_{rs} = 0$ are the reference patches themselves.

## 2.2 $dist_{rs}$ Distribution

For each reference patch in each image in the training set, we summarize the $dist_{rs}$ distribution as shown in Fig. 2. The relevant parameters above are set as follows[1], $Wsz = 31$, $Psz = 7$, $L = 5$.

---

[1] The values of the parameters are set to the same as those in TNLRD and other nonlocal methods using k Nearest Neighbors (kNN).

As shown in Fig. 2(a), we can find that about 77% of $dist_{rs}$ distributes within the $9 \times 9$ spatial range. As the distance increases, the distribution of similar patches in that distance gets smaller.

**Why Large $dist_{rs}$s Distribute Little? We Think this May Be Related with the Image Context of the Dataset.** To verify this, we manually assign a class to each image in the training set. The classes are chosen from WordNet [14] nouns which is used to label the ImageNet dataset [7]. We employ a simple yet robust way to assign a class to each image: (1) identify the main objects in the image; (2) match the main objects with the classes in WordNet nouns; (3) assign the matched class to the image. The 400 images in the training set are then divided into 5 classes, namely animal, artifact, natural object, person and plant. There are 134, 94, 42, 114 and 16 images in these 5 classes respectively.

The $dist_{rs}$ distribution for each class is shown in Fig. 2. In Fig. 2(b), we can see that in the artifact and natural object classes, the large $dist_{rs}$s distribute more than those in the other classes. It can be summarized as the right tail of the $dist_{rs}$ distribution of the artifact and natural object classes are heavier than those in the other classes. Taking a closer look at the images in the natural object class, one can find that these images are highly textured. Therefore, the similar patches are more likely far away distributed with respect to these highly textured reference patches. In the artifact class, one can find that a lot of periodical textures, e.g., the line structures in the building. The total images in the artifact and natural object classes are only about 34%. The large $dist_{rs}$s in artifact and natural object classes distribute larger than those in other classes, but they are still quite a few for the overall distribution. Therefore, the large $dist_{rs}$s distribute a little.

Note that, different from that in [26], we give an explanation from a semantic aspect instead of a pixel aspect.

## 3   The Proposed Work

In this section, we introduce the proposed model, and then give the detailed gradients of loss with respect to parameters for training the model.

### 3.1   Fast Nonlocal Diffusion by Stacking

Before introducing the proposed work, we will give a brief review of TNLRD. For more details about TNLRD, we recommend one refers to [15] due to the limited space.

In TNLRD, the energy function that embeds NSS prior is as follow

$$E(u, f) = \frac{\lambda}{2} ||u - f||^2 + \sum_{i=1}^{N_k} \rho_i(W_i K_i u), \tag{2}$$

where $u, f \in \mathcal{R}^p$, $p$ is the number of pixels in the image. $\rho_i$ is the nonlinear functions. $K_i \in \mathcal{R}^{p \times p}$ is the matrix form of the local filter $k_i$, and is a highly sparse matrix, such that $k_i * u \Leftrightarrow K_i u$. $W_i$ is a highly sparse matrix defined to model the NSS prior, which is related to the non-local filter.

Based on the $dist_{rs}$ distribution in Sect. 2, we can approximate nonlocal filters by a couple of local filters. In such a setting, in the proposed model there is no need to exploit kNN to get the NSS prior as TNLRD. To this end, we propose to accelerate the TNLRD by stacking two local filters instead of stacking a nonlocal filter on the top of a local filter in (2). The resulting energy function is formulated as

$$E(u, f) = \frac{\lambda}{2} ||u - f||^2 + \sum_{i=1}^{N_k} \sum_{j=1}^{N_h} \rho_{ij}(h_j * k_i * u), \tag{3}$$

where $k_i, h_j \in \mathcal{R}^{m^2}$ are the $1^{st}$layer and $2^{nd}$layer filters respectively. $m$ is the size of the filter and is an odd. Note that, the spatial range of $h_j$ can be different from that of $k_i$. $N_k$ and $N_h$ is the number of the $k_i$ and the $h_j$ respectively.

### 3.2   2layerTNRD

One gets a denoised image $u_T$ by conducting the gradient descent of the Eq. (3) with $T$ steps given an initial condition $u_0 = f$. This $T$ gradient descent steps of the Eq. (3) is the proposed model 2layerTNRD. It is formulated as

$$\begin{cases} u_0 = f, \quad t = 1, \cdots, T \\ u_t = u_{t-1} - \left( \sum_{i=1}^{N_k} \sum_{j=1}^{N_h} K_i^{t\top} H_j^{t\top} \phi_{ij}^t (H_j^t K_i^t u_{t-1}) + \lambda^t (u_{t-1} - f) \right), \end{cases} \tag{4}$$

where $\phi_{ij}(\cdot) = \rho'_{ij}(\cdot)$, is the influence function [2] or flux function [21]. The parameters in step $t$ are $\Theta_t = \{\lambda^t, \phi_{ij}^t, K_i^t, H_j^t\}$, where $i = 1, ..., N_k, j = 1, ..., N_h$ and $t = 1, ..., T$. Note that, the parameters are varying for each step $t$ and are learnt in a supervised manner. The $1^{st}$layer filter $k_i$, the nonlinear function $\rho_{ij}$ are parameterized in the way as TNRD. The $2^{nd}$layer filters $h_j$ are with unit length constraint, and are initialized as

$$h_j(l) = \begin{cases} 1 & l = (m^2 + 1)/2 \\ \mathcal{N}(0, 0.01) & otherwise \end{cases}. \tag{5}$$

### 3.3   Training

These parameters $\{\Theta_t\}_{t=1}^T$ are learned from training samples using the discriminative learning scheme. To be more specific, given pairs of degraded image $f$ and its ground-truth image $u_{gt}$, the parameters $\{\Theta_t\}_{t=1}^T$ are optimized by minimizing the loss function $\ell(u_T, u_{gt})$, where $u_T$ is the output given by the inference procedure (4). In summary, the training procedure is formulated as

$$
\begin{cases}
\Theta^* = \mathrm{argmin}_\Theta \mathcal{L}(\Theta) = \sum\limits_{s=1}^{S} \ell\left(u_T^s, u_{gt}^s\right) \\
\text{s.t.} \begin{cases} u_0^s = f^s \\ u_t^s = u_{t-1}^s - \left( \sum\limits_{i=1}^{N_k} \sum\limits_{j=1}^{N_h} {K_i^t}^{\top} {H_j^t}^{\top} \phi_{ij}^t (H_j^t K_i^t u_{t-1}^s) + \lambda^t (u_{t-1}^s - f^s) \right) \\ t = 1 \cdots T. \end{cases}
\end{cases}
\tag{6}
$$

The training problem (6) can be solved via gradient based algorithms, e.g., commonly used L-BFGS algorithm [11]. The gradients of the loss function w.r.t. $\Theta_t$ are computed using the standard back-propagation technique widely used in the neural networks learning [10].

$$
\frac{\partial \ell(u_T, u_{gt})}{\partial \Theta_t} = \frac{\partial u_t}{\partial \Theta_t} \cdot \frac{\partial u_{t+1}}{\partial u_t} \cdots \frac{\partial \ell(u_T, u_{gt})}{\partial u_T}.
\tag{7}
$$

Detailed derivation of the required gradients in (7) is similar to that in [4,15].

**Table 1.** Influence of the parameters. PSNR(dB)/time($s$).

| 2layer | | $m \times m$ | | | |
|---|---|---|---|---|---|
| TNRD | | $3 \times 3$ | $5 \times 5$ | $7 \times 7$ | $9 \times 9$ |
| $N_h$ | 8 | 28.74/2.75 | 28.95/8.41 | 29.01/18.51 | 29.00/75.63 |
| | 16 | 28.77/4.62 | 28.97/24.90 | 29.00/33.26 | -/- |
| | 24 | 28.77/6.22 | 28.96/23.98 | -/- | -/- |
| | 48 | 28.78/23.51 | 28.97/49.72 | -/- | -/- |
| TNRD [5] | | 28.38/0.29 | 28.78/0.81 | 28.91/1.80 | 28.95/3.23 |
| TNLRD [15] | | -/- | 28.92/41.41 | 29.01/58.53 | -/- |

## 4 Experiments

The training dataset is described in Sect. 2. The noisy image $f$ is generated via (1) with a specific noise level $\sigma$. We trained 2layerTNRD via a joint training after the greedy training to avoid the poor local minimizers of the loss function. We trained 2layerTNRD on a server with Intel(R) Xeon E5-2650 @ 2.00 GHz and NVIDIA GTX980. We ran 200 L-BFGS iterations for both greedy and joint training, when the training loss decreases marginally.

### 4.1 The Influence of the Parameters

In this subsection, we discuss the influence of the parameters $T$, $m \times m$ and $N_h$.

**The Steps $T$.** As illustrated in [5,15], setting $T$ to 5 is a reasonable trade-off between model performance and computation efficiency.

**The Local Filter Size** $m \times m$. $m \times m$ is set to $3 \times 3$, $5 \times 5$, $7 \times 7$ and $9 \times 9$. As shown in each row of Table 1, $7 \times 7$ is preferred, which is a better trade-off between image restoration performance and runtime. From Table 1, we found that via stacking local filters we can efficiently enlarge the equivalent filter size while achieving the same or even superior performance. By stacking local filters, we use few parameters to achieve the equivalent large filter size. For example, we can stack two $7 \times 7$ filter to achieve the equivalent filter size with $13 \times 13$, and use few parameters, $2 \times 7 \times 7 = 98 < 13 \times 13 = 169$. It is widely believed that fewer parameters mean more easier to train. This insight is consistent with relevant findings in VGG [18] for image classification, and DnCNN [23] for image restoration.

**The Number of $2^{nd}$ Layer Filter** $N_h$. In TNLRD, the nonlocal filter $W_i$ shares the same filter coefficients for different positions in the image. It means in TNLRD the nonlocal filter is enforced to use the same coefficients without a consideration of the reference patches context. To alleviate this strong assumption, we set $N_h$ to 8, 16, 24 and 48. As shown in each column of Table 1, we can find that with larger $N_h$, the performance gain is within 0.05 dB. Considering both restoration performance and computational efficiency, we tend to set $N_h$ to 8. The learned filters in stage $t = 1$ and $t = 5$ for 2layerTNRD$_{7 \times 7 \times 8}^{5}$ are shown in Fig. 3. Taking the $2^{nd}$ layer filters in stage $t = 5$ in Fig. 3(d) as an example, one can see that the $2^{nd}$ layer filters weight more on the nearby patches. Some $2^{nd}$ layer filters are learnt to capture directional line structures, which are corresponding to the edges in images.

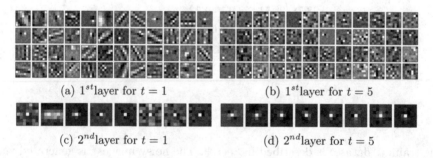

(a) $1^{st}$ layer for $t = 1$          (b) $1^{st}$ layer for $t = 5$

(c) $2^{nd}$ layer for $t = 1$          (d) $2^{nd}$ layer for $t = 5$

**Fig. 3.** The learned $1^{st}$ layer and $2^{nd}$ layer filters in stage $t = 1$ and $t = 5$ for 2layerTNRD$_{7 \times 7 \times 8}^{5}$.

## 4.2   Runtime Comparison

We compare the runtime among BM3D, WNNM, TNRD, TNLRD and 2layerT-NRD. The images are with resolution $256^2$, $512^2$, $1024^2$, $2048^2$ and $4096^2$. For the sake of fair comparison, the runtime of the comparison methods is measured via single thread implementation in Matlab on Intel(R) Core(TM) i5-4460 CPU @ 3.20 GHz. The runtime is summarized in Table 2.

**Table 2.** Runtime comparison for image denoising methods in seconds. GPU* runtime for TNRD is reported in [5] on NVIDIA GTX780Ti. GPU** runtime for 2layerTNRD is executed on NVIDIA GTX980.

| Method | $256^2$ | $512^2$ | $1024^2$ | $2048^2$ | $4096^2$ |
|---|---|---|---|---|---|
| BM3D [6] | 0.48 | 1.94 | 8.37 | 39.44 | 166.42 |
| WNNM [9] | 164.95 | 731.15 | 2962.17 | - | - |
| TNRD [5] | 1.00 | 2.65 | 8.89 | 31.65 | 128.55 |
| GPU* | 0.01 | 0.032 | 0.116 | 0.4 | - |
| TNLRD [15] | 25.87 | 97.30 | 385.76 | 1551.78 | 6557.73 |
| 2layerTNRD | 11.61 | 28.96 | 97.65 | 332.08 | 1129.40 |
| GPU** | 0.49 | 1.61 | 6.20 | 24.33 | 97.27 |

BM3D runs on par with TNRD, and outperforms other methods. However, the denoised images via BM3D is not as good as the others.

Compared with TNRD, 2layerTNRD runs 10 times slower in CPU runtime comparison; and about 60 times slower in GPU runtime comparison. The computational complexity of TNRD is $\mathcal{O}(T \cdot p \cdot N_k \cdot (2 \cdot ksz^2 + NumW)) = \mathcal{O}(38640 \cdot p)$. The computational complexity of 2layerTNRD is $\mathcal{O}(T \cdot p \cdot N_k \cdot (ksz^2 + 3 \cdot ksz^2 \cdot N_h + NumW \cdot N_h)) = \mathcal{O}(414960 \cdot p)$. In the above analysis, the parameters are set to $T = 5$, $N_k = 48$, $ksz = 7$, $NumW = 63$, $N_h = 8$, $p$ is the number of pixels in the image. Considering the computational complexity, 2layerTNRD is at least 10 times slower than TNRD, which is consistent with the runtime comparison above. However, the denoised image via 2layerTNRD is more visually plausible than that of TNRD.

Compared with TNLRD, 2layerTNRD achieves about 6 times speedup via the CPU implementation. The computational complexity of TNLRD is $\mathcal{O}(T \cdot p \cdot N_k \cdot (2 \cdot ksz^2 + 2 \cdot L + NumW) + p \cdot ksz^2 \cdot Wsz^2 \cdot L) = \mathcal{O}(276485 \cdot p)$, where $Wsz = 31$, $L = 5$. Considering the computational complexity, TNLRD runs slightly slower than TNRD and faster than 2layerTNRD. While in runtime, TNLRD runs quite slow. In TNLRD, kNN is time-consuming and is not easy to accelerate via the GPU implementation. The construction of nonlocal filters is also time-consuming. On the contrary, in 2layerTNRD there are merely convolutional operation and nonlinear functions, which are suitable to be accelerated via GPU implementation. The GPU implementation of 2layerTNRD runs about 10 times faster than the CPU implementation of 2layerTNRD.

When WNNM deonised images with resolution larger than $2048^2$, it caused an out-of-memory error. Therefore, we were not able to test WNNM for those resolutions. From the tested images, WNNM is quite time-consuming and memory-inefficient.

**Table 3.** Average PSNR(dB)/SSIM/time($s$) on 68 testset. The GPU runtime for 2layerTNRD is marked in bracket.

| Method | $\sigma$ | | | Time |
|---|---|---|---|---|
| | 15 | 25 | 50 | |
| BM3D [6] | 31.08/0.8717 | 28.56/0.8013 | 25.62/0.6864 | 1.21 |
| WNNM [9] | 31.37/0.8759 | 28.83/0.8084 | 25.83/0.6981 | 483.40 |
| TNRD [5] | 31.42/0.8821 | 28.91/0.8152 | 25.96/0.7024 | 1.80 |
| TNLRD [15] | **31.50**/0.8852 | **29.01**/0.8201 | 26.06/0.7094 | 58.53 |
| 2layerTNRD | 31.49/0.8847 | 29.01/0.8195 | **26.08**/0.7093 | 18.51(**0.95**) |

### 4.3   Denoising Comparison

We compared denoising performance using 68 testset as [5,16], which are widely used in image denoising. We evaluated the denoising performance using PSNR [5] and SSIM [20]. The codes of the compare methods were downloaded from the authors' homepage. We discussed the denoising performance for noise level $\sigma$=15, 25 and 50. The comparison result is summarized in Table 3. Visual comparison is shown in Fig. 4 and 5.

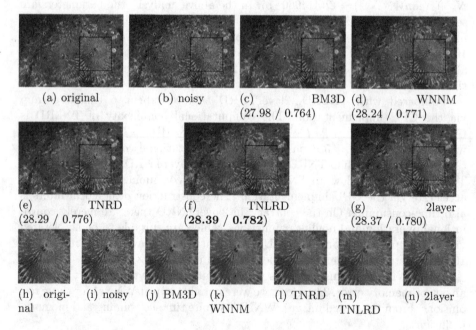

(a) original   (b) noisy   (c)       BM3D   (d)       WNNM
(27.98 / 0.764)   (28.24 / 0.771)

(e)       TNRD   (f)       TNLRD   (g)       2layer
(28.29 / 0.776)   (**28.39 / 0.782**)   (28.37 / 0.780)

(h)  origi-   (i) noisy   (j) BM3D   (k)       (l) TNRD   (m)       (n) 2layer
nal   WNNM   TNLRD

**Fig. 4.** Denoising results of the competing algorithms on a test image from the 68 test images for $\sigma = 25$. The PSNR/SSIM results in **bold** are best.

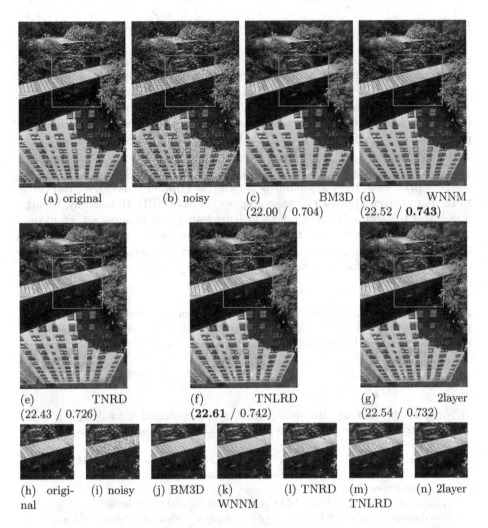

**Fig. 5.** Denoising results for $\sigma = 50$. The PSNR/SSIM results in **bold** are best.

As shown in Table 3, 2layerTNRD achieves competing denoising performance compared with TNLRD, and outperforms TNRD, WNNM and BM3D in both PSNR and SSIM. As shown in Fig. 5(h–n), we found 2layerTNRD produces fewer artifacts in the homogenous regions than TNLRD. When noise level gets larger, NSS prior obtained via kNN is getting less accurate. As a result, TNLRD may introduce artifacts in homogenous regions. On the contrary, the $2^{nd}$layer filters are learnt from training data, and the spatial location of these filters is less sensitive to the noise level. To this end, 2layerTNRD may create less artifacts than TNLRD. Limited by the kernel size, 2layerTNRD is not able to recover as much details as TNLRD.

# 5    Conclusion

Based on the analysis of the $dist_{rs}$ distribution from a semantic aspect, we found that in artifact and natural object images the large $dist_{rs}$s occur while their contribution to the overall distribution is small. Based on this observation, we proposed a fast nonlocal diffusion by stacking two local filters. Experimental results show that the proposed 2layerTNRD achieves competing denoising performance compared with TNLRD and is more computational efficiency than TNLRD.

To summarize the comparison results, we argue that exploiting kNN to get the NSS prior from the noisy input is not the only option at hand, we can turn to the data-driven approach like the proposed 2layerTNRD. In this way, we are more likely to capture the NSS prior which are learnt from training set. From the $dist_{rs}$ distribution, it suggests that we should use more efficient and effective way to embed the NSS prior into the image restoration methods. We will carefully regard the omitted far away similar patches corresponding to the large $dist_{rs}$s in our future work.

**Acknowledgements.** This work was supported by the National Key Research and Development Program of China under the Grant No. 2018YFB1003405, and National Natural Science Foundation of China under the Grant No. 61732018.

# References

1. Aharon, M., Elad, M., Bruckstein, A.: K-SVD: an algorithm for designing over-complete dictionaries for sparse representation. IEEE Trans. Sig. Process. **54**(11), 4311 (2006)
2. Black, M., Sapiro, G., Marimont, D., Heeger, D.: Robust anisotropic diffusion and sharpening of scalar and vector images. In: Proceedings of International Conference on Image Processing, vol. 1, pp. 263–266. IEEE (1997)
3. Buades, A., Coll, B., Morel, J.M.: A non-local algorithm for image denoising. In: Proceedings of the IEEE Conference on Computer Vision and Pattern Recognition, vol. 2, pp. 60–65 (2005)
4. Chen, Y.: Notes on diffusion networks. arXiv preprint arXiv: 1503.05768 (2015)
5. Chen, Y., Yu, W., Pock, T.: On learning optimized reaction diffusion processes for effective image restoration. In: Proceedings of the IEEE Conference on Computer Vision and Pattern Recognition, pp. 5261–5269 (2015)
6. Dabov, K., Foi, A., Katkovnik, V., Egiazarian, K.: Image denoising by sparse 3-D transform-domain collaborative filtering. IEEE Trans. Image Process. **16**(8), 2080–2095 (2007)
7. Deng, J., Dong, W., Socher, R., Li, L.J., Li, K., Fei-Fei, L.: ImageNet: a large-scale hierarchical image database. In: CVPR 2009 (2009)
8. Dong, W., Zhang, L., Shi, G., Li, X.: Nonlocally centralized sparse representation for image restoration. IEEE Trans. Image Process. **22**(4), 1620–1630 (2013)
9. Gu, S., Zhang, L., Zuo, W., Feng, X.: Weighted nuclear norm minimization with application to image denoising. In: Proceedings of the IEEE Conference on Computer Vision and Pattern Recognition, pp. 2862–2869 (2014)

10. LeCun, Y., Bottou, L., Bengio, Y., Haffner, P.: Gradient-based learning applied to document recognition. Proc. IEEE **86**(11), 2278–2324 (1998)
11. Liu, D.C., Nocedal, J.: On the limited memory BFGS method for large scale optimization. Math. Program. **45**(1), 503–528 (1989)
12. Mairal, J., Bach, F., Ponce, J., Sapiro, G., Zisserman, A.: Non-local sparse models for image restoration. In: 2009 IEEE 12th International Conference on Computer Vision, pp. 2272–2279. IEEE (2009)
13. Mao, X., Shen, C., Yang, Y.B.: Image restoration using very deep convolutional encoder-decoder networks with symmetric skip connections. In: Advances in Neural Information Processing Systems, pp. 2802–2810 (2016)
14. Miller, G.A., Beckwith, R., Fellbaum, C., Gross, D., Miller, K.J.: Introduction to WordNet: an on-line lexical database. Int. J. Lexicogr. **3**(4), 235–244 (1990)
15. Qiao, P., Dou, Y., Feng, W., Li, R., Chen, Y.: Learning non-local image diffusion for image denoising. In: Proceedings of the 2017 ACM on Multimedia Conference, pp. 1847–1855. ACM (2017)
16. Roth, S., Black, M.J.: Fields of experts. Int. J. Comput. Vis. **82**(2), 205–229 (2009)
17. Schmidt, U., Roth, S.: Shrinkage fields for effective image restoration. In: Proceedings of the IEEE Conference on Computer Vision and Pattern Recognition, pp. 2774–2781 (2014)
18. Simonyan, K., Zisserman, A.: Very deep convolutional networks for large-scale image recognition. arXiv preprint arXiv:1409.1556 (2014)
19. Sun, J., Tappen, M.F.: Learning non-local range Markov random field for image restoration. In: Proceedings of the IEEE Conference on Computer Vision and Pattern Recognition, pp. 2745–2752 (2011)
20. Wang, Z., Bovik, A.C., Sheikh, H.R., Simoncelli, E.P.: Image quality assessment: from error visibility to structural similarity. IEEE Trans. Image Process. **13**(4), 600–612 (2004)
21. Weickert, J.: Anisotropic Diffusion in Image Processing, vol. 1. Teubner, Stuttgart (1998)
22. Xu, J., Zhang, L., Zuo, W., Zhang, D., Feng, X.: Patch group based nonlocal self-similarity prior learning for image denoising. In: Proceedings of the IEEE International Conference on Computer Vision, pp. 244–252 (2015)
23. Zhang, K., Zuo, W., Chen, Y., Meng, D., Zhang, L.: Beyond a gaussian denoiser: residual learning of deep CNN for image denoising. IEEE Trans. Image Process. **26**(7), 3142–3155 (2017)
24. Zhang, K., Zuo, W., Gu, S., Zhang, L.: Learning deep CNN denoiser prior for image restoration. In: IEEE Conference on Computer Vision and Pattern Recognition, vol. 2 (2017)
25. Zhang, K., Zuo, W., Zhang, L.: FFDNet: toward a fast and flexible solution for CNN based image denoising. IEEE Trans. Image Process. **27**, 4608–4622 (2018)
26. Zontak, M., Irani, M.: Internal statistics of a single natural image. In: 2011 IEEE Conference on Computer Vision and Pattern Recognition (CVPR), pp. 977–984. IEEE (2011)
27. Zoran, D., Weiss, Y.: From learning models of natural image patches to whole image restoration. In: 2011 IEEE International Conference on Computer Vision (ICCV), pp. 479–486. IEEE (2011)

# Relationship Between Tyndall Light Path Attenuation and Concentration Based on Digital Image

Cunbo Jiang, Yaling Qin[✉], and Tiantian Zhu

College of Information Science and Engineering,
Guilin University of Technology, Guilin 541004, China
1129198514@qq.com

**Abstract.** When the red laser illuminates the lyosol, the Tyndall effect will form a light path with a certain distance, and the optical properties of the lyosol will have a certain influence on the Tyndall light intensity. This paper mainly aims at the theoretical and experimental studies on the change situation of the lyosol concentration and the attenuation characteristics of the light path when the red laser changes with the distance of the light path in the solution. In order to study the effect of lyosol concentration on the Tyndall light path, digital image technology was applied to the measurement of lyosol concentration. Due to the non-contact property of the image, the liquid concentration can be measured accurately in real time. The attenuation characteristics of the laser in the lyosol were obtained by image processing technology, and the quantitative relationship between the attenuation coefficient of the Tyndall light path and the lyosol concentration was obtained.

**Keywords:** Digital image · Tyndall effect · Attenuation characteristics · Quantitative relationship

## 1 Introduction

The concentration of the solution is related to many of its physical quantities, so in principle one of the changes in the amount can be used to characterize the change in concentration. Existing ultrasonic testing is sensitive to the temperature of the object being measured. The microwave method causes the object that to be measured be heated and cause a series of chemical and physical changes [1]. With the development of image processing technology, the detection of the concentration of the liquid to be tested by the Tyndall effect experimental phenomenon of the liquid dispersion system and the image processing technology has become the focus of research [2]. The detection based on image processing technology can avoid the deficiency of the traditional detection technology, realize non-contact measurement, and can measure the concentration of the lyosol accurately and in real time.

Since the absorption, reflection and scattering of light when it travels in solution, the intensity of light decreases with the increase of paths that it moves [3–5], and the intensity of the Tyndall light may also attenuate with the path. In this paper, the

© Springer Nature Singapore Pte Ltd. 2019
R. Mao et al. (Eds.): ICPCSEE 2019, CCIS 1059, pp. 102–109, 2019.
https://doi.org/10.1007/978-981-15-0121-0_7

relationship between the concentration of known samples and the attenuation of Tyndall light intensity is studied experimentally, and by which the relationship between the concentration of lyosol and the Tyndall light intensity attenuation, the concentration of lyosol and the attenuation characteristics of Tyndall light intensity is established. Computer can analyze the images of colloidal Tyndall phenomenon with certain precision to obtain experimental data, combining with the attenuation characteristics of laser in lyosol solution, analyze the mathematical model that can improve the accuracy of concentration detection.

## 2   Basic Principle

Particles with a diameter of 1 to 100 nm in the colloidal solution scatter the incident light to cause the Tyndall phenomenon [5]. When the particle diameter is between 5 and 100 nm, its scattered light intensity [7, 8]:

$$I = K \frac{\upsilon v^2}{\lambda^4} \left( \frac{n_1^2 - n_2^2}{n_1^2 + 2n_2^2} \right)^2 I_0 \tag{1}$$

Where, $I_0$ is incident light intensity; $\lambda$ is incident light wavelength; $\upsilon$ is number of particles per unit volume; V is the volume of single particle (V); $n_1$ is the refractive index of dispersed phase (colloidal particle); $n_2$ is the refractive index of dispersion medium; K is a constant. The intensity of scattered light is related to the concentration of the substance. When I0 is constant, I increases linearly with the increase of $\upsilon$. Studies have shown that light attenuates along the propagation path, and the attenuation characteristic of light along the path in aqueous solution have exponential properties that can be described by formula (2) [9, 10].

$$I_{(Td)} = I_{(T0)} e^{-kd} \tag{2}$$

Where, $I_{(T0)}$ is the light intensity of the incident point in the solution, d is the geometric distance in the direction of the light path in the solution, $I_{(Td)}$ is the light intensity at the distance d from the incident point in the solution, and k is the attenuation coefficient. Equations (1) and (2) show that by measured the path attenuation of Tyndall light intensity, the undetermined coefficient can be obtained by using the reference solution and the calibration can be carried out in the measurement process, so as to realize the measurement of the concentration of the solution under test. The principle of the measuring device is shown in Fig. 1.

In the Figs. 1, 2, 3, 4, 5 and 6 are reference patterns of known concentration, 7 is an empty sample, 8 is a sample to be tested, and eight samples can be sequentially rotated and positioned to a photographing station under the control of the control device; In the photographing station, a constant light source is vertically injected into the sample from the bottom of the sample, and a Tyndall light is generated perpendicular to the incident light path in the vertical direction. After the sample is accurately positioned and stabilized, the camera is controlled to take an image containing the Tyndall light.

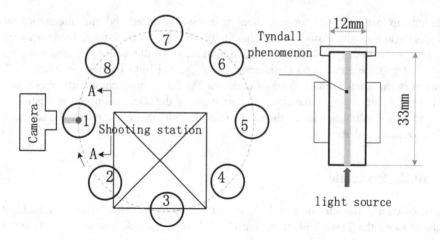

(a) Shooting station conversion principle    (b) A-A View

**Fig. 1.** Measurement schematic diagram.

# 3  Acquisition and Processing of Digital Images

## 3.1  Acquisition of Image

The nanocrystal contents are set as follows: 70%, 60%, 40%, 20%, 10%, 8%, 2% of the standard (33 mm high and 12 mm wide). In addition, 50%, 30% and 5% nanocrystalline solutions are prepared to test the expression of experimental results.

The standard samples are placed in the same full dark background environment to capture images of the Tyndall phenomenon using a camera. The source light wavelength is 650 nm (50 mW red laser), and the incident direction of the light is perpendicular to the bottom of the glass bottle as shown in Fig. 2(a) Concentration is 100% (b) Concentration is 70% (c) Concentration is 40% (d) Concentration is 8%.

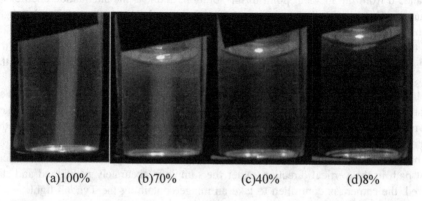

(a)100%        (b)70%        (c)40%        (d)8%

**Fig. 2.** Images of Tyndall phenomenon at several different concentrations. (Color figure online)

## 3.2  Image Preprocessing

Image thinning by computer will occur distortion phenomenon, that is, image noise, which affects the experimental results to a large extent, and brings huge error to the solution concentration detection method. Therefore, the image containing noise should be denoised first.

Median filtering [11, 12] is based on the theory of order statistics of a nonlinear signal processing technology can effectively restrain noise, median filtering is the basic principle of the digital image or the value of the point in the sequence used at various points in the field of a point at which the median value replace, let the surrounding pixel values close to the real value of order to remove isolated noise points. Median filtering is a classical noise smoothing method, which is often used to protect edge information in image processing. The purpose of image processing in this paper is to effectively extract the Tyndall light path, so the median filtering method is adopted to remove the noise, and the image with the concentration percentage of 100% is selected for processing to obtain the filtered effect picture e.

To capture the central part image of the Tyndall phenomenon light path, the image width is 20 * 80 mm. The image of the central portion of the light path is segmented by the image to extract the range of research required. Specific steps are as follows:

(1) Select a rectangle in the area with the light path (this rectangle should basically replace the color of the light path, such as selecting a rectangular area inside the light path)
(2) Calculate the mean value and variance within the rectangular area
(3) It is specified that if the variance larger than the mean value n times and less than the mean negative n times is considered to be the position of the light path (n value depends on the light intensity range).

Through the image segmentation effect picture g, the image in the analysis range is converted into a gray image picture h, and the tricolor value is read. The light intensity conversion measurement is realized according to formula 3.

$$V = 0.2989 \times R + 0.5870 \times G + 0.1140 \times B \tag{3}$$

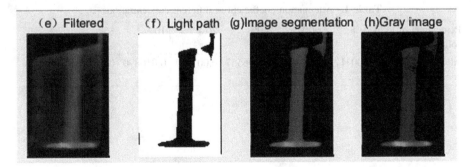

**Fig. 3.**  Image processing effect.

## 4  Tyndall Light Path Attenuation Analysis

Extract the intensity value matrix of the light path center in figure h, and the corresponding light intensity values on the unit pixel are obtained as $I_1, I_2 \ldots I_{20}$, calculate the average value $I_a$ of light intensity per unit distance. The obtained data are drawn into the scatter diagram as shown in Fig. 4, so as to further analyze the attenuation characteristics of the optical path.

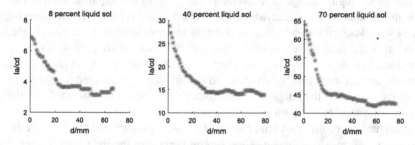

**Fig. 4.** Attenuation trend of concentration of 8%, 40%, 70%.

It can be seen from the figure above that the light intensity value of the Tyndall light path changes regularly with the distance and basically presents an exponential distribution. At the beginning and end of the light path, local highlights are generated by the refraction of the glass bottle and the horizontal plane, resulting in an increase in the light intensity value. Since the glass bottle of the same size and the same solution were used in the experiment, it can be considered that each highlighted area has a consistent impact on the results of the experiment, which is in the normal range.

Then, after smoothing the obtained data, it can be seen from Fig. 4 that with the increase of transmission distance, the light intensity of Tyndall shows a trend of exponential decline in whatever lyosol concentration. MATLAB is used to fit the obtained discrete data with formula 2. Record the attenuation coefficient K of the lyosol, and the data were shown in Table 1. The fitting effect diagram of attenuation trend is shown in Fig. 5.

**Table 1.** Attenuation coefficient at different concentrations

| Percentage of solution (c) | 70% | 60% | 40% | 20% | 10% | 8% | 2% |
|---|---|---|---|---|---|---|---|
| Attenuation coefficient (k) | 0.004177 | 0.003650 | 0.006557 | 0.007971 | 0.010890 | 0.012260 | 0.01501 |

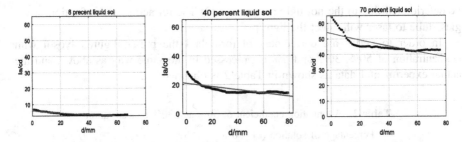

**Fig. 5.** Fitting image of attenuation trend.

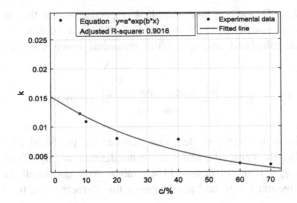

**Fig. 6.** Relationship between attenuation coefficient and concentration.

It can be seen from the data in Table 1 that the higher the concentration of the lyosol, the smaller the corresponding attenuation coefficient, and the better the correlation line. The results obtained by fitting the data in Table 1 are shown in Fig. 6.

That is, the equivalence relationship between the attenuation coefficient of the lyosol and the concentration is:

$$k = 0.0147 * \exp(-0.02325 * c) \tag{4}$$

The fitting correlation coefficient $R^2$ is 0.9016, and the fitting degree is relatively high.

## 5   Results and Analysis

The experimental results show that the attenuation coefficient of different percentages of lyosol is approximately exponential relationship with the concentration, indicating that there are still some experimental errors. The reasons for the errors may be: (1) The purity of the nano gold used in the preparation of the solution is not necessary so high as written on the label, causing the solution is not the standard concentration, and may be deteriorated due to the not tight packaging; (2) The refraction and reflection of the glass tube to laser will affect the light path.

In order to verify the correctness of formula 4 the pre-configured lyosol with concentration of 50%, 30% and 5% is processed in the same way as above, and the actual experimental data as shown in Table 2 is obtained:

**Table 2.** Attenuation coefficient of 50% and 30% concentration

| Percentage of solution (c) | 50% | 30% | 5% |
|---|---|---|---|
| Attenuation coefficient (k) | 0.004091 | 0.006557 | 0.01464 |

Put c = 50%, 30% and 5% into formula 4 respectively, and the attenuation coefficient of the lyosol is calculated for 0.004597, 0.007318, 0.01309. The relative error RE between the calculated result and the measured result can be calculated by formula 5

$$RE = \left| \frac{observed - predicted}{observed} \right| \times 100\% \tag{5}$$

RE can evaluate the degree of deviation between the predicted value and the true value. The smaller the value, the better the generalization ability of the formula. The calculated $RE_{(50\%)}$ = 12%, 11%, and 11% relative error are small, which basically meets the experimental needs, and also explains the correctness of formula 4.

## 6  Conclusion

In this paper, the variation of the intensity of scattered light along the Tyndall light path with the transmission distance is analyzed through the acquisition of the image of the Tyndall light path of the lyosol. The attenuation of the lyosol with different concentrations on the light path distance is subject to exponential distribution through non-linear fitting and analysis with a large number of experimental data, and the attenuation coefficient value is obtained. There is an obvious correspondence between the intensity attenuation coefficient along the light path and the concentration: k = 0.0147 * exp * (−0.02325 * c). Verify that the maximum relative error does not exceed 0.12, which meets the accuracy requirements of concentration measurement. The experimental results provide more accurate results for the measurement of outdoor liquid concentrations. Since the fixed light source is adopted in the selection of the light source in this paper, the influence of the incident wavelength is neglected, and the experiment may have some certain errors. It is hoped that in the follow-up studies, the errors in the experiment can be reduced by improving the choice of light source and the factor of refraction of the glass bottle to the light.

**Acknowledgements.** As the research of the thesis is sponsored by Development and application of rapid quantitative detection technology for heavy metals in water environment without instruments, Project number: guike AB17129003.

# References

1. Zhang, H., Wang, S.L., Jin, J.: Detection of suspended particle concentration in sewage based on image processing. J. Jiamusi Univ. (Nat. Sci. Ed.) **35**(1), 111–113 (2017)
2. Hu, W.J.: Research on Cd(II) concentration measurement of RGB image color detection. Sens. Microsyst. **35**(10), 17–19 (2016)
3. Wei, Z., Sheng, N., Zhang, X.J.: Improvement and experiment of transmission attenuation model of 532 nm laser in rain. Infrared Laser Eng. **47**(11), 43–50 (2018)
4. Barabanov, I.R., et al.: Influence of neodymium on the light attenuation length in a liquid organic scintillator. Phys. Part. Nuclei Lett. **15**(6), 630–636 (2018)
5. Lei, Y., Liu, C., Chen, B.: A TIE based technique for measurement of stress generated by laser induced damage. Chin. J. Lasers **45**(09), 262–268 (2018)
6. Alouini, M.S., Goldsmith, A.J.: Capacity of Rayleigh fading channels under different adaptive transmission and diversity-combining techniques. IEEE Trans. Veh. Technol. **48**(4), 1165–1181 (1999)
7. Li, J.S., Zhao, L., Sun, J., et al.: Research on the tyndall effect. Chem. Teach. (1), 44–47 (2014)
8. Wang, F.S.: Research on detection unit of liquid composition concentration based on optical scattering method. Harbin University of Science and Technology (2014)
9. Kirk, J.T.O.: Light and Photosynthesis in Aquatic Ecosystem. Cambridge University Press, New York (1983)
10. Morel, A., Maritorena, S.: Bio-optical properties of oceanic waters: a reappraisal. J. Geophys. Res. Oceans **106**(C4), 7163–7180 (2002)
11. Huang, M.T., Hu, Y.C.: Improved adaptive median filter denoising in smoke image of low illumination. Comput. Eng. Des. **39**(06), 1659–1663 (2018)
12. Chen, T., Ma, K.K., Chen, L.H.: Tri-state median filter for image denoising. IEEE Trans. Image Process. **8**(12), 1834 (1999). A Publication of the IEEE Signal Processing Society

# Systematic Framework of the All-for-One Tourism Digital Ecosystem

Dian He, Ying Liang$^{(\boxtimes)}$, Xiaolong Li, Yao Liu, and Jianglian Liu

College of Computer and Information Engineering,
Hunan University of Commerce, Changsha 410205, Hunan, China
yingl@126.com

**Abstract.** All-for-one tourism is a new mode of regional development, which can promote the coordinated development of economy and society. Due to diverse tourism information sources and scattered data services, the real-time and effectiveness of tourism information service is insufficient, and the "information island" is numerous. In order to integrate and share information resources effectively, and achieve the intelligentialized and real-time all-for-one tourism information services, this paper proposes a systematic framework of all-for-one tourism information system, where tourists and tourist destinations are considered as all-for-one tourism digital ecosystem. The whole mode of this framework was combined with four levels and two parts. Some advanced information technologies such as big data, cloud computing and mobile internet application development were utilized to construct tourism digital ecosystem. How to implement all-for-one tourism information collection, integration, processing and sharing was discussed and the corresponding solution was provided.

**Keywords:** All-for-one tourism · Digital ecosystem · Big data management · Cloud platform · Mobile information service

## 1 Introduction

### 1.1 Motivation

The all-for-one tourism refers to a new regional coordinated development concept and model. It takes tourism as the dominant industry, which is based on the comprehensive and systematic optimization of regional economic and social resources, especially tourism resources, tourism related industries, natural environment, and public service, to promote the economic and social development. To develop the all-for-one tourism, we must rely on advanced information technology, and adopt the Internet + tourism mode. The basis of all-for-one tourism informatization is the construction of cloud service platform for tourism resources. On this basis, the analysis and application of big data are carried out.

Due to the large number of tourism information platforms, diverse information sources, inconsistent data records, and scattered data services, the real-time and effectiveness of tourism information service is insufficient. It leads to serious and numerous "information island". Therefore, it is necessary to study the key technology

© Springer Nature Singapore Pte Ltd. 2019
R. Mao et al. (Eds.): ICPCSEE 2019, CCIS 1059, pp. 110–119, 2019.
https://doi.org/10.1007/978-981-15-0121-0_8

of cloud service software for all-for-one tourism and apply it to all-for-one tourism, effectively integrate and share data, so as to realize the intelligentialize, personalization, real-time and mobile of all-for-one tourism services.

## 1.2   Related Works

In terms of tourism informatization, Guilin, Zhengzhou and other cities proposed and built intelligent tourism; and some demonstration zones developed an all-for-one tourism APP (such as Libo). Ctrip, Qunar, Fliggy and other enterprises built e-business platform, through big data analysis to implement precision marketing, and to provide personalized services for users. Dingyou, Qinyi and other enterprise platforms provided ticketing, big data display, virtual tour and other services for scenic spots. However, it is difficult for these information platforms to monitor the interaction between tourists and tourist destinations, and it is difficult to meet the needs of timely analysis, diagnosis and optimization of the all-for-one tourism ecosystem.

In the macro research of smart tourism, Zheng [1] reviewed the research progress of information technology and tourism in the past two decades which was divided into digital era (1997–2006) and accelerated era (2007–2016). It was pointed out that due to the emergence of technologies such as smartphones, UAVs, wearable devices, new connections and big data, information technology in tourism research has changed from the main market-driven tools to knowledge created tools. Del Vecchio et al. [2] used multi-case analysis method to explore the tourism experience of Southern Europe and the regions related to intelligent tourist destinations, and obtained the mode and opportunity of creating value of tourism big data.

In tourism information platform technology research, Zhou et al. [3] constructed intelligent tourism cloud platform based on IaaS, PaaS and SaaS modes of cloud computing to virtualizes tourism resources: to integrate and distribute tourism resources through IaaS, to reuse and virtualize tourism resources through PaaS, and customize and rent tourism resources through SaaS.

In the tourist destination heat and tourist destination competitiveness, Ren et al. [4] took Huairou, Beijing as an example, by mining the Internet search data to analyze the heat of tourist destinations, using the principal component analysis method to build a search index to measure tourism demand, and by using the big data of Ctrip Huairou to analyze the basic characteristics of tourists and tourism behavior. Zhang et al. [5] studied the BP neural network model of tourism competitiveness evaluation, and used this model to evaluate and analyzed the tourism competitiveness of the above-mentioned intelligent tourism cities.

In the scenic area navigation and tourist location services, Wang et al. [6] established a navigation grid model to account the multi-dimensional dynamic environmental factors such as scenic area topography, meteorology and tourist density. An A* algorithm for intelligent tourism dynamic routing was designed and implemented.

To recommend travel routes and products, Zhu et al. [7] proposed SECT, a tourism product recommendation engine based on thematic sequential pattern. They attempted to generate recommendation by mining click logs on online tourism websites. Li et al. [8] proposed distance-sensitive recommendation strategy DRSS based on semantics and conformity recommendation strategy CRSS based on semantics. Jiang et al. [9]

proposed a hybrid density clustering method to identify the heat of scenic spots based on the geographic coordinate data and tour path data of real scenic spots. On the basis of Bayesian model and probability matrix decomposition model, Zhang et al. [10] proposed a travel distance-sensitive tourism recommendation model, GeoPMF, with the theme of user's time and constraint of cost. Guo et al. [11] proposed a multi-angle tourism information perception method based on multi-source social data fusion, preprocessed heterogeneous tourism data, combined image with text, proposed a cross-media multi-angle association method to connect fragmented tourism information.

In the aspect of tourism management and prediction, Liu et al. [12] constructed an intelligent management and control mode of "mobile signaling data, matching of tourists' location and behavior, precise management and control of scenic spots".

In the new information technology era, advanced communication technologies, such as 5G, mobile Internet and the analysis of big data, can also be used in smart cities and smart tourism [13]. Considering the interference of noise in prediction, Li et al. [14] proposed a CLSI-EMD-BP prediction model based on network search. Jing et al. [15] proposed a recommendation strategy based on taxi traces data to help passengers get taxis more easily. This idea can also be used in scenic spots and tourist transportation. Liu et al. [16] used vector autoregressive model to test the causality between the actual number of arrivals of tourist destinations and the number of searches on the Internet. Del Chiappa et al. [17] held that people generally recognize the role of ICT, Internet of Things and cloud computing in providing tools and platforms, promoting information and knowledge dissemination among stakeholders.

However, there are few studies on the process of information and knowledge transmission, sharing and transformation in smart tourism. Based on the massive, heterogeneous, multi-source, multi-dimensional and multi-granularity characteristics of all-for-one tourism spatial and temporal information resources, the traditional centralized architecture cannot be effectively processed and analyzed. Although the cloud data management method based on the new distributed architecture has achieved effective results, how to support the application of all-for-one tourism remains to be explored. In addition, in the real scene of all-for-one tourism mobile services, the tourism information resources are in a wide range of dynamic changes and high real-time requirements. How to collect and publish real-time dynamic data to meet the needs of all-for-one tourism is also in urgent need of research.

In summary, there is still a lack of cloud service framework suitable for the requirements of all-for-one tourism currently. Therefore, this paper will put forward a systematic framework of the all-for-one tourism digital ecosystem, and discuss its definition, function and characteristics. At the same time, this paper further addresses the problems such as the difficulty in the integration of all-for-one tourism information, the island of tourism information, the lack of intelligent, real-time and mobile information services, etc. In the framework of the all-for-one tourism digital ecosystem, the key technologies to solve these problems are proposed.

# 2  All-for-One Tourism Digital Ecosystem

From the perspective of all-for-one tourism, based on complex and self-organization system, tourists and tourist destinations can be regarded as an all-for-one tourism ecosystem. In order to monitor the interaction between tourists and tourist destinations, this paper proposes to build a digital all-for-one tourism ecosystem, to solve tourists and tourist destinations information collection, processing integration and sharing systematically and timely, which meet the needs of tourist on the real-time tourism information. The digital all-for-one tourism ecosystem provides a big data base for real-time analysis, diagnosis, early warning and optimization of the all-for-one tourism ecosystem.

The all-for-one tourism digital ecosystem is a general information platform for the application of new technologies including big data, cloud computing, mobile Internet and artificial intelligence to realize standardization, generalization, intelligentialize and real-time mobile service of information of tourists and tourist destinations. To construct an all-for-one tourism digital ecosystem, it can effectively integrate and share social resources, innovate travel services, satisfy the needs of the tourists, improve service satisfaction, and provide data support for travel monitoring and macro-control, to make the entire all-for-one tourism service intelligent, personalized, real-time, and mobile.

To realize a digital all-for-one tourism digital ecosystem, the following main problems must be solved:

- The definition, classification and standardization of information related to the all-for-one tourism ecosystem.
- A variety of tourism information resource discovery, real-time acquisition, identification, coding and storage technology.
- Tourism and transportation, meteorology, public security, hotel, environmental protection and other cross-sectoral, cross-industry data fusion and sharing technology.
- Real-time information dissemination, dynamic query, intelligent push and visualization display technology.

Through the big data analysis of the collected historical and real-time data, it provides basis for government supervision and decision-making, public opinion monitoring for tourism, big data analysis and prediction of tourism market, development and recommendation of tourism products. Specific examples include:

- Through multi-source of tourism related information collection, identification and real-time encoding and storage technology research and development, implement automatic collection of information resources, integration and management, under the background of all-for-one tourism ecological tourism industry. It forms the information resource base of the all-for-one tourism digital ecosystem.
- Using technologies such as data mining, intelligent recommendation of research and development implement tourism information resources, data analysis and automatic push, build an intelligent and personalized service system for digital all-for-one tourism ecological system.

- By mobile terminal and large screen, etc., in response to user inquiry or active pushing, the tourist information resources such as traffic, weather, hotel, current feature product and service, air health index, attraction evaluation, etc., can be used to provide a visual demonstration of the digital all-for-one tourism ecosystem for tourists, tour guides and tourism enterprises. It realizes the visual and mobile service of the digital all-for-one tourism ecosystem.

## 3   The Systematic Framework

In this paper, the digital all-for-one tourism ecological system architecture model which includes four levels and two plates was put forward. The four levels are data acquisition layer, data layer, application layer and presentation layer, the two plates are terminals and cloud computing environment. The overall systematic framework diagram is shown in Fig. 1.

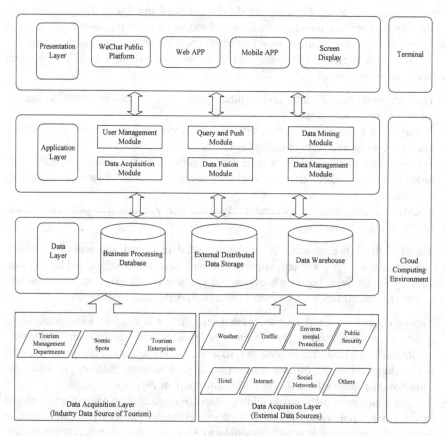

**Fig. 1.** All-for-one tourism digital ecosystem framework

The whole framework is divided into two parts: cloud computing environment and terminal. The terminal is the presentation layer, which is composed of WeChat public platform, Web APP, mobile APP and screen display. Cloud computing environment includes data acquisition layer (data source), data layer and application layer.

Data sources include industry data sources and external data sources. Industry data sources of tourism are divided into tourism management departments, scenic spots, tourism enterprises and so on. External data sources are divided into weather, traffic, environmental protection, public security, hotel, the Internet, social networks and so on. Data layer includes business processing database, external distributed data storage and data warehouse. The application layer consists of user management, query and push, data mining, data acquisition, data fusion, data management and other modules.

The framework integrates internal and external data sources in the tourism industry, comprehensively adopts multiple data storage methods and data management methods, realizes multiple application services with one platform, and achieves the goal of multi-platform self-adaptation.

The framework is the basis of constructing the all-for-one tourism digital ecosystem, and its main advantages are as follows: Based on the perspective of all-for-one tourism, the framework meets the requirements of all-for-one tourism for information service in full space-time, all-round and systematization, and realizes the definition, classification and standardization of digital ecological information of all-for-one tourism. Under the background of all-for-one tourism, the automatic collection, identification, coding and storage of eco-related information resources in the tourism industry are carried out in this framework to solve the problems of wide range of all-for-one tourism information resources, large amount of data, and difficult fusion of structured data and unstructured data. The framework contributes to realize the discovery, release, query and personalized recommendation of tourism information resources under the concept of all-for-one tourism, provide personalized services, improve user experience, and achieve the goal of real-time, intelligent and mobile tourism information resources services.

## 4 The Technical Solutions

To construct the all-for-one tourism digital ecosystem, the following major issues must be faced and addressed:

- Due to multiple sources of information, huge amount of data and inconsistent data structure, it is difficult to integrate data. Therefore, it is necessary to adapt to new data standards and adopt new data coding.
- A single traditional data management and access method cannot solve the serious problem of information fragmentation. New technology options should avoid creating new islands of information.
- The current tourism information service is not real-time and intelligent enough under the requirement of all-for-one tourism. New software technology and personalized service technology must be applied to improve the user experience of tourism information mobile service.

According to levels of the above framework diagram, the technical route adopted in this paper correspondingly includes four main parts which are data collection, data storage, data processing and data representation. Cloud computing and real-time computing are also important technologies used in the system framework proposed in this paper.

## 4.1  The Proposed Data Collection Method

The methods adopted in the data collection part include the technology of Web spider, Open API of data source, real-time data acquisition and the historical data import of the government and enterprises. Data sources are mainly from government, scenic spots, enterprises, Internet and social networks.

Data acquisition is the first step in the life cycle of big data. It obtains various types of structured, semi-structured and unstructured massive data by means of RFID radio frequency data, sensor data, social network data and mobile Internet data. The main technologies of data acquisition include system log acquisition, network data acquisition and database acquisition. The all-for-one tourism digital ecosystem mainly uses network data acquisition and database acquisition. After searching and fully exploring relevant data such as travel, trading, social networking, data resources are stable and secure. Network data acquisition refers to the process of acquiring data information from web pages, social networks and other platforms by means of web crawlers and website open API. In this way, unstructured and semi-structured data can be extracted from the platform and stored as a unified local data file in a structured way. It supports the collection of pictures, audio, video and other documents.

## 4.2  The Proposed Data Storage Method

The data storage of the data layer combines relational database, distributed file system, data warehouse and other methods.

The system based on this framework needs to process and analyze the distributed and massive data. Due to the lack of consistency in data obtained from multiple sources, standard processing and storage technologies are not feasible. Big data storage and management mainly includes distributed file storage, NoSQL database, NewSQL database and so on. The typical distributed file storage is the open source implementation HDFS of Google's GFS. Traditional relational databases such as MySQL and Oracle can also be used to store real-time business data and data analysis results. Cloud data management technology is mainly Google's BigTable data management technology and the open source data management module HBase developed by Hadoop team. In order to transplant traditional SQL database interface directly into cloud data management system, some technical tools provide RDBMS and SQL interfaces for cloud data management, such as Hive, a data warehouse tool based on Hadoop subproject, which is the basic framework of data warehouse built on Hadoop. It can map data files onto a database table, provide SQL query function, transform SQL statements to Map-Reduce tasks and run it. Open source Sqoop can import data from relational databases into Hadoop's HDFS, or HDFS data into relational databases.

As the all-for-one tourism digital ecosystem has many sources of information, different structures and complex uses, it should use a combination of data warehouse, big data management and database technology. Data warehouse manages historical data, mainly for data mining and personalized services. Big data technology is mainly used to collect, process and store unstructured data. Traditional databases are used to access general information and data analysis results of system queries and pushes.

### 4.3   The Proposed Data Processing Method

The data processing mainly studies and applies realization technologies of the tourism information definition, classification and standardization, multi-source data fusion technology, data mining technology, information search technology. In this layer, Data management technology is mainly suitable for Hadoop, HDFS, Hbase Hive, Sqoop, MySQL, Oracle, etc.

Data fusion technology refers to the information processing technology that uses computer to automatically analyze and synthesize some observation information acquired in time sequence under certain criteria in order to complete the required decision-making and evaluation tasks. In order to achieve big data fusion, the main technologies at present are pattern alignment, entity link, and conflict resolution and relationship deduction.

Massive Data Mining Technologies mainly include decision tree, Bayesian, support vector machine, k-means, apriori association rule and its improved algorithm, HotSpot association rule, regression analysis, BP neural network, RBF neural network, data mining methods based on artificial intelligence algorithm such as ant colony algorithm, etc. Using these algorithm models can achieve classification and regression, clustering analysis, time series analysis, association rule discovery, divergence detection and other common data mining and recommendation algorithms, which can be used to mine tourism information resources and user behavior patterns.

### 4.4   The Proposed Data Presentation Method

In the presentation layer, intelligent Personalized Mobile Service is realized. It mainly adopts the current mainstream (mobile) Internet application development technologies which include SOA Web Service, Web App, Android/IOS, HTML5, JavaScript MV Cross-platform framework and Python, PHP7.0.

The new software engineering technologies involved in the platform, such as the service-oriented architecture (SOA), which use a group of distributed services to construct and organize applications, and reuse business functions to construct new applications and modify existing applications; Web Service Bus technology can effectively integrate tourism business applications in a very low coupling way; WSGI development specification is implemented by development language. As a cross-platform development language, the system it developed can run on various main-stream operating systems. It can be used to improve the availability of the system and realize the cross-platform of the system. In mobile Internet software development, besides Android/IOS mobile terminal development technology, new technologies available also include some cross-platform frameworks such as JavaScript MV,

PhoneGap and Encha Touch, which are used to create App based on HTML5 technology; using SVG + JavaScript on Canvas to deal with complex transitions and animations; using mobile Web Apps technology to replace traditional Web sites or Apps; using faster dynamic Web page programming technology (such as PHP 7.0) to achieve back-end systems; using Python language to carry out big data acquisition and analysis programming.

### 4.5    The Proposed Cloud Computing and Real-Time Computing Method

The above parts need to apply technologies of cloud computing and real-time computing to implement source virtualization, parallel computing and real-time services.

Software applications can be isolated from underlying hardware by virtualization technology, including resource splitting and resource aggregation. According to the object, virtualization technology can be divided into storage virtualization, computing virtualization, network virtualization and so on. Virtualization technology can be applied in CPU, operating system, server and other aspects, and it is the best solution to improve service efficiency.

At present, the distributed parallel programming model commonly used in cloud computing is Map-Reduce and Apache Spark. Map-Reduce is a programming model and task scheduling model, which is mainly used for parallel operation of big data sets and scheduling of parallel tasks. Apache Spark is a big data processing framework built around speed, ease of use and complex analysis.

Tourism information resources and services have a strong real-time, so the real-time requirement of the system is higher. In the field of Internet, real-time computing mostly aims at massive data, and requires real-time response to the calculation results, generally requiring seconds. The real-time computing technology of all-for-one tourism digital ecosystem includes three aspects: data real-time acquisition, data real-time computing and real-time query service. At present, the main research object of real-time data acquisition is log data acquisition for petroleum, chemical, manufacturing, agriculture, forestry and other industrial and agricultural production. There are few studies on data acquisition of all-for-one tourism which belongs to the service industry. Due to the inconsistency of the distribution and structure of the all-for-one tourism information resources, it is difficult to achieve real-time data acquisition. Therefore, further breakthroughs need to be made in related technologies for real-time data acquisition.

## 5  Conclusions

In order to satisfy the needs of data resource management and information service from the perspective of all-for-one tourism, in view of the current difficulties in data integration and fusion, multiple data fragments and data islands, and low level of real-time intelligence of mobile services in the traditional smart tourism information system and platform, this paper gives the requirements of the all-for-one tourism digital ecosystem, puts forward the systematic framework of it, discusses the main problems to be solved, provides the corresponding solutions.

**Acknowledgment.** This project is supported by Hunan Province Key Research and Development Plan (Grant No. 2017GK2274), the Scientific Research Fund of Hunan Provincial Education Department (Grant No. 15B127), the Key Laboratory of Hunan Province for New Retail Virtual Reality Technology (Grant No. 2017TP1026).

# References

1. Zheng, X.: From digitization to the age of acceleration: on information technology and tourism. Tour. Manag. Perspect. **25**, 147–150 (2018)
2. Pasquale, D.V., Gioconda, M., Valentina, N., Giustina, S.: Creating value from social big data: implications for smart tourism destinations. Inf. Process. Manag. **54**(5), 847–860 (2018)
3. Zhou, X.B., Ma, H.J., Miao, F.: A solution to wisdom tourism cloud based on cloud computing. J. Southeast Univ. **42**, 261–264 (2012)
4. Ren, W.J., Li, X.: Tourism demand analysis based on internet big data: the case of Huairou, Beijing. Syst. Eng.-Theory Pract. **38**(02), 437–443 (2018)
5. Zhang, G.Y., Hu, Z.: Improved BP neural network model and its stability analysis. J. Cent. South Univ. (Sci. Technol.) **42**(01), 115–124 (2011)
6. Wang, S., Du, J.P., Gao, T.: Study and realization of multimedia intelligent tourism navigation system. J. Cent. South Univ. (Sci. Technol.) **40**(S1), 335–340 (2009)
7. Zhu, G.X., Cao, J.: A recommendation engine for travel products based on topic sequential patterns. J. Comput. Res. Dev. **55**(5), 920–932 (2018)
8. Li, X.X., Yu, Y.X., Zhang, W.C., Wang, L.: Mining coteries trajectory patterns for recommending personalized travel routes. J. Softw. **29**(3), 587–598 (2018)
9. Jiang, Z.A., Wang, M., Chen, Y.: Path recommendation based on geographic coordinates and trajectory data. J. Commun. **38**(05), 165–171 (2017)
10. Zhang, W., Han, L.Y., Zhang, D.L., Ren, P.J., Ma, J., Chen, Z.M.: GeoPMF: a distance-aware tour recommendation model. J. Comput. Res. Dev. **54**(02), 405–414 (2017)
11. Guo, T., Guo, B., Zhang, J., Yu, Z., Zhou, X.: Crowd travel: leveraging heterogeneous crowd sourced data for scenic spot profiling. J. Zhejiang Univ. (Eng. Sci.) **51**(04), 663–668 (2017)
12. Liu, G., Wang, X.M., Zhang, J.Q., Li, S.M.: Study on intelligent management and control of tourist attraction based on mobile signaling data. J. Univ. Electron. Sci. Technol. China **44** (05), 769–777 (2015)
13. Han, Q.L., Liang, S., Zhang, H.L.: Mobile cloud sensing, big data, and 5G networks make an intelligent and smart world. IEEE Netw. **29**(2), 40–45 (2015)
14. Li, X.X., Lv, B.F., Zeng, P.Z., Liu, J.X.: Tourism prediction using web search data based on CLSI-EMD-BP. Syst. Eng.-Theory Pract. **37**(01), 106–118 (2017)
15. Jing, W.P., Hu, L.K., Shu, L., Mithun, M., Takahiro, H.: RPR: recommendation for passengers by roads based on cloud computing and taxis traces data. Pers. Ubiquit. Comput. **20**(3), 337–347 (2016)
16. Liu, Y.Y., Tseng, F.M., Tseng, Y.H.: Big data analytics for forecasting tourism destination arrivals with the applied vector autoregression model. Technol. Forecast. Soc. Change **130**, 123–134 (2018)
17. Giacomo, D.C., Rodolfo, B.: Knowledge transfer in smart tourism destinations: analyzing the effects of a network structure. J. Destin. Mark. Manag. **4**(3), 145–150 (2015)

# Multi-focus Image Fusion Combined with CNN and Algebraic Multi-grid Method

Ying Huang[1,2], Gaofeng Mao[1(✉)], Min Liu[2], and Yafei Ou[2]

[1] School of Computer Science and Technology, Chongqing University of Posts and Telecommunications, Chongqing 400065, China
17783119842@163.com
[2] School of Software Engineering, Chongqing University of Posts and Telecommunications, Chongqing 400065, China

**Abstract.** The aim of the paper is to solve the problem of over-segmentation problem generated by Watershed segmentation algorithm or unstable clarity judgment by small areas in image fusion. A multi-focus image fusion algorithm is proposed based on CNN segmentation and algebraic multi-grid method (CNN-AMG). Firstly, the CNN segmentation result was utilized to instruct the merging process of the regions generated by the Watershed segmentation method. Then the clear regions were selected into the temporary fusion image and the final fusion process was performed according to the clarity evaluation index, which was computed with the algebraic multi-grid method (AMG). The experimental results show that the fused image quality obtained by the CNN-AMG algorithm outperforms the traditional fusion methods such as DSIFT fusion method, CNN fusion method, ASR fusion method, GFF fusion method and so on with some evaluation indexes.

**Keywords:** Image segmentation · Image fusion · Algebraic multi-grid · Clarity · Evaluation index

## 1 Introduction

Due to the limited depth of field of the optical lens, it is often difficult for the camera to capture the image in which all the objects are focused [1], while multi-focus image fusion can obtain a completely sharp image by fusing multi-focus images. Multi-focus image fusion can obtain more useful information from the source images through a certain fusion algorithm, so that the fused image is clearer, more accurate and more favorable for the human than a single image. Nowadays, a large number of image fusion technologies have been applied to computer vision, clinical medicine, remote sensing, military surveillance and digital imaging [2]. Image fusion can be roughly divided into three categories: pixel level image fusion, feature level image fusion and decision level image fusion [3]. Pixel level image fusion retains more useful information from the source images than the latter two levels. Pixel-level image fusion is the basis of the latter two methods, and it is also a hotspot in the current research [4]. At present, the commonly utilized image fusion methods mainly include linear weighting method, principal component analysis (PCA), HIS transform method, image fusion

© Springer Nature Singapore Pte Ltd. 2019
R. Mao et al. (Eds.): ICPCSEE 2019, CCIS 1059, pp. 120–129, 2019.
https://doi.org/10.1007/978-981-15-0121-0_9

based on Laplace pyramid (LP), image fusion based on wavelet transform, image fusion based on non-subsampled contourlet wavelet transform (NSCT), dual-tree complex wavelet transform, image fusion based on pulse coupled neural networks (PCNN), dense scale invariant feature transform (DSIFT) method, etc. [1, 5]. The concept of linear weighting method is relatively simple, the calculation amount is relatively small, and it is suitable for real-time processing. But the fused image contains some strong noises, which will lead to stitching marks and produce poor visual effect, especially when the gray level of the fused image is different. The image fusion method based on pyramid transform mainly decomposes the source images into sub-images with different scales and spatial frequency bands. These sub-images can preserve the details information of the source images very well. Although the fused image looks good, the algorithm also has shortcomings. Redundancy may be generated for the hierarchical data and part of high frequency information may be lost. The image fusion method based on wavelet transform mainly utilize filter banks to decompose the source images into a series of high and low frequency sub-images. Although the fusion image obtained by this method has good visual effect, the fused image will be distorted due to the existence of oscillation and lack of translation invariance of the wavelet method. The image fusion method based on non-subsampled contourlet wavelet transform and dual-tree complex wavelet transform are all multi-scale transform methods, and they have the shortcoming of spatial discontinuity. Pulse coupled neural networks image fusion method has high spatial continuity, but the details are relatively blurred [6]. The DSIFT method mainly obtains a significant image based on the summation of the DFIFT features of each pixel. On the basis of DSIFT features, the sliding window is utilized to obtain an active image, and the image is divided into clear regions, unclear regions and uncertain regions. Finally, the undefined regions are registered by using the normalized DFIFT, and the sharpness of the pixel is described with spatial frequency to determine the weight of the pixel in the uncertain region. Although the DSIFT method can solve the unregistered situation effectively, it is time-consuming and ineffective when processing the image without obvious contours.

Deep learning has been widely applied to computer vision and image processing recently, such as license plate recognition, face recognition, behavior recognition, speech recognition, image classification, semantic segmentation, Yu et al. [7] applies CNN to multi-focus image fusion and achieved good results. Algebraic multi-grid (AMG) method can be utilized to judge the clarity of the closed area, and will produce good effect. Watershed method can obtain the closed areas, but there is a serious phenomenon of over-segmentation. Therefore, CNN method is utilized to instruct the reduction of the over-segmentation effect in Watershed method, and AMG is utilized as the clarity evaluation index for multi-focus image fusion in the paper. The combination of CNN method, Watershed method and AMG index will propose a better fusion results for multi-focus image fusion and the strategy is proved with some experiences.

## 2  Related Work

Algebraic multi-grid method [8] is a multi-grid method utilizing some important principles and concepts of geometric multi-grid method, and it is independent of the actual geometric network. It can be utilized for more types of linear equations. At present, algebraic multi-grid method is mainly utilized in image reconstruction, image binarization, image restoration, image dryness, image fusion and contour analysis [9]. The algebraic multi-grid method is utilized to reconstruct the image, and mean square error (MSE) is obtained from the reconstructed image and original image to evaluate the clarity of the image [10].

The Watershed segmentation method [11] is a mathematical morphology segmentation method based on topological theory. The basic idea is to regard the images as geomorphological topography. The gray value of each pixel in the image indicates the altitude of the point. Each local minimum value and its affected area is called water collection basin, and the boundary of the water collection basin forms a Watershed. The advantage of Watershed segmentation algorithm is that it has obvious advantages in computational complexity, and it can also get a closed segmentation curve. However, Watershed segmentation method also has some shortcomings, such as it is sensitive to noise, easy to produce over-segmentation phenomenon, and there are too many closed regions in the segmentation results.

Convolutional neural network (CNN) is a kind of feed forward neural network with deep structure and convolution computation. It is one of the representative algorithms of deep learning [12, 13]. The structure of convolution neural network includes input layer, hidden layer and output layer. The hidden layer includes convolution layer, pooling layer and full connection layer. The precise image contour can be obtained by image segmentation based on CNN. The RCF [14] network architecture based on CNN image segmentation is adopted in this paper.

## 3  Image Fusion Algorithm

In order to avoid the over-segmentation of Watershed method, CNN method is utilized to instruct the merging of the regions produced by Watershed method. Although the contour obtained by CNN segmentation is more accurate, the edge of the image in CNN segmentation is multi-pixel wide, which is not conducive to the accurate extraction of image contour. The combination of CNN and Watershed method can extract more accurate single-pixel edges and closed regions. In this paper, Watershed segmentation and CNN segmentation are combined to extract precise and closed regions, so as to achieve better image fusion effect with the help of AMG index. The proposed algorithm can obtain a fused image that is clearer and contains more useful information. The frame diagram of this algorithm (CNN-AMG) is shown as Fig. 1. The approximate process of the algorithm is presented as follows:

1. The source images are reconstructed by AMG [10] method, and the corresponding reconstructed images are obtained.
2. The CNN segmentation method is utilized to segment the source images to obtain the corresponding segmentation results.
3. The Watershed segmentation method is utilized to segment the source images to obtain the corresponding segmentation results.
4. CNN segmentation results are utilized to guide the merging process of the areas produced by Watershed segmentation and the merged images are obtained.
5. For the image obtained from the above steps, the MSE between each segmented area of each source image and its corresponding reconstructed image area is calculated respectively, then the values of MSE are compared. The image region with a larger MSE value is selected and its source is marked to generate the initial decision map. Initial decision map $D1$ is obtained from formula (1):

$$D_1(x,y) = \begin{cases} 1, MSE_A > MSE_B \\ 0, MSE_A \leq MSE_B \end{cases} \tag{1}$$

The $MSE_A$ indicates the mean square error between the segmented region of source image A and the region of the corresponding reconstructed image, $MSE_B$ indicates the mean square error between the segmented region of source image B and the region of the corresponding reconstructed image. Figure 1 indicates that the image block is comes from the source image A, and 0 indicates that the image block comes from the source image B.

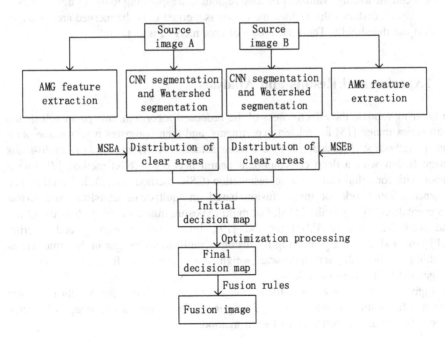

**Fig. 1.** Algorithmic framework diagram.

6. The generated initial region decision map will produce some small anomaly regions, a simple anomalous small region removal method is utilized to remove the region whose area is less than a certain threshold (which is adaptively set to one percent of the total pixels of the image). This method is utilized to adjust the initial decision map *D1*, the final decision map *D2* is generated, and then a fused image is obtained according to the decision map *D2*. The fused image can be obtained from formula (2):

$$F(x,y) = D_2(x,y) \times A(x,y) + (1 - D_2(x,y)) \times B(x,y) \tag{2}$$

Where F(x, y) represents the fused image, and A(x, y) and B(x, y) represent the source image A and the source image B respectively.

The specific implementation of the fourth step is as follows:

1. The CNN segmentation result and the Watershed segmentation result are binarized.
2. The overlapped portion of the two segmentation results after binarization is obtained and the region where the coincident portion is located and retained to obtain an image I.
3. At this time, the area of the partial area is small, and further optimization and merging are required. The two regions are merged when they are adjacent and the pixels in all the two regions are evenly distributed.
4. When the area of the two regions is less than a certain threshold t, the centroid distance of the two regions is less than a certain distance d and the difference between the average values of the two regions corresponding to the source images is less than a certain value u, then the region is merged until the merged area is larger than the threshold t. The final image of area merged is obtained.

# 4   Experimental Results and Analysis

In order to validate the effectiveness of the proposed algorithm, this paper selects the lytro series image [15] for related experiments, and then compares it with some other fusion methods such as adaptive sparse representation (ASR) method [16], multi-focus image fusion with a deep convolutional neural network (CNN) method [7], image fusion with convolutional sparse representation (CSR) method [17], DSIFT method [1], a general framework for image fusion based on multi-scale transform and sparse representation (MST) method [18], fast multi-exposure image fusion with median filter and recursive filter (FMMR) method [19], image fusion with guided filtering (GFF) method [20], image matting for fusion of multi-focus images in dynamic scenes (Matting) method [21] and multi-scale weighted gradient based fusion for multi-focus images (MWGF) method [22].

Figure 2 shows some multi-focus images. Figure 3 shows the results of region merging for source image named ly3. Figure 4 shows the fusion image of source images ly3 obtained with various fusion methods.

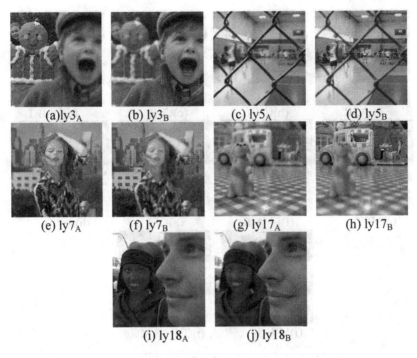

(a)ly3<sub>A</sub>          (b) ly3<sub>B</sub>          (c) ly5<sub>A</sub>          (d) ly5<sub>B</sub>

(e) ly7<sub>A</sub>          (f) ly7<sub>B</sub>          (g) ly17<sub>A</sub>          (h) ly17<sub>B</sub>

(i) ly18<sub>A</sub>          (j) ly18<sub>B</sub>

**Fig. 2.** Original image sets.

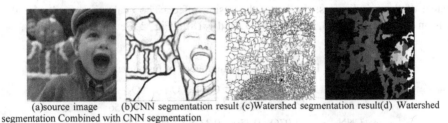

(a)source image      (b)CNN segmentation result (c)Watershed segmentation result(d) Watershed segmentation Combined with CNN segmentation

**Fig. 3.** Results produced by Watershed segmentation combined with CNN segmentation.

In this paper, the fusion image is evaluated by both subjective and objective methods. From the subjective point of view, (f) and (h) in Fig. 4 have obvious blurring phenomenon and artificial edges, while other methods have better fusion effects. At the same time, some commonly utilized objective evaluation indicators are utilized to evaluate the fusion results, such as $Q_{MI}$, $Q_{TE}$, $Q_{NCIE}$, $Q_G$, $Q_M$ and $Q_{SF}$ [23]. $Q_{MI}$ quantifies the overall mutual information between source images and fused images, $Q_{TE}$ estimates the amount of information present in the fused images, $Q_G$ is a gradient-based fusion metric witch evaluate the extent of gradient information injected into the fused images from the source images, $Q_M$ is a metric based on a multi-scale scheme and $Q_{SF}$ uses spatial frequency to measure the activity level of an image. The larger the

**Fig. 4.** Comparison of multiple image fusion method.

value of these indicators, the better the fusion effect. The evaluation results for the
fusion of the images ly3, ly5, ly7, ly17, and ly18 in Fig. 2 are shown as Table 1.

**Table 1.** Objective evaluation index of fusion results

| Image | Method | $Q_{MI}$ | $Q_{TE}$ | $Q_{NCIE}$ | $Q_G$ | $Q_M$ | $Q_{SF}$ |
|-------|--------|----------|----------|------------|-------|-------|----------|
| ly3 | ASR | 0.8245 | 0.3599 | 0.8226 | 0.6766 | 1.3456 | −0.0503 |
| | CNN | 1.0768 | 0.3701 | 0.8356 | **0.7117** | 2.6183 | −0.0251 |
| | CSR | 0.9244 | 0.3613 | 0.8272 | 0.6492 | 1.3568 | −0.0552 |
| | DSIFT | 1.1014 | 0.3714 | 0.8372 | 0.7117 | **2.7347** | **−0.0171** |
| | FMMR | 1.0376 | 0.3863 | 0.8410 | 0.0056 | 0.2258 | −0.9975 |
| | GFF | 1.0144 | 0.3657 | 0.8320 | 0.7094 | 2.5708 | −0.0222 |
| | Matting | 1.0733 | 0.3686 | 0.8354 | 0.0056 | 0.2257 | −0.9961 |
| | MST | 0.8631 | 0.3564 | 0.8244 | 0.6786 | 1.7768 | −0.0295 |
| | MWGF | 1.0211 | 0.3643 | 0.8324 | 0.0056 | 0.2256 | −0.9962 |
| | CNN-AMG | **1.1644** | **0.4071** | **0.8458** | 0.6993 | 2.6077 | −0.0275 |
| ly5 | ASR | 0.9604 | 0.3924 | 0.8339 | 0.6753 | 1.1203 | −0.0898 |
| | CNN | 1.0928 | 0.4010 | 0.8441 | 0.6817 | 1.6356 | −0.0864 |
| | CSR | 0.9975 | 0.3958 | 0.8351 | 0.5878 | 1.2681 | −0.0738 |

*(continued)*

**Table 1.** (*continued*)

| Image | Method | $Q_{MI}$ | $Q_{TE}$ | $Q_{NCIE}$ | $Q_G$ | $Q_M$ | $Q_{SF}$ |
|---|---|---|---|---|---|---|---|
| | DSIFT | 1.2000 | 0.4070 | 0.8526 | **0.6891** | **2.3989** | −0.0455 |
| | FMMR | 1.0567 | **0.4137** | 0.8445 | 0.0051 | 0.2569 | −0.9977 |
| | GFF | 1.0547 | 0.3990 | 0.8415 | 0.6846 | 1.6362 | −0.0763 |
| | Matting | 1.1279 | 0.4046 | 0.8508 | 0.0058 | 0.2570 | −0.9965 |
| | MST | 0.9755 | 0.3933 | 0.8346 | 0.6643 | 1.4824 | −0.0557 |
| | MWGF | 1.1100 | 0.4005 | 0.8508 | 0.0056 | 0.2571 | −0.9965 |
| | CNN-AMG | **1.2117** | 0.4114 | **0.8546** | 0.6817 | 2.3582 | **−0.0302** |
| ly7 | ASR | 0.9296 | 0.3924 | 0.8297 | 0.6625 | 1.5481 | −0.0585 |
| | CNN | 1.1061 | **0.4033** | 0.8399 | 0.6852 | 2.3155 | −0.0446 |
| | CSR | 0.9722 | 0.3960 | 0.8319 | 0.6161 | 1.3216 | −0.0690 |
| | DSIFT | **1.1425** | 0.3973 | 0.8424 | **0.6869** | **2.5381** | −0.0361 |
| | FMMR | 1.1364 | 0.3742 | 0.8322 | 0.0072 | 0.3127 | −0.9972 |
| | GFF | 1.0462 | 0.3985 | 0.8362 | 0.6778 | 2.2613 | −0.0411 |
| | Matting | 1.0935 | 0.3981 | 0.8392 | 0.0059 | 0.3128 | −0.9962 |
| | MST | 0.9437 | 0.3908 | 0.8305 | 0.6564 | 1.8040 | −0.0404 |
| | MWGF | 1.0363 | 0.3935 | 0.8356 | 0.0059 | 0.3128 | −0.9962 |
| | CNN-AMG | 1.1418 | 0.3961 | **0.8428** | 0.6829 | 2.4876 | **−0.0341** |
| ly17 | ASR | 0.7212 | 0.3374 | 0.8179 | 0.6889 | 1.4883 | −0.0355 |
| | CNN | 0.9870 | 0.3600 | 0.8313 | **0.7157** | 2.7635 | −0.0118 |
| | CSR | 0.8399 | 0.3515 | 0.8232 | 0.6786 | 1.4394 | −0.0339 |
| | DSIFT | 1.0311 | 0.3640 | 0.8333 | 0.7130 | **2.8137** | **−0.0086** |
| | FMMR | 0.9598 | 0.3540 | 0.8351 | 0.0048 | 0.2662 | −0.9981 |
| | GFF | 0.9111 | 0.3545 | 0.8274 | 0.7115 | 2.7096 | −0.0115 |
| | Matting | 0.9809 | 0.3589 | 0.8306 | 0.0055 | 0.2664 | −0.9961 |
| | MST | 0.7928 | 0.3459 | 0.8210 | 0.6875 | 2.1073 | −0.0142 |
| | MWGF | 0.9453 | 0.3554 | 0.8285 | 0.0055 | 0.2665 | −0.9961 |
| | CNN-AMG | **1.0406** | **0.3703** | **0.8354** | 0.7020 | 2.7280 | −0.0152 |
| ly18 | ASR | 1.2460 | 0.4349 | 0.8467 | 0.7023 | 1.4356 | −0.0676 |
| | CNN | 1.3642 | 0.4455 | 0.8545 | 0.7223 | 2.3236 | −0.0453 |
| | CSR | 1.2061 | 0.4387 | 0.8444 | 0.6196 | 1.3137 | −0.0831 |
| | DSIFT | 1.3753 | 0.4461 | **0.8553** | 0.7226 | 2.4767 | **−0.0378** |
| | FMMR | 1.2835 | 0.4458 | 0.8528 | 0.0106 | 0.2940 | −0.9974 |
| | GFF | 1.3384 | 0.4434 | 0.8526 | 0.7174 | 2.0862 | −0.0461 |
| | Matting | 1.3571 | 0.4444 | 0.8540 | 0.0103 | 0.2942 | −0.9962 |
| | MST | 1.2531 | 0.4351 | 0.8471 | 0.6970 | 1.6607 | −0.0511 |
| | MWGF | 1.3228 | 0.4403 | 0.8517 | 0.0101 | 0.2942 | −0.9962 |
| | CNN-AMG | **1.3769** | **0.4487** | 0.8550 | 0.6989 | 2.3855 | −0.0426 |

It can be seen from Table 1 that for the image ly3 and ly17, the three index values of $Q_{MI}$, $Q_{TE}$ and $Q_{NCIE}$ of CNN-AMG method are higher than other fusion methods. Although the three index values of $Q_G$, $Q_M$ and $Q_{SF}$ are not the highest, it is kept in the top five for all the fusion methods. For image ly5, the three index values of $Q_{MI}$, $Q_{NCIE}$ and $Q_{SF}$ of CNN-AMG method are the highest, and the three index values of $Q_{TE}$, $Q_G$ and $Q_M$ are kept in the top three positions. The $Q_{TE}$ index value of FMMR method is the highest, but the values of $Q_G$, $Q_M$, $Q_{SF}$ of FMMR are the lowest, which shows that the FMMR method is not stable. For image ly7, CNN-AMG method has the highest index values of $Q_{NCIE}$ and $Q_{SF}$. Although the four index values of $Q_{MI}$, $Q_{TE}$, $Q_G$ and $Q_M$ are not the best, they ranked in the top five among all the methods. For the image ly18, although the four index values of $Q_{NCIE}$, $Q_G$, $Q_M$, and $Q_{SF}$ are not as high as the DSIFT method, the $Q_{NCIE}$ and $Q_M$ index values are ranked the second, and the CNN-AMG method has the highest $Q_{MI}$ and $Q_{TE}$ values. Considering subjective evaluation and objective evaluation, the CNN-AMG method has certain advantages.

## 5  Conclusion

This paper proposed a multi-focus image fusion algorithm (CNN-AMG) combining CNN segmentation and algebraic multi-grid method. The CNN segmentation is utilized to guide the merging of the regions with Watershed segmentation to avoid the over-segmentation of Watershed method. Through the analysis of experimental results, the effective information of the source image can be preserved to the greatest extent in the process of image fusion with CNN-AMG method, which makes the fusion image clearer, richer in details and higher in fusion quality. For further research, the fusion of multi-focus images will utilize convolutional neural network and Watershed method to improve the fusion effect.

## References

1. Yu, L., Liu, S., Wang, Z.: Multi-focus image fusion with dense SIFT. Inf. Fusion **23**, 139–155 (2015)
2. Liu, Z., Yi, C., Yin, H., Zhou, J., Zhu, Z.: A novel multi-focus image fusion approach based on image decomposition. Inf. Fusion **35**, 102–116 (2016)
3. Li, H., Chai, Y., Ling, R., Yin, H.: Multifocus image fusion scheme using feature contrast of orientation information measure in lifting stationary wavelet domain. J. Inf. Sci. Eng. **29**(2), 227–247 (2013)
4. Wang, M., Zhou, S., Yang, Z., Liu, Z., Ren, S.: Image fusion based on wavelet transform and gray-level features. J. Mod. Opt. **66**(5), 1–10 (2018)
5. Qiang, Z., Yi, L., Blum, R.S., Han, J., Tao, D.: Sparse representation based multi-sensor image fusion for multi-focus and multi-modality images: a review. Inf. Fusion **40**, 57–75 (2017)
6. Li, H., Wu, X.-J.: Multi-focus image fusion using dictionary learning and low-rank representation. In: Zhao, Y., Kong, X., Taubman, D. (eds.) ICIG 2017. LNCS, vol. 10666, pp. 675–686. Springer, Cham (2017). https://doi.org/10.1007/978-3-319-71607-7_59

7. Yu, L., Xun, C., Hu, P., Wang, Z.: Multi-focus image fusion with a deep convolutional neural network. Inf. Fusion **36**, 191–207 (2017)
8. Brandt, A.: Algebraic multigrid theory: the symmetric case. Appl. Math. Comput. **19**(1–4), 23–56 (1986)
9. George, P., Petros, M.: Multigrid geometric active contour models. IEEE Trans. Image Process. **16**(1), 229–240 (2006)
10. Wang, D.: Hybrid fitting energy-based fast level set model for image segmentation solving by algebraic multigrid and sparse field method. IET Image Process. **12**(4), 539–545 (2018)
11. Beucher, S., Lantuéjoul, C.: Use of watersheds in contour detection. In: International Workshop on Image Processing (1979)
12. Schmidhuber, J.: Deep learning in neural networks: an overview. Neural Netw. **61**, 85–117 (2015)
13. Zhang, Q., Zhang, M., Chen, T., Sun, Z., Ma, Y., Yu, B.: Recent advances in convolutional neural network acceleration. Neurocomputing **232**, 37–51 (2019)
14. Liu, Y., Cheng, M.M., Hu, X., Wang, K., Bai, X.: Richer convolutional features for edge detection. IEEE Computer Society (2016)
15. Nejati, M., Samavi, S., Shirani, S.: Multi-focus image fusion using dictionary-based sparse representation. Inf. Fusion **25**, 72–84 (2015)
16. Yu, L., Wang, Z.: Simultaneous image fusion and denoising with adaptive sparse representation. IET Image Process. **9**(5), 347–357 (2014)
17. Yu, L., Xun, C., Ward, R.K., Wang, Z.J.: Image fusion with convolutional sparse representation. IEEE Sig. Process. Lett. **23**, 1882–1886 (2016)
18. Yu, L., Liu, S., Wang, Z.: A general framework for image fusion based on multi-scale transform and sparse representation. Inf. Fusion **24**, 147–164 (2015)
19. Li, S., Kang, X.: Fast multi-exposure image fusion with median filter and recursive filter. IEEE Trans. Consum. Electron. **58**(2), 626–632 (2012)
20. Shutao, L., Xudong, K., Jianwen, H.: Image fusion with guided filtering. IEEE Trans. Image Process. **22**(7), 2864–2875 (2013)
21. Li, S., Kang, X., Hu, J., Yang, B.: Image matting for fusion of multi-focus images in dynamic scenes. Inf. Fusion **14**(2), 147–162 (2013)
22. Zhou, Z., Sun, L., Bo, W.: Multi-scale weighted gradient-based fusion for multi-focus images. Inf. Fusion **20**, 60–72 (2014)
23. Liu, Z., Blasch, E.: Statistical comparison of image fusion algorithms: recommendations. Inf. Fusion **36**, 251–260 (2017)

# Fish Behavior Analysis Based on Computer Vision: A Survey

Yizhi Zhou[1,2], Hong Yu[1,2], Junfeng Wu[1,2,3(✉)], Zhen Cui[1,2], and Fangyan Zhang[1,2]

[1] School of Information Engineering,
Dalian Ocean University, Dalian 116023, China
wujunfeng@dlou.edu.cn
[2] Guangdong Province Key Laboratory of Popular High Performance Computers, Shenzhen University, Shenzhen 518060, China
[3] School of Computer Science and Technology, Tianjin University, Tianjin 300072, China

**Abstract.** Fish behavior refers to various movements of fish. Fish behavior is closely related to the ecology of fish, physiological changes of fish, aquaculture and so on. Related applications will be expanded if fish behavior is analyzed properly. Traditional analysis of fish behavior mainly relies on the observation of human eyes. With the deepening and extension of application and the rapid development of computer technology, computer vision technology is increasingly used to analyze fish behaviors. This paper summarized the research status, research progress and main problems of fish behavior analysis by using computer vision and made forecast about future research.

**Keywords:** Fish behaviour · Analysis · Computer vision · Fish behaviors monitoring

## 1 Introduction

The ocean is the largest body of water on the earth. Its total area is about 360 million square kilometers, accounting for 71% of the earth's surface area. At the same time, fish is the most abundant species in the ocean. According to statistics [1], there are about 28,000 known species of fish in the world. Furthermore, fish is also one of the main food sources for people. Therefore, all aspects of fish have a great impact on the ecological environment and human production, as well as human life.

Fish behavior [2] refers to the various movements of fish, which is the external response of fish to the changes of external and internal environment. Therefore, researching on the behavioral characteristics of these fishes and analyzing these behavioral characteristics are of great significance in forecasting weather, aquaculture, ecological protection, and other aspects for human.

The first research on analyzing fish behavior was started by scientists such as Ullman [3] from the view of animal behaviors. However, due to the immaturity of technology and theory at that time, insufficient data samples were provided. In addition, behavioral analysis technology did not receive proper attention at that time. Moreover,

R. Mao et al. (Eds.): ICPCSEE 2019, CCIS 1059, pp. 130–141, 2019.
https://doi.org/10.1007/978-981-15-0121-0_10

many of these fish behaviors are observed by human eyes, which consume a lot of time and energy. At the same time, the data observed by human is too subjective, which is not conducive to continuous, accurate and stable records.

Computer vision is also called machine vision. It mainly uses image sensors to obtain the image of object instead of human eyes. And then it will convert the image into digital image. After that, the image will be recognized by using the discriminate criterion of computer simulation, so as to achieve the purpose of analyzing and understanding the image in Fig. 1 [4]. In recent years, the application of computer vision technology in analyzing fish behavior and fish monitoring has been increasing. In addition, many experts and scholars have applied computer vision technology to fish behavior recognition. For example, the early warning system for abnormal behaviors of fish is designed through the phenomenon of fish turning over under unfavourable conditions to the surrounding environment [5]. At the same time, the swimming trajectory of fish can be used to judge whether the surrounding waters contain toxic chemicals, which is aimed to give an alarm to deal with them [6]. Moreover, the swimming vigor of carp is used to monitor the oxygen content in water in real time, and to monitor the oxygen content, as well as whether it is suitable for fish breeding [7] can be judged and so on.

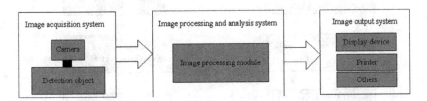

**Fig. 1.** Composition of computer vision inspection system

## 2 Analyzing Basic Behaviors of Fish by Using Computer Vision Technology

The monitoring of fish behaviors can provide a large amount of data for the analysis of fish behaviors. In the previous research, fish behaviors were monitored mainly based on direct observation and manual recording. Therefore, it consumes a lot of time and energy. Moreover, the subjective nature of the observed data is too strong, which is not conducive to accurate, stable and continuous recording. In recent years, scientists have made innovations in fish monitoring methods in order to analyze fish behavior better.

### 2.1 Analyzing the Center of Mass and Angular Velocity by Computer Vision Technology

At present, the basic implementation steps of fish behavior analysis system based on computer vision technology include image acquisition, image processing, background modeling, foreground detection, extraction of fish centroid, calculation of fish spatial position, calculation of fish velocity and angular velocity in Fig. 2. The core of the study is the extraction of fish centroid and analyzes its behaviors.

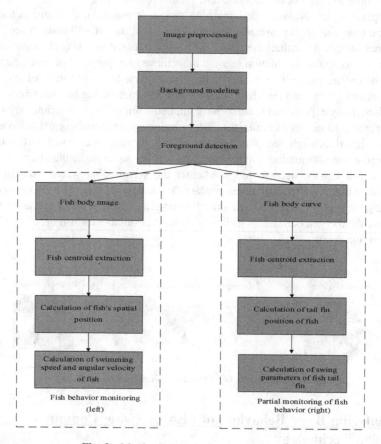

**Fig. 2.** Monitoring methods of fish behaviors

Fan [8] and other scientists have analyzed the monitoring of fish swarm and then they proposed a method that combines inter-frame difference with motion average background modeling. It overcomes the shortcomings of the inter-frame difference algorithm, such as the performance of the algorithm will be significantly degraded and the monitoring error rate will increase when the larger noise occurs in the video image. Moreover, it improves the serious shadows and cavitation phenomena in the foreground detection of the inter-frame difference algorithm. In addition, effective detection can be carried out in the case of background mutation. However, it is not ideal for too complex monitoring.

Based on the inter-frame difference method, Nguyen [9] and other scientists have detected the motion of fish is detected by combining the inter-frame difference algorithm and the mixed Gauss model. This method improves the tracking performance in different scenarios. In addition, it has higher tracking accuracy, if analyzing the performance of the algorithm in the use of mean square error. However, it is difficult to track the position or the centroid of the fish in some specific scenarios.

Zhao [10] and other scientists have put forward a method for detecting moving fish. Statistical background modeling algorithm was used to quickly build the background model. Then, background subtraction algorithm was used to separate the fish object from the background, so that the gray fish object image was obtained. Through maximum class variance algorithm, the adaptive segmentation threshold can be obtained. Then, two valued fish target image can be obtained through using the image segmentation. The centroid can be obtained by the method of connected region centroid detection, so as to monitoring fish based on centroid and fish characteristics. The experimental results show that the average accuracy of target fish detection is over 0.91. But in reality, fish live in three-dimensional space. Fish motion pictures will be collected in real time acquisition at the top and side, which will make the detection data more accurate.

Rodríguez [11] and other scientists have detected the fish by the use of the combination of background modeling, edge and region analysis, which replaces the artificial neural network described by Rodriguez [12]. Although artificial neural network provides satisfactory results, the detection speed is slow and the accuracy is low. In addition, it also uses a new method based on Kalman filter, which is aimed to obtain the trajectory of one or more individuals in the fish way. Compared with the standard method, it can effectively track the fish covered by the image or the fish in multiple scenes. In addition, it also allows the separation of different fish tracks and eliminates anomaly detection. Therefore, the balance between accuracy and analysis time is achieved.

Wang [13] studies the behaviors of American fish in natural habitats. In his research, the SURF algorithm is used to recognize and describe the image of the American fish. The collected pictures were detected by feature points, described by feature proximity description, paired by descriptors, and swimming of fish is analyzed frame by frame. Then, K-means clustering algorithm is combined with other outlier removal techniques to classify the collected fish samples. After that, the movement of American fish will be separated from the background and foreground. In the analysis of fish behavior, K classifier is combined with the similarity cycle detection method.

Xiao [14] have taken photos of fish targets in the use of one camera, based on monocular vision system. At the same time, he extracted the centroid of the target fish by using image segmentation and pixelization methods. Then, according to the principles of plane mirror imaging and refraction, the relationships between the target of mirror imaging and the actual target were studied. Then, the three-dimensional coordinates of real fish and imaginary fish can be calculated, based on this kind of relationship and underwater imaging process. Finally, the occlusion tracking of multitarget fish was realized by combining the three-dimensional coordinates with Kalman filter, which solves the difficulty of fish tracking under blocking conditions.

In accordance to the low tracking accuracy of two-dimensional fish tracking and difficulties to record individual's comprehensive motion state, Ren [15] and other scientists show that the structure, advantages and disadvantages of 3D fish tracking facilities including single camera with auxiliary facilities, top camera, binocular camera and multi-view camera are summarized. The single camera and the top camera are simple, but the operation accuracy is relatively low. In addition, binocular cameras and multi-view cameras have high accuracy, which can be used to track a large number of fish or in large-scale space.

Generally speaking, there are many studies on fish behaviors from the view of centroid and angular velocity. Although the target fish can be effectively monitored to a certain extent, the fish behaviors cannot be analyzed in the environment with a large number of fish populations. At the present stage, there are few studies on fish monitoring and analysis from three-dimensional perspective, which makes it difficult to fully express the spatial movement behaviors of fish.

## 2.2    Analyzing the Texture and Color Changes of Fish by Using Computer Vision Technology

At present, the analysis of fish behaviors mainly focuses on obtaining the direction, acceleration and speed of the movements of the fish, or extracting the trajectory of the fish through related algorithms, so as to analyze the behaviors of fish. However, the disadvantage of this method is that it can only be used in the case of fewer fish stocks. At the same time, it is not suitable for the ponds with great quantities of fish. Fish texture is also a very important visual attribute of fish. Therefore, texture features are easy to extract and observe. Moreover, it is suitable to analyze the behaviors of fish under the condition of large quantities of fish in Fig. 3. At present, most studies focus on how to extract fish texture features.

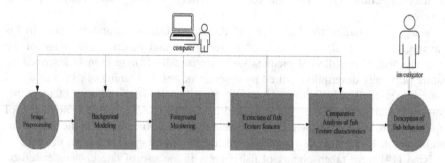

**Fig. 3.** Flow chart of fish behaviors in texture analysis

Li [16] and other scientists have selected 40 juvenile turbot fish with uniformly distributed pigments and no damage. Then, they placed them into a glass tank with milky white bottom for temporary cultivation. The aquaculture background is divided into nine different colors. At the same time, the water body is connected with each other, which means that the quality of water is consistent. Moreover, there is a crevice at the bottom, which can let the fish pass freely. The feeding behaviors and body color change rates of juvenile turbot under different backgrounds were observed under the computer vision system. In results, it showed that the fish stayed for a long time under the light purple background, and they consumed more food. At the same time, they had a little change in body color. However, under black and red background conditions, turbot had a shorter residence time, and they only have a smaller intake. At the same time, there is a larger body color change rate (The dispersion of body color change rate could be used as an indicator of stress).

Papadakis [17] and other scientists have studied the feeding behaviors of mirror carps, in accordance to the problem of large amount of fish behavior analysis and computation. Firstly, the image was preprocessed. Secondly, texture features were extracted by the methods of gray difference statistics, gray level co-occurrence matrix and Gauss-Markov random field model. In addition, the texture features extracted by three methods were analyzed and synthesized. After that, PCA algorithm is used to reduce the dimension, so as to extract the feature vectors, which can represent the texture of fish swarm to a large extent. Finally, SVM is used to classify and identify the motion state of the fish. The accuracy of classification is to 96.5%, so as to determine the movement state of the fish.

Zhao [18] and other scientists have observed tilapia and analyzed their feeding behaviors from the point of view of the changing characteristics of reflective aquaculture water surface. Firstly, the reflection area of the water surface was segmented and extracted in the HSV color space. Then, Lucas-Kanade optical flow method and information entropy are used to calculate and analyze the degree of irregularity of the changes of the reflective region, so as to measure the dispersion degree and the changes of the motion amplitude of the fish. At last, a kinetic energy model was constructed based on the variation amplitude of the reflecting area. In order to avoid foreground extraction and individual tracking of complex fish targets, the feeding intensity of fish was evaluated at the same time.

Hu [19] and other scientists have proposed a new method for fish classification. Firstly, the skin color and texture sub-images of fish were obtained from the original image. Then the color features, texture features are extracted, as well as the wavelet-based color and texture sub-image feature, so as to form six groups of feature vectors. After that, multi-support vector machine (MSVM) is used to classify the textures and colors. Finally, fish can be classified and analyzed better by the extractor based on HSV color space and the classifier algorithm based on DAGMSVM.

Generally speaking, fish texture is a very important visual attribute of fish. Furthermore, the texture feature information is easy to extract and observe. However, the fish texture photographed by camera is easy to be occluded in the underwater environment, which will lead to unstable feature extractions and large errors.

## 3   Analyzing Integrated Behaviors of Fish by Using Computer Vision

In the process of observing the fish behaviors, if monitoring fish for a long term under computer vision, it can not only intuitively show the living conditions of fish, but also play an important role in water quality monitoring, aquaculture, fishing industry and so on. Many researchers have carried out different comprehensive studies in Fig. 4.

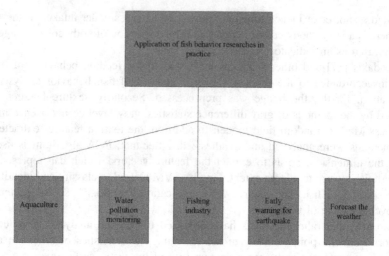

**Fig. 4.** Application of computer vision in fish behavior analysis

## 3.1 Analysis of Abnormal Behaviors of Fish

The motion characteristics and external morphological characteristics of fish will change if the environment changes where fish live in. And then, it will show abnormal changes in the motion characteristics of the center of mass and speed, as well as the acceleration of fish. Moreover, the differences in color, texture and shape in the external characteristics will be shown. Therefore, many researchers analyze the abnormal behaviors of fish from the perspective of computer vision.

Tang [20] and other scientists have proposed a fish behavior analysis system based on 3D computer vision. A planar mirror and a CCD camera were combined to monitor and analyze the fish in the tank in real time, so as to obtain the three-dimensional trajectory of fish movements. In the aspect of foreground extraction, background subtraction method is used, which can extract the foreground objects more completely in video images when compared with other algorithms. In foreground detection, LB Adaptive SOM is used, which is a self-organizing background modeling method. Compared with other algorithms, the segmentation accuracy of LB Adaptive SOM is to 91.2%. At the same time, after extracting the foreground two value images, we can get the three-dimensional data of fish space.

Liu [21] has improved the methods on the basis of Tang Yiping. In the target tracking algorithm, IMMJPDA tracking algorithm was used to track fish in three-dimensional, so as to obtain the three-dimensional coordinates of fish targets. Among it, the IMMJPDA tracking algorithm is an algorithm that combines the Interactive Multiple Model (IMM) algorithm with the Joint Probabilistic Association (JPDA) algorithm. In accordance to identifying and judging the abnormal trajectories of fish objects, the hidden Markov model is used to describe the behaviors of fish. At the same time, the correct recognition rate of the abnormal behaviors is up to 92.3%.

Yu [22] and other scientists first used threshold method to reduce the influence of noise from the image of original video sequence. Then the motion vectors under different space-time conditions are obtained by optical flow algorithm. After that, the behavior characteristics of fish motion are counted. Finally, standard mutual information (NMI) and local distance anomaly factor (LDOF) are used to detect abnormal behaviors of fish. Experiments show that the accuracy of NMI and LDOF methods in detecting abnormal behaviors of fish is up to 99.92% and 99.88% respectively, which shows that both of the two methods have good results. However, the optical flow algorithm has poor anti-noise performance, which requires special hardware to support. In addition, its real-time performance is poor, as well.

Xu [23] has proposed a method for identifying abnormal behaviors of spotted groupers. Fifty spotted groupers were first placed in aquaculture water for photography, which are 16.5 cm in length and 166.5 g in average weight. Then, a suitable and unsuitable environment for the survival of grouper was constructed by adjusting the dissolved oxygen concentration of water bodies artificially, while keeping the PH, ammonia nitrogen and temperature unchanged in the aquaculture water. Then, the collected image was segmented by Hue component. Then, the differences between normal and abnormal images are selected to do differential operation after the binarization, open operation, median filtering and other pre-processing, so as to obtain the differences of fish mouth areas in two states. The fisheye shape is used to get the center coordinates of fish eyes, and the target image only containing fish mouth will be cut out. According to the area of fish mouth, the opening and closing state and duration of fish mouth can be judged. Therefore, it will alarm if the fish has abnormal behaviors. At the same time, the accuracy of this method to judge the abnormal behaviors of spotted grouper is as high as 84%.

In general, fish are used as biological indicators. Through analyzing the abnormal behaviors of fish, the surrounding environment can be evaluated. This method is convenient and characterized. Moreover, it can play a certain role of early warning [24]. But at present, the researches on abnormal behaviors of fish are limited to empirical observation, and only a few related experimental researches were carried out.

## 3.2    Analysis of Feeding Behaviors

China is the largest aquaculture country in the world. In the current aquaculture industry, the aquaculture density is increasing. At the same time, the feeding of fish is the key to aquaculture fishes. Traditional fish feeding methods mainly rely on artificial feedings, which easily leads to overfeeding or inadequate feeding of fish. Furthermore, it will not only reduce the economic benefits of aquaculture fishes, but also will cause deterioration of water quality in aquaculture water when overfeeding, thereby affecting fish welfare. Therefore, traditional aquaculture methods cannot meet the requirements of modern intensive farming management. Therefore, a new non-destructive feeding control method is urgently needed nowadays. By using computer vision to analyze the feeding behaviors of fish, the saturation degree of fish can be well observed in the feeding process, so that the fish can be accurately fed. In addition, the loss of human resources can be reduced in the process of breeding.

Zhou [25] has collected images of fish feeding. Two-valued fish images were obtained, through a series of processes, such as image enhancement, background extraction and target extraction. In addition, a method of classification and removal of reflective frames based on support vector machine and gray gradient co-occurrence matrix has been proposed in order to eliminate the influence on the results caused by spatters and reflections. Finally, the Fishing Behavior Index (FIFFB) was calculated by Delaunay triangulation. At the same time, the FIFFB value, removed from reflective frame, is fitted by least squares polynomial fitting method. In the results, it showed that using FIFFB value could accurately quantify and analyze the changes of fish feeding behaviors.

Liu [26] and other scientists have taken Atlantic salmon as the research object. At the same time, they have also proposed a quantitative method to measure fish feeding activities based on differential frames. In accordance to the effects of overlapping problems in large-scale fish stocks on image analysis, an overlap coefficient has been proposed to correct the analysis results. At the same time, the reflective image of the water surface caused by fish feeding on the water surface will also lead to the inaccuracy of the calculation results. In this paper, an automatic detection and recognition method for the reflective image on the water surface is established. Then, delete the result calculated by the image and interpolate it again. Finally, we can quickly measure the feeding behaviors of the Atlantic salmon fishes based on this method.

Ye and other scientists [27] have put forward a new method of raising quality of Tilapia based on shoal behaviors. At first, the YCbCr color model is used to model the average background. Secondly, optical flow method was used to extract tilapia's behavior characteristics, in order to avoid complex tracking of individuals. Thirdly, standardize the collected behavior characteristics with the help of statistical analysis. Finally, the foraging intensity of fish was evaluated by combined entropy. This study provides feasibility for intelligent feeding of fish stocks in aquaculture.

Fish feeding has certain regularity. It is of great significance to distinguish the feeding status of fish in real time and to use it to control feeding quantities. However, there is a certain gap between the experimental environment and the aquaculture environment. At the same time, there are some problems such as water turbidity and different size of fish. Therefore, further research is needed on the analysis of feeding behaviors.

## 4  Problems in Analyzing Fish Behaviors

Since the 1980s, computer vision has developed from simple visual simulation to replacing and interpreting human visual information, which accelerates the development of visual information acquisition. But for the current technology, fish behavior analysis based on computer vision is difficult to achieve. The main problems are as the follows:

(1) Most of the current studies are based on two-dimensional images if we are observing the collected fish images. However, it is difficult to express the spatial motion behaviors of fish fully. A few researchers have analyzed fish images from three dimensions, but the amount of computation was very large. Some existing methods can track fish accurately from three-dimensional direction, but they are only applicable to some certain specific situations.

(2) Some researchers attempted to analyze the behaviors of fish from the aspect of texture. However, the matching of feature points is unstable and the error is large for the reason that the changes of fish texture color is not obvious. In addition, some fish texture has occlusion problem, which will make it difficult to carry out fish identification and extraction. Therefore, the focus of future research is to extract the characteristics of fish under the condition of partial occlusion.

(3) There are many kinds of fish. Current studies are all about the behavior analysis of individual fish. However, there is no in-depth study on the relationships between individual fish, which makes the results of the study have great limitations. Therefore, the studies can go deep into the characteristics of fish stocks in future.

# 5 Future Research Prospect

## 5.1 Researches on Computer Vision Algorithms

From the view of computer vision, there are a lot of problems in the researches of fish behavior analysis at present. However, more work could be made from the following aspects: (1) From the view of data processing, on the one hand, it may increase the complexity of calculation and reduce the computer processing speed while improving the accuracy of the algorithm. Therefore, we could try to use parallel computing to increase the practicability and improve real time performance. On the other hand, it can improve the method of image modeling, as well as the accuracy of 3D image acquisition. (2) From the view of fish behavior characteristics, we can combine the analysis of fish centroid, acceleration and angular velocity with the analysis of fish texture to improve the accuracy of fish behavior analysis. (3) Because of the long growth cycle of fish, the colour, texture and swimming speed of fish will change dramatically at each stage of its growth. However, the existing researches on fish behavior analysis were all monitoring and observing the same stage of a class of fish. Furthermore, there is little analysis on the special behaviors of fish at different stages. Therefore, the dynamic tracking and monitoring of the whole growth process of fish has important research significance.

## 5.2 Researches on Deep Learning

Deep learning [28] is an extension in the field of machine learning. It can simulate the structure of human brain, and realize efficient processing of complex input data, as well as learning different knowledge intelligently and solving a lot of problems effectively. Extracting the features through constructing in-depth learning neural network is a new research direction applied to behavior recognition. In future, fish behavior analysis will increasingly rely on deep learning model to achieve the real intelligence.

At present stage, some researchers use interactive information to improve the depth learning model by extracting the most representative image blocks. Combining with space-time convolution, the behaviors of the fish can be judged, so as to get the real-time life status of the current fish stocks. At the same time, the real-time behaviors of the fish will be sent, so as to deal with it pertinently [29]. Some researchers have proposed that optical flow can be self-learned by using in-depth learning networks as a motion feature, which can better identify fish behaviors and reduce a lot of operational force and time [30].

From the results of current researches, the next development of fish behavior recognition based on in-depth learning focuses on models (network architecture), tricks (introducing Res-net, new loss functions, input forms, etc.) and new data base. However, the most important thing is to deal with the characteristics of fish behaviors.

## 6 Conclusions

In conclusion, the analysis of fish behaviors is of great significance for further research on aquaculture and ecological protection. Although there are many studies on fish behavior analysis based on computer vision at present, the experimental environment is too simple and the number of samples is too small. In addition, a large number of fish behavior analysis cannot be carried out. Finally, the research of analyzing fish behaviors is still in its infancy. Therefore, new methods are needed to solve the current problems.

**Acknowledgements.** This work is supported by Guangdong Province Key Laboratory of Popular High Performance Computers (SZU-GDPHPCL201805), Institute of Marine Industry Technology of Universities in Liaoning Province (2018-CY-34), National Natural Science Foundation of China (61701070), Liaoning Doctoral Start-up Fund (20180540090) and China Postdoctoral Science Foundation (2018M640239).

## References

1. Henriques, S., Cardoso, P.: Processes underpinning fish species composition patterns in estuarine ecosystems worldwide. J. Biogeogr. **44**(3), 627–639 (2016)
2. Popper, A.N., Carlson, T.J.: Application of sound and other stimuli to control fish behavior. Trans. Am. Fish. Soc. **127**(5), 673–707 (1998)
3. Johansson, G.: Visual motion perception. Sci. Am. **232**(6), 76–89 (1975)
4. Favorskaya, M.N., Jain, L.C.: Development of mathematical theory in computer vision. In: Favorskaya, M.N., Jain, L.C. (eds.) Computer Vision in Control Systems-1. ISRL, vol. 73, pp. 1–8. Springer, Cham (2015). https://doi.org/10.1007/978-3-319-10653-3_1
5. Lu, H.D., Liu, Y.: Design and implementation of fish behavior automatic monitoring system based on computer vision. Fish. Mod. **21**(21), 39–50 (2011)
6. Kang, I.J., Moroishi, J.: Biological monitoring for detection of toxic chemicals in water by the swimming behavior of small freshwater fish. J.- Fac. Agric. Kyushu Univ. **54**(1), 209–214 (2009)
7. Israeli, D., Kimmel, E.: Monitoring the behavior of hypoxia-stressed Carassius auratus using computer vision. Aquacult. Eng. **15**(6), 423–440 (1996)
8. Fan, L., Liu, Y.: Fish motion detecting algorithms based on computer vision technologies. Trans. Chin. Soc. Agric. Eng. **27**(7), 226–230 (2011)
9. Nguyen, N.D., Huynh, K.N.: Fish detection and movement tracking. In: 2015 International Conference on Advanced Technologies for Communications, pp. 484–489. IEEE (2015)

10. Zhao, X., Peng, H.: A video image detection method for fish motion characteristics. J. Xi'an Univ. Posts Telecommun. (2), 7 (2017)
11. Rodríguez, Á.: Fish tracking in vertical slot fishways using computer vision techniques. https://doi.org/10.2166/hydro.2014.034/. Accessed 10 Oct 2018
12. Rodriguez, A., Rabuñal, J.R., Bermudez, M., Puertas, J.: Detection of fishes in turbulent waters based on image analysis. In: Ferrández Vicente, J.M., Álvarez Sánchez, J.R., de la Paz López, F., Toledo Moreo, F.J. (eds.) IWINAC 2013. LNCS, vol. 7931, pp. 404–412. Springer, Heidelberg (2013). https://doi.org/10.1007/978-3-642-38622-0_42
13. Wang, N.X.R.: Automated analysis of wild fish behavior in a natural habitat. In: Proceedings of the 2nd International Workshop on Environmental Multimedia Retrieval, pp. 21–26. ACM (2015)
14. Xiao, G., Fan, W.K.: Research of the fish tracking method with occlusion based on monocular stereo visions. In: 2016 International Conference on Information System and Artificial Intelligence, (ISAI), pp. 581–589 (2017)
15. Ren, Z., Chon, T.S.: The monitoring and assessment of aquatic toxicology. Biomed. Res. Int. **2017**, 1–2 (2017)
16. Li, X., Fan, L.: Preference habit of juvenile turbot for different color backgrounds based on computer vision. Trans. Chin. Soc. Agric. Eng. **28**(10), 189–193 (2012)
17. Papadakis, V.M.: A computer-vision system and methodology for the analysis of fish behavior. Aquacult. Eng. **46**, 53–59 (2017)
18. Zhao, J., Gu, Z.: Spatial behavioral characteristics and statistics-based kinetic energy modeling in special behaviors detection of a shoal of fish in a recirculating aquaculture system. Comput. Electron. Agric. **127**, 271–280 (2016)
19. Hu, J., Li, D.: Fish species classification by color, texture and multi-class support vector machine using computer vision. Comput. Electron. Agric. **88**, 133–140 (2012)
20. Tang, Y.P., Liu, S.S.: A study of fish behavior analysis based on 3D computer vision". High Technol. Commun. **25**(3), 249–256 (2015)
21. Liu, S.S.: Research on the key technology of the seismic behavior analysis of fish based on 3D computer vision. Zhejiang University of Technology (2016)
22. Yu, X., Hou, X.: Anomaly detection of fish school behavior based on features statistical and optical flow methods. Trans. Chin. Soc. Agric. Eng. **30**(2), 162–168 (2014)
23. Xu, S., Xing, K.Z.: A method for identifying abnormal behavior of point grouper based on machine vision. Fish. Mod. **43**(1), 18–23 (2016)
24. Cerdà, V., Estela, J.M.: Flow techniques in water analysis. Talanta **50**(4), 695–705 (1999)
25. Zhou, C., Zhang, B.: Near-infrared imaging to quantify the feeding behavior of fish in aquaculture. Comput. Electron. Agric. **135**, 233–241 (2017)
26. Liu, Z., Li, X.: Measuring feeding activity of fish in RAS using computer vision. Aquacult. Eng. **60**(3), 20–27 (2014)
27. Ye, Z., Zhao, J.: Behavioral characteristics and statistics-based imaging techniques in the assessment and optimization of tilapia feeding in a recirculating aquaculture system. Trans. ASABE **59**(1), 345–355 (2016)
28. Deng, L., Yu, D.: Deep learning: methods and applications. Found. Trends® Sig. Process. **7** (3–4), 197–387 (2014)
29. Neshat, M., Sepidnam, G.: Artificial fish swarm algorithm: a survey of the state-of-the-art, hybridization, combinatorial and indicative applications. Artif. Intell. Rev. **42**(4), 965–997 (2014)
30. Sevilla-Lara, L., Liao, Y., Güney, F., Jampani, V., Geiger, A., Black, Michael J.: On the integration of optical flow and action recognition. In: Brox, T., Bruhn, A., Fritz, M. (eds.) GCPR 2018. LNCS, vol. 11269, pp. 281–297. Springer, Cham (2019). https://doi.org/10.1007/978-3-030-12939-2_20

# Method for Recognition Pneumonia Based on Convolutional Neural Network

Xin Li[⊠], Dongdong Gao, and Haijiang Hao

Guangxi Key Laboratory of Embedded Technology and Intelligent System,
Guilin University of Technology, Guilin 541006, China
1996019@glut.edu.cn

**Abstract.** Pneumonia is one of the most common infectious diseases in clinical practice. In the field of pneumonia recognition, traditional algorithms have limitations in feature extraction and scope of application. To solve this problem, a pneumonia recognition is proposed based on convolutional neural network. Firstly, the morphological preprocessing operation was performed on the chest X-ray. Secondly, the convolutional layer containing the 1 * 1 convolution kernel was used instead of a fully connected layer in the convolutional neural network to segment the lung field and obtain the segmentation. The index Dice coefficient can reach 0.948. Finally, a pneumonia recognition model based on convolutional neural network was established. The segmented images were trained and tested. The experimental results show that the average accuracy of the proposed method for pneumonia is up to 96.3%.

**Keywords:** Convolutional neural network · Lung field segmentation ·
Pneumonia recognition

## 1 Introduction

In recent years, many experts have proposed a method for identifying pneumonia images using image recognition. The extraction of image features is the most basic work, Based on a large number of data samples on the current network, various learning algorithms are used to learn the internal features of medical images, eliminating the need for manual design, thereby simplifying the process of feature extraction [1]. Since 2006, various convolutional neural network algorithms for solving target recognition problems have been proposed, such as AlexNet, VGGNet, GoogleNet, and R-CNN [2] series, to extend these convolutional neural networks to different applications. Such as object recognition [3], face recognition [4], image classification [5], speech recognition [6] etc., have achieved relatively good results, especially in the medical field has achieved great success.

In 2017, Wang et al. used a self-built database, ChestX-ray8, to process clinical chest radiographs using natural language processing and deep convolutional neural networks, and to identify and locate 8 chest diseases such as atelectasis and cardiac hypertrophy. 63.3% [7]; In the same year, Yao et al. used statistical dependencies between tags to make more accurate predictions, which was better than 13 of the 14 groups in Wang et al., where the accuracy of pneumonia was 71.3% [8]. Subsequently,

© Springer Nature Singapore Pte Ltd. 2019
R. Mao et al. (Eds.): ICPCSEE 2019, CCIS 1059, pp. 142–156, 2019.
https://doi.org/10.1007/978-981-15-0121-0_11

Rajpurkar et al. proposed a chest X-ray pneumonia recognition algorithm (ChexNet), in which the accuracy of a pneumonia was 76.8% [9]. In China, in 2015, Wang in Taiwan used K nearest neighbor classifiers for pneumonia identification, and the final recognition was 75% [10]. Subsequently, Zhu of Capital Medical University used entropy texture based on wavelet transform. The classification of pneumoconiosis was characterized by an accuracy rate of 84.6% [11]. In 2017, Shao of East China Normal University used the improved cSVM algorithm to diagnose pneumonia, with an average correct rate of 90% [12].

The rest of the paper is organized as follows: In the second part of the work introduces X-ray chest image preprocessing, The third part and The fourth part introduce the experimental results and analysis of the segmentation and recognition phases, respectively, The last part is the conclusion.

## 2   X-Ray Chest Image Preprocessing

In the pneumonia recognition problem of X-ray chest radiograph studied in this paper, because the image map is in the process of formation, the noise will cause the edge of the image to be blurred, plus some overlaps of the image itself, and the background gray scale contrast is not prominent. Will affect the follow-up work, so this section first pre-processes the chest radiograph, using the filtering algorithm, Gauss-Laplace pyramid algorithm.

### 2.1   Image Preprocessing

Mean filtering is a typical linear filtering algorithm, and the main method used is the neighborhood averaging method. Its basic principle is to replace the individual pixel values in the original image with the mean value, that is, the current pixel point $(x, y)$ to be processed, and select a template consisting of several pixels of its neighbors, and find the mean of all the pixels in the template. And assigning the mean to the current pixel $(x, y)$ as the gray value $\mu(x, y)$ of the processed image at that point,

$$\mu(x, y) = \frac{1}{m} \sum f(x, y) \tag{1}$$

Where m is the total number of pixels in the template including the current pixel.

Median filtering is a non-linear image smoothing method that filters out impulse noise and protects the edges of the target image compared to averaging filters and other linear filters. It is a neighborhood operation, similar to convolution, but it is not a weighted summation, but the pixels in the neighborhood are sorted by gray level, and then the middle value of the group is selected as the output pixel value, the median value. Filtering can be defined as:

$$g(x, y) = median\{f(x - i, y - i)\}, \quad (i, j) \in W \tag{2}$$

Where $g(x, y)$ and $f(x - i, y - i)$ are the gray values of the output and input pixels, respectively, $W$ is the template window which can take a line shape, a square shape, a cross shape, a circle shape, a diamond shape, and the like.

For medical images, the Gaussian filter is a 2-dimensional convolution operator that uses Gaussian kernels for image blurring (removing detail and noise). The basic idea is to discretize the Gaussian function, and to use the value of the Gaussian function at the discrete points as the weight, and to perform a weighted average of a certain range of neighborhoods for each pixel of the acquired gray matrix, thereby effectively eliminating Gaussian noise. Its one-dimensional Gaussian distribution and two-dimensional Gaussian distribution are shown in Eqs. (3) and (4).

$$G(x) = \frac{1}{\sqrt{2\pi}\sigma} e^{-\frac{x^2}{2\sigma^2}} \tag{3}$$

$$G(x, y) = \frac{1}{2\pi\sigma^2} e^{-\frac{x^2+y^2}{2\sigma^2}} \tag{4}$$

The degree of image smoothing after Gaussian filtering depends on the standard deviation. Its output is the weighted average of the domain pixels, and the closer the pixel is to the center, the higher the weight. Therefore, the smoothing effect is softer and the edges are better preserved than the mean filtering.

Figure 1 shows the effect of three different filtering algorithms. After analysis, it can be known that the mean filtering is to give a template to the target pixel on the image, and then replace the original pixel value with the average value of the entire pixel in the template. Smoothing the image is faster, but you can't remove the noise and only weaken it weakly. The median filtering has a good effect on removing the salt and pepper noise. The probability that the noise-pollution point in the image is replaced by the noise point is relatively large, and the contour of the image is relatively clear. The overall effect of Gaussian filtering is good, not only effectively removes noise, but also preserves edge information. In summary, the method chosen in this paper is Gaussian filtering algorithm for image processing.

## 2.2 Gauss-Laplace Pyramid Enhancement Processing

The main idea of using Gaussian pyramid and Laplacian pyramid reconstruction is that the content in the Gaussian pyramid will lose information in the process of Gaussian blur and downsampling, and in order to reconstruct the image that is not much different from the original image, The idea of the Laplacian pyramid is used. Each layer in the Laplacian pyramid is a residual map of two adjacent layers in the Gaussian pyramid, that is, information lost between adjacent layers.

The algorithm is described in detail as follows: the original image is used as the first layer of the Gaussian pyramid, then Gaussian blur is used and downsampled to obtain the second layer of the Gaussian pyramid. The third layer of the Gaussian pyramid is Gaussian blur and downsampling the second layer. The resulting other layers of the Gaussian pyramid are also obtained by such operations. Then after getting the Gaussian pyramid, we need to find the residual between its two adjacent layers. The Gaussian

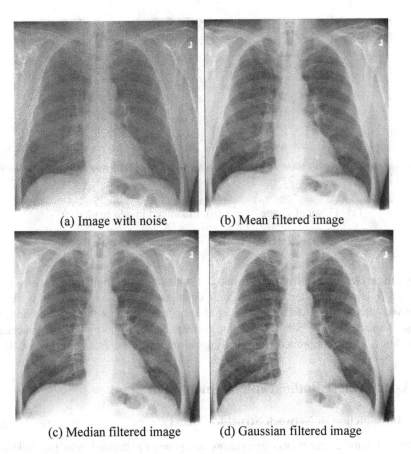

(a) Image with noise          (b) Mean filtered image

(c) Median filtered image     (d) Gaussian filtered image

**Fig. 1.** Comparison of several filtering algorithms on chest image

gold layer 2 image is upsampled and Gaussian blur is obtained to obtain the same resolution image as the Gaussian pyramid layer 1 image, and then (Layer 1 to Layer 2) to obtain the first layer image of the Laplacian pyramid, and then The second layer image is obtained from (the Gaussian pyramid layer 2 to the third layer of the Gaussian pyramid is upsampled and the Gaussian blur is obtained), and so on. It should be noted that the Laplacian pyramid generally has one layer less than the Gaussian pyramid. When the Laplacian pyramid was built, all the information was saved. Next, you can reconstruct the image, use the last layer of the Gaussian pyramid to upsample and smooth, add the residual of the last layer of the Laplacian pyramid, and then upsample and smooth the resulting image plus the Laplacian pyramid. The second-to-last layer, which has been reciprocating until the end of the lowest layer of the Laplacian pyramid, is the reconstructed image. The experimental results are shown in Fig. 2.

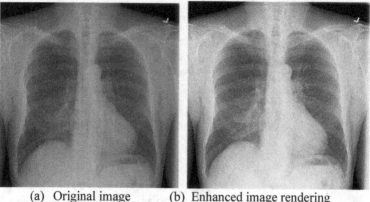

(a)  Original image        (b)  Enhanced image rendering

**Fig. 2.** Gauss-Laplace pyramid enhancement

It can be clearly seen from Fig. 2 that the contrast of the original image is low, the contrast is improved after the histogram equalization, and the contrast is further improved after the Gauss-Laplace pyramid enhancement algorithm. In addition, the Gauss- Laplace average contrast C value corresponding to the pyramid enhancement algorithm is greatly improved from the original image index, from 0.0023 to 0.0324.

## 3  X-Ray Chest Radiograph Segmentation

### 3.1  Introduction to Network Structure Model

Traditional lung segmentation algorithms are generally divided into the following categories, rule-based segmentation methods [13], pixel-based segmentation methods [14], deformation models based segmentation methods [15] and methods for combining multiple segmentation methods [16]. Etc. However, all of the above methods have problems that require manual setting of seed points, manual selection of features, and dependence on prior knowledge. Combining with the advantages of deep learning, this paper proposes an automatic segmentation method based on improved CNN, which transforms the image segmentation problem into image block classification. The experimental results are robust and can segment the target well and present the outline of the lung field.

In the X-ray chest segmentation experiment, this paper discards a fully connected layer with multiple parameters and changes it to a convolutional layer with a convolution kernel of 1 * 1. The benefit of this improved approach is that since the convolutional layer has the characteristics of parameter sharing, which greatly reduces the number of parameters, these convolutional networks can also be considered as convolutions with kernels covering their entire input area, and this improvement The network can be effectively trained on a block-by-block basis. In the image segmentation task, the position information of each point is crucial for each pixel category. The network model after the change is shown in Fig. 3.

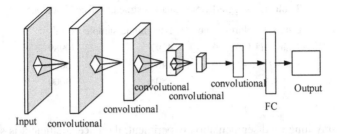

**Fig. 3.** Schematic diagram of an improved neural network model

The network has 8 layers, including 5 convolution layers and 1 full connection layer (Full Connection, FC). The first three convolution layers are followed by a maximum pooling layer. In order to improve the generalization ability of the network, The LRN layer is added after the first layer of the pooling layer, followed by a convolutional layer and a convolutional layer with a 1 * 1 convolution kernel, and finally a fully connected layer and output. The first three convolutional layers use a convolution kernel size of 3 * 3, a padding of 1, a step size of 1, a maximum pooling layer core size of 2 * 2, a step size of 2, and a fully connected layer of neurons. The number is 4096. Before the training starts, in order to be consistent with the pre-training network, the spatial resolution of all training sample images needs to be uniformly adjusted to 224 * 224 pixels. In the training process using the above network model, the input 224 * 224 RGB image is subtracted from the training. Sets the grayscale mean of the sample and sends it to the convolutional layer for processing. After the first layer of convolutional layer, the size of the feature map is still 224 * 224, and the obtained feature map is sent to the pooling layer through the activation function. Since the pooled layer core has a size of 2 * 2 and the stride is 1, the size of the calculated feature map becomes 112 * 112. By analogy, in this case, the size of the feature map obtained through the convolution layer is the same as the size of the previous layer, and the size of the feature map becomes the original quarter after the pooling. After the convolution layer, the activation function, and the pooling layer are alternately processed, the finally obtained 28 * 28 feature map is sent to the fully connected layer, and the category result is obtained and output.

## 3.2 Experiment

This experiment is carried out under Windows 7 and Ubuntu 16.04 operating system respectively. The Window 7 processor is Intel(R) Core(TM) i5-6500 CPU @3.20 GHz, memory is 8G; Ubuntu graphics card is NVIDIA GPU1070Ti, and the memory is DDR2 6 GB. The pre-processing of X-ray chest images in the previous part was carried out in Matlab 2016a environment. The network construction, training and debugging of subsequent deep learning were carried out in the deep learning framework Caffe environment.

The experimental data set of this paper is selected from the data set provided by the North American Radiological Society (RSNA®) in conjunction with Kaggle in the medical image pneumonia recognition competition in September 2018. The data set contains the training set and the test set. The specific allocation is shown in Table 1.

**Table 1.** Kaggle-RSNA data set structure distribution

| Data set | Normal sample | Pneumonia sample | Total |
|---|---|---|---|
| Train set | 16692 | 4297 | 20989 |
| Val set | 4518 | 1482 | 6000 |
| Test set | 2192 | 808 | 3000 |

In the X-ray lung field segmentation experiment, the Dice coefficient is selected as the criterion for judging the quality of image segmentation, also known as F1 score or Dice similarity coefficient (DSC).

The Dice coefficient was originally used to measure the presence or absence of data. The formula is:

$$QS = \frac{2|X \cap Y|}{|X| + |Y|} \tag{5}$$

Where $|X|$ and $|Y|$ represent the number of elements in the two samples. In this part of the experiment, the standard sample segmentation results and the results segmented by the algorithm are respectively indicated.

The Dice coefficient is not much different in form from the Jaccard index, but it has some different properties. The same point is that their range is 0 to 1, and the closer to 1, the higher the accuracy of the segmentation, close to 0, indicating image segmentation. The accuracy rate is low, unlike the Jaccard indicator, the corresponding difference function:

$$d = 1 - \frac{2|X \cap Y|}{|X| + |Y|} \tag{6}$$

It can be seen from Table 2 that the convolutional neural network used in this paper classifies the sample superpixel image, and then divides the X-ray chest image to obtain a relatively good segmentation effect, compared with the traditional Ostu segmentation. Algorithm, the average Dice coefficient of the method used in this paper is increased by 0.156; compared with the traditional CNN algorithm, the single picture test time of this method is shortened by 62 s; the combination shows that the proposed method is excellent in segmentation performance and consumption time. In the past, other traditional segmentation algorithms.

**Table 2.** Comparison of experimental results of different algorithms

| Experimental method | Ostu algorithm | Traditional CNN algorithm | Method of this paper |
|---|---|---|---|
| Average Dice coefficient | 0.792 | 0.893 | 0.948 |
| Test time per picture | 243 s | 440 s | 378 s |

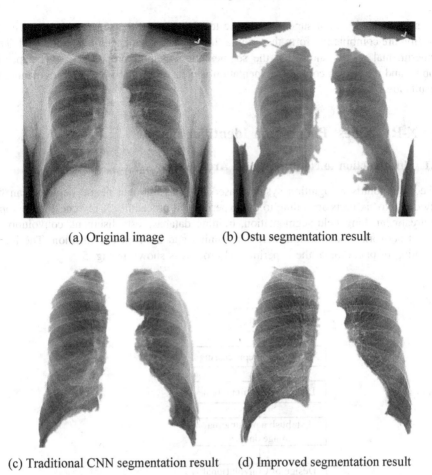

(a) Original image                    (b) Ostu segmentation result

(c) Traditional CNN segmentation result    (d) Improved segmentation result

**Fig. 4.** Comparison of segmentation results of different segmentation algorithms

Figure 4 reflects the influence of different segmentation algorithms on the segmentation results. From the perspective of segmentation effect, the Ostu segmentation algorithm is largely affected by the clavicular region and the thoracic angle region, and the segmentation effect is poor. The traditional CNN is more completely segmented. The outline of the lung field, but from the experimental results, it is still slightly affected by the ribs. In this paper, by constructing a convolutional neural network, the superpixels in the original image are classified, and then the X-ray chest is segmented. This segmentation method achieves better results and accurately segments the lung field.

In this paper, based on the advantages of deep learning, a lung field segmentation model based on convolutional neural network is proposed. In this model, the image block for the central pixel of the lung region is first obtained from the image, which is defined as super pixel, and then for each extracted A superpixel uses CNN to extract features, and then classifies the extracted features. After the model is trained, the model is used to classify each superpixel in the test image, and the superpixels after classification are combined to form the original image. The segmentation is completed. The experimental results show that the segmentation method proposed in this paper is robust and the region contour segmentation is clear, and relatively good segmentation results are obtained.

## 4  X-Ray Chest Pneumonia Identification

### 4.1  Introduction to Network Model Architecture

The pneumonitis recognition system based on convolutional neural network can be divided into six parts according to the experimental procedure, X-ray chest radiograph pretreatment, lung field segmentation, training database establishment, convolutional neural network model design, network training and network optimization. The identification of pneumonia, the experimental process is shown in Fig. 5.

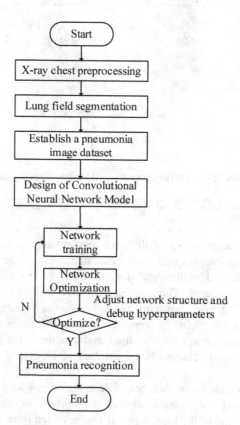

**Fig. 5.** Flow chart of pneumonia recognition system based on convolutional neural network

## 4.2   Preliminary Experimental Results Analysis

In the preliminary experiment, the preliminary experiment was carried out with the structure of the same hidden layer as VGG16. The network structure in the preliminary experiment has 13 convolution layers and 3 fully connected layers. According to Caffe's related theory, when making data in lmdb format. The category is marked with 0–1 numbers, the normal chest radiograph is marked as 0, and the pneumonia chest radiograph is marked as 1. The recognition of the image by the system is realized by the label. The performance of the system during the training phase is evaluated by the final accuracy value Acc and the loss value loss. Since the sample is less in the preliminary experiment, the learning rate is also set low (base_lr = 0.0001). The hyperparameter configuration of network training is shown in Fig. 6, and training accuracy, loss value and iteration number relationship are shown in Fig. 7.

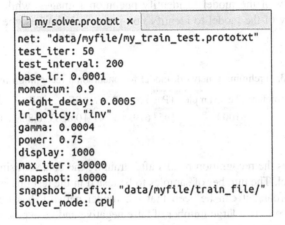

```
my_solver.prototxt  ×
net: "data/myfile/my_train_test.prototxt"
test_iter: 50
test_interval: 200
base_lr: 0.0001
momentum: 0.9
weight_decay: 0.0005
lr_policy: "inv"
gamma: 0.0004
power: 0.75
display: 1000
max_iter: 30000
snapshot: 10000
snapshot_prefix: "data/myfile/train_file/"
solver_mode: GPU
```

**Fig. 6.**  Hyperparameter configuration for network training

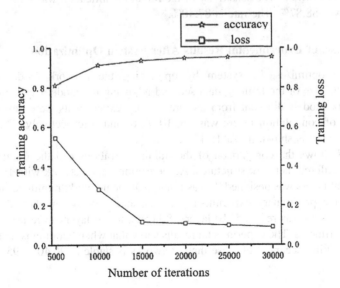

**Fig. 7.**  Training accuracy, loss value and iteration number relationship

In order to comprehensively measure the classification performance of the model, three model evaluation indicators were used in the experiment: accuracy rate (ACC), sensitivity (SE), and specificity (SP). The calculation formulas are as shown in (7), (8), (9):

$$ACC = (TP + TN)/(TP + TN + FN) \tag{7}$$

$$SE = TP/(TP + FN) \tag{8}$$

$$SP = TN/(TN + FP) \tag{9}$$

Among them, TP, TN, FP, and FN represent true positive, true negative, false positive, and false negative, respectively. The accuracy rate reflects the degree to which the model's prediction results are consistent with the actual test results. The sensitivity reflects the ability of the model to identify pneumonia images, while the specificity reflects the ability of the model to identify non-pneumonia images, the test results are shown in Table 3.

**Table 3.** Preliminary network model for pneumonia recognition results

| Network structure | Test sample | TP | TN | FP | FN | SE | SP | ACC |
|---|---|---|---|---|---|---|---|---|
| VGG-16 | 1000 | 195 | 679 | 85 | 41 | 0.8263 | 0.8887 | 0.8740 |

Table 3 shows the recognition results after training the data set using the VGG-16 pre-training model. The number of samples to be tested in the test phase is 1000. From the experimental data, it can be seen that the model network has not made actual adjustments. There were a large number of false negative and false positive cases in the identification. 41 cases were identified as normal chest images when the pneumonia images were identified. The sensitivity of the initial identification was 82.63% and the specificity was 88.87%. The rate is 87.40%.

## 4.3    Analysis of Experimental Results After System Optimization

This section optimizes the system by optimizing the network model structure, increasing the size of the training data set, and adjusting the model super parameters. Five network models different from the preliminary experiments were designed, and the number of convolution layers was 16, 10, 8, 6, and 6 respectively. The detailed network structure is shown in Table 4.

Figure 8 shows the comparison of the training iterations and the training model accuracy of different network structures. At the beginning, a network with 16 layers of convolutional layers was designed. It was found that the network training accuracy was lower than the preliminary experimental results, and then the number of layers of convolution layers was reduced. 10 layers, 8 layers, and 6 layers were used for comparative experiments. The experimental results show that when iteration is up to 30,000 times, the training accuracy of these three networks is stable at 96.40%, 95.80%, and

**Table 4.** CNN network model with different convolution layers

| Network structure | Number of convolution layers | Convolution and pooling layer distribution |
|---|---|---|
| VGG-19 | 16 | 2Conv* + Pool⊚ + 2Conv + Pool + 4Conv + Pool + 4Conv + Pool + 4Conv + Pool |
| VGG-16 | 13 | 2Conv + Pool + 2Conv + Pool + 3Conv + Pool + 3Conv + Pool + 3Conv + Pool |
| VGG-13 | 10 | 2Conv + Pool + 2Conv + Pool + 2Conv + Pool + 2Conv + Pool + 2Conv + Pool |
| VGG-11 | 8 | 1Conv + Pool + 1Conv + Pool + 2Conv + Pool + 2Conv + Pool + 2Conv + Pool |
| VGG-9 | 6 | 1Conv + Pool + 1Conv + Pool + 1Conv + Pool + 1Conv + Pool + 2Conv + Pool |
| VGG-9-LRN | 6 | 1Conv + Pool + LRN + 1Conv + Pool + 1Conv + Pool + 1Conv + Pool + 2Conv + Pool |

Conv* stands for convolutional layer; Pool⊚ stands for max pooling layer.

96.82%, respectively. It can be seen that the appropriate reduction of the number of layers of convolution is more suitable for the research of this topic, in order to further improve 6 The generalization capability of the layer convolutional layer network, the LRN layer is added to the network. The LRN is mainly a technical method for improving the accuracy in deep learning training, generally the normalization process after the activation and pooling layers. Methods to enhance the generalization ability of the model. Under the premise of not changing the original network parameters, the improved VGG-9-LRN network model is added after the first layer of the pooling layer, and the training accuracy of the final model is stable at 97.24%. Based on the above analysis, this paper finally chose VGG-9-LRN as the network model for pneumonia recognition. Table 5 shows the accuracy of different network models for pneumonia recognition.

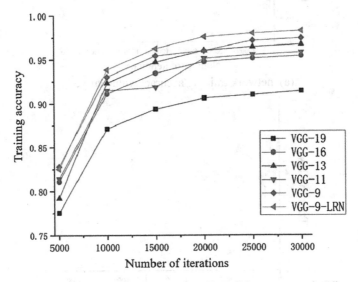

**Fig. 8.** Comparison of training iteration times and model accuracy of different network structures

**Table 5.** Comparison of different network training models for pneumonia recognition

| Network structure | Test sample | SE | SP | ACC |
|---|---|---|---|---|
| VGG-19 | 1000 | 0.7839 | 0.8639 | 0.8450 |
| VGG-16 | 1000 | 0.8263 | 0.8887 | 0.8740 |
| VGG-13 | 1000 | 0.8856 | 0.8966 | 0.8940 |
| VGG-11 | 1000 | 0.8390 | 0.8901 | 0.8780 |
| VGG-9 | 1000 | 0.9322 | 0.9202 | 0.9230 |
| VGG-9-LRN | 1000 | 0.9534 | 0.9332 | 0.9380 |

Then, the original data was amplified, and the amplified data set obtained by rotation and mirroring, 86967 training sets (including 60075 normal chest radiographs, 26892 chest radiographs), and 3000 test sets. (Includes 2,292 images of normal chest

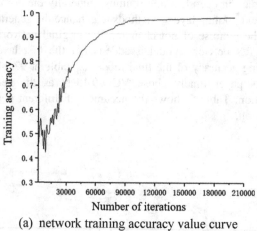

(a) network training accuracy value curve

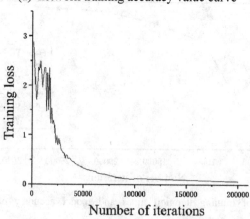

(b) network training loss graph curve

**Fig. 9.** The network training accuracy (Acc) and loss (Loss) curves

radiographs and 708 images of pneumonia). The convolutional neural network uses the VGG-9-LRN model with better model convergence in Sect. 4.3. The network training accuracy (Acc) and loss (Loss) curves of the augmented data set are shown in Fig. 9.

Compared with the previous part of the experiment, the training accuracy of the final network using the unamplified data training network is 97.24%, the training accuracy of the final network using the augmented data training network is 97.74%, and the recognition rate of pneumonia is also increased by 2.5%. It can be seen that increasing the number of training data samples can significantly improve the accuracy of the entire system, thereby reducing the loss value.

### 4.4 Compared with Previous Methods

In recent years, many researchers at home and abroad have done relevant research on such work. For example, Wang Jiaming mentioned in the introduction uses K nearest neighbor classifiers for pneumonia identification, because the data set is only 136. For the sample, the final recognition is only 75%; Zhu Biyun et al. used the wavelet transform-based entropy texture feature to classify pneumoconiosis, with an accuracy rate of 84.6%; Pranav Rajpurkar et al. proposed a chest X-ray pneumonia recognition algorithm (ChexNet), in which the accuracy of a pneumonia was 76.8%. Shao Xinwei used the improved cSVM algorithm to diagnose pneumonia, and the average correct rate Up to 90%. The comparison results are shown in Table 6. The accuracy and stability of the method in this paper are significantly higher than other methods. Here, the data set size and pneumonia recognition rate are compared.

**Table 6.** Accuracy comparison of different classification methods

| Years | Author | Data set size | Method | ACC |
|-------|--------|---------------|--------|-----|
| 2015 | Wang Jiaming | 136 | K nearest neighbor classifiers | 75% |
| 2016 | Zhu Biyun | 110 | Wavelet transform | 84.6% |
| 2017 | Pranav Rajpurkar | 5856 | CNN algorithm | 76.8% |
| 2017 | Shao Xinwei | 2614 | Improved cSVM algorithm | 90% |
| 2017 | Xiang Wenbo | 2000 | CNN algorithm | 86.75% |
| 2019 | Our method | 89967 | CNN algorithm | 96.30% |

## 5  Conclusion

In this paper, an improved convolutional neural network model is proposed to realize the identification of X-ray pneumonitis pneumonia. The network model structure is improved on the segmentation model, so that the segmentation sample obtains a higher Dice coefficient. At the same time, the classical network VGG-16 is improved. Increasing the number of training samples, fine-tuning the hyperparameters of network training, and finally achieving a 96.3% X-ray chest pneumonia recognition rate, showing better recognition and generalization ability than other recognition algorithms.

In the follow-up work, taking into account the particularity of medical images, other network structures can be used to conduct such research to achieve better recognition results.

**Acknowledgement.** This research work was supported by Guangxi key Laboratory Fund of Embedded Technology and Intelligent System (Guilin University of Technology) under Grant No. 2017-2-5.

# References

1. Su, X.: Classification of medical image mode based on deep learning. Zhejiang Normal University (2016)
2. Ren, S., He, K., Girshick, R., et al.: Faster R-CNN: towards real-time object detection with region proposal networks. In: International Conference on Neural Information Processing Systems, pp. 91–99 (2015)
3. Schwarz, M., Schulz, H., Behnke, S.: RGB-D object recognition and pose estimation based on pre-trained convolutional neural network features. In: IEEE International Conference on Robotics and Automation, pp. 1329–1335 (2018)
4. Li, H., Lin, Z., Shen, X., et al.: A convolutional neural network cascade for face detection. In: Computer Vision and Pattern Recognition, pp. 5325–5334 (2015)
5. Xiao, T., Xu, Y., Yang, K., et al.: The application of two-level attention models in deep convolutional neural network for fine-grained image classification. **40**(1), 842–850 (2014)
6. Abdel-Hamid, O., Mohamed, A.R., Jiang, H., et al.: Convolutional neural networks for speech recognition. IEEE/ACM Trans. Audio Speech Lang. Process. **22**(10), 1533–1545 (2014)
7. Wang, X., Peng, Y., Lu, L., et al.: ChestX-ray8: hospital-scale chest X-ray database and benchmarks on weakly-supervised classification and localization of common thorax diseases, 3462–3471 (2017)
8. Yao, L., Poblenz, E., Dagunts, D., et al.: Learning to diagnose from scratch by exploiting dependencies among labels. **23**(15), 1265–1269 (2017)
9. Rajpurkar, P., Irvin, J., Zhu, K., et al.: CheXNet: radiologist-level pneumonia detection on chest X-rays with deep learning. **69**(20), 239–245 (2017)
10. Wang, J.: Using pattern recognition to detect electronic nose pneumonia. Taiwan Med. J. **39**(01), 68–74 (2015)
11. Zhu, B., Chen, H., Chen, B., et al.: Early diagnosis of pneumoconiosis on digital radiographs based on wavelet transform-derived texture features. Beijing Biomed. Eng. **33**(02), 148–152 + 171 (2014)
12. Shao, X.: Early diagnosis of community-acquired pneumonia in children based on SVM + Algorihm. East China Normal University (2017)
13. Li, X., Peng, D., Wang, H.: High-resolution image classification based on optimal scale and rules. Surv. Eng. **26**(09), 14–22 (2017)
14. Jia, G., Zhao, H., Liu, F., et al.: Superpixel-based graph-based image segmentation algorithm. J. Beijing Univ. Posts Telecommun. **41**(03), 46–50 (2018)
15. Dong, X.: CT image liver tumor segmentation based on geometric deformation model. Shandong Normal University (2017)
16. Zhao, F., Zheng, Y., Liu, H., et al.: Multi-objective joint multi-objective evolutionary adaptive threshold image segmentation algorithm. Appl. Res. Comput. **35**(06), 1858–1862 (2018)

# Underwater Image Saliency Detection Based on Improved Histogram Equalization

Zhen Cui[1,2], Junfeng Wu[1,2,3], Hong Yu[1,2(✉)], Yizhi Zhou[1,2],
and Liang Liang[1,2]

[1] School of Information Engineering,
Dalian Ocean University, Dalian 116023, China
yuhong@dlou.edu.cn
[2] Guangdong Province Key Laboratory of Popular High Performance
Computers, Shenzhen University, Shenzhen 518060, China
[3] School of Computer Science and Technology,
Tianjin University, Tianjin 300072, China

**Abstract.** In order to solve the problem of unsatisfactory detection effect of underwater visual saliency map, an image saliency detection algorithm based on improved histogram equalization is proposed. Underwater images are often not clear enough because the refraction of light underwater causes insufficient image resolution. Therefore, in order to solve the existing problems of traditional histogram equalization algorithm, an improved histogram equalization method is proposed to enhance the quality of images, which makes the saliency regions smoother and clearer. In this paper, the simulation experiments were conducted on UIEBD dataset and DLOU_underwater dataset. The experimental results show the effectiveness, robustness and accuracy of the proposed algorithm.

**Keywords:** Underwater image · Histogram equalization · Saliency extraction

## 1 Introduction

The saliency detection of images can skillfully represent the effective objects or regions in complex scenes [1–3]. Therefore, it is applied to many computer vision researches such as image retrieval [4], image recognition [5], image segmentation [6–8], target detection and so on [8]. There are many algorithms for saliency detection, such as HC algorithm proposed by Cheng, This method is based on the contrast of global color histogram, comparing the color contrast of each pixel with all the others in the image to get its saliency value (The greater difference, the higher saliency). Thus, the full resolution saliency image can be obtained. However, it is not ideal for complex background images [9]. The AC algorithm proposed by Achanta uses low-level features of brightness and color to calculate the local contrast of a perception unit in different neighborhoods to achieve multi-scale saliency detection [10]. This method does not have a prominent place in considering the overall contrast. PCA saliency detection algorithm is based on principal component analysis (PCA) proposed by Ran M, combining two descriptions of morphology and Lab color space effectively, and certain results have been achieved [11].

© Springer Nature Singapore Pte Ltd. 2019
R. Mao et al. (Eds.): ICPCSEE 2019, CCIS 1059, pp. 157–165, 2019.
https://doi.org/10.1007/978-981-15-0121-0_12

As one of the important bases for ocean research, analysis and identification, underwater images have complex shooting conditions and poor quality. The obtained images are difficult to be used directly in marine engineering, scientific research, military and other fields, which brings great difficulties to practical operation and scientific research [12].

The existing saliency detection algorithm has achieved good results for common images. However, it is not suitable for the saliency detection of underwater images with complex environment. For such images, the following three methods will be used [13]. The first method is based on image features, which is representative of the histogram equalization method [14]. The second method is based on the visual effect of the image. The common used method is Retinex algorithm, but the algorithm has the disadvantage that there may be halo and weak color retention under strong light [15]. The third is based on the theory of image enhancement, which is integrated with mathematical theory, but the computational complexity is not suitable for images of complex underwater backgrounds [16]. In summary, histogram equalization has solved the preexisting problems of overexposure or blurring edges during extraction. However, traditional histogram equalization is not suitable for all fields, also needs to be adjusted according to specific fields. Therefore, this paper proposed an underwater image saliency detection method based on improved histogram equalization, which has achieved good results in underwater images.

The structure of this paper is as follows: The second part introduces the histogram equalization method for underwater images; The third part describes the underwater image detection algorithm based on improved histogram; The fourth part carried on the experimental verification, and analysis of experiment result; The fifth part summarizes the whole paper.

## 2 Histogram Equalization Algorithm Based on Underwater Features

The traditional histogram equalization works well for processing ordinary images. However, underwater images are often accompanied by the problem of strong attenuation and insufficient resolution [22]. Therefore, we propose a saliency extraction method based on improved histogram equalization for the specific characteristics of underwater images.

Histogram Equalization, also known as histogram flattening, essentially nonlinearly stretches the image so that the transformed histogram is evenly distributed [17], that enhances the contrast of the peak portion in the middle of the histogram, while the contrast at the bottom of the valley is reduced, so that the histogram of the output becomes a flat segmented histogram.

The histogram is a statistical representation of the occurrence frequency of each grayscale in digital images [18]. At the same time, histogram can also give an overview of the grayscale range of the image, the grayscale distribution of each grayscale, and the average brightness and contrast of the image. The gray histogram is a function of gray level [19], which reflects the number of pixels in the image with the gray level, the abscissa is the gray level r, and the ordinate is the frequency at which the gray level appears (i.e. the number of pixels) pr(r), the whole coordinate system describes the distribution of the gray level of the image, so that the gray distribution characteristics of the image can be showed, which means if the image is dark, most of the pixels are concentrated at a low level. If the image is bright, the pixels are distributed in the high grayscale area.

Figure 1 shows the histogram distribution of original image, and Fig. 2 shows the result after equalization, which is obviously different with before. The randomly distributed image histogram is modified into uniformly. The basic idea is to make mapping transformation on the pixel gray level of the original image, so that the probability density of the gray scale of the transformed image is evenly distributed. So that the dynamic range of image grayscale is increased, which improves the contrast of the image.

**Fig. 1.** Histogram distribution of original image

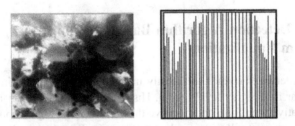

**Fig. 2.** Histogram distribution after equalization

Through this algorithm, the distribution of image brightness on the histogram can be clearly showed. The brightness of the image can be adjusted as needed. In addition, this method is reversible, it is possible to restore the original histogram, if the equalization function is known. Set the variable r representing the gray level of the pixels in the image. Normalize the gray level, then $0 \leq r \leq 1$, where r = 0 for black and r = 1 for white. For a given image, each pixel value is random at the gray level of [0, 1]. Probability density function $P_r(r^k)$ to represent the distribution of the gray level of the image.

In order to facilitate digital image processing, discrete forms are introduced. In discrete form, $r^k$ represents discrete gray levels, $P_r(r^k)$ represents $P_r(r)$, And the following formula is established

$$P_r(r^k) = \frac{nk}{n} \tag{1}$$

Where $0 \leq r^k \leq 1$, k = 0, 1, 2, ..., n − 1. In this formula, $n^k$ is the number of pixels that grayscale $r^k$ appears in the image, n is the total number of pixels in the image, and $\frac{nk}{n}$ is the frequency in probability theory. The function of histogram equalization is:

$$S_i = T(r_i) = \sum_{i=0}^{k-1} \frac{n_i}{n} \tag{2}$$

Where k is the number of gray levels. The corresponding inverse transformation is:

$$r^i = T^{-1}(S_i) \tag{3}$$

For the particularity of the underwater images in this paper, which are mostly blue-green, so we adjusted the image. Firstly, convert images to HSV channels. $p_i$ represents pixel,i represents the brightness value of the average pixel of the image, mi and m are the values after conversion to blue-green.

$$p_i = \left( \frac{((i - mi) * (255 - mi))}{(m - mi)} \right) + 0 \tag{4}$$

## 3   Saliency Detection Algorithm Based on Improved Histogram Equalization

Because the underwater images are mostly unclear and dark, pixels in the histogram generally appear in the gray level region. Histogram equalization can make the pixel distribution relatively uniform and improve the contrast. The higher contrast and more prominent saliency area will make the better effect of saliency extraction. In this paper, the underwater images were equalized and then saliency extracted. The saliency detection algorithm is based on Principal Component Analysis (PCA) [20], which considers both pattern difference and color difference, extracts image saliency [21, 22], effectively avoids the error caused by the traditional K-nearest neighbor algorithm and reduces the background interference (see Fig. 3).

For an image, an image block centered at pixel point i is represented by Pi and the average tile is defined by $P_a$ [17, 23]:

$$p_a = \frac{1}{N} \sum_{i=1}^{N} p_i \tag{5}$$

Next, calculate the distance between a certain block Pi and the average block Pv based on the principal component direction $d(P_i, P_a)$ [17, 23]:

$$d(p_i, p_v) = \frac{1}{N} \times \sum_{i=1}^{N} p_i \times \sqrt{\frac{(i_x - v_x)^2 + (i_y - v_y)^2}{S_{(i,v)}}} \qquad (6)$$

The position coordinates of each block is represented by its central pixel point $P_i(i_x, i_y)$, the average image block position is represented by $P_a(a_x, a_y)$, and the variance between the two image blocks is represented by $S^2_{(i,a)}$.

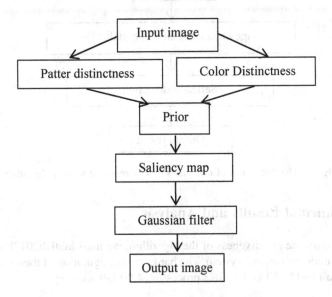

**Fig. 3.** PCA saliency detection process

When $d(P_i, P_a)$ is large, the block is a saliency block, otherwise, it is a normal block. From the mathematical theory, the detection of structural features is attributed to the $L_1$ norm under the PCA coordinate system, define the pattern feature $P(p_i)$ [17, 23]:

$$P(p_i) = ||p_i'||_1 \qquad (7)$$

Where $P_i$ is the coordinate of $P_i$ in PCA coordinate system. The $|| \cdot ||$ table demonstrates the arithmetic symbol of the $L_1$.

The first step of algorithm is to reduce the dimension of basic part with PCA and find the most representative feature as the basis for the saliency calculation. The image is divided into different regions by super pixel segmentation and the contrast between each region and the average value of the full image is calculated to obtain a saliency map Then, calculate the color contrast between the various areas under the Lab space. Add prior information, combine the morphological saliency map and the color saliency map to find the saliency region in the image, add a Gaussian filter, and highlight the saliency region to increase the brightness.

$$S(p_x) = G(p_x) \bullet P(p_x) \bullet C(p_x) \tag{8}$$

Combined with the improved histogram equalization algorithm and classification algorithm, the flow chart of the final saliency region detection algorithm is shown in Fig. 4.

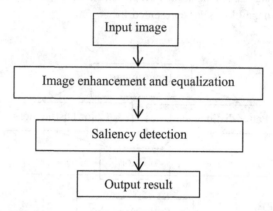

**Fig. 4.** The flow chart of the final saliency region detection algorithm

## 4 Experimental Results and Analysis

In order to verify the effectiveness of the algorithm, we used Matlab2012b to conduct experiments under windows7 system. The hardware configuration of the experiment is: Intel(R) Core i3-4150 CPU 3.5 GHz processor, 4.00 GB memory.

**Fig. 5.** UIEBD dataset

The UIEBD dataset [23, 24] was established by Tianjin University. It contains 890 original underwater images and corresponding high-quality reference images. The data is shown in Fig. 5. At the same time, DLOU_underwater dataset was also utilized, which was established by the Image Processing Research Group of Dalian Ocean University, containing 200 underwater images. The data is shown in Fig. 6.

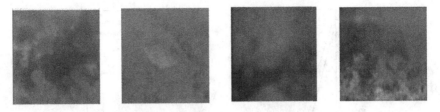

**Fig. 6.** DLOU_underwater dataset

In this paper, a comparison experiment between the original image and the experimental image was carried out:

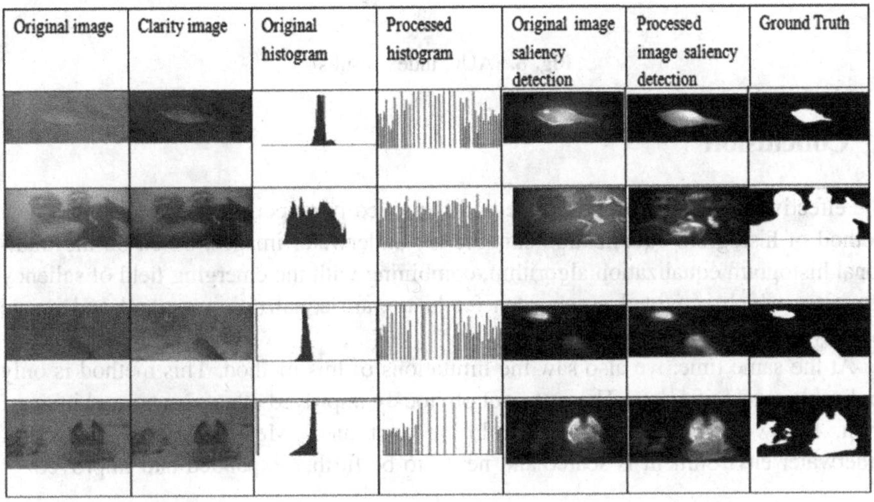

| Original image | Clarity image | Original histogram | Processed histogram | Original image saliency detection | Processed image saliency detection | Ground Truth |
|---|---|---|---|---|---|---|
| | | | | | | |
| | | | | | | |
| | | | | | | |
| | | | | | | |

**Fig. 7.** Experimental comparison

The experimental comparison chart (Fig. 7) includes original image, clarity image, original histogram, processed histogram and their saliency detection results. It is indicated that the algorithm proposed in this paper equalized the images, the effect of saliency detection is obviously improved, and the extracted information is more complete.

In order to verify the robustness and accuracy of the algorithm, AUC (Area-Under-The-Curve) was used as an evaluation index to evaluate the proposed algorithm (Fig. 8).

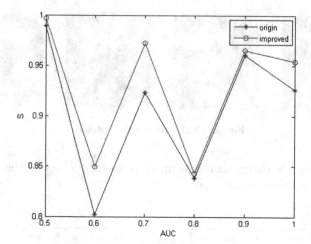

**Fig. 8.** AUC index results

## 5 Conclusion

The effectiveness of the proposed method is proved by experiments. We found a new method of histogram equalization suitable for underwater images, based on the traditional histogram equalization algorithm, combining with the emerging field of saliency detection, which provided a new idea for histogram equalization method and image saliency detection.

At the same time, we also saw the limitations of this method. This method is only applicable to specific data. There are not obviously improved effects on normal images. Then, follow-up research work should be continued. Meanwhile, the dataset of underwater environment is scarce and needs to be further expanded and improved.

**Acknowledgement.** This work is supported by Guangdong Province Key Laboratory of Popular High Performance Computers (SZU-GDPHPCL201805), Institute of Marine Industry Technology of Universities in Liaoning Province (2018-CY-34), China Postdoctoral Science Foundation (2018M640239).

## References

1. Wu, X.N., Wang, Y.J.: Cognitive and neurobiology models of visual attention. Adv. Psychol. Sci. **13**(3), 16–222 (1995)
2. Yang, L.E.: Saliency detection and application in complex scenes. Tianjin University, Tianjin (2016)
3. Shi, D.: A visual saliency tracking algorithm based on priori information. Microcomput. Appl. **35**(4), 46–49 (2016)
4. Datta, R.: Image retrieval: ideas, influences, and trends of the new age. ACM Comput. Surv. **40**(2), 1–60 (2008)

5. Shih, J.L., Lee, C.H.: An adult image identification system employing image retrieval technique. Pattern Recogn. Lett. **28**(16), 2367–2374 (2007)
6. Fowlkes, C.C., Arbelaez, M.P.: Contour detection and hierarchical image segmentation. IEEE Trans. Pattern Anal. Mach. Intell. **33**(5), 898–916 (2011)
7. Martin, D.R., Fowlkes, C.C.: Learning to detect natural image boundaries using local brightness, color, and texture cues. IEEE Trans. Pattern Anal. Mach. Intell. **26**(5), 530–549 (2004)
8. Papageorgiou, C.P., Oren, M.: A general framework for object detection. In: Sixth International Conference on Computer Vision (ICCV) (2002)
9. Cheng, M.M., Mitra, N.J.: Global contrast based salient region detection. IEEE Trans. Pattern Anal. Mach. Intell. **37**(3), 409–416 (2011)
10. Achanta, R., Estrada, F., Wils, P., Süsstrunk, S.: Salient region detection and segmentation. In: Gasteratos, A., Vincze, M., Tsotsos, J.K. (eds.) ICVS 2008. LNCS, vol. 5008, pp. 66–75. Springer, Heidelberg (2008). https://doi.org/10.1007/978-3-540-79547-6_7
11. Ran, M., Tal, A.: What makes a patch distinct. In: IEEE Conference on Computer Vision and Pattern Recognition, pp. 1139–1146 (2013)
12. Wang, B.: Research on the enhancement algorithms of under images. Ocean University of China, Qingdao(2008)
13. Wang, J.P., Li, J.: Development and prospect of image contrast enhancement. Electron Technol. **26**(5), 937–940 (2013)
14. Gonzalez, R.C., Wintz, P.: Digital Image Processing. Publishing House of Electronics Industry, Beijing (2007)
15. Jobson, D., Rahman, Z.: Properties and performance of a center/surround retinex. IEEE Trans. Image Process. **6**(3), 451–462 (1997). A Publication of the IEEE Signal Processing Society
16. Niu, H.M., Chen, X.J.: Image contrast enhancement based on wavelet transform and unsharp masking. High Technol. Lett. **21**(6), 600–606 (2011)
17. Wang, Y.L., Li, L.: Saliency detection based on hierarchical PCA technology. Comput. Sci. Appl. **008**(003), 398–409 (2018)
18. Acharya, T., Ray, A.K.: Image Processing: Principles and Applications, p. 610. Wiley, Hoboken (2005)
19. Wu, C.M.: Studies on mathematical model of histogram equalization. Acta Electron. Sin. **41**(3), 598–602 (2013)
20. Jin, H.L., Zhu, P.: The best thresholding on 2-D gray level histogram. Pattern Recogn. Artif. Intell. **3**, 83–87 (1999)
21. Margolin, R., Tal, A.: What makes a patch distinct. In: 2013 IEEE Conference on Computer Vision and Pattern Recognition, Portland (2013)
22. Ke, Y., Sukthankar, R.: A more distinctive representation for local image descriptors. In: IEEE Computer Society Conference on Computer Vision and Pattern Recognition, pp. 506–513. IEEE Computer Society (2004)
23. Wang, H., Wang, M.: Vision saliency detection of rail surface defects based on PCA model and color features. Process Autom. Instrum. **38**(1), 73–76 (2017)
24. Murthy, A.V., Karam, L.J.: A MATLAB-based framework for image and video quality evaluation. In: Second International Workshop on Quality of Multimedia Experience, pp. 242–247 (2010)
25. Li, C., Guo, C.: An underwater image enhancement benchmark dataset and beyond (2019)

# Kernelized Correlation Filter Target Tracking Algorithm Based on Saliency Feature Selection

Minghua Liu[1(✉)], Zhikao Ren[1], Chuansheng Wang[2],
and Xianlun Wang[2]

[1] College of Information Science and Technology,
Qingdao University of Science and Technology, Qingdao 266061, China
qustlmh@163.com
[2] College of Electromechanical Engineering,
Qingdao University of Science and Technology, Qingdao 266061, China

**Abstract.** To address the problem of using fixed feature and single apparent model which is difficult to adapt to the complex scenarios, a Kernelized correlation filter target tracking algorithm based on online saliency feature selection and fusion is proposed. It combined the correlation filter tracking framework and the salient feature model of the target. In the tracking process, the maximum Kernel correlation filter response values of different feature models were calculated respectively, and the response weights were dynamically set according to the saliency of different features. According to the filter response value, the final target position was obtained, which improves the target positioning accuracy. The target model was dynamically updated in an online manner based on the feature saliency measurement results. The experimental results show that the proposed method can effectively utilize the distinctive feature fusion to improve the tracking effect in complex environments.

**Keywords:** Kernel correlation filter · Feature selection ·
Patch-based target tracking · Saliency detection

## 1 Introduction

Visual target tracking aims to determine the location of the tracking target in the video or image sequence. It is one of the most important tasks in computer vision that finds numerous applications such as video surveillance, motion analysis, pattern recognition, image processing. It is also the core technology of video analysis and understanding, in intelligent video surveillance, Intelligent robots, unmanned aerial vehicles, human-computer interaction.

Although the research on visual tracking algorithms has made great progress in recent years [1–14], due to the interference of many complicated factors in the actual tracking scene, the design of stable and reliable target tracking system still faces great challenges.

The visual target tracking mainly includes two stages, the description of the target model and the target positioning. Choosing the right features to establish an accurate description of the target model is the key to determining the success of the tracking

© Springer Nature Singapore Pte Ltd. 2019
R. Mao et al. (Eds.): ICPCSEE 2019, CCIS 1059, pp. 166–178, 2019.
https://doi.org/10.1007/978-981-15-0121-0_13

algorithm. At present, the tracking algorithm can be divided into the generative model [3–5] and the discriminative model [6, 7]. The generative model is a processing method based on pattern matching. Firstly, the prior feature model of the target is constructed, and the final position of the target is obtained by searching the region that best matches the target template. For example, Ross et al. [3] proposed the Incremental Visual Tracking (IVT) algorithm, which uses low-dimensional subspace learning to model the apparent appearance of the target, and locates the tracking target according to the size of the candidate target and the center distance of the feature subspace. He et al. [4] proposed a Local Sensitive Histogram (LSH) target model description method, based on this model, a multi-region target tracking algorithm was implemented.

Discriminative model is a method based on target detection. The basic idea is to construct a discriminative model that can distinguish between target and background. Determine the target location by training and updating the online classifier to accurately classify the target from the target and background areas to be classified. For example, the Multiple Instance Learning (MIL) algorithm proposed by Babenko et al. [6] and the Tracking-Learning-Detection (TLD) algorithm proposed by Kalal et al. [7] belong to the classical discriminative model.

In recent years, the kernel correlation filtering target tracking algorithm based on the discriminative model has received wide attention from researchers because of its extremely high tracking efficiency [1, 2, 8–12]. For example, Bolme et al. [8] applied the correlation filter ideas to the target tracking domain for the first time, and proposed a correlation filter tracking algorithm with the goal of the minimum mean square errors. Henriques et al. [9] proposed a circulant structure of tracking with kernels (CSK) to locate targets by a kernel regularized least squares classifier. Henriques et al. [10] proposed a kernelized correlation filter (KCF) target tracking algorithm, which provides a more accurate description of the target model and obtains more higher tracking accuracy and robustness, by extended the single channel operation of CSK algorithm into multiple channels, and introduced HOG feature instead of gray.

Regardless of which target model is used, appearance of the target model will change significantly during the tracking process. If a fixed feature is used to create a single target apparent model, it will easily cause tracking drift or even loss of target. To this end, we propose a target tracking method based on saliency feature selection under the framework of kernel correlation filter target tracking. In the tracking process, a feature saliency detection mechanism is introduced to dynamically measure the saliency of different features and adaptively select saliency. Constructing target model using the feature of high saliency can effectively distinguish the target and background, and separate the target from the complex background, further improving the performance of the tracking based on the kernel correlation filter algorithm in complex environments.

The key benefits of the proposed method are (i) In the tracking process, the feature saliency measurement mechanism is introduced, the saliency weights of different features are dynamically calculated, and the feature with high saliency is adaptively selected to construct the target model, so that the target model has stronger adaptability to complex scenes; (ii) Under the correlation filter framework, the kernel correlation filter output response values of different feature target models are calculated, and the output saliency weights are used to weight the output response values to increase the contribution of the saliency feature model to the target position prediction. The target location will be more reliable; (iii) Adopting a dynamic update mode for the target model, selecting only the feature model with higher significance in each frame for updating, and applying it to the adaptive training and updating process of the classifier, effectively avoiding the tracking drift problem caused by using the single fixed feature representation model. The experimental results show that the proposed algorithm can adapt to the dynamic changes of targets and scenes effectively, and effectively improve the robustness of tracking in complex environments.

The rest of the paper is organized as follows. Section 2 introduces the feature saliency measure, Sect. 3 describes the kernelized correlation filter, the proposed method is presented in Sect. 4, experimental results are presented in Sect. 5. Finally, Sect. 6 summarizes the main conclusions from this work.

## 2 Feature Saliency Measure

During the tracking process of visual target, the target models established by different features (such as Color, Edge, and HOG) have different saliency and resolvability relative to the surrounding background and are dynamically changed. The visual saliency feature is a kind of feature that visually expresses the difference between the target area and the background area, and the visual saliency of the features which can distinguish the target and background area effectively is high, and vice versa. During the tracking process, the features with higher saliency should be selected. Hereby the target model established has better target-background separability, and the result of target location will be more accurate.

Defining the feature saliency $s_t^f$ is the difference between the target model and its surrounding background region at the $t$ frame, based on the feature $f$, it is measured by Kullback-Leibler relative entropy

$$s_t^f = K\_L(h_r, h_b) = \sum_i h_r(i) log \frac{h_r(i)}{h_b(i)} \tag{1}$$

where, $h_r(\cdot)$ and $h_b(\cdot)$ are the target histogram and the surrounding background area histogram respectively.

For the target model $P_t^{(f)} = \left(d_y^{(f)}, w, h\right)$ established based on feature $f$, the surrounding background area sampling range is $B_t^{(f)} = \left(d_y^{(f)}, \lambda_s w, \lambda_s h\right)$, where, $d_y^{(f)}$ is the sampling center, $\lambda_s$ is the parameters of sampling range control, we set $\lambda_s = 2$ in this paper.

In order to avoid the interference of background information and occlusion, further improve the robustness of the target model, we firstly perform superpixel segmentation on the target model, and then constructs the multi-feature representation of the segmented superpixel model to measure the saliency of its multi-feature model. Select a highly salient sub-block model to participate in the tracking and positioning of the target. Figure 1 shows a salient diagram of the different feature sub-block models, where (a), (c) is the target model and its sub-block model, (c), (d) is the corresponding salient map.

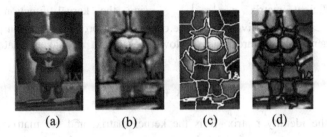

(a)        (b)        (c)        (d)

**Fig. 1.** The salient map of target and its sub-block model

## 3   Kernelized Correlation Filter

The Kernelized Correlation Filter (KCF) target tracking algorithm solves the tracking problem by learning the Kernelized Regularization Least Squares (KRLS) linear classifier. The algorithm introduces a dense sampling strategy, and all the training samples are obtained by cyclically shifting the standard target samples. At the domain space, locates the target by calculating the Gaussian kernel function response values of all training samples. The algorithm mainly includes three parts: classifier training, target detection and classifier adaptive update.

### 3.1   Classifier Training

First, the target image block $X_{0,0}$ of size $M * N$ is obtained, and the target image block is used as a reference sample, and all the shifted image blocks $X_{i,j}$, $i \in [0, M], j \in [0, N]$ are obtained as a training sample by cyclically shifting of $X_{0,0}$. A plurality of feature vectors for each training sample are extracted to characterize the image block.

For all training samples, KCF performs target tracking by learning a kernel regularized least squares (KRLS) linear classifier. The RLS classifier is to find the weight $w$ to minimize the mean square error of the response output and expected value of the classification function $f = w^T x$ for all training samples, that is

$$\min_{w} \left( \sum_{i,j} \left( f(x_{i,j}) - y_{i,j} \right)^2 + \lambda \|w\|^2 \right) \tag{2}$$

Where, $f$ is the classification function, $\lambda$ is the regularization parameter to prevent overfitting. In practical applications, mapping the feature space to a higher-dimensional space by nonlinear mapping can obtain better classification performance. Assuming the mapping function is $\varphi(x_{i,j})$, the weight vector of the classifier becomes

$$W = \sum_{i,j} \alpha_{i,j} \varphi\left(x_{i,j}\right) \tag{3}$$

At this time, the solution of the optimization objective function is transformed into the weight coefficient matrix $\alpha = \{\alpha_{i,j}\}_{M*N}$, using the kernel function $k(p,q) = <\varphi(p), \varphi(q)>$ can effectively solve the complex dot product operation problem of the high-dimensional feature matrix. Now, the weight coefficient matrix can be expressed as

$$\alpha = (K + \lambda I)^{-1} y \tag{4}$$

Where $I$ is the identity matrix, $K$ is the kernel matrix, and the matrix element is $K_{i,j} = k(x_i, x_j)$, $y$ is a regression label matrix composed of label elements $y_{i,j}$, $<$, $>$ represents the dot product operation. According to the dense sampling strategy of literature [10], the circulant matrix characteristics of $K$ can be used to transform into the Fourier frequency domain for the fast calculation of $\alpha$

$$\hat{\alpha} = \frac{\hat{y}}{\hat{k}(x_{0,0}, x_{0,0}) + \lambda} \tag{5}$$

Where, $x(0,0)$ is the feature matrix of the reference target sample, the kernel correlation vector $\hat{k}(x_{0,0}, x_{0,0})$ representing the first row of the kernel matrix, contains information of all training samples, and the symbol ^ represents the discrete Fourier transform of the matrix.

## 3.2    Target Detection

When a new frame arrives, the trained classifier is used to predict the position of the tracking target. The specific method is: the center of the tracking target detected in the previous frame is centered, and the area with the size $M * N$ is extracted as the reference candidate target, and the cyclic shift is performed. The obtained region is used as a candidate target sample, and the response output of the classifier is tested for all the test samples in the candidate region, and the position where the largest response output is found as the new position of the moving target. For a given single test sample, the response output of the classifier is

$$f(z) = w^T z = \sum_{i,j}^{M,N} k\left(z, x_{i,j}\right) \tag{6}$$

Since all samples need to be calculated, the calculation speed of Eq. (6) is very slow. The dense sampling and fast algorithm of literature [8–10] is used to perform the Fourier domain kernel correlation operation on a single test sample, and all test samples are calculated at one time. The classifier response output is

$$\hat{f}(z) = \hat{k}(z, x_{i,j}) \odot \hat{\alpha} \tag{7}$$

Where $\odot$ is the point multiplication operation, by Eq. (7), the position of the maximum response value of all sample response vectors $\hat{f}(z)$ is detected as the predicted position of the current frame target.

### 3.3 Classifier Update

In the tracking process, the target apparent model will be affected by the environment, and the target model needs to be dynamically updated to adapt to the apparent changes of the target. The update process mainly includes two parts: the apparent model of the target $\hat{x}$ and the matrix of the classifier coefficients $\hat{\alpha}$. The update method is as follows

$$\hat{x}_t = (1 - \eta)\hat{x}_{t-1} + \eta\hat{x} \tag{8}$$

$$\hat{\alpha}_t = (1 - \eta)\hat{\alpha}_{t-1} + \eta\hat{\alpha} \tag{9}$$

Where $t$ is the frame number, $\eta$ is the learning rate, $\hat{x}$ is the target apparent model for the predicted position, and $\hat{\alpha}$ is the classifier coefficient matrix obtained by the Eq. (5) for the training sample of the predicted position.

## 4  Proposed Method

### 4.1  Correlation Filter of Fusion Target Saliency Features

Firstly, the input target and its surrounding background area are superpixel segmented to obtain the superpixel sub-block model $\{SP_i\}_{i=1...N_0}$ of the target, and for each sub-block $SP_i$, its saliency $s_i^f$ is calculated based on the feature $f$ according to formula (1) (the color, texture, edge and HOG features are selected in this paper). According to the specified threshold, for each sub-block, the two higher value of $s_i^{f_j}(j = 1, 2)$ feature models are selected to participate in the target tracking framework of the correlation filter.

Assume that at the $t$ moment, during the training process of the classifier, the maximum response value $\max \hat{f}(z_{t,i}(f_j))$ obtained for all training samples corresponding to the feature model $z_i(f_j)$, then the filter response output after combining the feature saliency weights is

$$p_t = \sum_j \delta_j \max \hat{f}(z_{t,i}(f_j)) \tag{10}$$

Where, $\delta_j$ is the feature saliency weight of the corresponding sub-block model $z_i(f_j)$, and the calculation method is as follows

$$\delta_j = \frac{s_i^{f_j}}{\sum_j s_i^{f_j}} \tag{11}$$

By using Eqs. (10) and (11), the filter response value of sub-block model with higher saliency is given higher weight, and the high-saliency feature sub-block model is fused to locate the target under the correlation filter framework possessing higher tracking accuracy.

## 4.2  Target Model Update

In a complex scenario, the tracking target is prone to occlusion. If the target model is updated without distinction, it is easy to cause tracking drift. Therefore, it is necessary to detect the occlusion and use different update strategy with different target sub-block models accordingly. In this paper, each target sub-block is handled from the perspective of the change of the maximum response value of the correlation filter.

**Occlusion Detection.** It is assumed that at time $t$, the target sub-block model to be tested is $z_t(i)$ corresponding to the superpixel sub-block model $\{SP_i\}_{i=1...N_0}$ which can be performed occlusion determination by measuring the Maximum Response Variation (MRV) of the maximum response value, which is defined as follows

$$MRV = \frac{\max\left(\hat{f}(z_t(i))\right)}{\frac{1}{t-1}\sum_{n=1}^{t-1}\max\left(\hat{f}(z_t(i))\right)} \tag{12}$$

Where, the parameter MRV represents a ratio of the maximum response value of the sub-block $z_t(i)$ at time $t$ to the mean value of the maximum response value of the previous time $1...t-1, \hat{f}(z_t(i))$ represents the response value of all candidate target sub-blocks $z_t(i)$ at time $t$. Generally, when the target sub-block is occluded the maximum response value of the sub-block will be reduced significantly. When $MRV < T_o$, the target sub-block is considered to be occluded, $T_o$ is determined as an occlusion threshold, and dynamically adjusted according to the scene change.

**Model Update.** We adopt an asynchronous update strategy for different sub-block models. First, occlusion detection is performed on each sub-block, the sub-blocks under the occlusion state are not updated, otherwise, each target sub-block is updated according to formulas (13) and (14)

$$\hat{x}_t(i) = (1-\eta)\hat{x}_{t-1}(i) + \eta\hat{x}(i) \tag{13}$$

$$\hat{\alpha}_t(i) = (1-\eta)\hat{\alpha}_{t-1}(i) + \eta\hat{\alpha}(i) \tag{14}$$

Where $\hat{x}_t(i)$ and $\hat{\alpha}_t(i)$ are the target sub-block model and the corresponding classifier coefficient matrix at time $t$, and $\eta$ is the update factor, which controls the target sub-block updating rate.

### 4.3    Tracking Algorithm Flow

The specific process of the proposed algorithm is as follows:
Begin:

(1)  Initialization: $t = 1$

The tracking target is selected, superpixel segmentation is performed, the super-pixel sub-block model is obtained as $\{SP_i\}_{i=1...N_0}$, the multi-feature representation of the sub-block model is constructed, the saliency $s_i^f$ based on different features $f$ is calculated, and the two higher saliency values are saved and its fusion weights are calculated using Eq. (11).

(2)  Tracking process: $t = 2, 3, \cdots$

(a)  Input the two target sub-block models with higher saliency as the reference samples into the correlation filter framework, and obtain the maximum response value respectively. Determine the filter response output after the feature saliency weight fusion using the formula (10);

(b)  According to the result of the current frame target positioning, the occlusion judgment is performed by the Eq. (12), and the model update is performed by the Eqs. (13) and (14) for the not be occluded sub-block model.

End.

## 5    Analysis of Experimental Results

### 5.1    Experimental Setup

In order to verify the superiority of the proposed algorithm, the I7-8550 2.0 GHz CPU and 16G memory notebook are used to implement the algorithm in VS2015 +OPENCV3.1.0 platform. The KCF algorithm [10] framework is selected to implement kernel correlation filter tracking, and the SLIC algorithm [15] is used to achieve superpixel segmentation. The tracking target is automatically calibrated in the first frame, the update factor $\eta$ is set $\eta = 0.03$, the occlusion threshold $T_o$ is set $T_o = 0.6$.

We select the six video sequences commonly used in the target tracking data set for testing [16]. These sequences cover the main challenges of target tracking, including occlusion, complex background and similar target interference, fast target movement, target pose, scale and illumination changes etc. Table 1 gives a detailed description of the characteristics of the selection test sequence, where Occ DF Clu FM SV and IV is abbreviation of occlusion deformation clusters fast-motion and scale-variation respectively.

Three kinds of correlation filter target tracking algorithms (KCF [10], Staple [12], CN [13]) and the Distractor Aware Tracking algorithm (DAT) [14] with excellent performance are selected to compare qualitatively and quantitatively with the proposed algorithm.

**Table 1.** Experimental sequences.

| Sequences | Resolution | Length | Object | Description |
|---|---|---|---|---|
| Basketball | 576*432 | 725 | 34*81 | Occ, DF, Clu |
| Lemming | 640*480 | 1336 | 61*103 | Occ, FM, SV |
| Faceocc1 | 352*288 | 892 | 114*162 | Occ |
| Woman | 352*288 | 597 | 21*95 | Occ, IV |
| Tiger2 | 640*480 | 365 | 68*78 | Occ, Clu, FM |
| Car2 | 320*240 | 913 | 64*52 | FM, IV |

## 5.2   Qualitative Analysis

In this paper, the effectiveness of the proposed algorithm is analyzed from the aspects of occlusion, target deformation, similar targets and complex background interference, and the algorithm is compared with the above method. Figure 2 (a)–(f) shows the comparison of tracking results on the frame numbers of corresponding test sequences.

Sequence (a) shows the tracking results of fast-moving targets under many similar targets and complex background interference conditions. There are multiple consecutive occlusion situations in the tracking process; sequence (b) shows the tracking results for fast moving targets, tracking There are long-term occlusion and target scale changes in the process; sequence (c) shows the tracking results of continuous occlusion of different directions of the tracking target; sequence (d) shows the tracking of the vehicle continuously occluded during pedestrian walking. The effect; sequence (e) shows the tracking effect on fast moving toy tigers under continuous occlusion conditions; sequence (f) shows the tracking of fast moving cars.

It can be seen from the tracking results that the contrast tracking algorithm shows better tracking performance on different tracking sequences, but under complex tracking conditions, these methods appear different levels of tracking offset or missing targets, for example: The Staple and DAT method shows strong capture ability for fast moving objects, but it is easy to be disturbed by local similar targets and occlusion interference. CNT and DAT methods have better tracking result on similar target interference and target deformation and illumination changes, but in the case of continuous occlusion, the tracking robustness is low; the KCF algorithm is sensitive to occlusion and target deformation, but displayed the better tracking result on rigid targets. By dynamically merging the target sub-block model with high saliency, the proposed algorithm performs target localization under the correlation filter framework, which can effectively resist the interference of continuous occlusion, similar targets and complex background. In complex scenarios, the proposed algorithm shows stronger robustness compared with other algorithms.

(a) Basketball sequence:77,198,400,619,723

(b) Lemming sequence:23,131,326,500,737

(c) Faceocc1 sequence:37,170,407,600,888

(d) Woman sequence:31,157,306,420,533

(e) Tiger2 sequence:17,100,200,298,365

(f) Car2 sequence:29,173,458,647,799

——— Ours    ——— Staple    ——— KCF    ——— CN    ——— DAT

**Fig. 2.** Comparison of tracking results on different sequence

## 5.3 Quantitative Analysis

In this paper, the average center point error and the tracking accuracy rate are used to quantitatively compare and track the tracking results. The tracking accuracy rate (*TA*) and the average overlap rate (*AOR*) given by the PASCAL VOC [17] are used as the evaluation indicators for tracking success. When *AOR* is greater than the threshold the specified threshold, the tracking is considered successful. In this paper, we select $AOR > 0.5$, Assuming $R$ is the tracking result box, $T$ is set to the real target bounding box. $M$ is the total number of frames in the test sequence, $M_0$ is the number of successfully tracked frames, then the *TA* and *AOR* are calculated as follows

$$AOR = \frac{R \cap T}{R \cup T} \tag{15}$$

$$TA = \frac{M_0}{M} \tag{16}$$

The tracking correct rate (*TA*) describes the number of successfully tracked frames in the video sequence. The larger the value, the more frames are successfully tracked, and the accuracy of the tracking result is higher. The average center point error reflects the mean of the deviation between the tracking result and the actual target center point. The smaller the value, the higher the tracking accuracy. Table 2 gives a comparison of the tracking accuracy of the algorithm and other methods. Table 3 gives the comparison of the average center point error between the proposed algorithm and other methods (red is the best result; blue is the second-best result).

**Table 2.** Comparison of *TA*.

| Sequences | Proposed | Staple | KCF | CN | DAT |
|---|---|---|---|---|---|
| Basketball | 0.90 | 0.88 | 0.89 | 0.79 | 0.96 |
| Lemming | 0.95 | 0.27 | 0.96 | 0.31 | 0.91 |
| Faceocc1 | 1.00 | 1.00 | 1.00 | 0.90 | 0.10 |
| Woman | 0.96 | 0.94 | 0.48 | 0.26 | 0.21 |
| Tiger2 | 0.97 | 0.89 | 0.96 | 0.64 | 0.62 |
| Car2 | 0.94 | 1.00 | 1.00 | 1.00 | 0.07 |

**Table 3.** Comparison of the average center location errors.

| Sequences | Proposed | Staple | KCF | CN | DAT |
|---|---|---|---|---|---|
| Basketball | 10.7 | 19.1 | 40.3 | 15.3 | 7.2 |
| Lemming | 9.2 | 118.0 | 9.7 | 90.7 | 22.3 |
| Faceocc1 | 5.4 | 19.7 | 5.4 | 13.2 | 97.9 |
| Woman | 5.7 | 6.4 | 27.9 | 89.0 | 58.5 |
| Tiger2 | 10.1 | 10.3 | 13.5 | 18.3 | 14.3 |
| Car2 | 7.8 | 3.5 | 3.6 | 2.1 | 57.2 |

# 6 Summary

We proposed a target tracking algorithm based on the correlation filter framework and saliency feature selection in this paper. The method performs superpixel segmentation on the target area, constructing multi-feature descriptions of target sub-block models, and calculating the saliency of different feature sub-block models to establish a feature saliency weight measurement mechanism, under the correlation filter framework, the sub-blocks with higher weights are combined to calculate the filter response value, and the response result of the higher saliency feature model is assigned higher weights and the results of target location are more accurate. To further improve the robustness of tracking, the proposed algorithm provides effective target occlusion detection and model update mechanism, which adopts the method of asynchronous update of the target sub-block, by measuring the change of the highest filter response value of the target sub-block, timely determine which target sub-block models need to be updated, and ensure that the model update can adapt well to the target local changes and occlusion effectively. Experimental results on a variety of complexities scenario show that the proposed algorithm is more robust.

**Acknowledgments.** This work was partially supported by the National Natural Science Foundation (61472196, 61672305); Natural Science Foundation of Shandong Province (BS2015DX010, ZR2015FM012) and Key Research and Development Foundation of Shandong Province (2017GGX10133).

# References

1. Zhang, T.Z., Xu, C.S., Yang, M.H.: Learning multi-task correlation particle filters for visual tracking. IEEE Transact. Pattern Anal. Mach. Intell. **41**(2), 365–378 (2019)
2. Ma, C., Yang, X., Zhang, C.: Long-term correlation tracking. In: IEEE Conference on Computer Vision and Pattern Recognition (CVPR), pp. 5388–5396. IEEE (2015)
3. Ross, D.A., Lim, J., Lin, R.S., Yang, M.H.: Incremental Learning for Robust Visual Tracking. Adv. Neural. Inf. Process. Syst. **77**(1–3), 125–141 (2008)
4. Kwon, J., Lee, K.M.: Visual Tracking Decomposition. In: IEEE Conference on Computer Vision and Pattern Recognition (CVPR), pp. 1269–1276. IEEE (2010)
5. He, S., Yang, Q., Lau, R.W.H.: Visual tracking via locality sensitive histograms. In: IEEE Conference on Computer Vision and Pattern Recognition (CVPR), pp. 2427–2434. IEEE (2013)
6. Babenko, B., Yang, M.H., Belongie, S.: Robust object tracking with online multiple instance learning. IEEE Transact. Pattern Anal. Mach. Intell. **33**(8), 1619–1632 (2011)
7. Kalal, Z., Mikolajczyk, K., Matas, J.: Tracking-learning-detection. IEEE Transact. Pattern Anal. Mach. Intell. **34**(7), 1409–1422 (2011)
8. Bolme, D.S., Beveridge, J. R., Draper, B.A.: Visual object tracking using adaptive correlation filters. In: IEEE Conference on Computer Vision and Pattern Recognition, pp. 2544–2550. IEEE (2010)
9. Henriques, João F., Caseiro, R., Martins, P., Batista, J.: Exploiting the circulant structure of tracking-by-detection with kernels. In: Fitzgibbon, A., Lazebnik, S., Perona, P., Sato, Y., Schmid, C. (eds.) ECCV 2012. LNCS, vol. 7575, pp. 702–715. Springer, Heidelberg (2012). https://doi.org/10.1007/978-3-642-33765-9_50

10. Henriques, J.F., Caseiro, R., Martins, P., et al.: High-speed tracking with kernelized correlation filters. IEEE Transact. Pattern Anal. Mach. Intell. **37**(3), 583–596 (2015)
11. Ning, W., Wengang, Z., Qi, T.: Multi-cue correlation filters for robust visual tracking. In: IEEE Conference on Computer Vision and Pattern Recognition, pp. 4844–4853. IEEE (2018)
12. Bertinetto, L., Valmadre, J., Golodetz, S.: Staple: complementary learners for real-time tracking. In: IEEE Conference on Computer Vision and Pattern Recognition, pp. 1401–1409. IEEE (2016)
13. Danelljan, M., Shahbaz, K.F., Felsberg, M.: Adaptive color attributes for real-time visual tracking. In: IEEE Conference on Computer Vision and Pattern Recognition, pp. 1090–1097. IEEE (2014)
14. Possegger, H., Mauthner, T., Bischof, H.: In defense of color-based model-free tracking. In: IEEE Conference on Computer Vision and Pattern Recognition, pp. 2113–2120. IEEE (2015)
15. Radhakrishna, A., Appu, S., Kevin, S.: SLIC superpixels compared to state-of-the-art superpixel methods. IEEE Transact. Pattern Anal. Mach. Intell. **31**(7), 1195–1209 (2009)
16. Wu, Y., Lim, J., Yang, M. H.: Online object tracking: a benchmark. In: IEEE Conference on Computer Vision and Pattern Recognition, pp. 2411–2418. IEEE (2013)
17. Everingham, M., Gool, L.V., Williams, C.K.I.: The pascal visual object classes (VOC) challenge. Int. J. Comput. Vis. **88**(2), 303–338 (2010)

# Image Edge Detection Method Based on Ant Colony Algorithm

Qirong Lu[1], Qianmin Liang[2(✉)], Jiqiu Chen[2], and Jiwei Xia[2]

[1] Modern Education Technology Center, Guilin University of Technology,
Guangxi 541004, China
[2] College of Mechanical and Control Engineering,
Guilin University of Technology, Guangxi 541004, China
664159671@qq.com

**Abstract.** Ant colony algorithm has good results in finding the optimal solution in a certain field; and image edge detection is an essential foundation for all kinds of image processing. How to improve image edge detection becomes a hot topic in image processing. In this paper, the ant colony algorithm is applied to image edge detection, and the ant colony algorithm's discreteness, parallelism and positive feedback are fully utilized. Through repeated iteration, pheromone acquisition and pheromone matrix were continuously updated to search for images step by step. The experimental results show that the ant colony algorithm can effectively detect the edge of the image, and the detection effect of the algorithm is significantly improved compared with the Roberts algorithm.

**Keywords:** Ant colony optimization · Edge detection · Pheromones · Pheromone matrix

## 1 Preface

The edge of the image is one of the most important features of the image, it is the basis of image segmentation, target region recognition, region shape extraction, etc. The edge of the image often shows local range grayscale changes, and most algorithms use this feature of the image edge to edge recognition. The common methods for image edge detection are Robert operator, Sobel operator [1], Prewitt operator, Canny operator, etc. But these methods are only effective for images with high definition, and cannot suppress image noise [2]. While the method used in this paper is ant colony algorithm. It is a random search method with the characteristics of discreteness, parallelism, robustness and positive feedback. This method has a good effect in finding the optimal solution in a certain field, such as business travel problems, workshop task scheduling, etc. Image edge detection can also be seen as the process of finding the "optimal solution".

© Springer Nature Singapore Pte Ltd. 2019
R. Mao et al. (Eds.): ICPCSEE 2019, CCIS 1059, pp. 179–185, 2019.
https://doi.org/10.1007/978-981-15-0121-0_14

# 2 Ant Colony Algorithm

## 2.1 Bionics Foundation

The method used by ants to search for food is teamwork, they always can find out an optimal path to get food. The acquisition of the optimal path is due to a special pheromone. Ant released the pheromone when they searching for the path [3]. The optimal path is founded by a large number of ants through many tests. When an ant passes a new intersection, it randomly selects a path to advance and releases the pheromone. The concentration of the pheromone is determined by the length of the path. If the path is shorter than others, the pheromone concentration released by ants will be higher, and vice versa. The following ants can select the path by the concentration of pheromone. The probability that the ant chooses a path with high concentrations of pheromone is much higher than the path with low concentrations. This action will form a positive feedback, then the hormone concentration on the optimal path will be getting higher. Finally, the entire ant colony can use this method to weed out the far foraging route and find the optimal path. This foraging process of ants show the following two characteristics for us. First characteristic is diversity: each ant in the ant colony has different routes, their initial direction is random, and the coverage of routes is wide. All these make this method have diversity [4]. Second characteristic is positive feedback: the better foraging path is constantly choosing by ants; and the probability of its selection is increased because of the concentration of pheromone become higher.

## 2.2 Ant Colony Algorithm Mathematical Description and Implementation Steps

Ant colony algorithm is used to find the optimal solution of the target problem by repeating the search step over and over again. It searches the solution space of target problem by building up pheromone information [5]. Suppose K ants are used to find the optimal solution in the solution space consisting of N1 × N2 nodes. The specific optimization steps can be summarized as follows:

(1) Initialize the starting position and pheromone matrix $\tau^{(0)}$ of K ants.
(2) Loop execution step (the number of loops N is determined by setting the maximum number of iterations).
   ① According to the probability conversion matrix $p^{(n)}$ (n is the number of executions, and the matrix is N1N2 × N1N2-dimensional.), the kth ant is continuously taken L steps ($1 \leq k \leq K$). The probability conversion matrix $p^{(n)}$ is as shown in Eq. (1).

$$p_{i,j}^{(n)} = \frac{(\tau_{i,j}^{(n-1)})^{\alpha}(\eta_{i,j})^{\beta}}{\sum_{j \in \Omega_i} (\tau_{i,j}^{(n-1)})^{\alpha}(\eta_{i,j})^{\beta}}, if \ j \ \in \Omega_i \tag{1}$$

$\tau_{i,j}^{(n-1)}$ in Eq. (1) is the radian pheromone value of connecting node i and node j; ant $a_k$ is at node i, $\Omega_i$ is its neighbor node; $\alpha$ and $\beta$ are constants, $\alpha$ represents the importance of the pheromone information, and $\beta$ represents the importance of other information; $\eta_{i,j}$ represents the heuristic information from node i to node j, and its value is fixed for each execution step.

② The pheromone matrix $\tau^{(n)}$ is updated on each loop, and the pheromone matrix is updated twice in the ant colony optimization process. The first update occurs after each ant moves in each execution step. The update formula is as shown in Eq. 2.

$$\tau_{i,j}^{(n-1)} = \begin{cases} (1-\rho) \bullet \tau_{i,j}^{(n-1)} + \rho \bullet \Delta_{i,j}^{(k)}, (i,j) \in U \\ \tau_{i,j}^{(n-1)}, others \end{cases} \tag{2}$$

In Eq. (2), $\rho$ represents the evaporation rate; U represents the point set on the optimal path.

The second update occurs after all ants move in each execution step, and the update formula is as shown in Eq. 3.

$$\tau^{(n)} = (1-\psi) \bullet \tau^{(n-1)} + \psi \bullet \tau^{(0)} \tag{3}$$

In Eq. (3), $\psi$ is the pheromone attenuation coefficient.

(3)  Calculate the result using the final pheromone matrix $\tau^{(N)}$.

## 3    Ant Colony Algorithm

The ant colony algorithm has been introduced in the previous section is applied to image edge detection in this section. In other words, It means the target image will convert to undirected graph, and each pixel in this picture will be treated as a node [6]. Then, the "ant" is randomly placed on a node, and the field of the pixel points adopts an 8-connected field (see in Fig. 1).

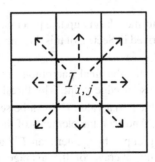

**Fig. 1.** 8-connected field of pixels. $I_{i,j}$ indicates the current position coordinates of the ant.

Then, the selection of the next crawling node is based on the pheromone intensity α and the heuristic guiding function of the pixels in the field. At the same time, the "ant" releases the corresponding pheromone in the node. By repeating this step over and over again, most ants concentrate on the edge of the image. Then image edge detection is completed.

## 4  Algorithm Implementation Steps

(1) Initialization process

$K$ ants are randomly distributed on the image I of size $M_1 \times M_2$ and each pixel of the image can be regarded as one node. Then, set the initial value of the pheromone matrix $\tau_{i,j}^{(n-1)}$.

(2) Implementation process

During the execution of the nth step, one of the K ants which is randomly selected is continuously moved L steps on the image. Its movement follows the definition of Eq. 4.

$$p_{(l,m),(i,j)}^{(n)} = \frac{(\tau_{i,j}^{(n-1)})^\alpha (\eta_{i,j})^\beta}{\sum_{(i,j)\in\Omega_{(l,m)}} (\tau_{i,j}^{(n-1)})^\alpha (\eta_{i,j})^\beta} \tag{4}$$

In Eq. 4, $\Omega_{(l,m)}$ represents a point adjacent to the node $(l, m)$.

In the execution process, the allowable range of ant motion adopts 8- connected fields, and the determination of heuristic information is determined by the local statistic at pixel point $(i, j)$. The definition formula is as shown in Eq. 5.

$$\eta_{i,j} = \frac{1}{Z} V_C(I_{i,j}) \tag{5}$$

In Eq. 5, $Z = \sum_{i=1:M_1} \sum_{j=i:M_2} V_C(I_{i,j})$, Z is a normalization factor; $I_{i,j}$ is the intensity value of the pixel at $(i, j)$; $V_C(I_{i,j})$ is a function of a set of local pixels which is called a "pixel group". The value of the function is determined by the change in image intensity of the pixel group C.

(3) Update process

Two updates need be performed during update process. This step is the same as the previous step which described in Sect. 2 of this document, so it is not repeated here.

(4) Decision-making process

By using the threshold T for the final pheromone matrix $\tau^{(N)}$, it is possible to judge a pixel whether It is an edge point. The threshold T is calculated by the following method. First, the initial value $T^{(0)}$ can be selected as the average value of the pheromone matrix. And then the elements of the pheromone matrix can be divided into two groups, grouped by greater than $T^{(0)}$ and less than $T^{(0)}$. Then the new threshold value is the average of the average values of the two sets of elements. This process is repeated until the threshold no longer changes [7]. Finally, the following formula 6 is used as a reference to judge whether the pixel position $(i, j)$ is an edge.

$$E_{i,j} = \begin{cases} 1, \tau_{i,j}^{(N)} \geq T^{(l)} \\ 0, \text{others} \end{cases} \tag{6}$$

If the pixel position $(i, j)$ is the edge the value of $E_{i,j}$ is 1; otherwise it is 0.

## 5  Experimental Results and Analysis

This paper uses MATLAB software to verify the effect of ant colony algorithm in image edge detection. All experimental simulations in this paper are done in MATLAB (R2016b).

The experimental image is Animal (see in Fig. 2), the image size is $128 \times 128$ pixels, and the experimental parameter values are set as $\alpha = 1.2$, $\beta = 0.2$, $\rho = 0.1$, $\psi = 0.05$, N = 300.

**Fig. 2.**  Animal

The experimental result is shown in Fig. 3. In order to prove the superiority of this algorithm, Roberts algorithm is used to detect the edge of the same image. The result is shown in Fig. 4.

**Fig. 3.**  Ant colony algorithm result

**Fig. 4.** Roberts algorithm result

The experimental results show that the Roberts algorithm has edge information loss in the region where the gray level change is not obvious in the image (such as the eyes and lower right side of animal), and multiple edges are discontinuous. In contrast, the ant colony algorithm is more sensitive to grayscale changes in the image, and detects edge images with better continuity and integrity.

In order to verify the stability and universality of the algorithm, different types of pictures were tested. Figure 5 is a simple real photograph, the test result is shown in Fig. 6. And Fig. 7 is a more complicated cartoon picture, and the detection result is shown in Fig. 8. It can be seen from the results (Figs. 3, 6 and 8) that the algorithm can obtain better edge detection result whether it is simple or more complicated pictures or real photos. (All of these edge images have good continuity and integrity).

**Fig. 5.** Apple

**Fig. 6.** Apple's test result

**Fig. 7.** Bear

**Fig. 8.** Bear's test result

# References

1. Nausheen, N., Seal, A., Khanna, P., Halder, S.: A FPGA based implementation of Sobel edge detection. Microprocess. Microsyst. **56**(2), 84–91 (2018)
2. Tang, Z., Zhu, L., Ding, Y., He, M., Yingqi, L.: Research on the optimization algorithm of image edge detection. Technol. Innov. Prod. **2**, 71–74 (2019)
3. Zhang, W., Haijun, X., Ni, Z.: Research on ant colony algorithm and its application in navigation. J. Guangzhou inst. Navig. **26**(4), 66–70 (2018)
4. Ning, J., Zhang, Q., Zhang, C., Zhang, B.: A best-path-updating information-gided ant colony optimization algorithm. Inf. Sci. **4**, 142–162 (2018)
5. Ghimatgar, H., Kazemi, K., Helfroush, M.S., Aarabi, A.: An improved feature selection algorithm based on graph clustering and ant colony optimization. Knowl.-Based Syst. **159** (11), 270–285 (2018)
6. Han, Y., Shi, P.: An image segmentation method based on ant colony algorithm. Comput. Eng. Appl. **18**, 5–7 (2004)
7. Shahdoosti, H.R., Tabatabaei, Z.: MRI and PET/SPECT image fusion at feature level using ant colony based segmentation. Biomed. Sig. Process. Control **47**(2), 63–74 (2019)

# Biological Network Modeling Based on Hill Function and Hybrid Evolutionary Algorithm

Sanrong Liu and Haifeng Wang[(✉)]

School of Information Science and Engineering, Zaozhuang University,
Zaozhuang, China
batsi@126.com

**Abstract.** Gene regulatory network inference helps understand the regulatory mechanism among genes, predict the functions of unknown genes, comprehend the pathogenesis of disease and speed up drug development. In this paper, a Hill function-based ordinary differential equation (ODE) model is proposed to infer gene regulatory network (GRN). A hybrid evolutionary algorithm based on binary grey wolf optimization (BGWO) and grey wolf optimization (GWO) is proposed to identify the structure and parameters of the Hill function-based model. In order to restrict the search space and eliminate the redundant regulatory relationships, $L_1$ regularizer was added to the fitness function. SOS repair network was used to test the proposed method. The experimental results show that this method can infer gene regulatory network more accurately than state of the art methods.

**Keywords:** Gene regulatory network · Hill function · Grey wolf optimization · Hybrid evolutionary algorithm · Ordinary differential equation

## 1 Introduction

The biological functions of cells are realized through the interactions between genes and their products, which form a complex gene regulatory network (GRN). Understanding the interactions and functions of genes in the life system is the core of system biology [1, 2]. Although many computational and statistical methods have been proposed to analyze regulatory networks in order to systematically understand the molecular mechanisms of life activities, due to the complexity of transcriptional regulatory mechanisms and the inherent noise of high-throughput data, the complete understanding of genetic regulatory networks remains an unresolved problem [3–5].

Ordinary differential equation (ODE) model could model the known observation data and predict the future development. Because its characteristics of forward fitting, backward prediction with little error and high accuracy, ODE model has been applied for gene regulation network inference [6]. In order to model GRN accurately, Hill and sigmoidal functions, which could well identify biochemical mechanism, have been added into the ODE model. With the proper parameters, Hill curves could have the sigmoidal shape. Santillán analyzed the biochemical reactions and discussed the feasibility and constrains of gene regulation process identification with Hill function [7]. Baralla et al. presented Michael Menten-type Models, Mendes model based on Hill function and S-system

© Springer Nature Singapore Pte Ltd. 2019
R. Mao et al. (Eds.): ICPCSEE 2019, CCIS 1059, pp. 186–194, 2019.
https://doi.org/10.1007/978-981-15-0121-0_15

model to describe GRN and particle swarm optimization (PSO) was presented to evolve the parameters of models [8]. Polynikis et al. compared the performance of different ordinary differential equations for two-gene network identification and proved that the choice of Hill coefficients was extremely important [6]. Elahi and Hasan proposed the generalized profiling method to infer continuous models based on Hill functions, which was used to described gene regulatory network [9].

The past proposed methods have proved that Hill functions are very suitable for describing gene regulatory network. But these methods must fix the activation and suppression gene sets according to the original gene regulatory networks. Some evolutionary algorithms were proposed to optimize the parameters of the model based on Hill functions in order to model GRN. However, most of the gene regulatory net-works in organisms are unknown. The expression levels of each gene can only be measured by high-throughput sequencing technologies. Moreover, the purpose of gene regulatory network inference models is to identify the unknown network structures instead of using the known networks to infer models. Therefore, to infer the unknown gene regulatory networks with ODE based on Hill functions is extremely difficult task due to the unknown activation and suppression gene sets of each target gene.

In this paper, a novel hybrid evolutionary algorithm is proposed to identify the ODE model based on Hill functions for gene regulatory network inference. In the hybrid evolutionary algorithm, binary grey wolf optimization (BGWO) is proposed to select automatically the excitation and suppression genes in Hill functions, and grey wolf optimization (GWO) is utilized to optimize the parameters of ODE model.

## 2  Proposed Method

### 2.1  Ordinary Differential Equation Based on Hill Functions

The regulations of each target gene are identified by one ODE based on Hill function. The number of ODEs is equal to the size of GRN. The $i - th$ Hill function-based ODE-Hill is described as follows [9].

$$\frac{dx_i}{dt} = p_i \prod_{j \in \Omega_i} h^-(x_j, Q_{ij}, R_{ij}) \prod_{k \in \Upsilon_i} h^+(x_k, Q_{ik}, R_{ik}) - \lambda_i x_i. \tag{1}$$

Where

$$h^-(x_j, Q_{ij}, R_{ij}) = \frac{Q_{ij}^{R_{ij}}}{x_j^{R_{ij}} + Q_{ij}^{R_{ij}}}, \tag{2}$$

and

$$h^+(x_k, Q_{ik}, R_{ik}) = 1 + \frac{x_k^{R_{ij}}}{x_k^{R_{ik}} + Q_{ik}^{R_{ik}}}. \tag{3}$$

Where Eqs. (2) and (3) are inhibiting and activating functions, respectively. $x_i$ is the expression level of $i - th$ gene, $p_i$ represents a rate constant, $\Omega_i$ is a subset of all inhibiting genes of $i - th$ gene, $\Upsilon_i$ is a subset of all activating genes of $i - th$ gene, $\lambda_i$ denotes degradation coefficient, $R_{ij}$, $Q_{ij}$, $R_{ik}$ and $Q_{ik}$ are parameters of Hill function.

## 2.2  Hybrid Evolutionary Algorithm

Grey wolf optimization (GWO) algorithm is a novel swarm intelligence optimization algorithm based on wolf hunting behavior, which was proposed by Mirjalili et al. in year 2014 [10]. GWO algorithm has the characteristics of fewer parameters, fast convergence, strong global search ability and simple implementation. It has the better optimization performance than the traditional evolutionary methods such as particle swarm optimization (PSO) and differential evolutionary (DE), so GWO has been widely applied for feature selection [11], load frequency control of power system [12], economic load dispatch [13], and vehicular ad-hoc networks [14].

GWO algorithm was proposed by imitating the hierarchy and predation mechanism of grey wolf populations in nature. The grey wolf populations have the strict hierarchy, which are divided into four classes ($\alpha$, $\beta$, $\delta$ and $\omega$) according to the leadership level. In wolf populations, the wolf with $\alpha$ level is called the dominant one, which has the best fitness value in general. It is mainly responsible for making decisions about hunting behavior, habitat and food distribution. The wolf with $\beta$ level has the second best fitness value, which is the best successor of the wolf with $\alpha$ level. Its main responsibility is to help the wolf with $\alpha$ level make decisions and conduct other collective activities. The wolf with $\delta$ level mainly carries out the orders of the wolves with $\alpha$ and $\beta$ levels, but it could order the wolves with $\omega$ level. The wolves with $\omega$ level have the lowest level, which are responsible for balancing population relations and carries out the orders of the wolves with $\alpha$, $\beta$ and $\omega$.

**Real-Encoding Grey Wolf Optimization.** Grey wolves could encircle the prey when hunting. The mathematical model of encircling behavior is described as follows.

$$D = \left| C \cdot X_p(t) - X(t) \right|. \tag{4}$$

$$X_p(t+1) = X_p(t) - A \cdot D. \tag{5}$$

$$A = 2\alpha \cdot r_1 - \alpha. \tag{6}$$

$$C = 2 \cdot r_2. \tag{7}$$

Where $D$ denotes the distance between grey wolf and prey, $X_p(t)$ is the current position of prey, $X(t)$ denotes the current position of grey wolf, $C$ and $A$ are algorithm coefficients, and $\alpha$ is a convergence factor, which decreases linearly from 2 to 0. $r_1$ and $r_2$ are random variables in the interval [0, 1].

Due to that the position of the prey always changes, in order to search the better prey, suppose the positions of the wolves with $\alpha$, $\beta$ and $\omega$ levels as the ones closer to the prey. Therefore, during each iteration, save the positions of the three best wolves ($\alpha$, $\beta$ and $\delta$), which are used to update the positions of other wolves with Eq. (8).

$$X(t+1) = \frac{X_1(t) + X_2(t) + X_3(t)}{3}.$$
(8)

Where $X_1(t)$, $X_2(t)$ and $X_3(t)$ are the calculated position results when the positions of the wolves with $\alpha$, $\beta$ and $\delta$ levels are considered as the prey positions ($X_p$), respectively.

**Binary-Encoding Grey Wolf Optimization.** Binary grey wolf optimization (BGWO) algorithm is similar to the real encoding GWO. Due to that the location of each grey wolf is binary encoding, the following modifications are needed with Eq. (9) when updating the locations of the grey wolves [11].

$$X(t+1) = \begin{cases} 1, & if \ S(\frac{X_1(t)+X_2(t)+X_3(t)}{3}) \geq r_3 \\ 0, & otherwise \end{cases}.$$
(9)

Where $r_3$ is a random variable in the interval [0, 1], and $S(\cdot)$ denotes the conversion function as follows.

$$S(x) = \frac{1}{1 + e^{-10(x-0.5)}}.$$
(10)

## 2.3 Optimization of ODE Based on Hill Functions

In this paper, a hybrid evolutionary algorithm is proposed, which can automatically select inhibiting and activation gene sets, and optimize parameters. The hybrid evolutionary algorithm uses binary encoding and real encoding grey wolf optimization algorithm to optimize the structure and parameters of ODE model based on Hill functions. Suppose that gene regulatory network contains $n$ genes. The corresponding Hill functions-based ODE of each gene could be derived independently. For the $i - th$ gene, the optimization algorithm of ODE based on Hill functions is described as follows. The flowchart is depicted in Fig. 1.

(1) Binary grey wolf optimization algorithm is used to automatically select inhibiting and activation gene sets. In BWGO, initialize the binary grey populations. Each grey wolf contains two parts: inhibiting gene set and activating gene set, whose structure is depicted in Fig. 2. For GRN with $n$ genes, the dimension of each wolf is set to $2n$. Each bit of grey wolf ($p_i$, $i = 1, 2, \ldots, 2n$) is binary. If $p_j$ is set to 1, the $j - th$ gene could inhibit the expression of the $i - th$ gene. In order to infer the regulations of the $i - th$ gene, inhibiting gene set and activating gene set in Hill functions could not contain the $i - th$ gene. Therefore, in each grey wolf, $p_i$ and $p_{n+i}$ are set as 0 in the optimization. Then calculate the fitness values of binary wolf populations. Search the positions of the wolves with $\alpha$, $\beta$ and $\delta$ levels. Update the positions of other wolves with Eq. (9). Repeat this process until that stopping criterion is met.

(2) In the optimization process of BGWO, grey wolf algorithm is used to optimize the parameters of ODE based on Hill functions. During this process, the structure of

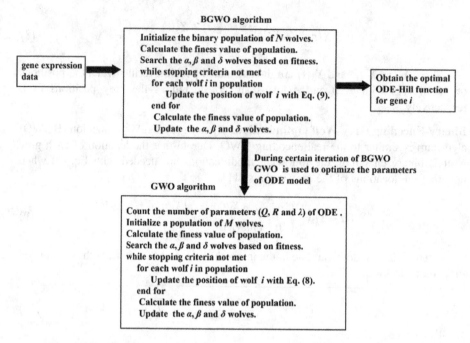

**Fig. 1.** The flowchart of ODE-Hill functions optimization for the $i-th$ gene.

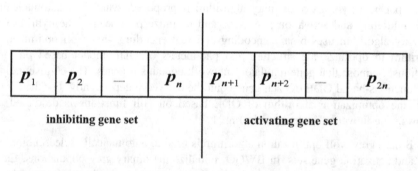

**Fig. 2.** Chromosome structure of binary grey wolf optimization algorithm.

ODE is fixed. According to the structure, the number of parameters is counted. If $p_j$ is 1 in the structure, two kinds of parameters ($R_{ij}$ and $Q_{ij}$) are given. Suppose that the number of 1 is counted as $m$, the dimension of grey wolf in GWO algorithm is set as $2m+2$.

(3) Fitness function. In order to restrict the structure of Hill functions and eliminate the redundant regulatory relationships, $L_1$ regularizer is added to the fitness function of the hybrid evolutionary method. The fitness function of the $i-th$ gene is described as follows.

$$f_i = \sum_{k=1}^{T} (y_{ik} - y'_{ik})^2 + \lambda |R_{ij} + Q_{ij}|. \tag{11}$$

Where $T$ denotes the number of gene expression sample points, $y_{ik}$ is the expression level of the $i - th$ gene at $k - th$ sample point, $y'_{ik}$ is the predicted expression level of the $i - th$ gene at $k - th$ sample point and $\lambda$ is a sparse coefficient.

## 3  Experiments

SOS DNA repair network is used to evaluate the performance of our proposed algorithm. SOS repair is that cells give the stress response when DNAs are seriously injured. SOS DNA repair network includes 8 genes (**uvrD**, **lexA**, **umuD**, **recA**, **uvrA**, **uvrY**, **ruvA** and **polB**) and 9 regulatory relationships, which is depicted in Fig. 3 [15]. The dataset contains four experiments and each experiment has 50 sample points. Sensitive, Precision and F-score are utilized to test the performance of SOS network inference by our proposed method, which are defined as follows.

$$Sensitivity = \frac{TP}{TP + FN} \tag{12}$$

$$Precision = \frac{TP}{TP + FP} \tag{13}$$

$$F - score = 2\frac{Sensitivity * Precision}{Sensitivity + Precision} \tag{14}$$

Where TP, FP, and FN are defined in Fig. 4.

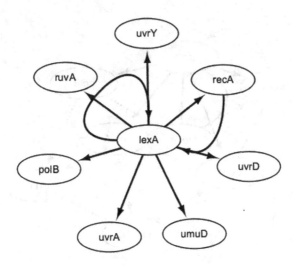

**Fig. 3.** The true SOS network.

**True output**

|  |  | Positive | Negative |
|---|---|---|---|
| **Predicting output** | Positive | TP | FP |
| | Negative | PN | TN |

**Fig. 4.** Introduction of TP, FP and FN.

With gene expression data, the inferred SOS repair network is described in Fig. 5. The solid line denotes the true regulations and the dotted line represents the false-positive regulatory relationships. Our method could identify 6 true edges and 9 false-positive edges. The performances of SOS network inference by seven methods are listed in Table 1. From Table 1, it can be seen that RNN with BA has the highest *Sensitive* among seven methods, which reveals that RNN with BA could identify more true-positive regulatory relationships. In terms of *Precision*, DBN performs best, which is 0.4444 and shows DBN method could infer less false-positive regulations. Our method has the second highest *Precision* and *Sensitive*, so our method could gain the highest $F - score$. In sum, our method could identify more accurately SOS network among seven methods.

We also investigate the effect of $L_1$ regularizer with SOS network. The results of SOS network inferred by ODE based on Hill functions with and without $L_1$ regularizer are also listed in Table 1. From the results, we can see that these two methods has the

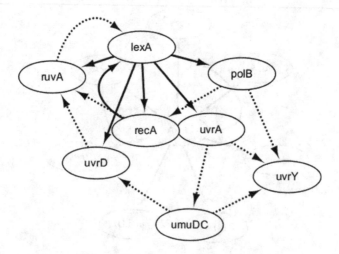

**Fig. 5.** The constructed SOS network.

same *Sensitive*, but ODE based on Hill functions with $L_1$ regularizer has higher $F-score$ and *Precision*.

**Table 1.** The performance of SOS network inferred by seven GRN inference methods.

| Methods | Sensitive | Precision | F-score |
| --- | --- | --- | --- |
| S-system [18] | 0.556 | 0.3333 | 0.41667 |
| DBN [19] | 0.444 | 0.4444 | 0.4444 |
| ODE [20] | 0.6667 | 0.28571 | 0.4 |
| RNN+PSO [16] | 0.3333 | 0.2308 | 0.2727 |
| RNN+BA [17] | 0.7778 | 0.3043 | 0.4375 |
| ODE based on Hill functions without $L_1$ regularizer | 0.6667 | 0.23077 | 0.3428 |
| ODE based on Hill functions | 0.6667 | 0.4 | 0.5 |

## 4 Conclusions

In order to infer gene regulatory network accurately, this paper proposes a novel hybrid evolutionary algorithm to identify the ODE based on Hill functions for GRN inference. In the hybrid evolutionary algorithm, binary grey wolf optimization is proposed to select automatically the excitation and suppression genes in Hill functions, and grey wolf optimization is used to optimize the parameters of ODE model. The inference performance of SOS repair network reveals that our method could predict gene expression data and infer gene regulatory network more accurately than some state of the art models.

## References

1. Li, M., Belmonte, J.C.: Ground rules of the pluripotency gene regulatory network. Nature Rev. Genet. **18**(3), 180 (2017)
2. Chan, T.E., Stumpf, M.P.H., Babtie, A.C.: Gene regulatory network inference from single-cell data using multivariate information measures. Cell Syst. **5**(3), 251–267 (2017)
3. Omranian, N., Eloundou-Mbebi, J.M.O., Mueller-Roeber, B., Nikoloski, Z.: Gene regulatory network inference using fused LASSO on multiple data sets. Sci. Rep. **6**, 20533 (2016)
4. Yeung, K.Y., Raftery, A.E., Young, W.C.: A posterior probability approach for gene regulatory network inference in genetic perturbation data. Math. Biosci. Eng. **13**(6), 1241–1251 (2017)
5. Lam, K.Y., Westrick, Z.M., Müller, C.L., Christiaen, L., Bonneau, R.: Fused regression for multi-source gene regulatory network inference. PLoS Comput. Biol. **12**(12), e1005157 (2016)
6. Polynikis, A., Hogan, S.J., Bernardo, M.D.: Comparing different ODE modelling approaches for gene regulatory networks. J. Theor. Biol. **261**(4), 511–530 (2009)
7. Santillán, M.: On the use of the hill functions in mathematical models of gene regulatory networks. Math. Model. Nat. Phenom. **3**(2), 85–97 (2008)

8.  Baralla, A., Cavaliere, M., de la Fuente, A.: Modeling and parameter estimation of the SOS response network in E.coli, MS thesis, University of Trento, Trento, Italy (2008)
9.  Elahi, F.E., Hasan, A.: A method for estimating hill function-based dynamic models of gene regulatory networks. Royal Soc. Open Sci. 5(2), 171226 (2018)
10. Mirjalili, S., Mirjalili, S.M., Lewis, A.: Grey wolf optimization. Adv. Eng. Softw. 69(7), 46–61 (2014)
11. Emary, E., Zawbaa, H.M., Hassanien, A.E.: Binary grey wolf optimization approaches for feature selection. Neurocomputing 172C, 371–381 (2016)
12. Guha, D., Roy, P.K., Banerjee, S.: Load frequency control of interconnected power system using grey wolf optimization. Swarm Evol. Comput. 27, 97–115 (2016)
13. Pradhan, M., Roy, P.K., Pal, T.: Grey wolf optimization applied to economic load dispatch problems. Int. J. Electr. Power Energy Syst. 83, 325–334 (2016)
14. Fahad, M., Aadil, F., Rehman, Z.U., Khan, S., Shah, P.A.: Grey wolf optimization based clustering algorithm for vehicular ad-hoc networks. Comput. Electr. Eng. 70, 853–870 (2018)
15. Zhang, Y., Pu, Y., Zhang, H., Cong, Y., Zhou, J.: An extended fractional Kalman filter for inferring gene regulatory networks using time-series data. Chemom. Intell. Lab. Syst. 138, 57–63 (2014)
16. Kentzoglannakis, K., Poole, M.: A swarm intelligence framework for reconstructing gene networks: searching for biologically plausible architecture. IEEE/ACM Trans. Computat. Biol. Bioinf. 9(2), 358–372 (2012)
17. Mandal, S., Saha, G., Pal, R.K.: Recurrent neural network based modeling of gene regulatory network using bat algorithm. J. Adv. Math. Comput. Sci. 23(5), 1–16 (2017)
18. Noman, N., Iba, H.: Reverse engineering genetic networks using evolutionary computation. Genome Inf. 16(2), 205–214 (2005). International Conference on Genome Informatics
19. Perrin, B.E., Ralaivola, L., Mazurie, A., Bottani, S., Mallet, J., d'Alche–Buc, F.: Gene networks inference using dynamic Bayesian networks. Bioinformatics 19, 138–148 (2003)
20. Kimura, S., Sonoda, K., Yamane, S., Maeda, H., Matsumura, K., Hatakeyama, M.: Function approximation approach to the inference of reduced NGnet models of genetic networks. BMC Bioinf. 9(1), 23 (2008)

# Design of Five-Axis Camera Stabilizer Based on Quaternion Untracked Kalman Filtering Algorithm

Xiaohui Cheng[1](✉), Yu Zhang[2], and Dezhi Liu[2]

[1] College of Information Science and Engineering,
Guilin University of Technology, Jiangan Road No.12, Guilin 541000, China
cxiaohui@glut.edu.cn
[2] Guangxi Key Laboratory of Embedded Technology and Intelligent System,
Jiangan Road No.12, Guilin 541000, China
969673766@qq.com

**Abstract.** A five-axis camera stabilizer based on quaternion unscented Kalman filter algorithm is designed. It combined the unscented Kalman filter algorithm with the quaternion attitude solution and was solved by attitude sensor. By attitude algorithm, the motor in three directions of pitch, heading and roll in the stabilizer was accurately adjusted to control the movement of the three electronic arms. In order to improve the three-axis hand-held camera stabilizer's performance, and to solve the jitter problem of up-and-down movement not being eliminated, two mechanical anti-shake arms were loaded under the stabilizer to balance the camera's picture in pitch, roll, heading, and above and below five directions. Movement can maintain a stable effect. The simulation results show that the algorithm can effectively suppress the attitude angle divergence and improve the attitude calculation accuracy.

**Keywords:** Attitude sensor · Quaternion · Attitude fusion ·
Untracked Kalman filter · Five-axis stabilizer

## 1 Introduction

In recent years, the development and progress of Micro-Electro-Mechanical System and automatic control technology promote the level of artificial intelligence to reach a new height. In the process of photography, the photographer's inability to keep the camera's picture stable while moving has always plagued each photographer, so how to keep the camera balance has become a hot topic. At present, the three-axis stabilizer technology has become more and more mature, but the traditional three-axis stabilizer can only solve the three axial vibrations of Course, roll, pitch three directions jitter, and cannot solve the vibration from the up, down, left and right directions, so when shooting, The photographer can only effectively remove the up and down jitter by slowly moving the controller to obtain a satisfactory picture. To solve this problem and consider the lightweight standard that the stabilizer handheld needs to achieve, this paper designs a five-axis camera handheld stabilizer to solve this problem. During the balance control of the handheld camera stabilizer, the attitude sensor inputs the detected

© Springer Nature Singapore Pte Ltd. 2019
R. Mao et al. (Eds.): ICPCSEE 2019, CCIS 1059, pp. 195–213, 2019.
https://doi.org/10.1007/978-981-15-0121-0_16

attitude signals of the three axes into the STM32 chip of the attitude sensor, and the algorithm calculates the control amount to drive the motor to control the roll or pitch of the stabilizer shaft or The heading changes and the two mechanical axes move up and down, left and right, so that the camera position is always balanced. It can be seen from this that obtaining the real-time posture position of the camera is a prerequisite for maintaining hand balance. Therefore, it is especially important to implement a real-time and reliable camera position and attitude estimation algorithm, which directly affects the stable self-balancing performance of the handheld.

To solve this problem, this paper uses the method of untracked kalman filter to fuse the data of gyroscope and accelerometer, and corrects the random drift component in the output of gyroscope. Zou et al. Propose an imaging, inertial and altitude combined navigation for a quad copter. Using Kalman filter to fuse the data of vision, inertial measurement system and ultrasonic range finder, estimate the relative height of the aircraft by similar triangle theorem, and estimate the position and displacement speed of the aircraft by visual metric method [1]. Mu et al. In order to improve the target positioning accuracy, a target positioning algorithm based on extended Kalman filter is proposed [2]. Zhu et al. Based on the correlation filter classifier, a visual target tracking method with motion state estimation and target scale estimation is proposed [3]. For the problem that the inertial device used in the small quad rotor in attitude calculation is susceptible to non-gravity acceleration during take-off and descent of the aircraft, Cui et al. Proposed the adaptive error attitude calculation method based on quaternion solution for unscented Kalman filter [5]. Zheng et al. Proposed a UAV attitude calculation method to enhance the attitude solution of the UAV based on the extended Kalman filter theory. The robustness of the calculation improves the control performance of the drone [6]. Zhou et al. In order to solve the problem that the heading angle error is easy to divergence, a zero integrated heading rate (ZIHR) correction method is proposed [7]. Li et al. Proposed a gyroscope error compensation algorithm based on improved Kalman filter for the inaccurate measurement of variance of measurement noise in MEMS gyroscope error compensation algorithm based on Kalman filter [8]. In addition, the application of Kalman filtering to control drones or other devices that need to maintain stability is a more optimized method [9–12].

Because the Kalman filter algorithm can feedback future values through prediction and actual measurement feedback, it has high precision and can be applied to nonlinear systems. There are a variety of algorithms for attitude solution, and the commonly used method is the quaternion attitude solution algorithm, because the quaternion method can represent rotation, and the amount of calculation is small, which can avoid singularity.

The main purpose of this study is to:

- Develop the required Kalman filtering algorithm, whose the weights are updated by the improved UKF method.
- To propose a stable system for UKF design handheld camera stabilizers using quaternion solution.
- Trajectory tracking of the load-bearing part of the camera stabilizer.
- Experimentally verified that the quaternion solution Kalman filter can be applied to handheld camera stabilizers.

The structure of this paper is divided into six sections: The Sect. 2 briefly introduces the model and use of the five-axis handheld camera stabilizer. The following sections highlight the design of the stabilizer-controlled camera stabilization system. In Sect. 3, the principle of quaternion solving is introduced. Section 4 introduces Kalman filtering, with a focus on the UKF algorithm. In Sect. 5, the unscented Kalman filter algorithm is solved by quaternion. In Sect. 6, the simulation experiment is carried out to verify that the unscented Kalman filter can be used in the stabilizer system and the conclusion is given. Finally, the conclusion is given.

## 2 Design of Five-Axis Stabilizer

### 2.1 Design Idea of Five-Axis Stabilizer

The five-axis stabilizer ensures that the camera position is stabilized mainly by the dynamic balance principle of the sensor and motor inside the device. When the photographer's hand or body moves, the corresponding posture information is output through the gyroscope and the acceleration sensor in the posture sensor. After the controller senses this information, the three motors in the stabilizer are controlled to rotate in the corresponding directions. The attitude sensor continuously measures the attitude of the vehicle at a certain frequency, and outputs the attitude information to the controller. The controller continuously adjusts the rotation direction and the rotation speed of the motor, so that the dynamic balance of the camera position can be ensured.

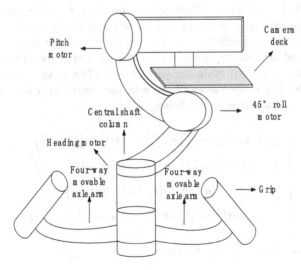

**Fig. 1.** Five-axis stabilizer conceptual model.

## 2.2   Hardware Design of Five-Axis Stabilizer

The conceptual model of the five-axis balance stabilizer is shown in Fig. 1. The hand grip has a control chip embedded in it, which controls the heading motor, roll motor and pitch motor in three different positions of the stabilizer. Each motor has a controllable rotary axis on the outside. In use, push the camera forward into the deck and connect it to the stabilizer through the port. Open the switch, the three axis will adjust the camera position steadily under the command of the internal algorithm. When the position of the camera changes, the chip monitors the change of the camera position, and the external rotation axis is controlled by the circuit drive motor to adjust the unbalance amount, and the difference between the feedback value and the preset value is calculated by the quaternion and Kalman filter. The parameters are adjusted, and the system parameters are adjusted to control the position of the three motors so that the camera is always in the equilibrium position, achieving the goal of smooth camera shooting.

## 3   Quaternion Attitude Solving Algorithm

The mathematical concept of quaternion was first proposed by Irish scientist Hamilton in 1843. It consists of a real unit and three imaginary units i, j, k, the formula is: $q = q_0 + q_1 i + q_2 j + q_3 k$. The quaternion can represent rotation, its scalar part represents the magnitude of the rotation angle, and the vector part represents the direction of the rotation axis. Normalization must be performed when applying the quaternion, even if its modulus is 1. The hand-held position rotates clockwise around the roll axis, i.e., around the x-axis of the carrier coordinate system, resulting in a roll angle $\alpha$. The x-axis is the direction-forward axis; the clockwise rotation about the pitch axis, that is, around the y-axis of the carrier coordinate system, gives the pitch angle $\beta$. The y-axis is the axis that is oriented forward; the yaw axis is rotated clockwise around the z-axis of the carrier coordinate system to obtain the yaw angle $\gamma$. The z-axis is the axis that is oriented upwards. The initial values of the four parameters are calculated from the initial values of the Euler angles as follows:

$$
\begin{cases}
q_0 = \cos\left(\frac{\alpha}{2}\right)\cos\left(\frac{\beta}{2}\right)\cos\left(\frac{\gamma}{2}\right) + \sin\left(\frac{\alpha}{2}\right)\sin\left(\frac{\beta}{2}\right)\sin\left(\frac{\gamma}{2}\right) \\
q_1 = \cos\left(\frac{\alpha}{2}\right)\cos\left(\frac{\beta}{2}\right)\sin\left(\frac{\gamma}{2}\right) - \sin\left(\frac{\alpha}{2}\right)\sin\left(\frac{\beta}{2}\right)\cos\left(\frac{\gamma}{2}\right) \\
q_2 = \cos\left(\frac{\alpha}{2}\right)\sin\left(\frac{\beta}{2}\right)\cos\left(\frac{\gamma}{2}\right) + \sin\left(\frac{\alpha}{2}\right)\cos\left(\frac{\beta}{2}\right)\sin\left(\frac{\gamma}{2}\right) \\
q_3 = -\cos\left(\frac{\alpha}{2}\right)\sin\left(\frac{\beta}{2}\right)\sin\left(\frac{\gamma}{2}\right) + \sin\left(\frac{\alpha}{2}\right)\cos\left(\frac{\beta}{2}\right)\cos\left(\frac{\gamma}{2}\right)
\end{cases}
\tag{1}
$$

The quaternion differential equation is:

$$
\dot{q} = \frac{1}{2}q * \omega
\tag{2}
$$

The quaternion differential equation can be updated with quaternions, where $\omega$ is the angular velocity measured by the gyroscope. The fourth-order Runge-Kutta method can be used to solve the quaternion differential equation. The specific process is as follows:

$$\text{Assume}\begin{cases} K_1 = \frac{1}{2}q(t) * \omega(t) \\ K_2 = \frac{1}{2}\left(q(t) + \frac{K_1}{2}T\right) * \omega\left(t + \frac{T}{2}\right) \\ K_3 = \frac{1}{2}\left(q(t) + \frac{K_2}{2}T\right) * \omega\left(t + \frac{T}{2}\right) \\ K_4 = \frac{1}{2}\left(q(t) + K_3T\right) * \omega(t + T) \end{cases} \tag{3}$$

Then:

$$q(t+T) = q(t) + \frac{T}{6}(K_1 + 2K_2 + 2K_3 + K_4) \tag{4}$$

In this equation T is the update period, $q(t)$ is the initial quaternion, $q(t+T)$ is the updated quaternion, $K_1$ is the slope at the beginning of time period T, and $K_2$ is the slope of the midpoint of the time period. $K_3$ is the slope of the midpoint too, and $K_4$ is the slope of the end of the time period. After calculating the four parameters, the Euler angle can be updated by the attitude matrix T:

$$T = \begin{bmatrix} q_0^2 + q_1^2 - q_2^2 - q_3^2 & 2(q_1q_2 + q_0q_3) & 2(q_1q_3 - q_0q_2) \\ 2(q_1q_2 - q_0q_3) & q_0^2 - q_1^2 + q_2^2 - q_3^2 & 2(q_2q_3 + q_0q_1) \\ 2(q_1q_3 - q_0q_2) & 2(q_2q_3 - q_0q_1) & q_0^2 - q_1^2 - q_2^2 + q_3^2 \end{bmatrix} \tag{5}$$

Through the different forms of the attitude matrix and the Euler angle, it can be concluded:

$$\alpha = \arctan\left(\frac{2(q_1q_2 + q_0q_3)}{q_1^2 + q_0^2 - q_2^2 - q_3^2}\right) \tag{6}$$

$$\beta = -\arcsin(2(q_1q_3 - q_0q_2)) \tag{7}$$

$$\gamma = \arctan\left(\frac{2(q_2q_3 + q_0q_1)}{-q_1^2 + q_0^2 - q_2^2 - q_3^2}\right) \tag{8}$$

Ideally, after the axis of the five-axis camera stabilizer is automatically adjusted to the balance state, the output angle should be 0°, but the Euler angle will accumulate due to the random drift error of the gyroscope. In order to solve this problem, this paper first uses the data of the gyroscope to obtain a quaternion attitude, as the attitude estimation value in the unscented Kalman filter, and then calculates another set of attitude data as the measured value based on the accelerometer and the direction data. Then, through the unscented Kalman filter, the estimated value and the measured value are compared and calculated, and an accurate attitude data is obtained to adjust the axis arm of the camera stabilizer.

# 4  Design of Unscented Kalman Filter Algorithm

## 4.1  Introduction to Kalman Filtering Algorithm

Kalman filter is characterized by processing the input and observation signals with Gaussian white noise interference in the dynamic linear discrete state space of uncertain information to accurately predict the system state or real signal. Therefore, the Kalman filter algorithm is an algorithm that feeds back future values through prediction and actual measurement feedback.

Since the basic Kalman filter is applied to a stochastic linear Gaussian dynamic system in which both dynamic equations and measurement equations are linear, most of the systems are nonlinear, and many are strongly nonlinear systems. Therefore, Bucy and Y. Sunahara et al. proposed an extended Kalman filter (EKF) for discrete nonlinear stochastic systems, which uses the Bayesian estimation principle to expand the non-linear model with Taylor polynomial or difference and cut it at the first order. The approximation term is used as the approximate expression form of the original state equation and the measurement equation to realize local linearization, and the system is approximated to a linear system, and then Kalman filtering is performed. Although extended Kalman filtering has been widely used, it still has many theoretical limitations that cannot be overcome. For example, extended Kalman filtering requires that non-linear system state functions and measurement functions must be continuously differentiable, and the first-order linearization approximation accuracy of the linear function is low. In particular, when the system has strong nonlinearity, the estimation accuracy of the extended Kalman filter is seriously degraded and even diverged. In addition, it is necessary to calculate the Jacobian matrix of the nonlinear function, which is easy to cause the numerical stability of the extended Kalman filter to be poor and the calculation divergence, and the calculation of the Jacobian and the Hesse matrix is large.

In order to overcome the shortcomings of extended Kalman filter, the filtering problem of nonlinear Gaussian system can be processed with higher precision and faster calculation speed. According to the basic idea of deterministic sampling, the untracked kalman filter is proposed based on UT transform from Julier et al. UT transform is a new method for calculating the statistical characteristics of random variables in nonlinear transforms and is the core of UKF. Similar to EKF, UKF still inherits the basic structure of the Kalman filter, except that UKF replaces local linearization in EKF with UT transform. The UKF still assumes that the state of the stochastic system must obey the Gaussian distribution but does not require the system to be approximately linear. At the same time, the UKF does not need to calculate the Jacobian matrix, so the state function and the measurement function are not required to be continuously differentiable, and it can even be applied in a discontinuous system. Experiments show that regardless of the degree of nonlinearity of the system, the unscented transformation can theoretically approximate the posterior mean and covariance of any nonlinear Gaussian system state with third-order Taylor accuracy. Therefore, the theoretical estimation accuracy of the unscented Kalman filter is better than extended Kalman filtering.

Because the five-axis camera stabilizer designed in this paper is randomly moving with the photographer's hand, the environment is a nonlinear dynamic environment, so the design will use Unscented Kalman Filter (UKF).

## 4.2   Unscented Kalman Filter Algorithm Flow

The process of the unscented Kalman filter algorithm is shown in Fig. 2.

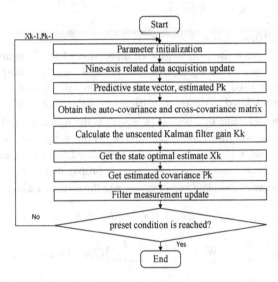

**Fig. 2.** Flow chart of unscented Kalman filter

The main steps are as follows: the first step is to initialize the parameters; the second step is to collect and update the nine-axis data; the third step is to predict the state vector and estimate the covariance $P_k$; the fourth step is to guide the observation equation The predicted value and the autocorrelation matrix and the cross-covariance matrix are obtained; the fifth step is to calculate the gain of the unscented Kalman filter; the sixth step is to obtain the state optimal estimate; the seventh step is to obtain the estimated covariance; the last step, filter update.

The UKF method first constructs a sigma scatter set, setting the state vector to n-dimensional, $\hat{x}_{k-1}$ as the state vector estimate at time k−1, and $p_{k-1}$ as the covariance matrix of the state vector at that time. The 2n+1-dimensional sigma point set can be expressed as:

$$\begin{cases} x_{0,k-1} = \hat{x}_{k-1} \\ x_{i,k-1} = \hat{x}_{k-1} + \left(\sqrt{(n+k)p_{i,k-1}}\right)_i & i = 1, 2, \ldots \ldots, n \\ x_{i+n,k-1} = \hat{x}_{k-1} - \left(\sqrt{(n+k)p_{i,k-1}}\right)_i & i = n+1, n+2 \ldots \ldots, 2n \end{cases} \quad (9)$$

The first-order and second-order weight coefficients corresponding to the sigma point set $\zeta_i$ are:

$$w_i^m = \begin{cases} \frac{\lambda}{n+\lambda} & i = 0 \\ \frac{1}{2(n+\lambda)} & i \neq 0 \end{cases} \tag{10}$$

$$w_i^c = \begin{cases} \frac{\lambda}{(n+\lambda)} + 1 + \varepsilon - \lambda^2 & i = 0 \\ \frac{1}{2(n+\lambda)} & i \neq 0 \end{cases} \tag{11}$$

among them, $\lambda = \delta^2(n+k) - n$.

The parameter $\delta$ determines the degree of dispersion of the $i^{th}$ sigma point around the state mean $\hat{x}_{k-1}$, usually taking a positive value less than 1, k being a redundancy, usually taking 0; $\varepsilon$ is used to describe the distribution information of x In Gaussian case, $\varepsilon$ is a parameter related to the prior distribution of the state vector. For a Gaussian distribution, the optimal value of $\varepsilon$ is 2. $\sqrt{(n+\lambda)P_x}$ is the square root of the matrix, the i-th column $w_i^m$ is the weight of the mean, and $w_i^c$ is the weight of the variance.

The sigma point set $x_{i,k-1}(i = 0, 1\ldots, L)$ is calculated from $\hat{x}_{k-1}$ and $P_{k-1}$ at time k−1, through a nonlinear function $f_{k-1}(\bullet) + q_{k-1}$ is propagated as $x_{i,k/k-1}$, and $x_{i,k/k-1}$ is predicted from the available state vector $\hat{x}_{k/k-1}$ and the error covariance matrix $P_{k/k-1}$.

$$x_{i,k/k-1} = f_{k-1}(x_{i,k-1}) + q_{k-1} \quad i = 0, 1, \ldots\ldots, L \tag{12}$$

$$\hat{x}_{k/k-1} = \sum_{i=0}^{L} W_i^m x_{i,k/k-1} = \sum_{i=0}^{L} W_i^m f_{k-1}(x_{i,k-1}) + q_{k-1} \tag{13}$$

$$P_{k/k-1} = \sum_{i=0}^{L} W_i^c (x_{i,k/k-1} - \hat{x}_{k/k-1})(x_{i,k/k-1} - \hat{x}_{k/k-1})^T + Q_{k-1} \tag{14}$$

In the above equation, $Q_{k-1}$ is the process noise covariance matrix.

Similarly, $\hat{x}_{k-1}$ and $P_{k-1}$ are used to calculate the sigma point set $x_{i,k-1}(i = 0, 1, \ldots\ldots.L)$ according to the previous sampling strategy. The linear measurement function $h_{k-1}(\bullet) + r_{k-1}$ propagates as $\chi_{ik/(k-1)}$, $\chi_{ik/(k-1)}$ from the available output prediction value $\hat{z}_{k/k-1}$ and the auto-covariance matrix $P_{\tilde{z}_k}$ and the cross-covariance matrix $P_{\tilde{x}_k \tilde{z}_k}$.

$$\chi_{i,k/k-1} = h_k(x_{i,k/k-1}) + r_k \tag{15}$$

$$\hat{z}_{k/k-1} = \sum_{i=0}^{L} W_i^m \chi_{i,k/k-1} = \sum_{i=0}^{L} W_i^m h_k(x_{i,k-1}) + r_k \tag{16}$$

$$P_{\tilde{z}_k} = \sum_{i=0}^{L} W_i^c (\chi_{i,k/k-1} - \hat{z}_{k/k-1})(\chi_{i,k/k-1} - \hat{z}_{k/k-1})^T + R_k \tag{17}$$

$$P_{\tilde{x}_k \tilde{z}_k} = \sum_{i=0}^{L} W_i^c (X_{i,k/k-1} - \hat{X}_{k/k-1})(\chi_{i,k/k-1} - \hat{Z}_{k/k-1})^T \tag{18}$$

In the above equation, $R_k$ is the measurement noise covariance matrix.

After the new output prediction value $z_k$ is obtained, the filter measurement update is performed.

$$\begin{cases} \hat{x}_k = \hat{x}_{k/k-1} + K_k(z_k - \hat{z}_{k/k-1}) \\ K_k = P_{\tilde{x}_k \tilde{z}_k} P_{\tilde{z}_k}^{-1} \\ P_k = P_{k/k-1} - K_k P_{\tilde{z}_k} K_k^T \end{cases} \quad (19)$$

# 5  Attitude Algorithm Fusion

In this paper, the quaternion pose algorithm and Kalman filter algorithm are used to pose fusion, which is applied to the five-axis handheld camera balance stabilizer to achieve the purpose of quickly predicting and adjusting the camera position. The method of algorithm fusion is as follows: in the first step, the rotation quaternion is updated by the data recorded by the gyroscope, and the external force acceleration is updated; and in the second step, the prediction covariance $P_k$ and the filter gain $K_k$ are estimated. In the third step, the inverse gravity vector in the carrier coordinate system is calculated by updating the obtained rotated quaternary value; finally, the optimal estimated value of the state variable is obtained according to the measured value and the covariance is updated. The method [4] studied by Qiao et al. can be used in this design to achieve the purpose of algorithm fusion.

Because the selection of the perturbed quaternary sigma point is limited by the normalization when the quaternion attitude algorithm is combined with the UKF algorithm, in the design of this paper, the quaternion state variable represented by vector is used. Therefore, the state variable of the system can be expressed as:

$$x^a = \begin{bmatrix} x^T & x_n^T \end{bmatrix}^T = \begin{bmatrix} \rho^T & b^T & \eta_a^T & \eta_b^T & \eta_s^T \end{bmatrix}^T \quad (20)$$

In the above equation, $x_n$ is the noise part and $x$ is the n-dimensional state variable, which consists of quaternion and gyro drift. Initialization state

First, initialize the system:

$$\hat{x}_0^a = E(x_0^a) = \begin{bmatrix} \hat{\rho}_0^T & \hat{b}_0^T & 0 & 0 & 0 \end{bmatrix}^T \quad (21)$$

$$P_0^a = E\left[ (x_0^a - \hat{x}_0^a)(x_0^a - \hat{x}_0^a)^T \right] = \begin{bmatrix} P_0 & 0 & 0 & 0 \\ 0 & \sigma_a^2 & 0 & 0 \\ 0 & 0 & \sigma_b^2 & 0 \\ 0 & 0 & 0 & \sigma_s^2 \end{bmatrix} \quad (22)$$

Second, make a time update:

The state variable is selected as:

$$x_{k-1}^{ax} = \begin{bmatrix} \rho_{k-1}^T & b_{k-1}^T & \eta_{ak-1}^T & \eta_{bk-1}^T \end{bmatrix}^T \tag{23}$$

Disturbed sigma point is selected as:

$$\delta X_{i,k-1}^{ax} = \left( \sqrt{(L_1 + \lambda)P_{k-1}^{ax}} \right)_i \tag{24}$$

$L_1 = 2n$, $P_{k-1}^{ax}$ is a dimensionally extended variance matrix composed of a state error variance matrix $P_{k-1}$ and a process noise matrix $Q_{k-1}$ at time k−1. Since the filter update process is transmitted by the quaternion, the quaternary point is selected as the sigma point of the pose part, which can be expressed as:

$$X_{i,k-1}^q = \begin{bmatrix} \hat{q}_{i,k-1} & \delta\hat{q}_{i,k-1} \otimes \hat{q}_{i,k-1} & \delta\hat{q}_{i,k-1}^{-1} \otimes \hat{q}_{i,k-1} \end{bmatrix} \quad i = 0, 1, \ldots, 2L_1 \tag{25}$$

In the above equation, $\hat{q}_{k-1}$ is the quaternion estimate at time k-1, and $\delta\hat{q}_{i,k-1}$ is the perturbed quaternion at time k−1.

$$\delta\hat{q}_{i,k-1}(i) = \begin{bmatrix} \delta X_{i,k-1}^{q0} & \left( \delta X_{i,k-1}^\rho \right)^T \end{bmatrix}^T \tag{26}$$

$\delta X_{i,k-1}^{q0}$ is the scalar part of the perturbed quaternion $\delta\hat{q}_{i,k-1}$,

$$\delta\hat{q}_{i,k-1} = \sqrt{1 - \left( \delta X_{i,k-1}^\rho \right)^T \delta X_{i,k-1}^\rho} \tag{27}$$

The mean weights corresponding to the Sigma points are:

$$x_{k-1}^{ax} = \begin{bmatrix} \hat{x}_{k-1}^T & \hat{w}_{k-1}^T \end{bmatrix}^T \tag{28}$$

The variance weight is:

$$P_{k-1}^{ax} = \begin{bmatrix} P_{k-1} & 0 \\ 0 & Q_{k-1} \end{bmatrix} \tag{29}$$

In the case of a mean and a weight of 1,

$$K = 4M - I_{4 \times 4} \tag{30}$$

$$M = \sum_{i=1}^{N} \omega_i q_i q_i^T \tag{31}$$

The one-step prediction mean for quaternions is:

$$\hat{q}_{k/k-1} = argmax\chi^q_{i,k/k-1} K\left(\chi^q_{i,k/k-1}\right)^T \tag{32}$$

$$\delta\chi^q_{i,k/k-1} = \hat{q}_{k/k-1} \otimes K\left(\chi^q_{k-1}\right)^{-1} \tag{33}$$

Therefore, the attitude prediction error variance matrix is:

$$P^\rho_{k/k-1} = \sum_{i=0}^{2L_1} \omega^c_i \delta\chi^\rho_{i,k/k-1} \left(\delta\chi^\rho_{i,k/k-1}\right)^T \tag{34}$$

In the above equation, $\delta\chi^\rho_{i,k/k-1}$ is the vector portion of the prediction error quaternion.

Then, make a measurement update:

In the part of the measurement update, since the process noise is not involved, the switching system status is:

$$x^{az}_{k/k-1} = \left[\hat{\rho}^T_{k/k-1} \hat{b}^T_{k/k-1} \hat{\eta}^T_{sk-1}\right]^T \tag{35}$$

The selection of the disturbance sigma point during the measurement update process is:

$$\delta\chi^{az}_{i,k/k-1} = \sqrt{(L_2 + \lambda)P^{az}_{k/k-1}} \tag{36}$$

In the above equation, $L_2 = n + m$, $P^{az}_{k/k-1}$ is the dimensionality variance of the state prediction error variance matrix $P_{k/k-1}$ and the measurement noise array $R_k$ at time k. Array. The attitude quaternion sigma point is selected as:

$$\chi^q_{i,k/k-1} = \left[\hat{q}_{k/k-1} \quad \delta\hat{\chi}_{i,k-1} \otimes \hat{q}_{k,k-1} \quad \delta\hat{\chi}_{i,k-1} \otimes \hat{q}_{k,k-1}\right] \tag{37}$$

The quaternion quantity update is updated to:

$$Z^q_{i,k/k-1} = \chi^{q_{\eta_s}}_{i,k} \otimes \chi^q_{i,k/k-1} \tag{38}$$

The ternary quantity measurement prediction mean value is calculated according to formula (31):

$$\hat{z}^q_{k/k-1} = argmaxZ^q_{i,k/k-1} K\left(Z^q_{i,k/k-1}\right)^T \tag{39}$$

Since the quantity measurement is only quaternion, the system state measurement error variance matrix is:

$$\delta Z_{i,k/k-1}^q = \hat{z}_{k/k-1}^q \otimes \left( Z_{i,k/k-1}^q \right)^{-1} \tag{40}$$

$$P_{zz,k} = P_{z,kk-1}^\rho = \sum_{i=0}^{2L_2} \omega_i^c \delta Z_{i,k/k-1}^\rho \left( \delta Z_{i,k/k-1}^\rho \right)^T \tag{41}$$

Since the white noise estimate is preferably 0, the state update only considers the pose portion and the gyro drift portion, so that the cross-covariance matrix can be calculated by:

$$P_{xz,k} = \left[ \sum_{i=0}^{2L_2} \omega_i^c \delta x_{k/k-1}^\rho \left[ \delta Z_{k/k-1}^\rho \right]^T \quad \sum_{i=0}^{2L_2} \omega_i^c \left[ \chi_{i,k/k-1}^b - \hat{x}_{k/k-1}^b \right] \left[ \delta Z_{k/k-1}^\rho \right]^T \right] \tag{42}$$

Thus the state gain matrix is:

$$K_k = P_{xz,k} \tag{43}$$

The corresponding state error amount is updated to:

$$\delta \hat{x}_k = K_k \delta \hat{z}_{k/k-1}^\rho \tag{44}$$

In the above equation, $\delta \hat{x}_k = \left[ \delta \hat{\rho}_k^T \quad \delta \hat{b}_k^T \right]^T$, $\delta \hat{\rho}_k$ is the quaternion vector part of the attitude estimation error, $\delta \hat{z}_{k/k-1}^\rho$ is the vector part of the measurement error quaternion $\delta \hat{z}_{k/k-1}^\rho = q_s \otimes \left( \hat{z}_{k/k-1}^q \right)^{-1}$.

Thus the quaternion status is updated to:

$$\hat{\rho}_k = \delta \hat{\rho}_k \otimes \hat{q}_{k/k-1} \tag{45}$$

In the above equation, $\delta \hat{q}_k = \left[ \sqrt{1 - \delta \hat{\rho}_k^T \delta \hat{\rho}_k} \quad \delta \hat{\rho}_k^T \right]^T$.
The gyroscope drift is updated to:

$$\hat{b}_k = \hat{b}_{k/k-1} + \delta \hat{b}_k \tag{46}$$

The state error variance matrix is:

$$P_k = P_{k/k-1} - K_k P_{zz,k} K_k^T \tag{47}$$

The state prediction error equation matrix in which $P_{k/k-1}$ is composed of the attitude portion and the gyro drift portion can be expressed by the following formula:

$$P_{k/k-1} = \begin{bmatrix} P_{k/k-1}^{\rho} & \sum_{i=0}^{2L_1} \omega_i^c \delta\chi_{i,k/k-1}^{\rho} N^T \\ \sum_{i=0}^{2L_1} \omega_i^c N\left(\delta\chi_{i,k/k-1}^{\rho}\right)^T & P_{k/k-1}^b \end{bmatrix} \tag{48}$$

In the above equation, $P_{k/k-1}^b = \sum_{i=0}^{2L_1} \omega_c^i NN^T$, $N = \hat{x}_{i,k/k-1}^b - \hat{b}_{k/k-1}$.

## 6 Experimental Results

### 6.1 Software Application

Considering the performance and cost of the five-axis stabilizer, the MCU of this experiment selects the 32-bit microcontroller STM32 of ARM Cortex-M core, and the sensor uses WT931 module. The high-precision gyroscope, accelerometer and geomagnetic field are integrated in the module package sensor. It can connect the MCU through the IIC interface and output the acceleration signal and angular velocity signal in digital form. The axial direction of the module is on the back of the figure, the X-axis to the right, the Y-axis to the right, and the Z-axis to the vertical. The direction of rotation is defined by the right-hand rule, that is, the right thumb is pointing in the axial direction, and the direction in which the four fingers are bent is the direction of rotation around the axis. The X-axis angle is the angle of rotation around the X-axis, the Y-axis angle is the angle of rotation around the Y-axis, and the Z-axis angle is the angle of rotation around the Z-axis. The chip transmits the nine-axis data to the host computer through the serial port. After the analog-to-digital conversion, the data can be simulated by MATLAB and the data graph is output. The module has an output frequency of 500 Hz, a baud rate of 921600 kbps, and an operating voltage of 3.3 V. The sensor coordinate system is the five-axis handheld stabilizer coordinate system.

### 6.2 Simulation Results

The hardware device of the experiment is shown in Fig. 3. In the experiment, the Wit Smart hardware product will be redeveloped. The module uses a BMI160 chip with integrated attitude solver high-precision gyroscope, accelerometer, geomagnetic sensor and high-performance microprocessor. Through the combination of the designed dynamic solution and the Kalman dynamic filtering algorithm, the real-time motion pose of the current module can be quickly determined and displayed on the host computer. The main data acquisition process is as follows: First, the program is programmed for the chip, connected to the computer through the USB-TTL conversion interface, the driver software is installed on the computer to set the serial port and baud rate, and then the calibration of the three coordinate axes is started. Test the button, then have the target mobile chip position, record the data, and observe whether the chip can accurately record the position, angle, acceleration, and angular velocity values.

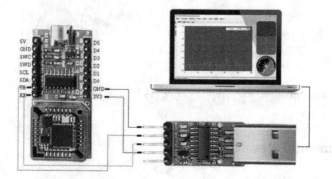

**Fig. 3.** Experimental hardware device diagram

The parameters of the module are as follows:

1. Voltage: 3.3 V
2. Current: < 25 mA
3. Volume: 12 mm × 12 mm × 2 mm
4. Pad spacing: up and down 1.27 mm, left and right 12 mm
5. Measurement dimension:
   acceleration: 3 dimensions, angular
   velocity: 3 dimensions, magnetic
   field: 3 dimensions, angle: 3 dimensions
6. Range: Acceleration: ± 2/4/8/16 g (optional), angular velocity: ± 250/500/1000/2000°/s (optional), angle X Z ± 180°
7. Y-axis ± 90°
8. Stability: Acceleration: 0.01 g, angular velocity 0.05°/s
9. Attitude measurement accuracy: X Y dynamic 0.1°, static 0.05°, Z axis 1° (no magnetic interference and calibration)
10. Data output content: relative time, acceleration, angular velocity, angle, magnetic field
11. The data output frequency 0.1 Hz–500 Hz, the default 500 Hz
12. Data interface: serial port (TTL level, baud rate support 2400, 4800, 9600, 19200, 38400, 57600, 115200, 230400, 460800, 921600 (default)).

The collected data is the initial value, and then the Kalman filter algorithm is used to calculate the difference between the preset value and the initial value, and the motor torque is adjusted to return the motor-controlled camera deck to the initial position, so that the camera position can be kept stable.

## 6.3 Simulation Results

The host computer software reads the serial data stream through the TTL connector, checks the frame header and reads the valid data. By decoding the hexadecimal floating point data by forced type conversion, and then outputting the angle values of the stabilizer pitch angle and the roll angle and the heading angle in real time through the control, the .txt file can be output. Input the read data into MATLAB to visually see the angle size and trend. Real-time animation can also be achieved with 3D control displays.

Because the five-axis camera stabilizer designed in this paper does not require a large range of activities, the experiment is carried out indoors, and the temperature change of the test is shown in Fig. 4.

The serial port address set during the experiment is $0 \times 50$, and the experimental part data is shown in Tables 1, 2 and 3.

**Table 1.** Partial experimental data sheet

| Time(s) | ax(g) | ay(g) | az(g) |
|---------|-------|-------|-------|
| 17.457 | −0.0469 | 0.0610 | 1.0127 |
| 17.457 | −0.0469 | 0.0610 | 1.0127 |
| ... | ... | ... | ... |
| 17.620 | −0.0249 | 0.0845 | 0.9463 |
| 17.622 | −0.0249 | 0.0845 | 0.9463 |
| ... | ... | ... | ... |
| 23.473 | 0.0034 | −0.0977 | 0.7676 |
| 23.475 | 0.0034 | −0.0977 | 0.7676 |
| ... | ... | ... | ... |
| 27.710 | −0.3071 | −0.1865 | 1.1792 |
| 27.711 | −0.3071 | −0.1865 | 1.1792 |

**Table 2.** Partial experimental data

| Time(s) | wx(deg/s) | | wy(deg/s) | | wz(deg/s) | |
|---------|-----------|----------|-----------|-----------|----------|-----------|
| 17.457 | 9.8267 | 7.9346 | 5.4321 | 6.3226 | 7.8442 | −59.4141 |
| 17.457 | 9.8267 | 7.9346 | 5.4321 | 6.3391 | 7.8552 | −59.4141 |
| ... | ... | ... | ... | ... | ... | ... |
| 17.620 | −7.3242 | 61.1572 | −40.8936 | 3.5376 | −5.2844 | −71.2793 |
| 17.622 | −7.3242 | 61.1572 | −40.8936 | 3.5321 | -5.1471 | −71.3232 |
| ... | ... | ... | ... | ... | ... | ... |
| 23.473 | −20.8740 | −97.4121 | −129.9438 | −10.9753 | 1.4722 | −63.1989 |
| 23.475 | −20.8740 | −97.4121 | −129.9438 | −11.0193 | 1.2305 | −63.3856 |
| ... | ... | ... | ... | ... | ... | ... |
| 27.710 | −104.9805 | −126.3428 | 160.4004 | 3.5651 | −9.3713 | −54.8328 |
| 27.711 | −104.9805 | −126.3428 | 160.4004 | 3.3014 | −9.6075 | −54.5691 |

**Table 3.** Partial experimental data

| Time(s) | Angle$_X$(deg) | Angle$_Y$(deg) | | Angle$_Z$(deg) |
|---------|---------------|---------------|------|---------------|
| 17.457 | 15.7700 | −253 | 100 | 0 |
| 17.457 | 15.7300 | −253 | 101 | 0 |
| ... | ... | ... | ... | ... |
| 17.620 | 15.7800 | −244 | 232 | 19 |
| 17.622 | 15.7400 | −243 | 232 | 19 |
| ... | ... | ... | ... | ... |
| 23.473 | 16.0700 | −478 | 647 | 53 |
| 23.475 | 16.0700 | −478 | 647 | 53 |
| ... | ... | ... | ... | ... |
| 27.710 | 16.2800 | −386 | 15 | −215 |
| 27.711 | 16.2400 | −387 | 18 | −217 |

Among the three different styles of lines in the figure, the dotted line is the x-axis, the one-by-one horizontal dotted line is the y-axis, and the solid line is the z-axis, which represents the parameter changes in the three axes.

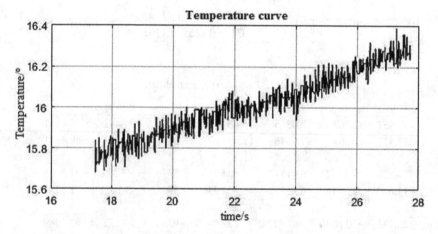

**Fig. 4.** Working temperature curve

When the handheld stabilizer moves, the chip record data is exported by software, and the experimental data shown in Tables 2 and 3 of 1 can be obtained. According to the experimental data, the MATLAB experimental platform is used to reflect the prediction and change of each measurement.

The change curve of acceleration is shown in Fig. 5. The changes of angular velocity and angle are shown in Figs. 6 and 7 respectively. According to the information in the figure, we can accurately obtain the acceleration change of the stabilizer

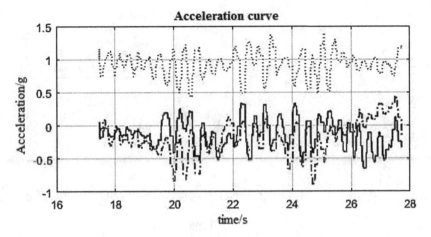

**Fig. 5.** Acceleration data record

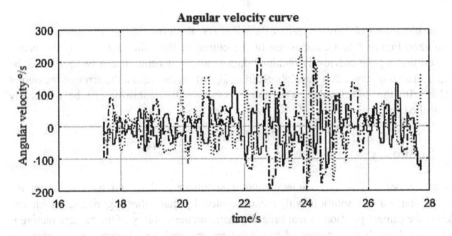

**Fig. 6.** Angular velocity data record

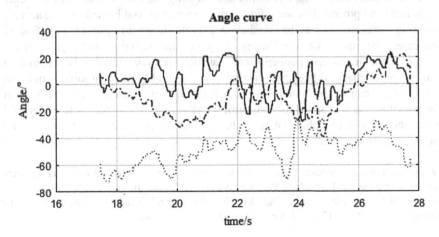

**Fig. 7.** Angle change data record

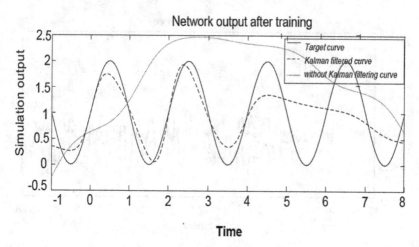

**Fig. 8.** Target tracking comparison chart

to calculate the stabilizer to predict the stabilizer. Position, adjust the shaft arm of the stabilizer. Figure 8 is a comparison of the output of the Kalman filter and the preset value by using the previously obtained data as an input value. It can be seen from the experimental diagram that the Kalman filter algorithm can adjust the error of the output value with the preset value. The motor output of the stabilizer can be effectively controlled.

# 7  Summary

In this paper, a five-axis camera balance stabilizer is designed by combining the quaternion attitude solution with the unscented Kalman filter algorithm, which can detect the camera position in real time. To maintain the stability of the camera picture is achieved through the control of the stabilizer manipulator. Advantage of using the Kalman filter algorithm is that it does not need to design any mathematical model, only need to learn and process data according to past experience, and is used for various data processing occasions such as data prediction, classification, and pattern recognition. In most cases, the results obtained using the unscented Kalman filter algorithm are superior to the traditional mathematical statistics method, and can be applied to occasions with many influencing factors and large changes. Through simulation experiments, it can be explained that the realization of the quaternion-based Kalman filter algorithm for fast and stable adjustment can achieve the expected function, and achieve the purpose of stable stabilizer shake when the photographer shoots, but the shooting picture is stable.

**Acknowledgements.** The research of this thesis is supported by the Guangxi Key R&D projects of Guangxi Science and Technology Program (AB17195042) and the Guangxi Innovation Drive Development Special Fund Project (AA1811009). We sincerely thank you. Special thanks to Shenzhen Wit Intelligent Co., Ltd. for their assistance in this design.

# References

1. Zou, Q., Fu, C., Mo, S.: Imaging, inertia and altitude combined navigation of four-axis aircraft based on Kalman filter. J. Transduct. Technol. **32**(1), 1–7 (2019)
2. Mu, S., Qiao, C.: Ground target localization method based on extended Kalman filter in small airborne photovoltaic platform. Acta Optica Sin. **39**(05), 0528001 (2019)
3. Zhu, H., Zhu, F., Pan, Z.: Object Tracking Method for Motion State and Scale Estimation. Comput. Sci. **44**(S2), 193–198 (2017)
4. Qiao, X.W., Zhou, W.D., Ji, Y.R.: Aircraft attitude estimation based on quaternion state switching unscented Kalman filter. Control Theory Appl. **29**(1), 97–103 (2012)
5. Cui, P.L., Zhou, Y.H., Lu, P., Hu, B.: Adaptive error quaternion unscented Kalman filter four-rotor attitude solution method. J. Xi'an Jiaotong Univ. **53**(3), y1–y7 (2019)
6. Zheng, J., Wang, H.Y., Yan, B.N.: Research on attitude calculation of four-rotor aircraft in wind disturbance environment. Comput. Simul. **35**(6), 41–45 (2018)
7. Zhou, G.T., Wang, Q.Q., Gao, Y.: Pedestrian navigation algorithm based on ZIHR heading angle correction method. Syst. Eng. Electr. **11**(1), 1–11 (2018)
8. Li, L., Zhao, L.Y., Tang, X.H., He, W., Li, F.R.: Gyro error compensation algorithm based on improved Kalman filter. J. Transduct. Technol. **31**(4), 538–544 (2018)
9. Huang, Y.G., Fan, Y.D., Chen, Z.R.: Attitude solution of four-rotor aircraft based on information fusion. Comput. Simul. **35**(9), 59–64 (2018)
10. Zhang, X.: Research on the algorithm of attitude and navigation information fusion for multi-rotor UAV. Chinese Academy of Sciences (2015)
11. Wu, Y.: Four-axis aircraft DIY-based STM32 microcontroller. Beijing Aerospace Publishing House, Beijing (2016)
12. Peng, X.D., et al.: Small UAV attitude estimation algorithm based on sensor correction and fusion. Acta Automatica Sin. Sin. Sin. **41**(4), 854–860 (2015)

# Mutual Learning Model for Skin Lesion Classification

Yanan Wang, Haiwei Pan[⊠], Bin Yang, Xiaofei Bian,
and Qianna Cui

Harbin Engineering University, Harbin, People's Republic of China
panhaiwei2006@hotmail.com

**Abstract.** Skin lesion classification in the dermoscopy images exerts an enormous function on the improvement of diagnostic performance and reduction of melanoma deaths. This skin lesion classification task remains a challenge. Deep learning requires a lot of training data, and the classification algorithms of skin lesions have certain limitations. These two points make the accuracy of the skin lesion classification needs to be further improved. In this paper, a mutual learning model was presented to separate malignant from benign skin lesions using the skin dataset. This model enabled dual deep convolutional neural networks to mutually learn from each other. Experimental results on the ISIC 2016 Skin Lesion Classification dataset indicate that the mutual learning model obtains the most advanced performance.

**Keywords:** Skin lesion classification · Deep learning · Mutual learning model

## 1  Introduction

As the maximal organ in the human body, the skin is usually directly exposed to the air, making skin disease one of the most common diseases in humans [1]. In the United States, there are 5.4 million new cases of skin cancer each year [2]. Melanoma, the most malignant skin tumor with the highest mortality rate, causes more than 9 000 deaths per year [3]. Early detection can enhance the 5-year survival rate of melanoma from 14% to 99%. Hence, early diagnosis and treatment of skin diseases are essential.

In clinic, besides general visual screening, dermatosis diagnosis based on skin images is the most commonly used means of diagnosis and treatment. Compared with other skin images, dermatoscope can visualize deeper lesions that are indistinguishable to the eye by eliminating reflections on the surface of the skin. Compared with visual inspection, the diagnostic sensitivity can be improved by 10% to 30% [4]. The biopsy rate has been reduced to some extent. However, artificial dermatoscopy image analysis is not only time-consuming and laborious, but the diagnosis results of that are susceptible to subjective factors such as doctor experience. With the help of the Computer-Aided Diagnosis (CAD) system, the automatic recognition and classification of skin diseases based on dermoscopy images can enhance the accuracy and efficiency of diagnosis.

However, the automatic skin lesion classification based on dermoscopy images remains a complex task. On the one hand, various classification algorithm models have

© Springer Nature Singapore Pte Ltd. 2019
R. Mao et al. (Eds.): ICPCSEE 2019, CCIS 1059, pp. 214–222, 2019.
https://doi.org/10.1007/978-981-15-0121-0_17

certain limitations, and the overall accuracy needs to be further improved; On the other hand, dermoscopic images will contain noise such as uneven illumination, black frame, hair, skin texture, and so on, which will also affect the classification of skin lesions to a certain extent.

Studies on the automatic classification of skin lesions images have emerged in the literature [5] since 1987. With the development of dermoscopic techniques and machine learning algorithms, dermatological classification based on dermoscopic image analysis and traditional machine learning algorithms has gradually become a trend. Literature [6] extracted feature information such as shape, color and texture after segmentation of lesions, and used integrated classifier to detect melanoma. Barata et al. [7] compared the different roles of color and texture features in melanoma detection. Sheha et al. [8] extracted chromaticity features from the segmented regions of interest, and selected the optimal features using Fisher and t-test methods to achieve the diagnosis of skin diseases. It is worth noting that these methods involve a series of complicated and inefficient problems such as image preprocessing, feature extraction and selection. The design of features is limited by the expertise of domain experts, and it is often necessary to extract distinguishing high quality features. There are certain difficulties. In the meantime, meaningful and representative high quality features are the key factors for the success of skin disease classification.

With the progress of deep learning, especially the successful application of the deep convolutional neural networks (DCNNs) in the field of computer vision [9], it laid the foundation for the application of DCNN in the field of medical image processing. Deep learning is an "end to end" model, which directly learns the mapping from the original input to the expected output. In this way, deep learning can automatically extract features compared with traditional machine learning algorithms. Besides it can extract the representative high-level features, especially in the aspect of fine-grained image recognition with great advantages and potential. Ge et al. [10] presented a new deep convolutional neural network model with a significant feature descriptor. The Stanford University Artificial Intelligence Laboratory adopted deep learning algorithms to classify dermoscopic images and clinical lesion images, and utilized the advantages of fine-grained labels to train the CNN-PA (CNN-Partitioning Algorithm) model for 129,450 dermoscopic and clinical skin lesion images. The relevant research results [11] were published in Nature in January 2017. Literature [12] combined with deep learning and machine learning methods to design a system for segmentation and classification of skin lesions. Mishra et al. [13] performed melanoma detection on the basis of summarizing and comparing various segmentation algorithms based on dermoscopy images. Adria Romero Lopez et al. [14] used VGG16 network and three different migration learning strategies to achieve the dichotomy task of the dermascopic image. Literature [15–18] adopted the commonly used deep learning model to achieve the detection of melanoma. In addition, some scholars considered skin lesion segmentation and classification as a consolidated issue [19]. Li et al. [20] realized the segmentation and classification of skin lesions simultaneously by using two full convolution residual networks. Zhang et al. [21] presented synergic deep learning model to classify skin lesion images. Liao et al. [22] presented a deep multi-task learning (MTL) framework to classify skin lesion images.

We present a mutual learning model to separate malignant from benign skin lesions using dermoscopy images. The characteristic of the proposed model is that dual DCNNs can learn image representations at the same time and can mutually learn from each other to improve the performance of each DCNN. We performed experiments on the ISIC 2016 skin lesion dataset [23] with a classification accuracy of 86.5% and an average precision of 67.3%.

## 2  Data and Materials

The ISIC 2016 Skin Lesion Classification dataset [23] is applied to this study and is published by the International Society for Skin Imaging (ISIC), consisting of 900 training sets and 379 test sets. The training set consists of 727 images of benign lesions and 173 images of malignant lesions. The test set includes 304 images of benign lesions and 75 images of malignant lesions.

## 3  Algorithm

The architecture of the mutual learning model can be seen in Fig. 1. During training, the input is the same skin image, and the image enters both SE-Inception-Resnet-V2 networks [25] at the same time. SE-Inception-Resnet-V2 network and SE-Inception-Resnet-V2$^2$ network both have a mimicry loss based on Kullback Leibler Divergence and a supervised learning loss in the course of training [24]. We will now go into the following sections in detail.

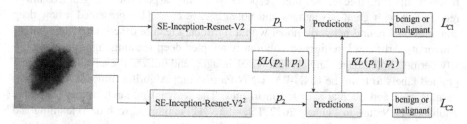

**Fig. 1.** Architecture of the proposed model.

### 3.1  Preprocessing and Data Augmentation

To unify the image size, we used the bicubic interpolation to resize each image to $299 \times 299 \times 3$ for SE-Inception-Resnet-V2 [25]. Deep neural networks generally need plenty of training data to obtain ideal results. In the case of limited data, Data Augmentation (DA) can be used to raise the diversity of dataset, improve the robustness of the model, avoid over-fitting, and obtain a stable model. To prevent overfitting, we generate four enhanced copies of each training image by vertical flipping, horizontal flipping, rotating and shearing. We add four enhanced copies to the training set, which can enrich the dataset.

## 3.2  SE-Inception-Resnet-V2

Convolutional Neural networks (CNNs) have achieved tremendous breakthroughs in plenty of areas recently. The heart of CNN is convolution kernel. Convolution kernels are treated as information polymerization. CNN is made up of a great many of convolutional and downsampling layers. Hence, CNN can catch plenty of features of an image and provide a better description of the image. However, it is quite difficult to learn a very powerful network. The difficulty lies in many aspects.

Recently, a lot of research has enhanced the network performance in the spatial dimension, such as embedding multi-scale information in the Inception structure [26]. ResNet [27] demonstrated a deeper, more powerful network by using identity-based skip connections. These jobs have achieved quite good results.

Squeeze-and-Excitation (SE) block [25] is a new architectural unit. It can improve the quality of the representation produced by the network. It can specifically simulate the interdependencies between channels of convolutional features to improve the quality of the representation produced by the network. SE block includes a lightweight gating mechanism that works to enhance the characterization capabilities of the network by modeling channel relationships in a computationally efficient manner. The SE block can be embedded in many network structures to improve performance. The SE-Inception-Resnet-V2 model [25] is derived by embedding the SE module into the Inception-Resnet-V2.

We use the SE-Inception-Resnet-V2 [25] model that has been pre-trained on the ImageNet dataset. In order to make SE-Inception-Resnet-V2 [25] adapt to our requirements, We fine-tune SE-Inception-Resnet-V2 [25]. The optimizer is the Mini-batch Stochastic Gradient Descent algorithm. The mini-batch size is 32.

## 3.3  Mutual Learning Model

The mutual learning model we proposed is shown in Fig. 1. For the training set $\{(x^{(1)}, y^{(1)}), \ldots, (x^{(m)}, y^{(m)})\}$, we have $y^{(i)} \in \{1, 2, \ldots, k\}$. $y^{(i)}$ indicates the label corresponding to sample $x^{(i)}$. $k$ indicates the number of classes. $z_1^j$ indicates the logit of the SE-Inception-Resnet-V2 network, which is the input of the softmax layer. The probability that the sample belongs to class $j$ is

$$p_1^j(x^{(i)}) = \frac{e^{z_1^j}}{\sum_{j=1}^{k} e^{z_1^j}} \tag{1}$$

For multiple classifications, the cross entropy loss of the SE-Inception-Resnet-V2 network is

$$L_{C1} = -\sum_{i=1}^{m} \sum_{j=1}^{k} 1\{y_i = j\} \log(p_1^j(x^{(i)})) \tag{2}$$

For the SE-Inception-Resnet-V2$^2$ network, $z_2^j$ represents the logit of the SE-Inception-Resnet-V2$^2$ network, which is the input of the softmax layer. The probability that the sample belongs to class $j$ is

$$p_2^j(x^{(i)}) = \frac{e^{z_2^j}}{\sum_{j=1}^{k} e^{z_2^j}} \tag{3}$$

The cross entropy loss of the SE-Inception-Resnet-V2$^2$ network is

$$L_{C2} = -\sum_{i=1}^{m}\sum_{j=1}^{k} 1\{y_i = j\} \log(p_2^j(x^{(i)})) \tag{4}$$

Relative entropy, also known as Kullback-Leibler divergence [24], measures the difference between two probability distributions in the same event space. The KL distance [24] from $p_1$ to $p_2$ is calculated as

$$KL(p_2 \parallel p_1) = \sum_{i=1}^{m}\sum_{j=1}^{k} p_2^j(x^{(i)}) \log \frac{p_2^j(x^{(i)})}{p_1^j(x^{(i)})} \tag{5}$$

The KL distance [24] from $p_2$ to $p_1$ is calculated as

$$KL(p_1 \parallel p_2) = \sum_{i=1}^{m}\sum_{j=1}^{k} p_1^j(x^{(i)}) \log \frac{p_1^j(x^{(i)})}{p_2^j(x^{(i)})} \tag{6}$$

The total loss of the SE-Inception-Resnet-V2 network is

$$L_1 = L_{C1} + KL(p_2 \parallel p_1) \tag{7}$$

The total loss of the SE-Inception-Resnet-V2$^2$ network is

$$L_2 = L_{C2} + KL(p_1 \parallel p_2) \tag{8}$$

By adding a mimicry loss based on Kullback Leibler Divergence [24] to the loss of each network, both networks can learn from each other to predict the correct label.

During training, we used two GPUs for distributed training. The two network models are optimised collaboratively and jointly in the optimization process. We calculate the predicated values of each model separately and update the parameters of the two networks based on the predictions of the other model in each iteration. The optimisation details are divided into the following steps:

1. determine the training sample and its corresponding label, and set two learning rates;
2. Then perform different initialization conditions on the two neural networks;
3. The process of gradient update is divided into the following steps:

(1) Select some samples randomly from the training samples as mini-batch, and then send the samples to SE-Inception-Resnet-V2 network and SE-Inception-Resnet-V2$^2$ network to get their corresponding outputs;
(2) On the first GPU, calculate the loss $L_1$ of the SE-Inception-Resnet-V2, and update the parameters of the SE-Inception-Resnet-V2 according to the gradient descent algorithm, and then send the samples of the mini-batch to the first network to obtain the corresponding predicted values.
(3) On the second GPU, calculate the loss $L_2$ of the SE-Inception-Resnet-V2$^2$, and update the parameters of the SE-Inception-Resnet-V2$^2$ according to the gradient descent algorithm, and then send the samples of the mini batch to the SE-Inception-Resnet-V2$^2$ network to obtain the corresponding predicted values.
(4) Cycle through steps (2) and (3) until it converges.

During the testing phase, the prediction of the SE-Inception-Resnet-V2 is $P^{(1)} = (p_1^{(1)}, p_2^{(1)}, \ldots, p_k^{(1)})$. The prediction of the SE-Inception-Resnet-V2$^2$ is $P^{(2)} = (p_1^{(2)}, p_2^{(2)}, \ldots, p_k^{(2)})$. The class label of the model we propose is defined as:

$$\arg\max_j \{ \sum_{i=1}^2 p_1^{(i)}, \ldots, \sum_{i=1}^2 p_j^{(i)}, \ldots, \sum_{i=1}^2 p_k^{(i)} \}, i = 1, 2 \tag{9}$$

## 4 Experimental Results

In Table 1, we can see AP, AUC, Acc values of the SE-Inception-Resnet-V2, SE-Inception-Resnet-V2$^2$ and the proposed mutual learning model. AP indicates average precision. Acc indicates classification accuracy. It indicates that DA can improve the performance of SE-Inception-Resnet-V2, SE-Inception-Resnet-V2$^2$ and the proposed mutual learning model. It demonstrates that the mutual learning model performs better than SE-Inception-Resnet-V2 no matter adopting or not adopting DA.

**Table 1.** Performance of SE-Inception-Resnet-V2, SE-Inception-Resnet-V2$^2$ and mutual learning model with or without DA.

| Methods | DA | AP | AUC | Acc |
|---|---|---|---|---|
| SE-Inception-Resnet-V2 | N | 0.641 | 0.806 | 0.854 |
| | Y | 0.654 | 0.814 | 0.856 |
| SE-Inception-Resnet-V2$^2$ | N | 0.643 | 0.807 | 0.855 |
| | Y | 0.656 | 0.815 | 0.857 |
| Proposed | N | 0.660 | 0.817 | 0.858 |
| | Y | 0.673 | 0.825 | 0.865 |

We compare our algorithm with four excellent algorithms on this dataset. Table 2 indicates the performance of the mutual learning model and four excellent algorithms.

CUMEND adopted the segmentation algorithm to segment the lesions and then classified them. The proposed algorithm achieved excellent results without adopting segmentation algorithms. The mutual learning algorithm obtained the best performance.

**Table 2.** Performance of five skin lesion classification methods. AP is the unique criterion.

| Methods | AP | AUC | Acc |
|---|---|---|---|
| BF | 0.598 | 0.826 | 0.834 |
| GTDL | 0.619 | 0.802 | 0.813 |
| CUMED | 0.637 | 0.804 | 0.855 |
| SDL | 0.664 | 0.818 | 0.858 |
| Proposed | 0.673 | 0.825 | 0.865 |

## 5  Conclusion

We present the mutual learning model to separate malignant from benign skin lesions using dermoscopy images. We use two pre-trained SE-Inception-Resnet-V2 networks. SE-Inception-Resnet-V2 network and SE-Inception-Resnet-V2$^2$ network both have a mimicry loss based on Kullback Leibler Divergence and a supervised learning loss in the course of training. Hence, the two models can mutually learn from each other to improve the performance of each DCNN. Experimental results on the ISIC 2016 Skin Lesion Classification dataset indicate that the mutual learning model obtains the most advanced performance.

**Acknowledgements.** This work was supported by the National Natural Science Foundation of China under Grant No. 61672181, No. 51679058, Natural Science Foundation of Heilongjiang Province under Grant No. F2016005. We would like to thank our teacher for guiding this paper. We would also like to thank classmates for their encouragement and help. We acknowledged the International Skin Imaging Collaboration (ISIC) for the publication of the ISIC 2016 Skin Lesion Classification Dataset. In the meantime, We would like to thank the scholars cited in this paper for their support and answers.

## References

1. Hay, R.J., et al.: The global burden of skin disease in 2010: an analysis of the prevalence and impact of skin conditions. J. Invest. Dermatol. **134**(6), 1527–1534 (2014)
2. Rogers, H.W., Weinstock, M.A., Feldman, S.R., Coldiron, B.M.: Incidence estimate of nonmelanoma skin cancer (keratinocyte carcinomas) in the U.S. population. Jama Dermatol. **151**(10), 1 (2015)
3. Siegel, R.L., et al.: Colorectal cancer statistics, 2017. Ca A Cancer J. Clin. **67**(3), 104–117 (2017)
4. Kittler, H.: Diagnostic accuracy of dermoscopy. Lancet Oncol. **3**(3), 159–165 (2002)
5. Cascinelli, N., Ferrario, M., Tonelli, T., Leo, E.: A possible new tool for clinical diagnosis of melanoma: the computer. J. Am. Acad. Dermatol. **16**(2), 361–367 (1987)

6. Schaefer, G., Krawczyk, B., Celebi, M.E., Iyatomi, H.: An ensemble classification approach for melanoma diagnosis. Memetic Comput. **6**(4), 233–240 (2014)
7. Barata, C., Ruela, M., Francisco, M., Marques, J.S., Mendonça, T.: Two systems for the detection of melanomas in dermoscopy images using texture and color features. IEEE Syst. J. **8**(3), 965–979 (2017)
8. Sheha, M.A., Sharwy, A.: Pigmented skin lesion diagnosis using geometric and chromatic features. In: 7th Cairo International Biomedical Engineering Conference, pp. 115–120 (2015)
9. Lecun, Y., Bengio, Y., Hinton, G.: Deep learning. Nature **521**(7553), 436–444 (2015)
10. Ge, Z., Demyanov, S., Chakravorty, R., Bowling, A., Garnavi, R.: Skin disease recognition using deep saliency features and multimodal learning of dermoscopy and clinical images. In: Descoteaux, M., Maier-Hein, L., Franz, A., Jannin, P., Collins, D.L., Duchesne, S. (eds.) MICCAI 2017. LNCS, vol. 10435, pp. 250–258. Springer, Cham (2017). https://doi.org/10.1007/978-3-319-66179-7_29
11. Esteva, A., et al.: Dermatologist-level classification of skin cancer with deep neural networks. Nature **542**(7639), 115–118 (2017)
12. Codella, N.C.F., et al.: Deep learning ensembles for melanoma recognition in dermoscopy images. IBM J. Res. Dev. **61**, 5:1–5:15 (2017)
13. Mishra, N.K., Celebi, M.E.: An overview of melanoma detection in dermoscopy images using image processing and machine learning. CoRR abs/1601.07843 (2016)
14. Lopez, A.R., Giro-i-Nieto, X., Burdick, J., Marques, O.: Skin lesion classification from dermoscopic images using deep learning techniques. In: 2017 13th IASTED International Conference on Biomedical Engineering (BioMed), pp. 49–54. IEEE, Innsbruck, Austria (2017)
15. Mirunalini, P., Aravindan, C., Gokul, V., Jaisakthi, S.M.: Deep learning for skin lesion classification. CoRR abs/1703.04364 (2017)
16. González-Díaz, I.: Incorporating the knowledge of dermatologists to convolutional neural networks for the diagnosis of skin lesions. CoRR abs/1703.01976 (2017)
17. Bi, L.: Dermoscopic image segmentation via multi-stage fully convolutional networks. IEEE Transact. Biomed. Eng. **64**(9), 2065–2074 (2017)
18. Devries, T., Ramachandram, D.: Skin lesion classification using deep multi-scale convolutional neural networks. CoRR abs/1703.01402 (2017)
19. Yang, X., Zeng, Z., Yeo, S.Y., Tan, C., Tey, H.L., Su, Y.: A novel multi-task deep learning model for skin lesion segmentation and classification. CoRR abs/1703.01025 (2017)
20. Li, Y., Shen, L.: Skin lesion analysis towards melanoma detection using deep learning network. Sensors **18**(2), 556 (2018)
21. Zhang, J., Xie, Y., Wu, Q., Xia, Y.: Skin lesion classification in dermoscopy images using synergic deep learning. In: Frangi, Alejandro F., Schnabel, Julia A., Davatzikos, C., Alberola-López, C., Fichtinger, G. (eds.) MICCAI 2018. LNCS, vol. 11071, pp. 12–20. Springer, Cham (2018). https://doi.org/10.1007/978-3-030-00934-2_2
22. Liao, H., Luo, J.: A deep multi-task learning approach to skin lesion classification. CoRR abs/1812.03527 (2018)
23. Gutman, D.: A challenge at the international symposium on biomedical imaging (ISBI). CoRR abs/1605.01397 (2016)
24. Ying, Z., Tao, X.: Deep mutual learning. In: The IEEE Conference on Computer Vision and Pattern Recognition (CVPR), pp. 4320–4328. IEEE, Salt Lake City (2018)

25. Jie, H., Shen, L.: Squeeze-and-excitation networks. In: The IEEE Conference on Computer Vision and Pattern Recognition (CVPR), pp. 7132–7141. IEEE, Salt Lake City (2018)
26. Christian, S., Wei, L.: Going deeper with convolutions. In: The IEEE Conference on Computer Vision and Pattern Recognition (CVPR), pp. 1–9. IEEE, Boston (2015)
27. Kaiming, H., Xiangyu, Z.: Deep residual learning for image recognition. In: The IEEE Conference on Computer Vision and Pattern Recognition (CVPR), pp. 770–778. IEEE, Las Vegas (2016)

# System

# Business-Oriented Dynamic Reconfiguration Model of Cloud Computing Network

Xin Lu, Lifeng Cao$^{(\boxtimes)}$, Xuehui Du, and Zhanbing Zhu

He'nan Province Key Laboratory of Information Security,
Zhengzhou 450001, Henan, China
1209774364@qq.com

**Abstract.** The complexity and diversity of the cloud business and the continuous growth of new services put forward higher requirements for business-oriented adaptive reconstruction of cloud networks. Therefore, by introducing the construction idea of reconfiguration network into cloud network, this paper designs a business-oriented dynamic reconfiguration model of cloud computing network. In the design process of the model, the formal description of the model reconfigurable goal, the target-tree decomposition method and the target order-graph relation representation method were proposed. A rapid-reconfiguration method based on similar node transformation, a specific reconfiguration process of the model and reconfiguration optimization algorithm were also presented in detail. The model provided an effective resolution to better realize the flexibility, scalability, security and self-adaptability of the network in the cloud environment, which ensures the reconfiguration continuity of the cloud network to meet ever-changing business requirements. Finally, the performance of the model is verified, which proves the high efficiency of the model the dynamic reconfiguration.

**Keywords:** Reconfiguration cloud network · Business-oriented ·
Reconfigurable virtual routing node · Meta-component ·
Reconfiguration optimization

## 1 Introduction

With the continuous development of the information age, cloud computing has become a new innovative concept in recent years, standing at the forefront of information technology. Cloud computing is based on data computing, storage, and networking [1]. As one of the important cornerstones of cloud computing, cloud network architecture design and evolution is an important part of cloud computing development. Nevertheless, researching cloud computing network architecture has involved many aspects such as reliability, scalability, security, migration flexibility, and communication performance. The traditional data center network for cloud computing has many limitations, such as low flexibility, poor scalability, service capability and basic capability of cloud network, which cannot meet the characteristics of the existing cloud computing [2].

With the development of cloud data center network, big two-tier network [3, 4] was a new network concept after the introduction of cloud computing which was once

© Springer Nature Singapore Pte Ltd. 2019
R. Mao et al. (Eds.): ICPCSEE 2019, CCIS 1059, pp. 225–245, 2019.
https://doi.org/10.1007/978-981-15-0121-0_18

defined as the next generation data center network. However, in the face of the increasing demand of cloud computing, problems such as broadcasting storm, low latency and poor flexibility of the two-tier network have arisen. Large-scale two-tier network technology was proposed to solve the above problems, such as TRILL/SPB/Fabric Path, which made the two-tier network expand well, but it had high requirements and high cost for network equipment. With the advancement of virtualization technology, new network technologies such as Overlay Virtualization Network Architecture [5] and Software Definition Network Architecture [6] have been proposed one after another. Overlay is a mode of virtualization technology superimposed on the network architecture that is mainly used for large-scale interconnection within the cloud data center. It achieves high flexibility and scalability but brings security and compatibility problems. SDN architecture was first proposed by ONF and has become a widely recognized network architecture in academia and industry. At present, OpenFlow is mainly used [7]. Inspired by SDN conceptual model, SDN-based cloud network architecture solutions [8–12] have been proposed. Although these solutions have shown good advantages in scalability, flexibility and performance, but some of them only support certain types of network environments or specific models of cloud network layout. In addition, security, the design of SDN switches and controllers, network management and other issues [13] have been gradually exposed.

It is worth mentioning that the development of reconfiguration network provides a new technology concept and development direction for the compatible evolution of cloud network architecture. Relevant research abroad includes the reference model of IEEE P1520, MSF, NPF, Soft Router of Lucent and so on. Relevant domestic research includes reconfiguration routing switching node platform operating on the novel network [14] and the architecture model—universal carrying network (UCN) [15] proposed by PLA Information Engineering University and scalable router structure [16] proposed by Tsinghua University. What's more, to ensure the autonomy, reactivity and initiative in the construct and reconfiguration process of the reconfiguration network, the construction principle of the component based hierarchical reconfiguration networks was proposed in [17] which also introduced the reconfiguration principle, model, function entity cooperate scheme, and the specific implement scheme. A service-oriented network architecture as a network convergence primitive and a prototype of reconfiguration router with detailed reconfiguration process were given in [18] so that provided more flexible and high-speed multi-service networks. Based on the thinking of network reconfiguration, the reconfigurable information communication basal network architecture was founded in [19] with the network atomic capacity theory, polymorphic addressing and routing mechanism, network reconfiguration mechanism and the secure manageability and controllability mechanism to present a resolution for current and future applications network architecture. The research on retrieval and classification methods of reconfiguration components [20, 21] provides an effective way for efficient retrieval of reconfiguration components. The resource reallocation problem was mathematically formulated as an optimization problem with minimum network reconfiguration overhead subject to QoS requirements of the applications flows and a forwarding table compression technique is devised decreasing the network reconfiguration overhead dramatically in [22]. A QoS satisfaction aware VNE algorithm was designed and then improved by network reconfiguration mechanisms that enable the reconfiguration of not only virtual networks but also substrate network in [23]. Among the existing research results, reconfiguration network exhibits good characteristics such as high flexibility and scalability, trustworthiness, manageability and controllability, convergence and compatibility, which can dynamically meet the complex and changing business needs and meet the characteristics of cloud computing network. However, there is no in-depth study on the framework model and key implementation mechanism of the network reconfiguration in cloud environment.

To solve the above problems, guided by the construction idea of reconfiguration network and referring to the overall architecture design of existing reconfiguration network, this paper designs a business oriented dynamic reconfiguration model of cloud computing network, which satisfies the characteristics of cloud computing network and the complex and diverse business requirements. In the design process of the model, the element definition, establishment thoughts, and formal description of the model is given. And then the formal representation of the model reconfigurable goal, the target-tree decomposition method and the target order-graph relation representation method are presented. What's more, a node rapid-reconfiguration method based on similar node transformation, the model specific reconfiguration method and reconfiguration optimization algorithm are also presented in detail. At last, the performance verification results of the model are given.

## 2  Overview of Models

Facing complex and diverse cloud businesses, the on-demand assembly of cloud network resources as well as the secure and flexible establishment of cloud communication networks are the key to ensuring continuous and reliable cloud services. Thus, based on the advantages of reconfiguration technology in flexibility, scalability and adaptability to network businesses, a business-oriented dynamic reconfiguration model of cloud network (RV_NetM) is proposed. This section mainly completes the model related concepts and ideas to lay a good foundation for the model's detailed design.

### 2.1  Model Element Definition and Relational Description

**Definition 1.** Reconfiguration cloud network (**RVNet**) refers to a dynamic cloud network that graphs the sub-graph of the underlying physical network in structure, is a part of the underlying resources, and achieves the functional objectives by adding or removing functional entities such as network meta-component in the cloud network, and then configures them accordingly to meet the changing needs of cloud businesses with less resource consumption. As shown in Fig. 2(a).

**Definition 2.** Meta-component refers to a functional unit that can complete an independent function in the process of cloud network reconfiguration, is reused in different environments, and can be reconfigured and easily extended in structure granularity, hierarchy and different dimensions according to the needs of cloud business.

It can be expressed as a five-tuple as $\mathbf{RU} = \{\text{ID}, \text{F}, \text{I}, \text{O}, \text{C}\}$, wherein, "ID" represents a unique identifier of the RU; "F" represents a function set of the RU,

$F = \{f_{u_1}, f_{u_2}, \ldots f_{u_i} \ldots, f_{u_k}\}$; "I" and "O" respectively denote an input and output interface set of the RU; $C$ denotes a trigger set of the RU. Meta-component is an encapsulated functional unit with black-box property, which interacts with external information through its own interface, that is, input from the input interface of meta-component, output from the output interface of meta-component after processing, and then input the input interface of the next meta-component. As shown in Fig. 1.

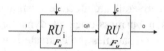

**Fig. 1.** Schematic diagram of Meta-component model.

**Definition 3.** Reconfigurable virtual routing nodes (Re − VR) refer to virtual routing nodes supported by underlying physical routing nodes, which are based on virtualization technology and component-based ideas. And they simulate the functions of physical routers in both hardware and software layers.

**Definition 4.** Reconfigurable virtual routing & switching platform (Re − VRP) refers to a virtualization platform that supports the reconfiguration cloud network logical to carry upper layer service and provides a Re − VR operating environment, which can ensure interconnected communication of the reconfiguration cloud network.

Re − VRP = {VMP, VCP, VEP, VTP, VIP, Link}, wherein, "VMP" is a reconfigurable management plane, which has global management control rights to the RVNet, and manages the meta-component library; "VCP" is a virtual controlling plane; "VEP" is a virtual switching plane; "VTP" represents a virtual forwarding plane; "VIP" represents a virtual accessing plane, and "Link" is a reconfigurable transmission link between different planes. As shown in Fig. 2(b).

**Definition 5.** The security tunnel (ST) refers to the secure transmission of data between different Re − VR nodes mainly through encapsulation technology. Both sides of communication need to negotiate Security Association SA. As shown in Fig. 2(a).

$ST_k = \ <T_k(Re − VR_i, Re − VR_j), RSA_k >$, wherein, $T_k$ represents the data transmission channel established by both sides of communication; $RSA_k$ is an agreement between the two sides of communication on security elements for the data's safe transmission, such as encryption and authentication algorithm, session key, etc.

**Definition 6.** Virtual forwarding table (VRF) refers to the exchange information table designed for data forwarding and exchanging in a secure tunnel to implement regular forwarding between different ST, ensuring the safe exchange within RVNet and the safe forwarding between cloud networks, which is used to determine the safe access relationship between reconfigurable virtual routing nodes and guides the network-level reconfiguration of nodes. As shown in Fig. 2(a).

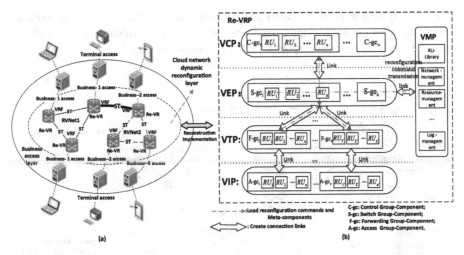

**Fig. 2.** Schematic diagram of reconfiguration cloud network architecture. The interaction relationship and deployment location among the elements of the model are described.

In Fig. 2(a), business-oriented reconfiguration cloud network is divided into three layers, the outer layer is the application layer to complete the terminal access, providing users with a visual access platform; the business-access layer mainly clusters and abstracts the characteristics and requirements of the cloud network tenants' businesses, and extracts basic network service elements required for each business to provide the basis for the reconfiguration; the cloud network dynamic reconfiguration layer mainly reconfigures the corresponding cloud network for the upper layer businesses and provides the cloud service pertinently while ensuring the maximization of the utilization of the underlying resources.

In Fig. 2(b), reconfigurable topology of the Re − VR on Re − VRP is described. Each Re − VR$_i$ contains a switching group-component as the core of the node, a controlling group-component, multiple forwarding group-components, and multiple accessing group-components. "VMP" interactively loads meta-components and reconfigurable commands through reconfigurable links and the "VEP". The multiple meta-components on the same reconfigurable plane are combined to form functional group-components. Functional group-components on different planes are connected by "Link". Thereby, a cloud network dynamic reconfiguration hierarchical architecture of "'RVNet' - 'Re − VR' - 'functional group – components' - 'RU'" is proposed.

## 2.2   The Establishment Thoughts of RV_NetM

The cloud network dynamic reconfiguration model is designed using a top-down approach. The upper-layer business requirements guide the dynamic reconfiguration of the cloud network to determine the network service capability of the cloud network; According to the network service capability requirements of cloud network, the allocations of the underlying basic transmission capacity and resources are determined. The functions of service quality assurance, flexible reconfiguration, security, trustworthiness, manageability and scalability are embedded in the architecture of cloud network to achieve business-oriented adaptive and on-demand services. According to the above ideas, the establishment thoughts of RV_NetM is shown in Fig. 3.

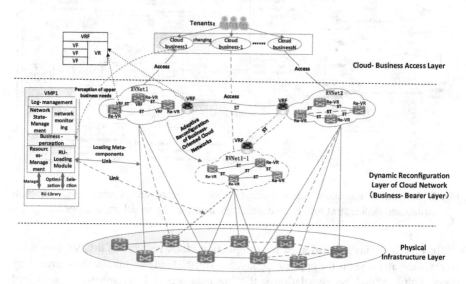

**Fig. 3.** Schematic diagram of RV_NetM Model Establishment. Reload to form a new reconfigurable cloud network RVNet1-1 for the changes of upper businesses and underlying resources.

1. By virtualizing the underlying physical infrastructure, RV_NetM model utilizes the hierarchical reconfiguration idea of "'**RVNet**' - 'Re − VR' - 'functional group – components' - 'RU'" to achieve the goal of reconfiguring the proprietary service bearer network for cloud businesses.
2. By perceiving the change of businesses demands, the model needs to determine the goals of cloud network reconfiguration, designs the reconfiguration methods, reconfigures the functional structure, number and connection relationship of cloud network nodes to better meet the needs of upper businesses.
3. By establishing virtual forwarding tables and security tunnels, and formulating secure communication rules, the connection relationship and secure communication among reconfigurable nodes can be determined to achieve network-level reconfiguration, while ensuring the safe transmission of data within and between reconfigurable cloud networks.
4. The optimization algorithm of model reconfiguration is designed to achieve the purpose of resources efficient utilization and improve the efficiency of reconfiguration.

# 3  Research on "RV_NetM" Model Construction Method

## 3.1  Formal Representation of Model

In the dynamic business-oriented reconfiguration of cloud networks, from the engineering point of view, the reconfiguration process of this model can be divided into: a. Determine the reconfigurable goals of the model for business needs; b. Decompose the reconfigurable goal, reduce the level of reconfiguration, and find the relationship between the goals; c. select a set of qualified meta-components that meet the criteria; d. Design and implement reconfiguration methods; e. Optimal reconfiguration Model. Thus, the formal definition of the model is as follows:

$$RV\_NetM = \sum_{\mapsto} < RG,RGT,RUT,REM,R\_opt > = \{(rg_i, RU_i)|rg_i \in RG \mapsto$$
$$< tree\_a(RG), graph(rg_i) > \in RGT \mapsto RU_i \in RU \mapsto REM < RU_i >$$
$$\mapsto RG \hookleftarrow R\_opt(RU)\}$$

Wherein, "$\mapsto$" indicates the direction of execution; "RG" represents the reconfigurable goal of the model; "RGT" represents the processing of reconfigurable goals, including the target-tree decomposition method "tree_a(RG)" and the order-graph relation representation method "graph(rg_i)"; "RUT" represents the candidate set of meta-components; "REM" represents the reconfiguration implementation of model, that is, model reconfiguration scheme; "R_opt" represents the reconfiguration optimization of the model. The formal description is as follows:

**1. Model Reconfigurable Goal (RG):** refers to the high-efficiency service state that the cloud network enters through reconfiguration to meet the new requirements of cloud business when the demand of cloud business changes.

RG = $<FG \triangleright EG <GT \Diamond GR> >$ , wherein, "FG" is the functional goal, "EG" is the performance goal, "$\triangleright$" means that the latter is satisfied under the premise of satisfying the former, and "$\Diamond$" means to satisfy both.

**Function Goal (FG):** refers to the set of goal functions of the reconfiguration cloud network after the reconfiguration is completed, which is consisted of sub-target functions, namely FG $= \{fg_1, fg_2, \ldots fg_i \ldots, fg_n\}$, $fg_i \cap fg_j = \emptyset$.

**Efficiency Goal (EG):** refers to the performance limitation of the reconfiguration cloud network, that is, the performance index, which is divided into "GR" and "GT", on the premise that reconfiguration networks have the required goal functions. "GR" represents the minimum system resources consumed by reconfigurable entities in the process of realizing reconfiguration functions, that is, GR $= \left[\min \sum r(e_{ij})\right] \Diamond$ $\{j \in [1, n], \forall i \in 1, \sum_{j=1}^{n} r(e_{ij}) \leq R(e_i)\}$, "$e_{ij}$" denotes the amount of resource "i" consumed by sub-function "j"; "$R(e_i)$" represents the total amount of resources "i"; "GT" represents the minimum time consumed to realize the goal function.

**2. Reconfigurable goal decomposition and relational representation (RGT):** refers to refine the reconfigurable goal, replace the high-level reconfigurable goal in the model with low-level reconfiguration sub-targets, and find the dependencies between the targets to guide the detailed design and implementation of the reconfiguration cloud network. There are tree_a(RG) and graph(rg$_i$) to represent the decomposition and relationship between the targets respectively.

- **The target-tree decomposition method. "tree_a(RG)":** hierarchical representation of the model's reconfigurable goal decomposition.

Let $RG = \{rg_i | i \in [1, n]\}$. If the realization of any sub-target is equal to the realization of the parent-target, it is called the "$\cup$" realization of RG, that is, $\bigcup_1^n rg_i \succ RG$; If the realization of the parent-target must require the realization of all the sub-targets, it is called the "$\cap$" realization of RG, that is, $RG \succ \bigcap_1^n rg_i$. "$\succ$" represents equivalent realization.

According to the above definition, RG can be decomposed into $\forall i \in [1, n], \forall j \in [1, m], RG = \left(\bigcup_1^n rg_i\right) \cup \left(\bigcap_1^m rg_j\right)$, so the tree_a(RG) can be expressed as follows:

$$\text{Tree\_a(RG)} = \sum L_k \left[\left(\bigcup_1^n rg_i\right) \cup \left(\bigcap_1^m rg_j\right)\right], \text{k represents the level.} \quad (1)$$

The feature description of Tree_a(RG): (a) the root node is the total goal RG, and the leaf nodes is the final sub-targets after the decomposition; (b) When the child nodes of the parent node are "$\cup$" relationship, they are called "$\cup$" node, and when they are "$\cap$" relationship, they are called "$\cap$" node. Different parent nodes can allow the existence of the same child nodes and exist independently of each other; (c) Each intermediate nodes of the target-tree can be further divided into smaller target-tree nodes. The complex reconfiguration process of cloud network needs the cooperation of various reconfigurable sub-targets. There may be resource competition and interactive communication among these sub-targets. Therefore, the relationship between reconfigurable sub-targets after decomposition needs to be expressed.

- **The target order-graph relation representation method. "map(rg$_i$)":** the representation of the relationship between the model reconfiguration sub-targets.

When the sub-nodes of a node in the tree_a(RG) are "$\cap$" relations, there are corresponding execution order problems among sub-nodes. This paper stipulates that the relationships among the sub-nodes of the "$\cap$" node are order, Parallel, cyclic and irrelevant.

According to the relationship between sub-nodes and the structure of the "tree_a (RG)", the corresponding order-graph of sub-targets is constructed. Each "$\cap$" node corresponds to an independent "graph$_k$(rg$_i$)", "k" is the level of the tree, "i" is the i$^{th}$ node of the k-layer from left to right, and the node in the graph is the sub-target of the "$\cap$" node, constructed as follows:

a. **Order.** It represents that the completion between rg$_i$ and rg$_j$ is sequential, divided into positive order and disordered trigger. As shown in Fig. 4. Positive order indicates that rg$_i$ must be completed before rg$_j$, which is transitive, and the connection of nodes is one-way solid line. Disordered trigger means that the completion of rg$_i$ and rg$_j$ is not in order, but they need to be executed separately. The connection of nodes is directed solid, and the direction depends on the specific needs. The formal description is as follows:

$$\text{Positive Order: } rg_i\_pre \wedge rg_j\_post == true \tag{2}$$

$$\text{Disordered Trigger: } select(rg_i, rg_j)\_pre \wedge remain(rg_i, rg_j)\_post == true \tag{3}$$

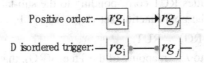

**Fig. 4.** Diagram of the order relationship.

**b. Parallel.** It means that $rg_i$ and $rg_j$ must be carried out simultaneously. The node connection is a two-way solid line. As shown in Fig. 4. The formal description is as follows (Fig. 5):

$$rg_i\_toge \wedge rg_j\_toge == true \tag{4}$$

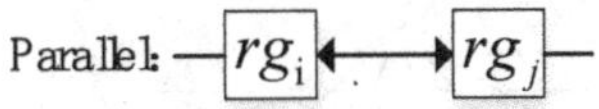

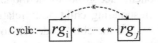

**Fig. 5.** Diagram of the parallel relationship.

**c. Cyclic.** It means that multiple sub-targets or single sub-target need to be completed many times in a cycle, and that the node connection is a dotted line in the direction of the cycle. As shown in Fig. 4. The formal description is as follows:

$$\exists RG_i, RG_j \subseteq RG, (RG_i \cup \ldots \cup RG_j) \wedge (r = 1) == true \tag{5}$$

"r" is a cyclic condition, and "∪" represents a cyclic execution (Fig. 6).

Cyclic:— $rg_i$ ←×— ... ←×— $rg_j$ —

**Fig. 6.** Diagram of the cyclic relationship.

**d. Irrelevant.** There is no relationship between the completion of the sub-targets, and the nodes are connected as undirected dashed line.

**3. Determining Candidate Sets of Meta-component (RUT):** The implementation of decomposed reconfigurable sub-targets is mapped to specific reconfigurable functional entities. Here, "RUT" is determined by the inverse of the goal function ($\lambda$ − mapping) and the reconfiguration of the entity tree "Tree_R(RU)".

- $\lambda -$ **mapping**: It can be seen from the Definition 2 of Sect. 2.1 that the meta-component $RU_i$ is input in the input interface "I", and is output from the output interface "O" under the trigger of condition "C", and then the function $F_U$ is completed to realize the sub-target $rg_i$, that is, $\lambda\left(RU_i : I \xrightarrow{c} o \Rightarrow F_U\right) = rg_i$. Here, the corresponding $RU_i$ is obtained by inverse mapping of the goal function $F_U$ under the sub-target $rg_i$, namely, $\lambda(F_U \overset{\sim}{\Leftarrow} rg_i) = RU_i$. Therefore, the candidate set of reconfigurable entities RUT corresponding to the sub-target can be obtained by inverse mapping of the goal function set $(F = \{F_{u_1}, \ldots F_{u_i} \ldots, F_{u_k}\})$ under the goal RG, namely, $\lambda(F \overset{\sim}{\Leftarrow} RG) = RUT$.
- **Tree_R(RU)**: refers to $\lambda -$ mapping of the Tree_a(RG), that is, $\lambda F \overset{\sim}{\Leftarrow} \widetilde{Tree}\_a(RG) = Tree\_R(RU)$, as shown in Fig. 7. Starting from the root node, the node is $\lambda -$ mapping layer by layer to obtain the reconfigurable entity tree Tree_R(RU).

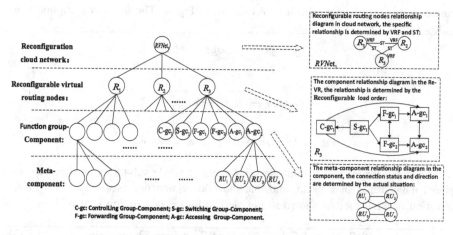

**Fig. 7.** Diagram of the Tree_R(RU).

Tree_R(RU) is divided into four layers of "cloud network–routing node–functional group-component–meta-component" according to the model reconfigurable target-tree; According to graph($rg_i$), a corresponding child node relationship graph is added for each layer node in Tree_R(RU), and the child node relationship establishment rule is the same as graph($rg_i$). The elements of the control node relationship should be represented in the graph, for example, the reconfigurable routing node relationship graph is a connection topology diagram of the routing node in the reconfiguration cloud network, the forwarding relationship is controlled by VRF, and the secure transmission relationship is controlled by ST, thus, ST and VRF should be reflected on the graph.

In addition, in the process of selecting meta-components, multiple RUs may be satisfied by $\lambda\left(F_U \overset{\sim}{\Leftarrow} rg_i\right) = RU_i$ and redundant meta-components are easily appearing in the RUT. So there needs to be a process of the reconfiguration optimizing scheme, that is, R_opt in the model (see Sect. 3.3).

## 3.2  Detailed Reconfiguration Scheme of the RV_NetM

In Sect. 3.1, the establishment of reconfigurable objectives, decomposition, and selection and analysis of reconfigurable functional entities have been described in detail. Therefore, this section designs the specific implementation scheme (REM) of the model. The reconfiguration scheme of the model is divided into node-level reconfiguration and network-level reconfiguration. The detailed design is as follows:

### Node-Level Reconfiguration of the Model

The node-level reconfiguration mainly represents the dynamic reconfiguration of the reconfigurable virtual routing node, which is realized by designing the Re-VR node reconfiguration representation method and the specific reconfiguration process. The reconfigurable virtual routing & switching platform includes multiple Re-VR nodes to support independent business-cloud networks. In the process of reconfiguration of cloud network, the number of Re-VR nodes will change dynamically. When it is necessary to reduce the number of nodes, it can be accomplished by deleting the corresponding nodes directly and adjusting the functions and relationships of other nodes. Therefore, this section mainly designs the method of adding reconfigurable nodes and realizes the rapid reconfiguration of nodes by using the reconfigurable method of similar node transformation.

### 1. RE-VR Node Rapid-Reconfiguration Method Based on Similar Node Transformation (SimNTM)

This method realizes the reconfiguration of new nodes by splitting and deforming similar components to reduce the impact of repeated retrieval meta-components on the reconfiguration efficiency. And the complex reconfiguration process of nodes is represented by an abstract tree transformation to guide the actual reconfiguration of nodes.

Suppose that a two-unit reconfigurable virtual routing node $Re - VR_k$ with two forwarding components is needed. Similarity function is defined as $\text{sim}(\text{Tree\_a}(Re - VR_k), \text{Tree\_a}(Re - VR_i))$ and uses cosine similarity method. $Re - VR_i$ is an existing node. The similarity function mainly finds the existing node $Re - VR_i$ with the largest similarity by using the reconfigurable target-tree as the node $Re - VR_k$. The formulas for calculating similarity is as follows:

$$\text{sim}(\text{Tree\_a}(Re - VR_k), \text{Tree\_a}(Re - VR_i)) = \frac{\sum_{j=1}^{n}(\mathbf{R}_{k,j} \times \mathbf{R}_{i,j})}{\sqrt{\sum_{j=1}^{n} \mathbf{R}_{k,j}^2} \times \sqrt{\sum_{j=1}^{n} \mathbf{R}_{i,j}^2}} \quad (6)$$

Wherein, $n = \text{number\_max}(Re - VR_k, Re - VR_i)$, that is, the maximum of leaf nodes in the two target-trees. Insufficient to use "0" to complement that the corresponding position has no leaf nodes, and the vectorization of sub-targets can be achieved by numbering sub-targets.

Assume that the similar node of $Re - VR_k$ is calculated by the similarity is $R_3$ in Fig. 7. Through the relation graph of the $R_3$ node and its child nodes, the tree-shaped split representation method of similar node transformation can be used to realize the logical representation of $Re - VR_k$ node reconfiguration. As shown in Fig. 8.

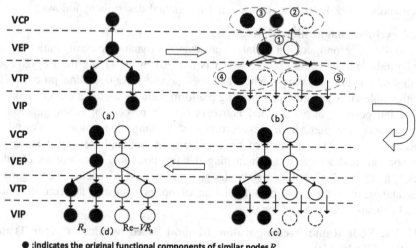

● :indicates the original functional components of similar nodes $R_3$
○ :indicates functional components of $Re-VR_k$
◌ :indicates the functional components to be loaded

**Fig. 8.** Diagram of the tree-shaped split representation of similar node transformation.

(a) According to the group-component relationship diagram of the $R_3$ node in Fig. 7, the $R_3$ tree-sharp structure is initialized. (b) Decouple the upper and lower nodes connections to facilitate nodes' reconfiguration and interconnection: ① is to Reconfigure Switching group-component Node; ②③ represent that a new control group-component node is reconfigured by splitting the control group-component node in $R_3$ and receiving the meta-component by the switching group-component node; ④⑤ represents reconfiguring the new forwarding group-components by separately splitting two forwarding group-component nodes in $R_3$ and receiving meta-component by the switching group-component node. (c) Reconfigure the connection relationship between group-component nodes. (d) Connect accessing group-component nodes, and finally complete the logical representation of tree reconfiguration of $Re - VR_k$.

## 2. Specific Process Design for Dynamic Reconfiguration of $Re - VR_k$

Under the guidance of the SimNTM, the actual reconfigurable process of the $Re - VR_k$ is shown in Fig. 9:

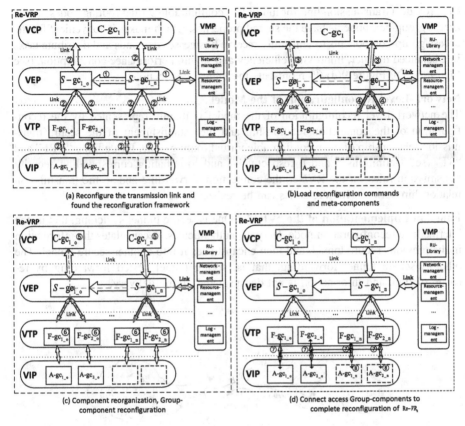

**Fig. 9.** Diagram of the specific process for dynamic reconfiguration of $Re - VR_k$

(a) According to the reconfiguration command of the "VMP", (see Fig. 9(a)-①) the switching group-component of the original node is split, and the meta-component is reloaded to form the switching group-component 1_o and the switching group-component 1_n. The transmission links with $S - gc_{1-o}$ and $S - gc_{1-n}$ as the core are established respectively, and the links between the upper and lower group-components are reconfigured to form a new reconfiguration framework. (see Fig. 9(a)-②).

(b) The "VMP" transmits reconfiguration commands and meta-components to the "VCP" (see Fig. 9(b)-③) and "VTP" (see Fig. 9(b)-④) through switching group-components along two sets of transmission links.

(c) Under the guidance of reconfiguration command, the existing control group-component are split, and the meta-components are reloaded to form $C - gc_{1-o}$ and $C - gc_{1-n}$. (see Fig. 9(c)-⑤). And the existing forwarding group-components are split and the meta-components are reloaded to form $C - gc_{1-o}$, $C - gc_{2-o}$ and $C - gc_{1-n}$ and $C - gc_{2-n}$. (see Fig. 9(c)-⑥).

(d) Though the new reconfiguration configuration requirements of the "VMP", connect the new access group-components to the nodes, and finally complete the reconfiguration. According to the link multiplexing principle, the same network link can provide a transmission link for multiple reconfigurable virtual routing nodes, so the access group-components have the shared properties, which makes it convenient for the access group-components to uniformly receive businesses demands and distribute tasks in a balanced manner, improving efficiency and saving performance consumption. (see Fig. 9(d)-⑦). Figure 9(d)-⑧ represents the logical connections.

### Network-Level Reconfiguration of the Model

The network-level reconfiguration of the model is mainly represented by the reconfiguration of the connection relationship and the establishment of the secure transmission relationship of the reconfigurable virtual routing node. By defining "VRF" and "ST" (see Definitions 5 and 6 respectively) and formulating corresponding security transmission rules for the RVNet, the connection, forwarding and secure transmission relationship between Re-VR nodes can be realized.

**Function Implementation of the "VRF":** One is secure tunnel switching, that is, the intermediate reconfiguration node decrypts, authenticates, decapsulates the information in the previous secure tunnel, and then encapsulates it into another secure tunnel; The other is secure tunnel forwarding, that is, secure tunnel nested forwarding by re-encapsulation, and information is not decapsulated during delivery.

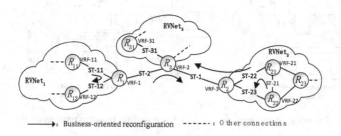

————▶: Business-oriented reconfiguration    ------: Other connections

**Fig. 10.** Diagram of the secure transport in reconfiguration cloud networks

As shown in Fig. 10, secure transmission mainly covers three categories, including the switching and forwarding between the intranet and the intranet tunnel, the extranet and the extranet tunnel, as well as the intranet and extranet tunnel. The intranet security transmission refers to the exchange or forwarding of the tunnels between the two communicating parties in the same cloud network, such as ST-11 and ST-12; intranet and extranet tunnel security transmission refers to the exchange or forwarding of the tunnels between the two communicating parties within the network and outside the network, such as ST-22 and ST-1; Extranet tunnel security transmission is also called cross-network transmission, such as ST-2 and ST-1. In general, cross-network transmission is carried out by means of secure tunnel forwarding.

**Security Transmission Rule 1. Switching Rule.** Assume that the communication ends of the secure tunnel ST-11 are $R_{11}$ and $R_1$, the communication ends of ST-12 are $R_1$ and $R_{12}$, and the delivery information is Data, then the security exchange rule is:

$$if\,((Data \xrightarrow{\text{from } R_{11} \text{ to } R_{12} \text{ according to VRF}} R_{12})\,\&\&\,R_1 \leftarrow visible(Data))$$

then {

$$ST\text{-}11(Data) \xrightarrow{VF} \overset{\substack{decap(ST-11(Data)) \\ ST-12(Data)=encap(data)}}{R_1} \rightarrow ST\text{-}12(Data) \xrightarrow{VRF} \overset{decap(ST-12(Data))}{R_{12}}$$

}

Wherein, "Visible()" is a visual function, and "visible(data)" means visible to data. "Encap" represents a secure channel encapsulation, "decap" represents a secure channel decapsulation, the secure switching is shown, for example, in RVNet$_1$ in Fig. 10. The essence of secure tunnel switching is that the data is protected by a secure tunnel, the data is visible at the reconfigurable switching node, and is re-encrypted.

**Security Transmission Rule 2. Forwarding Rule.** Assume that the communication ends of the secure tunnel ST-22 are $R_{21}$ and $R_2$, the communication ends of ST-23 are $R_2$ and $R_{22}$, the communication ends of ST-21 are $R_{21}$ and $R_{22}$, and the delivery information is Data, then the security exchange rule is:

$$f\,((Data \xrightarrow{\text{from } R_{21} \text{ to } R_{22} \text{ according to VRF}} R_{22})\,\&\&\,R_2, R_{21} \leftarrow invisible(Data))$$

then {

$$ST\text{-}21(Data) \rightarrow \overset{ST-22(ST-21(Data))=encap(ST-21(Data))}{R_{21}} \xrightarrow{\frac{ST-22(ST-21(Data))}{VRF}}$$

$$\overset{\substack{ST-21(Data)=decap(ST-22(ST-21(Data))) \\ ST-23(ST-21(Data))=encap(ST-21(Data))}}{R_2} \xrightarrow{\frac{ST-23(ST-21(Data))}{VRF}} \overset{\substack{ST-21(Data)=decap(ST-23(ST-21(Data))) \\ m=decap(ST-21(Data))}}{R_{22}}$$

}

Wherein, "invisible()" is an invisible function, and "invisible(data)" means invisible to data. The secure switching is shown, for example, in RVNet$_2$ in Fig. 10. The essence of secure tunnel forwarding is that one channel is nested and encapsulated by another channel, and then data is forwarded along the new channel.

### 3.3 Reconfiguration Optimization Algorithm (R_opt) of the RV_NetM

According to the optimization requirements of the model, the candidate set of meta-components is optimized to achieve the goal of minimizing the cost of resources and time and improve the efficiency of reconfiguration. Thus, a quadratic optimization algorithm for weighted coverage set of meta-components is proposed.

Suppose that the goal function set is $E = \{f_{e_1}, f_{e_2}, \ldots\ldots, f_{e_m}\}$, the function set of meta-component $RU_i$ is $F = \{f_{u_1}, f_{u_2}, \ldots f_{u_i}\ldots, f_{u_k}\}$, The set of meta-components to be optimized $RUT = \{f_{RU_1}, f_{RU_2}, \ldots\ldots, f_{RU_L}\}$. Orthogonality between Functional Elements. The total available resources are $R = \{r_1, r_2, \ldots r_i\ldots, r_n\}$. $RU_i$ consumes the resource cost vector is $\mathbf{r_{RU_i}} = (r_1', r_2', \ldots\ldots, r_h')$, the time cost vector is $\mathbf{T_{RU_i}}$. In order to weigh the resource cost and time cost, the weighting coefficients $\delta_r$ and $\delta_t$ are introduced, then the total cost vector of $RU_i$ is $V_i = \delta_r r_{RU_i} + \delta_t T_{RU_i}$, wherein, $\delta_r + \delta_t = 1$, $r_{RU_i} = \varepsilon_1 r_1' + \varepsilon_2 r_2' + \ldots + \varepsilon_h r_h'$, $\varepsilon_1 + \varepsilon_2 + \ldots + \varepsilon_h = 1$. The algorithm is as follows:

**Step 1.Select(F,E$_o$):**
**Input:** $RU_i$ function set F, the goal function set E.
**Output:** Approximate set of goal functions $E_o$.

1.  $E_o = \emptyset$;
2.  $U = E$;                // Initialization.
3.  While $U \neq \emptyset$ do Select $f_{RU_i} \in RUT$ and $f_{RU_i} \cap U \neq \emptyset$ that minimize
    $$select(RU_i) = \frac{V_i}{|f_{RU_i} \cap U|}$$
4.     $U = U - f_{RU_i}$;
5.     $E_o = E_o \cap \{f_{RU_i}\}$;
/*Select a $RU_i$ with more goal functions and lower weights until "E" is covered. */
6.  Return $E_o$;

**Step 2.Amend(E$_o$,E$_a$):**        // Check and remove redundant $RU_i$ in $E_o$,
**Input:** $E_o$, E
**Output:** Goal function optimization set $E_a$.

7.  $M = |E_o|, \forall i \in [1,k]$,counter[i] = 0,goal[i] = 0;
/* k is the number of all functions in $E_o$, and M is the number of RU in $E_o$.*/
8.  for(i = 1,i ≤ M,i ++)
9.    for(j = 1,j ≤ k,j ++)
10.      if $(f_j \in f_{RU_i})$ counter[j] = counter[j] + 1;
11.  for(j = 1,j ≤ k,j ++)
12.    if ($f_j \in E$)   goal[j] = goal[j] + 1;
13.  Sort $RU_i(\frac{V_i}{|f_{RU_i} \cap U|})$;

/* Sort the $RU_i$ in ascending order by $\frac{V_i}{|f_{RU_i} \cap U|}$ using the quick sort method.*/
14.  for(i = 1,i ≤ M,i ++)
15.    select(RU$_i$);
16.    for(j = 1,j ≤ k,j ++)
17.      tag[j] = 0;
18.      if(f$_j \in f_{RU_i}$)  tag[j] = tag[j] + 1;
19.    if all counter − tag ≥ goad
/* Determine whether $RU_i$ is a redundant meta-component.*/
20.      Delete $RU_i$ from $E_o$;
21.      counter = counter − tag;// Delete redundant meta-component.
22.  Return $E_o$;

The optimization algorithm is divided into two parts, the first step is to use a weighted set coverage algorithm to select a set of meta-components that can cover the target function and have less resource and time cost; the second step re-optimizes the set of meta-components by using the idea of heuristic optimization set to remove redundant meta-components.

# 4    Performance Verification of the Model

This experiment platform uses python3.6.1, Intel(R) Core(TM) i5-4200 CPU @ 1.60 GHz 2.30 GHz, 4 GB memory, and the system environment is Windows 10.

## 4.1    Performance Verification of the SimNTM

According to the component information in the reconfigurable routing exchange platform component library [20], Assume that the number of meta-components in the meta-component library is 3500, and numbers 1, 2, ..., the meta component retrieval time is about 0.82 s [22]. In order to facilitate the calculation, the number of leaf nodes of Tree_a(Re − VR) when the reconfigurable virtual routing node is reconfigured in the experiment is set to [50, 120]. The leaf nodes set of Tree_a(Re − VR) randomly is generated. The numbers of the leaf nodes are consistent with the number of meta-components in component library by "λ − mapping".

Using the "cosine_similarity()" function in "sklearn.metrics.Pairwise" library to calculate similarity. Based on the above assumptions and experimental environment, the number of leaf nodes in the reconfigurable target-tree "Tree_a(Re − $VR_k$)" that is to be reconfigured is generated randomly by programming. Ten kinds of experimental nodes (Re − $VR_i$) with similarity degree of [0, 0.1], (0.1, 0.2], (0.2, 0.3]......, (0.9, 1] to the number of leaf nodes in the Tree_a(Re − $VR_k$) are separately generated. Randomly generate 80 experimental nodes for each type of node Re − $VR_i$. And then the reconfiguring node Re − $VR_k$ are reconfigured according to experimental nodes Re − $VR_i$. Figure out separately time consumption of the calculating node similarity and Meta-component retrieval. Calculate the time saved by using the similar node transformation method (SimNTM). The formula is as follows:

$$T_S = \sum_{i=1}^{n} T_i - \left( \sum_{i=1}^{m} T_i + \sum_{j=1}^{z} K_i \right) \tag{7}$$

Wherein, $T_S$ is the time saved, $\sum_{i=1}^{n} T_i$ is the time (Time level–"s") consumed to retrieve the meta-component directly. $\sum_{i=1}^{m} T_i$ is the time consumption of meta-component retrieval by SimNTM. $K_i$ is the similarity calculation time (Time level–"ms") (Fig. 11).

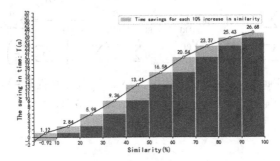

**Fig. 11.** Diagram of saving time for node reconfiguration using "SimNTM".

In this experiment, when the similarity is about 3%, the saving time $T_S$ begins to exceed 0. According to the characteristics of routing nodes with partial functional similarity, the "SimNTM" method can effectively save the reconfiguration time of new nodes and improve the reconfiguration efficiency.

## 4.2  Performance Test of the Model Optimization Algorithm (R_opt)

The relative approximation ratio (RAR) is an effective criterion for the degree of optimization of the set. Therefore, the performance of the R_opt algorithm is verified by obtaining the RAR.

**1. The RAR of the results by Select(F, E$_o$):** The set coverage problem has proven to be an NP-complete problem, The Select(F, E$_o$) algorithm in R_opt proposed is a polynomial approximation algorithm, which can find an approximate optimal set of the RUT. The proof is as follows:

Suppose that $E^*$ is the optimal solution set of the RUT, E is the goal function set, and $f_{RU_i}$ is the $i^{th}$ function set selected in step 3 of the algorithm, and the weight is $V_i$. Assign $V_i$ equally to the first selected goal functional element $f_i \in E$ in $f_{RU_i}$.

$$V_{f_i} = \frac{V_i}{|f_{RU_i} - (f_{RU_1} \cup f_{RU_2} \cup ... \cup f_{RU_{i-1}})|},$$ Let $V_{E_0} = \sum_{f_i \in E} V_{f_i}$ denote the weight generated after executing Select(F, E$_o$), that is, the preliminary approximate optimal solution OPT$_{1st}$. Because the function set of each component in $E^*$ may intersect and $f_i$ in E doesn't crossover. The following formula is obtained:

$$V_{E_0} \leq \sum_{f_{RU_i} \in E^*} \sum_{f_i \in f_{RU_i}} V_{f_i} = \text{OPTIMAL}(V) \tag{8}$$

Valuate the $\sum_{f_i \in f_{RU_i}} V_{f_i}$: $\forall f_{RU} \in F_{RU}$, Let $\beta_i = |f_{RU} - (f_{RU_1} \cup f_{RU_2} \cup ... \cup f_{RU_i})|$ be the number of goal functions that are not covered in $f_{RU}$ after $f_{RU_1}, f_{RU_2}, ..., f_{RU_i}$ is selected first. Let $\beta_0 = |f_{RU}|$ and "k" be the first subscript satisfying $\beta_k = 0$. Where each element in $f_{RU_k}$ is covered by $\{f_{RU_1}, f_{RU_2}, ..., f_{RU_i}\}$. Obviously, $\beta_{i-1} \geq \beta_i$, $\beta_{i-1} - \beta_i$ is the number of elements covered by $f_{RU_i}$ for the first time in $f_{RU}$. Let $H(n) = \sum_{i=1}^{n} \frac{1}{i}$.

$$
\begin{aligned}
\sum_{f_i \in f_{RU_i}} V_{f_i} &= \sum_{i=1}^{k} (\beta_{i-1} - \beta_i) \frac{V_i}{|f_{RU_i} - (f_{RU_1} \cup f_{RU_2} \cup ... \cup f_{RU_{i-1}})|} \\
&\leq \sum_{i=1}^{k} (\beta_{i-1} - \beta_i) \frac{V_{RU}}{|f_{RU} - (f_{RU_1} \cup f_{RU_2} \cup ... \cup f_{RU_i})|} = \sum_{i=1}^{k} \sum_{j=\beta_i+1}^{\beta_{i-1}} \frac{V_{RU}}{\beta_{i-1}} \\
&\leq \sum_{i=1}^{k} \sum_{j=\beta_i+1}^{\beta_{i-1}} \frac{V_{RU}}{j} = V_{RU} \sum_{i=1}^{k} \left( \sum_{j=1}^{\beta_{i-1}} \frac{1}{j} - \sum_{j=1}^{\beta_i} \frac{1}{j} \right) \\
&\leq V_{RU} \sum_{i=1}^{k} (H(\beta_{i-1}) - H(\beta_i)) = V_{RU}(H(\beta_0) - H(\beta_k)) = V_{RU} H(\beta_0) \\
&= V_{RU} H(|f_{RU}|) \leq V_{RU} H(\max\{|f_{RU}| : f_{RU} \in F_{RU}\})
\end{aligned}
$$

$$\tag{9}$$

Therefore, it can be known from Eqs. (8) and (9):

$$V_{E_0} \leq \sum\nolimits_{f_{RU_i} \in E^*} \sum\nolimits_{f_i \in f_{RU_i}} V_{f_i} \leq \sum\nolimits_{f_{RU_i} \in E^*} V_{RU} H(\max\{|f_{RU}| : f_{RU} \in F_{RU}\})$$

$$= H(\max\{|f_{RU}| : f_{RU} \in F_{RU}\}) \sum\nolimits_{f_{RU_i} \in E^*} V_{RU} \qquad (10)$$

$$= OPTIMAL(V) * \sum\nolimits_{i=1}^{|f_{RU}|} 1/i, (\max\{|f_{RU}| : f_{RU} \in F_{RU}\})$$

Meta-component under the fine-grained division has the single function, making $|f_{RU}|$ relatively small. Therefore, $OPT_{1st}$ can generally be controlled at 1–3 times.

**2. Amend($E_0$, $E_a$) optimization experiment proof:** Assume that the number of meta-components is 40 and the total number of atomic functions is 100 in the Meta-component Library. Programming randomly generates a time cost between 0.000–1.999 for each atomic function, accurate to three decimal places. And the total number of system resources is 20. Assume that each resource is divided into ten parts, the cost of each resource is 1. The weights of time and resource cost are 0.5 respectively. The total cost is the algebraic sum of time and resource costs. Randomly select x as an atomic function number of the meta-component in the interval [1, 15], and randomly select x atomic functions from $E_0$. Randomly generate a number y as the number of resources required for each atomic function in the interval [1, 10], and the number of resources used by all Meta-components must not exceed the total number of resources, that is, meet the reconfigurable performance goal in Sect. 3.1. The goal function set of each experiment is randomly generated by the system. The preliminary approximate optimal solution set $E_0$ is obtained by Amend($E_0$, $E_a$).

Here, 15 optimization experiments on $E_0$ using Amend($E_0$, $E_a$), repeated 10 times for each experiment(The goal function set in each experiment needs to be regenerated). Average the RAR for each experiment, as shown in Fig. 12. After optimization, the RAR can be controlled between 1 and 1.17. Obviously, the experimental results are approximately equal to the optimal solution and reach the desired effect.

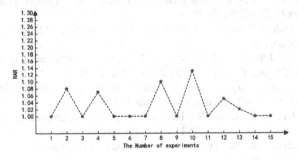

**Fig. 12.** Diagram of RAR comparison results between experimental optimization solutions and optimal solutions

# 5 Conclusion

Based on the idea of business-oriented service-on-demand in cloud network, by introducing the construction idea of reconfiguration network into the reconfigurable process of cloud network, this paper designs a business-oriented dynamic reconfiguration model of cloud computing network, gives the element definition, establishment thoughts and formal description of the model. In the design process of the model, the formal representation of the model reconfigurable goal, the target-tree decomposition method and the target order-graph relation representation method are proposed. What's more, a node rapid-reconfiguration method based on similar node transformation, the model specific reconfiguration method and reconfiguration optimization algorithm are also presented in detail. By skillfully integrating reconfigurable technology into the adaptive reconstruction process of cloud network facing environmental and business requirements, not only the flexibility, expansibility and adaptablity of cloud network facing business changes are improved, the manageability and controllability of cloud network are realized, but also the consumption of network resources is reduced, as well as the continuity and security requirements of multi-tenant business services are guaranteed. Finally, the performance of the model is verified, and the efficiency of dynamic reconfiguration of the model is proved. However, this paper only detailed designs and preliminarily validates the dynamic reconfiguration model of cloud network and lacks the overall realization of the model in engineering. Therefore, the next step is to build an experimental network of reconfigurable cloud network model to verify the application effect of the model.

**Acknowledgements.** This work is supported by the National Natural Science Foundations of China (grant No. 61502531 and No. 61702550) and the National Key Research and Development Plan (grant No. 2018YFB0803603 and No. 2016YFB0501901).

# References

1. Cui, Y., Song, J., Miao, C.C.: Mobile cloud computing research progress and trends. Chin. J. Comput. **40**(2), 273–295 (2017)
2. Li, D.S.: Research on Structure of Scalable Data Center Networks. Master, Dalian University of Technology (2017)
3. Wang, B.F., Su, J.S., Chen, L.: Review of the design of data center network for cloud computing. J. Comput. Res. Dev. **53**(9), 2085–2106 (2016)
4. Zhang, G.P.: Data center network for cloud computing based on SDN and overlay technology. China New Telecommun. (3), 109–111 (2015)
5. Hang, Z.L., Fan, Y.B., Chen, N.: Key technologies of heterogeneous compatibility in overlay SDN. Telecommun. Sci. **32**(11), 112–118 (2016)
6. Zhang, S.M., Zou, F.M.: Survey on software defined network research. Appl. Res. Comput. **30**(8), 2246–2251 (2013)
7. Li, C.S., Liao, W.: Software defined networks. IEEE Commun. Mag. **51**(2), 113–114 (2013)
8. Rao, S.Y., Chen, Y.Q., Feng, M.: Cloud data center based on SDN. Telecommun. Sci. (8), 33–41 (2014)

9. Bankazemi, M., Olshfski, D., Shaikh, A., et al.: An SDN platform for cloud network services. IEEE Commun. Mag. **51**(2), 120–127 (2013)
10. Zhang, C.K., Cui, Y., Tang, H.H.: State-of-the-art survey on software-defined networking (SDN). J. Softw. **26**(01), 62–81 (2015)
11. Yan, M.L., Zheng, L., Chen, Q.C.: Cloud computing networks based on SDN. Inf. Commun. Technol. (2), 60–66, 77 (2015)
12. Li, Y., Zhu, X.G.: Future network architecture based on SDN and NFV technologies. Commun. Technol. **50**(5), 956–961 (2017)
13. Wang, M.M., Liu, J.W., Chen, J.: Software defined networking: security model, threats and mechanism. J. Softw. **27**(4), 969–992 (2016)
14. Wang, B.Q., Wu, J.X.: Development trends and associated countermeasures analysis for NGN. J. Inf. Eng. Univ. **10**(1), 1–6, 17 (2009)
15. Wang, H.X., Wang, B.Q., Yu, J.: Research on architecture of universal carrying network. Chin. J. Comput. **32**(3), 371–376 (2009)
16. Zhang, X.P., Liu, Z.H., Zhao, Y.J., et al.: Scalable router. J. Softw. **19**(2), 1452–1464 (2008)
17. Liu, Q., Wang, B.Q., Xu, K.: Construction and reconfiguration scheme of the hierarchical reconfiguration network based on the components. Chin. J. Comput. **33**(9), 1557–1568 (2010)
18. Hu, Y., Lan, J., Wu, J.: Providing personalized converged services based on flexible network reconfiguration. Sci. China Inf. Sci. **54**(2), 334–347 (2011)
19. Lan, J.L., Chen, D.N., Hu, Y.X.: Research on reconfigurable information communication basal network architecture. J. Commun. (1), 128–139 (2014)
20. Qu, J.: Research on Component Retrieval of Reconfigurable Routing Switching Platform. Master, PLA Information Engineering University (2010)
21. Wang, W.X.: A component retrieval method based on classified policy and cluster index tree. Comput. Technol. Dev. **26**(4), 110–113 (2016)
22. Qian, X.J., Du, S.H.: Component retrieval method based on identification of faceted classification and cluster tree. J. Comput. Appl. **37**(10), 2973–2977 (2017)
23. Tajiki, M.M., Behzad, A., Nader, M.: Optimal QoS-aware network reconfiguration in software defined cloud data centers. Comput. Netw. **120**, 71–86 (2017)
24. Han, P.C., Liu, Y.J., Guo, L.: QoS satisfaction aware and network reconfiguration enabled resource allocation for virtual network embedding in Fiber-Wireless access network. Comput. Netw. **143**, 30–48 (2018)

# Secure, Efficient and Searchable File System on Distributed Clouds

Ximing Li, Weizhao Chen, Yubin Guo$^{(\boxtimes)}$, Sha Ma, and Qiong Huang

College of Mathematics and Informatics, South China Agricultural University,
Guangzhou, China
`guoyubin@scau.edu.cn`

**Abstract.** Many enterprises and personals are inclining to outsource
their data to public clouds, but security and privacy are two critical problems cannot be ignored. The door of cloud provider may be broken, and
the data may also be dug into by providers to find valuable information.
In this paper, a secure and efficient storage file (SES_FS) system is proposed to distribute files in several clouds and allows users to search the
files securely and efficiently. In the proposed system, keywords were transformed into integers and secretly shared in a defined finite field, then
the shares were mapped to random numbers in specified random domain
in each cloud. Files were encrypted with distinct secret key and scattered within different clouds. Information about keyword/file was secretly
shared among cloud providers. Legal users can search in the clouds to
find correct encrypted files and reconstruct corresponding secret key. No
adversary can find or detect the real file information even they can collude all the servers. Manipulation on shares by one or more clouds can
be detected with high probability. The system can also detect malicious
servers through introduced virtual points. One interesting property for
the scheme is that new keywords can be added easily, which is difficult
and usually not efficient for many searchable symmetric encryption systems. Detailed experimental result shows, with tolerable uploading delay,
the scheme exhibits excellent performance on data retrieving aspect.

**Keywords:** Secret sharing · Secure searchable file system ·
Distributed clouds

## 1 Introduction

For economic, efficiency and security reasons, small and medium-sized economic
and social organizations are intending to outsource the fast-growing business

This work is supported by Demonstration on the Construction of Guangdong Survey and Geomatics Industry Technology Innovation Alliance (2017B090907030), The
Demonstration of Big Data Application for Land Resource Management and Service
(2015B010110006). Qiong Huang is supported by Guangdong Natural Science Funds
for Distinguished Young Scholar (No. 2014A030306021), Guangdong Program for Special Support of Top-notch Young Professionals (No. 2015TQ01X796), Pearl River Nova
Program of Guangzhou (No. 201610010037), and the National Natural Science Foundation of China (Nos. 61472146, 61672242).

© Springer Nature Singapore Pte Ltd. 2019
R. Mao et al. (Eds.): ICPCSEE 2019, CCIS 1059, pp. 246–266, 2019.
https://doi.org/10.1007/978-981-15-0121-0_19

data to cloud service providers, and select multiple cloud service providers for better data security availability [27]. The departure of data owner and data users accelerates the generation of secure distributed storage and retrieval requirements. However, one of the top concerns of cloud users and would-be users remains security.

According to the work in [19,25], quantum computing is becoming more and more close to practicality. The related secret sharing technology represented by Shamir's secret sharing method is one of the best choices for post-quantum cryptography for its property of fast computing speed and anti-quantum attack.

On the other side, to handle data privacy, this can be achieved via an encryption scheme that allows evaluating conditions on encrypted data, i.e., homomorphic encryption scheme. Unfortunately, however, the practical applicability of homomorphic schemes to the cloud has been limited by their complexity. Conceptually, homomorphic encryption technology can almost completely solve the encryption problem on cloud database. However, most of full homomorphic encryption schemes [5,12,13,26] generally have low computational efficiency and even some security problems.

In some sense, our method is similar to Searchable Symmetric Encryption (SSE) scheme [7] which allows a party to outsource the storage of his data to another party in a private manner, while maintaining the ability to selectively search over it. But basically, we do not need to encrypt all the files with one same master key, that is to say, our scheme is more flexible. Also, our proposition has better communication cost.

## 1.1 Related Work

Recent work focuses on cloud storage data security, which includes Shamir's secret sharing algorithm [23] and concentrated on the confidentiality of outsourced data against untrusted servers.

Some studies focus on cloud database security based on secret sharing. Hadavi et al. [14–17] explored the DAS model and proposed a series of solutions using $(k, n)$ threshold secret sharing for preserving data confidentiality, which execute queries over encrypted data in the outsourcing database. Emekci et al. [10] proposed an order preserving polynomial building technique which uses hash functions to generate random coefficients. Ferretti et al. [11] focused on protecting database structure. They proposed Secure DBaas architecture that integrates cloud database services with data confidentiality and the possibility of executing concurrent operations on encrypted data. Secure DBaaS is designed to allow multiple and independent clients to connect directly to the untrusted cloud DBaaS without any intermediate server. Attasena et al. [2] analyzed and compared existing methods based on secret sharing technology in cloud data security. They proposed a novel approach to securing cloud data warehouses by flexible verifiable secret sharing (fVSS). The main novelty of fVSS is that, to optimize shared data volume and cost, they share a piece of data fewer than $n$ times by using pseudo shares to construct a polynomial and then enforce data integrity with the help of both inner and outer signatures. Dautrich et al. concluded in [8], that some secret sharing outsourcing schemes are not simultaneously secure and practical in the honest-but-curious server model, where servers are not trusted to keep data private. Their attack is to align shares based on their order and then map the aligned shares onto actual secret values relying upon servers' prior knowledge of original data distribution.

Zhu et al. [28] pointed out that the current study of cloud security storage mainly considered some low-level rules, such as data integrity check, identity-based cryptography, etc, while he proposed a consideration of potential malicious cloud server provider scheme. Doyel et al. [21] proposed a solution whose core content is multi-threshold secret sharing, which enhances the security of the secret key in a distributed cloud environment. Dolev et al. [3,9] proposed the accumulating automata structure to solve the problem of data hiding in the distribution phase. They also proposed the idea that uses the concept of accumulating automata to construct secret sharing data tables, as well as some simple ideas to verify data architecture. Benaloh et al. [4] proposed schemes by which a secret can be divided into many shares which can be distributed to mutually suspicious agents. Cai et al. [6] proposed a SEDB scheme, which supports a set of commonly used operations, such as addition, subtraction, comparison, multiplication, division, and modular. Tian et al. [24] present a secret share based scheme to guarantee the confidentiality of delegated data, while they construct a privacy preserving index to accelerate query and to help return the exactly required data tuples.

Some research focuses on data outsourcing scheme based on Asmuth–Bloom secret sharing (ABSS) [1], which is a threshold scheme based on the Chinese remainder theorem (CRT). Kaya et al. [18] present three function sharing schemes for RSA, ElGamal and Paillier cryptosystems and investigate how threshold cryptography can be conducted with the ABSS. Muhammad et al. [20] proposes a secure data outsourcing scheme based on ABSS which tries to address the issues in data outsourcing such as data confidentiality, availability and order preservation for efficient indexing.

## 1.2  Our Contribution

In this paper, we propose a secret shared based scheme to distribute files in several clouds, which allows users to search the files securely and efficiently. For simplicity, we refer Secure, Efficient and Searchable File System as SES_FS in the remainder of this paper. Basically, as other similar works based on secret sharing, all the participants in our scheme are assumed honest but curious. The proposed scheme can detect partial corruption through verification process.

SES_FS is an outsourcing file system on distributed clouds based on $(k, n)$ threshold secret sharing (similar to schemes in [2,10,14–17]), which is depicted in Fig. 1, where the data owner secretly shares data and sends the encrypted files to public clouds separately. The authorized data user submits the query share to each cloud and gets the result share from each cloud which works on the file shares and the query share independently to calculate its result share. At the

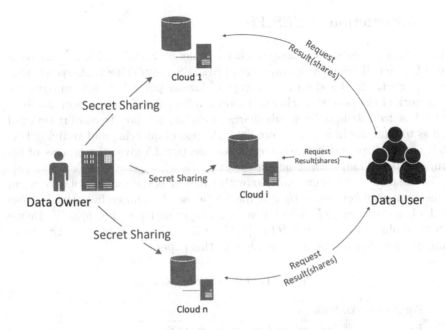

**Fig. 1.** SES_FS framework

end, the data user can reconstruct the correct result from the result shares while the clouds know neither the database nor the query results.

In SES_FS, keywords are transformed to integer values and secretly shared in a defined finite field, then the shares are mapped to random numbers in specially appointed random domain in each cloud. Files are encrypted with distinct secret key and scattered within clouds. With keywords, legal users can search in the clouds to find correct shares and reconstruct the corresponding file numbers and secret keys. The system can also detect malicious servers by introducing virtual points.

SES_FS can be applied to the following scenario. Suppose there is a modern medical research center which keeps large-scale of medical cases. Then, the center utilizes SES_FS to secretly encrypt and share these medical materials on clouds. Thus, the research center could authorize some hospitals to access and make use of these medical material.

## 1.3 Organization

The remainder of this paper is organized as follows. Section 2 briefly reviews some basic concepts and proposes a preliminary framework, and then details our secret sharing and reconstruction mechanisms. In Sect. 3 we show how the query and update is processed. Section 4 summarizes the results of our empirical studies and discusses about the security and computing cost issues in SES_FS. Section 5 summarizes and concludes the paper.

## 2     Construction of SES_FS

This section presents our proposed scheme in detail. Section 2.1 shows the basic idea of Secret Sharing. As a core part of SES_FS, Key/Value table structure is given in Sect. 2.2. Based on the concept of former parts, detailed construction framework of the proposed scheme is shown in Sect. 2.3. The next Sect. 2.4 illustrates how to distribute keywords shares to clouds, and how to convert keyword strings to numbers in a given finite field. Process of querying and verifying keyword is also displayed in the same section. Section 2.5 gives the process of file number sharing and reconstruction.

We assume $p$ is a prime number with $l$ bits, $n$ is the number of servers, $m$ is the number of keywords. $H_0$ is defined as a pseudo-random function. We also need a hash function $HF_1$ which maps a string to an interval $[a, b) \in \mathbb{Z}_p$. Define a two-variable hash function $HF_2$: $\{0,1\}^* \times \mathbb{Z} \to \mathbb{Z}_p$. In Table 1 we give some parameters which will be used throughout this paper.

**Table 1.** SES_FS parameters

| Parameters | Definitions |
|---|---|
| $kw_i$ | $i^{th}$ keyword used to search among files |
| $key_i$ | $i^{th}$ encryption key of keyword be used in $HF_2$ |
| $share_{kw_i}^{s_i}$ | Share of $kw_i$ stored in cloud server $s_i$ |
| $s_1, \cdots, s_n$ | Horizontal ordinate for each cloud server |
| $s^*$ | Horizontal ordinate for constructing polynomial |
| $fn_j$ | File number of $j^{th}$ file |
| $share_{fn_j}^{s_i}$ | Share of $fn_j$ stored in cloud server |
| $DOM_i$ | $i^{th}$ domain |
| $HF_{SToI}$ | Function transform each random string to a random integer |

### 2.1     Secret Sharing in SES_FS

Shamir's secret sharing scheme [23] is designed to share a single secret value $V$ among $n$ servers such that shares from at least $k$ servers must be obtained in order to reconstruct $v$. Such a scheme is called a $(k, n)$ threshold scheme.

We utilize a simple $(k, n)$ threshold scheme [23] which is based on Lagrange polynomial interpolation: given $k$ points in the 2-dimensional plane $(x_1, y_1), \ldots, (x_k, y_k)$ with each distinct $x_i$, there is one and only one polynomial $f(x)$ of degree $k - 1$ such that $f(x_i) = y_i$ where $i \in [1, k]$. To divide the data $V$ into $n$ pieces, we randomly pick $k$ points to construct a random $k - 1$ degree polynomial $f(x) = a_{k-1}x^{k-1} + a_{k-2}x^{k-2} + \cdots + a_1x + a_0$ in which $f(x_0) = V$, and evaluate: $Share_1 = f(s_1), \cdots, Share_n = f(s_n)$. Given any subset of $k$ of these $Share_i$ values (together with their identifying indices), one can find the coefficients of $f(x)$ by interpolation and then evaluate $V = f(x_0)$. Knowledge of at most $k - 1$ of these values does not suffice to calculate $V$.

In SES_FS, the data owner is the distributor of secrets among $n$ servers $server_1, server_2, ..., server_n$. A secret $V$ is split into $n$ shares $share_v^{s_1}$, $share_v^{s_2}, ..., share_v^{s_n}$, which are stored on $server_1, server_2, ..., server_n$, respectively. $share_v^{s_i}$ for each attribute value $v$ is the result of computing $f(x) = a_{k-1}x^{k-1} + a_{k-2}x^{k-2} + ... + a_1x + a_0$. The assigned random values $x$ (each for one cloud), which are known only to the data owner (and also to authorized users), form an $n$ dimension vector $S = \{s_1, s_2, ..., s_n\}$ upon which secrets are distributed and reconstructed. When an authorized user receives at least $k$ shares, the secret can be reconstructed by Lagrange polynomial interpolation.

## 2.2 Key/Value Table of SES_FS

SES_FS is a secure, efficient and searchable file system on distributed clouds. The data owner encrypts and splits each file into many shares which then be distributed to several public clouds. In SES_FS, each file is searched with a keyword which corresponds to one or more files. All files are cited by file numbers. Keyword and file numbers are mapped into a key/value table. Table 2 is a sample of key/value table, which shows that files with number 5, 12 and 23 contain keyword $kw_3$, file 12 contains both keywords $kw_1$ and $kw_3$. Therefore, we can search the key/value table according to a keyword to retrieve the file numbers, with which we can download files from clouds. We should say, all the files here are encrypted with distinct secret keys under symmetric encryption algorithms, such as AES. All the corresponding secret keys can be treated as random numbers and secretly shared and distributed in the same way as file numbers. So, in the paper we will not discuss about how to share and reconstruct secret keys independently.

Table 2. A sample of key/value table

| Keyword | File number |
|---------|-------------|
| $kw_1$ | 1, 12, 17 |
| $kw_2$ | 3, 14, 15 |
| $kw_3$ | 5, 12, 23 |
| ... | ... |
| $kw_i$ | ..., $fn_j$, ... |
| ... | ... |
| $kw_m$ | ... |

## 2.3 Detailed Construction Framework of SES_FS

The detailed construction framework of SES_FS is depicted in Fig. 2, in which we utilize Shamir's threshold scheme to encrypt and secretly share the key/value table among several cloud servers. To distribute a keyword(or file number)

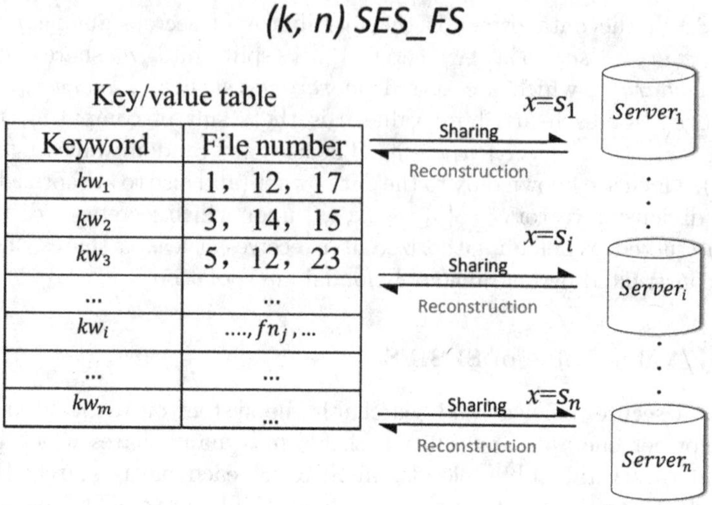

# (k, n) SES_FS

Key/value table

| Keyword | File number |
|---------|-------------|
| $kw_1$ | 1, 12, 17 |
| $kw_2$ | 3, 14, 15 |
| $kw_3$ | 5, 12, 23 |
| ... | ... |
| $kw_i$ | ...., $fn_j$, ... |
| ... | ... |
| $kw_m$ | ... |

$x{=}s_1$  Server₁
Sharing →
Reconstruction

$x{=}s_i$
Sharing →
Reconstruction
Server_i

Sharing $x{=}s_n$ →
Reconstruction
Server_n

**Fig. 2.** Detailed construction framework of SES_FS

between $n$ servers, the data owner constructs a random polynomial and randomly selects a distributed vector $\{s_1, s_2, ..., s_i, ..., s_n, s^*\}$, in which $s_1$ to $s_n$ represent respectively to $server_1$ to $server_n$, where $s^*$ is used to construct distribution polynomial. The randomly selected polynomial is $f(s_i) = a_{k-1}s_i^{k-1} + a_{k-2}s_i^{k-2} + ... + a_1 s_i + a_0$. Value $share_{kw}^{s_i}$ represents share of the keyword $kw$ which is stored on the $server_i$, as shown in Fig. 3.

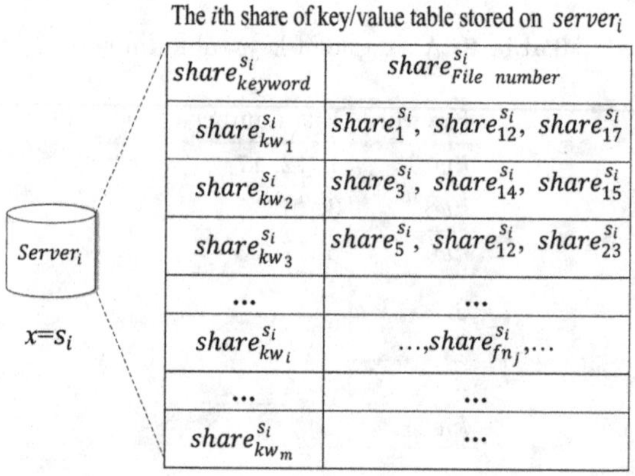

The $i$th share of key/value table stored on $server_i$

| $share_{keyword}^{s_i}$ | $share_{File\ number}^{s_i}$ |
|---------|-------------|
| $share_{kw_1}^{s_i}$ | $share_1^{s_i}$, $share_{12}^{s_i}$, $share_{17}^{s_i}$ |
| $share_{kw_2}^{s_i}$ | $share_3^{s_i}$, $share_{14}^{s_i}$, $share_{15}^{s_i}$ |
| $share_{kw_3}^{s_i}$ | $share_5^{s_i}$, $share_{12}^{s_i}$, $share_{23}^{s_i}$ |
| ... | ... |
| $share_{kw_i}^{s_i}$ | ..., $share_{fn_j}^{s_i}$, ... |
| ... | ... |
| $share_{kw_m}^{s_i}$ | ... |

$Server_i$

$x{=}s_i$

**Fig. 3.** Sharing values of key/value relation with secret sharing scheme

Different methods are used to share and reconstruct keywords and file numbers correspondingly. We secretly share keywords and query on the shares of keywords. Shares of file numbers are used to reconstruct original file number only, when it is retrieved accordingly with keyword shares. For convenience, in the rest of the paper, we assume a 3-server system and a (3,3) threshold secret sharing scheme. We randomly select a master key $masterKey$ which will be maintained in the data owner side, and randomly select a vector $X = \{s_1, s_2, s_3, s^*\}$ as horizontal ordinate for each cloud server.

## 2.4  Keyword Sharing and Verification

Assume all keywords are strings, and we need a function to convert them into integers in set $[0, p - 1]$. Assume $HF_{STol}$ is the function which can transform each random string to a random integer. It may be collision-free or not. A simple example is given in Appendix A. Suppose there are $m$ $(1 \leq i \leq m)$ keywords totally. We divide $DOM_p = [0, p-1]$ into $m$ equal consecutive partitions, i.e.

$$[0, \frac{p}{m} - 1], [\frac{p}{m}, 2\frac{p}{m} - 1], ..., [(i - 1)\frac{p}{m}, i\frac{p}{m} - 1], ..., [(m - 1)\frac{p}{m}, p - 1]$$

Define $DOM_i = [(i-1)\frac{p}{m}, i\frac{p}{m} - 1]$, function $HF_1$ is defined to return an interval for each keyword $kw_i$ as below:

$$HF_1(kw_i) = DOM_j \ (HF_{STol}(kw_i) \text{ is the } j^{th} \text{ biggest number })$$

$HF_1$ can be shown as a Fig. 4. Here we assume all the values transformed to integers are different.

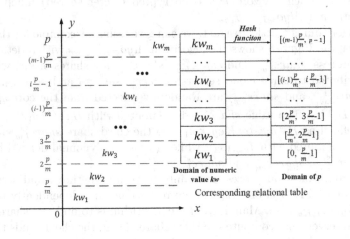

**Fig. 4.** Process of dividing finite field

To encrypt and share a number between $n$ cloud servers, we need $n$ points to construct a polynomial $f(x)$ with order $k - 1$ such that some value of $f(x)$ be equal to the number. Choosing values of $k$ and $n$ for a threshold secret sharing $(k, n)$ has some influences on the availability and fault tolerance aspects

---

**Algorithm 1.** Algorithm $Algo_{ssKeyword}$, to secret share each keyword independently.

---

**Input:**

    Keyword used to search among files, $kw_i$;

    Maintained in the data owner side $masterKey$;

**Output:**

    Shares of keyword stored in each cloud server, $share_{kw_i}^{s_1}$, $share_{kw_i}^{s_2}$, $share_{kw_i}^{s_3}$;

1: Randomly select two values $\{share_{kw_i}^{s_1}, share_{kw_i}^{s_2}\}$, from the corresponding intervals $DOM_i$;

2: Compute $key_i = H_0(masterKey, kw_i)$;

3: Compute $share_{kw_i}^{s^*} = HF_2(key_i, kw_i)$;

4: Construct a polynomial $f_{kw_i}(x)$ of degree 2 by Lagrange interpolation:

$$f_{kw_i}(x) = \sum_{\alpha=1}^{3} \prod_{1 \leq \beta \leq 3, \alpha \neq \beta} \frac{x - x_\beta}{x_\alpha - x_\beta} \cdot y_\alpha ,$$

    Where

$$((x_1, y_1), (x_2, y_2), (x_3, y_3)) = ((s_1, share_{kw_i}^{s_1}), (s_2, share_{kw_i}^{s_2}), (s^*, share_{kw_i}^{s^*}))$$

5: Compute $share_{kw_i}^{s_3} = f_{kw_i}(s_3)$;

6: **return** $share_{kw_i}^{s_1}, share_{kw_i}^{s_2}, share_{kw_i}^{s_3}$.

---

of the system. Each keyword $kw_i$ is encrypted (secret shared) independently with Algorithm 1 $Algo_{ssKeyword}$.

The algorithm is a little abstract, we can give a simple example for the three-cloud situation. Figure 5 shows the algorithm $Algo_{ssKeyword}$ with a detailed calculation process. $share_{kw_i}^{s_1}$, and $share_{kw_i}^{s_2}$, $share_{kw_i}^{s_3}$ are shares of $kw_i$ which are then distributed to three cloud servers: $server_1$, $server_2$ and $server_3$, respectively. $share_{kw_1}^{s_1}$, and $share_{kw_1}^{s_2}$ are Randomly selected from the corresponding intervals: $DOM_1$. The value $share_{kw_1}^{s^*}$ is computed with $HF_2(key_1, kw_1)$, while $key_1$ is defined as $H_0(masterKey, kw_1)$. So the third share $share_{kw_1}^{s_3}$ for sever $server_3$ is computed with $f_{kw_1}(s_3)$, while $f_{kw_1}$ is a polynomial defined by three valued $share_{kw_1}^{s_1}$, $share_{kw_1}^{s_2}$ and $share_{kw_1}^{s^*}$.

Keyword verification process is mainly to verify whether cloud servers are malicious or not. A keyword $kw_i$ can be reconstructed independently by Algorithm 2 $Algo_{VeriKeyword}$. Main idea of the algorithm is to find two shares in the former two servers and compute the third share. Then, the client sends the computed share to the third server. If there is correct share in the proper position, the algorithm will return success, otherwise failure. Failure indicates the data has been tampered by adversaries.

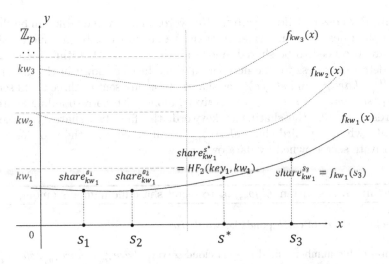

**Fig. 5.** Process of sharing keywords

---

**Algorithm 2.** Algorithm $Algo_{VeriKeyword}$, to verify the correctness of keyword.

**Input:**

Keyword corresponding to the file number, $kw_i$;

Maintained in the data owner side, $masterKey$;

**Output:**

If keyword is found, return $correct$, otherwise return $fail$;

1: Retrieve $share_{kw_i}^{s_1}, share_{kw_i}^{s_2}$ in domain $[(i-1)\frac{p}{m}, i\frac{p}{m} - 1]$ from $server_1$ and $server_2$;

2: Compute $key_i = H_0(masterKey, kw_i)$;

3: Compute $HF_2(key_i, kw_i)$;

4: Construct a polynomial of $f_{kw_i}(x)$ with the pairs of $\{(s_1, share_{kw_i}^{s_1}), (s_2, share_{kw_i}^{s_1}),$ $(s^*, HF_2(key_i, kw_i))\}$;

5: Compute $f_{kw_i}(s_3)$ and send it to $server_3$ to search $share_{kw_i}^{s_3}$;

6: **if** searching success **then**

7:     **return** $success$;

8: **else**

9:     **return** $fail$.

10: **end if**

---

### 2.5   File Number Sharing and Reconstruction

File numbers in the key/value table (e.g. Table 2) are also secretly shared within clouds, which are stored together with shares of the corresponding keyword. Therefore, sharing file number is carried out with sharing keyword, as well as searching over keywords can also retrieve the corresponding shares of file number, and correct files can be retrieved.

**Sharing Process of File Number.** Keywords in key/value table can be shared with method described in former sections. Values in the table are file numbers, which are also need to be shared among all the clouds. Algorithm 3 $Algo_{ssFile}$, shows detailed process for file number sharing. Just as shown in the algorithm, $Share_{fn_j}^{s_1}$, $share_{fn_j}^{s_2}$ and $share_{fn_j}^{s_3}$ as shares of $fn_j$ are sent to three cloud servers $server_1$, $server_2$ and $server_3$, respectively, along with corresponding shares of keywords. Just like the sharing of keyword, the first two shares are selected randomly, while the third share $share_{fn_j}^{s_3}$ is computed with the two shares and some private secret owned by data owner.

---

**Algorithm 3.** Algorithm $Algo_{ssFile}$ to secret share file numbers among clouds.

---

**Input:**

    The file number of corresponds keyword, $fn_j$;

**Output:**

    Shares of file number stored in each cloud server, $share_{fn_j}^{s_1}$, $share_{fn_j}^{s_2}$, $share_{fn_j}^{s_3}$;

1: Randomly select two values $share_{fn_j}^{s_1}$ and $share_{fn_j}^{s_2}$ from $\mathbb{Z}_p$;

2: Construct a polynomial $f_{fn_j}(x)$ of degree 2 by Lagrange interpolation:

$$f_{fn_j}(x) = \sum_{\alpha=1}^{3} \prod_{1 \leq \beta \leq 3, \alpha \neq \beta} \frac{x - x_\beta}{x_\alpha - x_\beta} \cdot y_\alpha ,$$

Where

$$((x_1, y_1), (x_2, y_2), (x_3, y_3)) = ((s_1, share_{fn_j}^{s_1}), (s_2, share_{fn_j}^{s_2}), (s^*, fn_j))$$

3: Compute the third share $share_{fn_j}^{s_3} = f_{fn_j}(s_3)$;

4: **return** $share_{fn_j}^{s_1}, share_{fn_j}^{s_2}, share_{fn_j}^{s_3}$.

---

**Reconstruction Process of File Number.** When user searches keywords, files which contain the keywords will be retrieved. Actually, the proposed system only sends the correspondent file numbers back, so the user can download files by himself. File number reconstruction algorithm is shown in Algorithm 4.

Figure 6 shows computation process of file number sharing and reconstruction process, where the polynomial $f_{fn_j}()$ is computed by data user with two shares in server 1 and server 2, along with private information owned by himself.

## 3 Query and Update Data in Remote Clouds

On the data owner side, keyword/filenubmer table are secretly shared and uploaded to clouds with algorithm $Algo_{ssKeyword}$ and $Algo_{ssFile}$. Storage maps of files in file bitmap table $Tab_{FileMap}$ is divided and secretly shared just as the algorithm $Algo_{ssFile}$. For data user, with querying algorithm and proper keywords, correct file names and the corresponding storage map can be achieved,

**Algorithm 4.** Algorithm $Algo_{ReconFile}$, to reconstruct file number from cloud shares.

**Input:**

Shares of $kw_i$ stored in each cloud server, $share_{kw_i}^{s_1}, share_{kw_i}^{s_2}, share_{kw_i}^{s_3}$;

**Output:**

The file number corresponds to the keyword, $fn_j$;

1: Retrieve $share_{fn_j}^{s_1}$, $share_{fn_j}^{s_2}$ and $share_{fn_j}^{s_3}$ from $server_1, server_2$ and $server_3$, respectively, in corresponding record of keywords;

2: Construct a polynomial $f_{fn_j}(x)$ with degree 2 by Lagrange interpolation with $(s_1, share_{fn_j}^{s_1}), (s_2, share_{fn_j}^{s_2}), (s^*, fn_j)$;

3: Compute $fn_j = f_{fn_j}(s^*)$ by the Lagrange interpolation;

4: **return** $fn_j$.

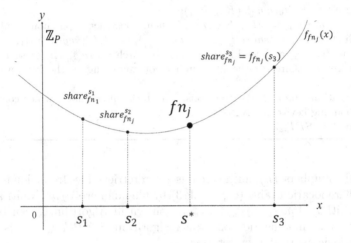

**Fig. 6.** Process of sharing file number

then data users retrieve wanted files from correct clouds. For our principal scheme, users can do exact query. With minor modification they can do more flexible searches, such as multi-keyword query. Query algorithm $Algo_{QueryExact}$ is shown in Sect. 3.1. Details in process of update, including inserting and deleting, are shown in Sect. 3.2.

## 3.1   Query Process of Data Users

Exact Query is to find a single keyword and return all the file numbers which contain this keyword. Exact Query algorithm $Algo_{QueryExact}$ is given in Algorithm 5. The input of $Algo_{QueryExact}$ is keyword and secret key of data user, the output is a set with all the file numbers. $FN\_SET_{kw_i}$ is used to describe filenumber set while searching keyword $kw_i$.

**Algorithm 5.** Algorithm $Algo_{QueryExact}$, to retrieve filenumber set shares back from clouds with assigned keyword.

**Input:**

   Authorized user submits a exact query request to data owner, '$Keyword == kw_i$';

   Encryption key for keyword be used in $HF_2$, $key_i$;

**Output:**

   a set with all the file numbers of corresponding files which contains the keyword, $FN - SET_{kw_i}$;

1: Data owner translate condition '$Keyword == kw_i$' into '$(i - 1)\frac{p}{m} < share_{kw_i}^{s_1, s_2} < i\frac{p}{m} - 1$', and send to $server_1$, $server_2$. Then, get $share_{kw_i}^{s_1}$, $share_{kw_i}^{s_2}$ together with shares of the file number corresponding to the keyword $kw_i$ on $server_1$ and $server_2$;

2: Compute $key_i = H_0(masterKey, kw_i)$;

3: Compute $share_{kw_i}^{s^*} = HF_2(key_i, kw_i)$. Then, construct a polynomial of $f_{kw_i}(x)$ with the help of $\{(s_1, share_{kw_i}^{s_1}), (s_2, share_{kw_i}^{s_1}), (s^*, HF_2(key_i, kw_i))\}$;

4: Compute $f_{kw_i}(s_3)$ and send it to $server_3$ to search $share_{kw_i}^{s_3}$. If the search is successful, return shares of the file number corresponding to the keyword $kw_i$ on $server_3$;

5: Use algorithm $Algo_{ReconFile}$ to reconstruct all the file numbers which contains the corresponding keyword $FN\_SET_{kw_i}$.

6: **return** $FN\_SET_{kw_i}$.

With file numbers $fn_j$, authorized user can retrieve files from clouds according files share location table (e.g. Table 3) by file number(s) $fn_j$. Multi-keyword query algorithm $Algo_{QueryMulti-keyword}$ (shown in Algorithm 6) can be easily gotten by minor modification on former algorithm. $EXP(kw_a, \cdots)$ is used to describe a logic expression on set $\{kw_a, \cdots\}$.

**Algorithm 6.** Algorithm $Algo_{QueryMulti-keyword}$, to retrieve filenumber set shares back from clouds with assigned keyword combination under regular expression.

**Input:**

   Authorized user submits a query request to data owner, '$EXP(kw_a, \cdots)$';

   Encryption key of keyword be used in $HF_2$, $key_i$;

**Output:**

   The file number of corresponds keyword, $FN\_SET_{EXP(kw_a, \cdots)}$;

1: Perform algorithm 5 $Algo_{QueryExact}$ of step 1 to step 5, for $kw_a, \cdots$, respectively, while get file number set: $FN\_SET_{kw_a}, FN\_SET...$;

2: Perform set operations on $FN\_SET_{kw_a}, FN\_SET...$, and get result set $EXP(FN\_SET_{kw_a}, FN\_SET...)$;

3: **return** $EXP(FN\_SET_{kw_a}, FN\_SET...)$.

## 3.2   Insertion and Deletion Process

Updates mainly means insertion and deletion. Without considering adding or delete new keywords, current insertion algorithm considers adding new files to present system. Here in Algorithm 7, the insertion algorithm $Algo_{Insert}$ is given, which can insert one new file to current system. Actually, to add one new file means to distribute all the keywords/filenumber pairs to clouds in secret sharing method.

---

**Algorithm 7.** Algorithm $Algo_{Insert}$, to insert a new file with several keywords to SES_FS.

---

**Input:**

An authorized user submits a insertion request to data owner, while $fn_j$ is file number of file $file_a$, 'INSERT $file_a$ TO $keyword_i$';

Encryption key of keyword be used in $HF_2$, $key_i$;

**Output:**

$Success$ or $Fail$;

1: Data owner translate condition into find '$(i-1)\frac{p}{m} < share_{kw_i}^{s_1,s_2} < i\frac{p}{m} - 1$', and send to $server_1$, $server_2$. Then, obtain $share_{kw_i}^{s_1}$, $share_{kw_i}^{s_2}$;

2: Compute $share_{kw_i}^{s^*} = HF_2(key_i, kw_i)$.

3: Construct a polynomial of with the help of $\{(s_1, share_{kw_i}^{s_1}), (s_2, share_{kw_i}^{s_1}), (s^*, HF_2(key_i, kw_i))\}$;

4: Perform algorithm 3 $Algo_{ssFile}$ for $fn_j$;

5: **return**

---

Algorithm 8 gives the algorithm $Algo_{delete}$ which can delete files from current system. Main operation of the algorithm is to delete corresponding shares of file number from keyword/filenumber pair table.

---

**Algorithm 8.** Algorithm $Algo_{Delete}$, to delete a file from clouds.

---

**Input:**

An authorized user submits a delete request to data owner while $fn_j$ is file number of file $file_a$, 'DELETE $file_a$ FROM $keyword_i$';

Encryption key of keyword be used in $HF_2$, $key_i$;

**Output:**

$Success$ or $Fail$;

1: Perform algorithm 5 $Algo_{QueryExact}$ of step 1) to step 5), then delete $fn_j$ and file $file_a$;

2: Delete $share_{fn_j}^{s_1}$, $share_{fn_j}^{s_2}$, $share_{fn_j}^{s_3}$ from $server_1$, $server_2$, $server_3$, respectively;

3: **return**

---

# 4  Security, Computation Analysis and Expreimental Results

## 4.1  Security and Computation Analysis

In the proposed SES_FS scheme, assume all the servers are honest but curious, the proposed algorithm $Algo_{QueryExact}$ is complete and sound. The proposed SES_FS scheme is secure in two aspects: (1) no adversary can find or detect real file information even if they can collude all the servers; (2) manipulation on keyword shares by one or more servers will be detected with high probability.

For the first claim. Because, all the corrupted servers together cannot reconstruct the original polynomials, so they cannot achieve the keyword/filenumber table. Even all servers are malicious, through the verification algorithm, manipulation of keyword shares can be detected easily. The best way for bad servers to get real information is to calculate possible polynomials on all the combination of shares.

As to the second claim. If malicious severs manipulate shares of file numbers, data users will retrieve wrong files back and detect the illegal manipulation. The proposed scheme is not only secure, but also efficient. Data users can achieve correct result within one round, no matter how many files are stored in clouds. The deep reason is that within our system all the data are indexed although they are encrypted in some way.

SES_FS preserves the order of $share_{kw}$ in finite field. It is a good property for searching, but adversaries may find real values of the shares by listening to the communication channel. We can utilize approaches described in [16] to use an order-obfuscated mapping function to improve the security of the proposed scheme.

## 4.2  Implementation of SES_FS

In this section, the feasibility of our method is demonstrated through evaluating the performance of whole sharing and query execution for files outsourcing to the cloud. We implemented all file system and examined computation cost.

We implemented whole system in Python on Win10 with Intel Core i7-6700 3.40 GHz processor and 8GB of memory. This is a lean Python program executed on behalf of the data owner over the original relation. The following experimental results only count the time of SES_FS, excluding the AES encryption time and network transmission time. The test data in the experiment are all randomly generated.

The information about how the files are split is stored in a file bitmap table $Tab_{FileMap}$. As shown in Table 3, the bitmap corresponding to $File_1$ is 101, with a 1 value at position $i$ representing share storage at $server_i$. Thus, the first row shows $File_1$ has been split into two parts which are shared on $server_1$ and $server_3$.

**Table 3.** A sample of file bitmap table $Tab_{FileMap}$

| File number | Share location | | |
|---|---|---|---|
| | $Server_1$ | $Server_2$ | $Server_3$ |
| $File_1$ | 1 | 0 | 1 |
| $File_2$ | 1 | 1 | 0 |
| $File_3$ | 0 | 1 | 1 |
| ... | ... | ... | ... |

## 4.3  Performance and Comparison

We first evaluate time cost for sharing processing scenario. All the files and keywords used in the experiments are randomly generated with python program. We need to check how much time it will take in the initial stage of the system. Apparently, computing shares for keywords/filenumber table will take more time then generating random keys used in the system. Firstly, we focus on the time cost of computing shares. Figures 7 and 8 describes the client-side sharing process times of different files in the case of different keywords. It can be seen, with increase of number of keywords, sharing processing time is gradually increasing.

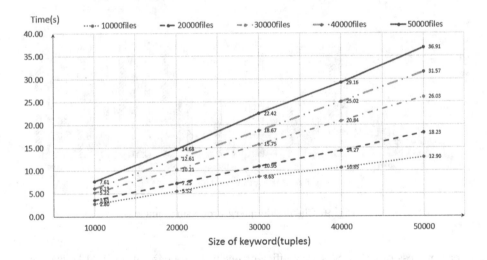

**Fig. 7.** Client-side sharing processing time

The second series of experiments evaluate time cost of query processing scenario. We issued equality query to the system and obtained query processing times of different files in the case of different keyword. Each query processing time is taken after 120 tests, where we remove the longest 10 records and remove the shortest 10 records, and thus get the average. The Fig. 9 suggests that the

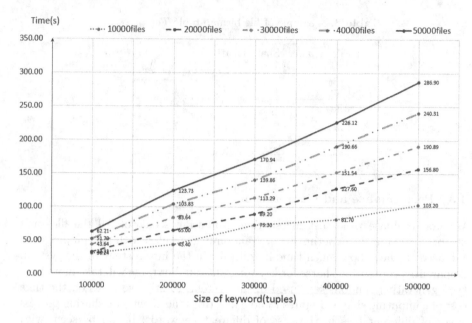

**Fig. 8.** Client-side sharing processing time

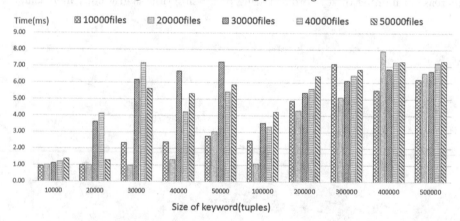

**Fig. 9.** Client-side query processing time

query process of SES_FS is very efficient, even in the case of 500,000 keywords, and the query time is still maintained at milliseconds level.

Compared to [22] which extend the notion of traditional SSE to SSE over multiple servers, our scheme have the same number of rounds required to retrieve the documents/blocks. Their implementation was run using a test file with file size: 1,252,862,080 bytes (i.e., about 1.25 GB). Each chunk of block was the size of 50,000 bytes (50 kB). A total of number of 25,058 blocks was generated. The time taken to generate all five server keys and Build the index blocks, which includes block segmentation and encryption, took about 122 seconds. While in our scheme, it cost us about 127s to share 20,000 files (with about 400,000 keywords). Compared to their work, our scheme nearly has the same performance in initial stage. For searching they spend 81 ms, while our scheme spends less than 9ms on all the situation. We can finish searching operation in one tenth of the time.

## 5    Discussion and Conclusion

In this paper, we proposed a new secure scheme to distribute files in several clouds, which allows users to search the files securely and efficiently. Keywords are secretly shared in finite field, which makes the system more versatile. The system can also detect malicious servers by using virtual points.

For our proposal, one problem is that, the function $HF_{SToI}$ may not be collision resistant. This may obsess data user on searching with the prescribed keywords. Data user will get more than one shares when searching the collision strings, for the algorithm $Algo_{ssKeyword}$ has put them in the same domain. A simple idea to solve this problem is use the fact that $Algo_{ssKeyword}$ gives distinct share $share_{kw_i}^{s_3}$ for different keyword. When querying with algorithm $Algo_{QueryExact}$, data user can judge which file should be retrieved. Also, the data user will not take wrong files back, for it does not know the correct $share_{kw_i}^{s_3}$ for other keywords.

Another problem for us is how to update keywords which have been put into the system at the initial phase. In this situation, new files can be added in the system and uploaded to clouds. The second columns in lines associated with specified keyword, say $keyword^*$, in Table 2 will be updated. This can be done through that data owner query keyword $keyword^*$ in clouds and add extra file number shares for the newly added files.

An interesting property for our scheme is that new keywords can be appended to the SEC_FS scheme easily by minor modification, while it is difficult and usually low efficient for many traditional Searchable Symmetric Encryption (SSE) schemes. Actually, we can get this flexibility by dividing $DOM_p = [0, p-1]$ into $\theta$ (where $\theta >> m$) equal partitions, let $HF_{SToI}()$ work on $\mathbb{Z}_\theta$, define $HF_1$ as:

$$HF_1(kw_i) = DOM_j \quad (j = HF_{SToI}(kw_i))$$

We have not discussed how the files are encrypted and uploaded to clouds because one can borrow techniques from traditional SSE schemes. We have not discussed the key distribution process for authorized users which may have independent interest. Basically, PKI scheme can be used to deal with key distribution problem.

## A    Algorithm of Transforming String to Integer

We here give a function $HF_{SToI}$ to convert a keyword to a distinct number in $\mathbb{Z}_p$ (where $p$ is a prime number). Below, we give the detailed process.

(1) Convert each character of the keyword $kw$ into a binary data $bkw$.
(2) For each binary data $bkw$, if its length is not an exact multiple, pad digit 1 in the front of $bkw$ until its length is a multiple which is named $fbkw$.
(3) For each $l$ bit block of $fbkw$, convert it to a decimal digit.
(4) Compute the product of all decimal blocks $fbkw$ and get an integer number $Ikw$.
(5) Compute $Ikw \mod p$.

In the function, there is a small probability that two or more strings are converted to the same integer in $\mathbb{Z}_p$. Figure 10 gives an example that converts string $kw$ to a numeric value where $p = 2017$. In the example the string $kw$ is converted to bitstring 0110101101110111, and then padded with 111111. Finally, the number of $kw$ is 559 in finite field $\mathbb{Z}_{2017}$.

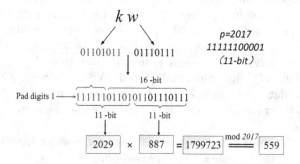

**Fig. 10.** Sample data type conversion process

# References

1. Asmuth, C., Bloom, J.: A modular approach to key safeguarding. IEEE Trans. Inf. Theory **29**(2), 208–210 (1983)
2. Attasena, V., Harbi, N., Darmont, J.: fVSS: a new secure and cost-efficient scheme for cloud data warehouses. In: Proceedings of the 17th International Workshop on Data Warehousing and OLAP, pp. 81–90. ACM (2014)
3. Avni, H., Dolev, S., Gilboa, N., Li, X.: SSSDB: database with private information search. In: Karydis, I., Sioutas, S., Triantafillou, P., Tsoumakos, D. (eds.) ALGO-CLOUD 2015. LNCS, vol. 9511, pp. 49–61. Springer, Cham (2016). https://doi.org/10.1007/978-3-319-29919-8_4
4. Benaloh, J.C.: Secret sharing homomorphisms: keeping shares of a secret secret (extended abstract). In: Odlyzko, A.M. (ed.) CRYPTO 1986. LNCS, vol. 263, pp. 251–260. Springer, Heidelberg (1987). https://doi.org/10.1007/3-540-47721-7_19
5. Brakerski, Z., Vaikuntanathan, V.: Efficient fully homomorphic encryption from (standard) LWE. SIAM J. Comput. **43**(2), 831–871 (2014)

6. Cai, Q., Lin, J., Li, F., Wang, Q.: SEDB: building secure database services for sensitive data. In: Hui, L.C.K., Qing, S.H., Shi, E., Yiu, S.M. (eds.) ICICS 2014. LNCS, vol. 8958, pp. 16–30. Springer, Cham (2015). https://doi.org/10.1007/978-3-319-21966-0_2

7. Curtmola, R., Garay, J., Kamara, S., Ostrovsky, R.: Searchable symmetric encryption: improved definitions and efficient constructions. J. Comput. Secur. **19**(5), 895–934 (2011)

8. Dautrich, J.L., Ravishankar, C.V.: Security limitations of using secret sharing for data outsourcing. In: Cuppens-Boulahia, N., Cuppens, F., Garcia-Alfaro, J. (eds.) DBSec 2012. LNCS, vol. 7371, pp. 145–160. Springer, Heidelberg (2012). https://doi.org/10.1007/978-3-642-31540-4_12

9. Dolev, S., Gilboa, N., Li, X.: Accumulating automata and cascaded equations automata for communicationless information theoretically secure multi-party computation. In: Proceedings of the 3rd International Workshop on Security in Cloud Computing, pp. 21–29. ACM (2015)

10. Emekci, F., Methwally, A., Agrawal, D., El Abbadi, A.: Dividing secrets to secure data outsourcing. Inf. Sci. **263**, 198–210 (2014)

11. Ferretti, L., Colajanni, M., Marchetti, M.: Distributed, concurrent, and independent access to encrypted cloud databases. IEEE Trans. Parallel Distrib. Syst. **25**(2), 437–446 (2014)

12. Gentry, C.: Fully homomorphic encryption using ideal lattices. In: Proceedings of the 41st Annual ACM Symposium on Theory of Computing, STOC 2009, Bethesda, MD, USA, 31 May–2 June 2009, pp. 169–178 (2009)

13. Gentry, C., Halevi, S.: Implementing gentry's fully-homomorphic encryption scheme. In: Paterson, K.G. (ed.) EUROCRYPT 2011. LNCS, vol. 6632, pp. 129–148. Springer, Heidelberg (2011). https://doi.org/10.1007/978-3-642-20465-4_9

14. Hadavi, M.A., Damiani, E., Jalili, R., Cimato, S., Ganjei, Z.: AS5: a secure searchable secret sharing scheme for privacy preserving database outsourcing. In: Di Pietro, R., Herranz, J., Damiani, E., State, R. (eds.) DPM/SETOP 2012. LNCS, vol. 7731, pp. 201–216. Springer, Heidelberg (2013). https://doi.org/10.1007/978-3-642-35890-6_15

15. Hadavi, M.A., Jalili, R.: Secure data outsourcing based on threshold secret sharing; towards a more practical solution. In: Proceedings of the VLDB PhD Workshop, pp. 54–59 (2010)

16. Hadavi, M.A., Jalili, R., Damiani, E., Cimato, S.: Security and searchability in secret sharing-based data outsourcing. Int. J. Inf. Secur. **14**(6), 513–529 (2015)

17. Hadavi, M.A., Noferesti, M., Jalili, R., Damiani, E.: Database as a service: towards a unified solution for security requirements. In: 2012 IEEE 36th Annual Computer Software and Applications Conference Workshops (COMPSACW), pp. 415–420. IEEE (2012)

18. Kaya, K., Selçuk, A.A.: Threshold cryptography based on asmuth-bloom secret sharing. Inf. Sci. **177**(19), 4148–4160 (2007)

19. Monz, T., et al.: Realization of a scalable shor algorithm. Science **351**(6277), 1068–1070 (2016)

20. Muhammad, Y.I., Kaiiali, M., Habbal, A., Wazan, A.S., Sani Ilyasu, A.: A secure data outsourcing scheme based on asmuth-bloom secret sharing. Enterp. Inf. Syst. **10**(9), 1001–1023 (2016)

21. Pal, D., Khethavath, P., Thomas, J.P., Chen, T.: Multilevel threshold secret sharing in distributed cloud. In: Abawajy, J.H., Mukherjea, S., Thampi, S.M., Ruiz-Martínez, A. (eds.) SSCC 2015. CCIS, vol. 536, pp. 13–23. Springer, Cham (2015). https://doi.org/10.1007/978-3-319-22915-7_2

22. Poh, G.S., Mohamad, M.S., Chin, J.J.: Searchable symmetric encryption over multiple servers. Crypt. Commun. **10**(1), 139–158 (2018)
23. Shamir, A.: How to share a secret. Commun. ACM **22**(11), 612–613 (1979)
24. Tian, X.X., Sha, C.F., Wang, X.L., Zhou, A.Y.: Privacy preserving query processing on secret share based data storage. In: Yu, J.X., Kim, M.H., Unland, R. (eds.) DASFAA 2011. LNCS, vol. 6587, pp. 108–122. Springer, Heidelberg (2011). https://doi.org/10.1007/978-3-642-20149-3_10
25. Trummer, I., Koch, C.: Multiple query optimization on the D-Wave 2X adiabatic quantum computer. Proc. VLDB Endow. **9**(9), 648–659 (2016)
26. Wang, W., Hu, Y., Chen, L., Huang, X., Sunar, B.: Exploring the feasibility of fully homomorphic encryption. IEEE Trans. Comput. **64**(3), 698–706 (2015)
27. Wu, C., Zapevalova, E., Chen, Y., Li, F.: Time optimization of multiple knowledge transfers in the big data environment. Comput. Mater. Continua **54**(3), 269–285 (2018)
28. Zhu, H., Liu, T., Zhu, D., Li, H.: Robust and simple N-party entangled authentication cloud storage protocol based on secret sharing scheme. J. Inf. Hiding Multimed. Signal Process. **4**(2), 110–117 (2013)

# Design of Landslide Warning System

Xiaoping Yang[1,2(✉)], Jixuan Du[1,2], Zhaoyu Su[1], Pubin Nong[1],
Zhirong Qin[1], and Bailin Chen[1]

[1] College of Information Science and Engineering,
Guilin University of Technology, Guilin 541004, China
707726045@qq.com
[2] Guangxi Key Laboratory of Embedded Technology and Intelligent System,
Guilin 541004, China

**Abstract.** Aiming at geological disaster monitoring and prevention work, a
real-time monitoring and early warning system is proposed for low-power
consumption of landslides to meet the needs of landslide monitoring in remote
mountainous areas. The inclination angle of the mountain body was detected by
a mechanical inclination sensor, and a plurality of inclination sensors were
placed on each landslide body to form an array distribution. The landslide body
was stereoscopically monitored. Each mesh node had a different node address,
and different inclination thresholds were set in advance. When the sensor
detection value reached the threshold, an alarm message was sent to the system
main control end, and the main control end generated an audible and visual
alarm, and an alarm message was sent at the same time. Compared with the
current landslide warning system on the market, the system achieves expected
results and its power consumption is extremely low. The sensor terminal of the
mountain monitoring is powered by dry battery and can work for 6 years in the
field without external power supply. It avoids the damages made by weather,
livestock and human, and has broad application prospects.

**Keywords:** Landslide · Warning · Tilt sensor · Wireless communication

## 1 Introduction

China is a country with a vast territory, and the mountain area accounts for a large
part. The mountain area accounts for about two-thirds of the country's total area. It is
one of the countries in the world that is more affected by landslides. As one of the most
common geological hazards in mountainous areas, landslides seriously threaten people's lives and property, damage engineering facilities, affect normal production and
life, and cause huge economic losses [1]. The purpose of monitoring and alarming the
landslide is to grasp the sliding law of the landslide, make judgments on the displacement of the landslide in time, and make predictions or alarms for possible slippage. Therefore, landslide monitoring is of great significance for disaster mitigation and
prevention.

In China, prior to the 1980s, conventional geodetic methods were used for deformation monitoring, the plane displacement was measured by theodolite wire or triangulation. After the total station appeared in the mid-1980s, deformation monitoring was

© Springer Nature Singapore Pte Ltd. 2019
R. Mao et al. (Eds.): ICPCSEE 2019, CCIS 1059, pp. 267–277, 2019.
https://doi.org/10.1007/978-981-15-0121-0_20

performed using the total station wire and the electromagnetic wave ranging triangle elevation method. However, all of the above methods require people to observe on the spot, and the workload is large, especially in the southern mountainous areas [2]. The trees are overgrown, the operation is very difficult, and it is difficult to achieve unattended monitoring. After the emergence of GPS satellite positioning system, GPS monitoring method can be used for long-distance monitoring without unattended, and has been widely used in landslide monitoring [3]. For example, Lijiaxia landslide, Sichuan Ya'an Xiakou landslide, Huanglaishi landslide and other landslide monitoring have adopted GPS technology. However, GPS positioning technology also has the disadvantages of low degree of freedom of point selection, GPS signal is susceptible to surrounding occlusion and more error sources. In foreign countries, aerial photography and color infrared photography and thermal infrared scanning are commonly used to investigate landslides, as well as new synthetic aperture radar interference (INSAR) measurements [4]. However, most methods of monitoring and alarming generally have the disadvantages of complicated structure, high cost, and difficulty in promotion.

To this end, this paper studies the tilting sensing unit and wireless monitoring and early warning device for monitoring landslides, and makes a wireless monitoring and array distribution warning system for landslides with simple structure and low power consumption.

## 2   System Structure and Working Principle

### 2.1   Array Distributed Low Power Sensor

The schematic diagram of the composition of the real-time monitoring and early warning system for landslides is shown in Fig. 1. The landslide real-time monitoring and early warning system includes the sensor end, the main control end and the client end. The tilt sensor consists of a mercury ball, a glass column and two wiring legs. The sensor is tilted at a certain angle with glass glue. The mercury ball sinks to the bottom of the glass column and is separated from the wiring pin. At the set angle, the mercury ball will turn on both pins and the circuit will trigger an alarm. In the absence of landslides, the circuit does not turn on and the power consumption is almost zero, so the sensor can achieve very low power consumption and is maintenance-free within six years of the shelf life of the dry battery.

The angle sensors of various angles are buried in the ground according to the array distribution, and are numbered separately. When the mountain displacement occurs, the array-distributed sensors can monitor more landslide sections, and the corresponding sensor ends send alarms to the master. After the main control receives the alarm, the regional alarm signal of the landslide is sent to the mobile phone user as a short message. Further, the number of tilt angles $\alpha$ of the sensor can be known from the sensor number distributed in the array, and the displacement distance x of the landslide is calculated from the depth d when the sensor is buried, that is $S = h \times \tan \alpha$. Schematic diagram of soil displacement calculation is shown in Fig. 2. It is therefore possible to monitor different degrees of landslide displacement by different depths of the buried sensor.

Through the data fed back by the array of sensors, the overall displacement of the landslide section can be known, rather than just warning the landslide.

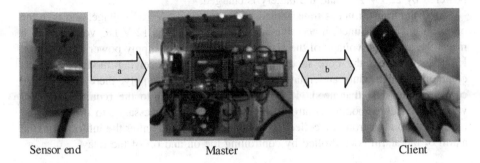

Sensor end                    Master                         Client

**Fig. 1.** Schematic diagram of landslide real-time monitoring and early warning system

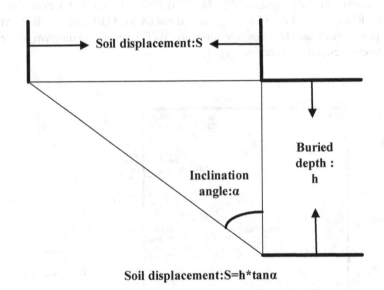

**Soil displacement:S=h*tanα**

**Fig. 2.** Soil displacement derivation

## 2.2  Power Supply Design

A major feature of the design of the landslide monitoring main control terminal is real-time monitoring and prevention. To prevent power outage caused by sudden situations, a multi-function access control power module is used, and the battery is used as a backup power source. The entire power module is mainly powered by the external 220 V AC voltage. Under normal circumstances, it will not be powered off, which basically ensures the long-term normal operation of the entire system. The transformer part reduces the 220 V AC voltage to 12 V AC voltage. In order to ensure safety, a fuse is connected between the external 220 V AC voltage and the transformer. In the case of an emergency, the fuse is blown to ensure the safety of the system and on the other hand to ensure personal safety. Under normal circumstances, the entire system is powered by 220 V AC, and the battery is charged.

After the transformer is transformed, the output is 12 V AC voltage, and then the output of the multi-function access control power module is 12 V DC voltage, which meets the operating voltage of the alarm, so it can directly supply power to the alarm. However, because the core module, ZW20-J wireless receiving module, wireless communication module and relay need 5 V DC voltage supply, there is a voltage conversion module that needs 12 V DC voltage. Convert to the required 5 V DC voltage. The core module controls SIM900 to send text messages to the user, and receives the output from the collection terminal. Immediately after the information, the alarm of the alarm is controlled by controlling the on and off of the relay.

## 2.3   Core Control Module

The core control module consists of STM32F103C8T6. It is a 32-bit microcontroller based on ARM Cortex-M core STM32 series. It has 64 kB Flash and 20 kB RAM, and has low power mode and IAP upgrade function [5]. The schematic diagram of the core control module circuit is shown in Fig. 3.

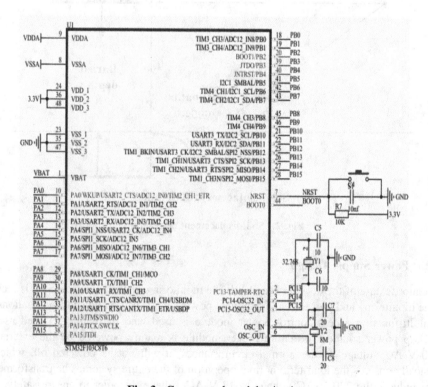

**Fig. 3.**  Core control module circuit

## 2.4    SMS Communication Module

The system main control terminal uses SIM900 as the short message communication module to realize the alarm function of sending text messages to users, and at the same time, it has the functions of adding and removing user numbers, setting alarm messages, and querying added user information. The SIM900 communication module supports 5 V and 3.3 V voltage supply, so the VCC pin is directly connected to the 5 V pin of the core module STM32, and the RXD and TXD pins of the module support serial communication with the STM32. The system selects USART1, so the system uses USART1. The RXD and TXD pins of the SIM900 module are connected to the A9 and A10 pins of the STM32 for data communication. The schematic diagram of the connection is shown in Fig. 4.

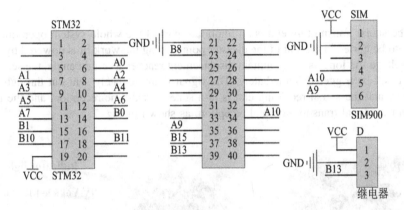

**Fig. 4.** Connection schematic

## 2.5    Wireless Communication Module

The wireless receiving module adopts ZW20-J, which can decode eV1527 and 2262 encoding chips and is widely used. The schematic diagram of the ZW20-J module is shown in Fig. 5. The main function of ZW20-J wireless receiving module in this design system is to receive the signal sent by the transmitting module, and then transmit it to the core module STM32 to judge the processing. At the same time, like the core module, it uses DC 5 V voltage to supply data. The final schematic diagram of the wiring with the core module is shown in Fig. 6.

**Fig. 5.** ZW20-J module

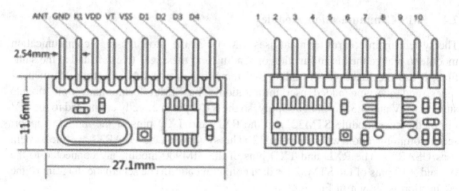

ANT GND K1 VDD VT VSS D1 D2 D3 D4

2.54mm

11.6mm

27.1mm

**Fig. 6.** Wiring principle

The signal transmitting end is an important part of the whole system operation. It needs to be buried in the soil of the disaster point during the work. It is powered by DC 9 V voltage. As long as the mountain displacement reaches a certain angle, the sensing circuit will be triggered immediately. The signal is immediately sent out through the antenna, and the signal receiving end is in a state of real-time reception, and the two implement signal transmission and reception, as shown in Fig. 7.

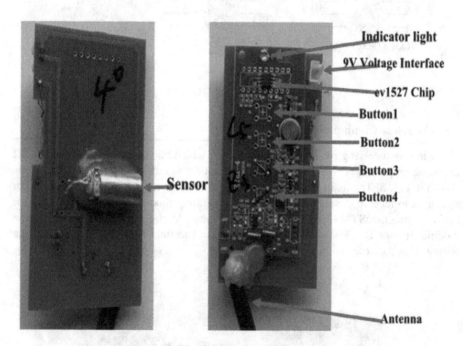

Sensor

Indicator light

9V Voltage Interface

ev1527 Chip

Button1

Button2

Button3

Button4

Antenna

**Fig. 7.** Signal sender

The ev1527 chip is a learning code encoding IC that can be pre-fired by CMOS. It can be decoded by software and contains an oscillating circuit. Four key inputs can be combined to obtain a variety of buttons. This design requires only six buttons. Button 1 to button 4 have a one-to-one correspondence with D1 to D4 of ZW20-J, and different button combinations have a one-to-one correspondence with sensor number and LED lamp number. When there is a key press, ev1527 corresponds. There is a high level on the pin, so the ev1527 chip is triggered, and the TXD pin outputs the corresponding data string, which is transmitted through the amplifier and the oscillating circuit. When the ZW20-J receives the signal, the VT pin will output a high level, and D1, D2, D3, and D4 will output the corresponding signals to the microcontroller. See Tables 1 and 2 for the correspondence and function of each button.

**Table 1.** Correspondence between buttons, ZW20-J, STM32

| Button | ZW20-J pin | STM32 pin |
|--------|-----------|-----------|
| Button 1 | D1 pin | PA1 pin |
| Button 2 | D2 pin | PA3 pin |
| Button 3 | D3 pin | PA5 pin |
| Button 4 | D4 pin | PA7 pin |

**Table 2.** Correspondence between the combination button and the sensor terminal and LED lamp

| Combination button | Sensor number | LED number |
|--------------------|---------------|------------|
| Button 1&&Button 4 | Sensor 1 | LED1 |
| Button 1&&Button 3 | Sensor 2 | LED2 |
| Button 1&&Button 2 | Sensor 3 | LED3 |
| Button 2&&Button 3 | Sensor 4 | LED4 |
| Button 2&&Button 4 | Sensor 5 | LED5 |
| Button 3&&Button 4 | Sensor 6 | LED6 |

## 3  Software Design

After powering up the system, the STM32F103C8T6 core module first performs a series of initializations. After the preparation work is finished, check whether the system can work normally. If it cannot work normally, it will alarm through the LED work indicator. If it can work normally, set the SMS command, modify the alarm information, and then judge whether the receiving module receives the signal from the acquisition terminal. If there is no wireless module, it is always in the state of receiving signal. If the signal of the collecting end is received, it is judged that it is the signal sent by the first collecting end, the corresponding LED warning light is lit, and the STM32F103C8T6 core control module will give the relay. A high level closes the alarm, the alarm is in a closed loop, the alarm starts to alarm, and then the short message program is executed. The core control module sends an alarm message to the user who has stored the mobile phone number by controlling the communication module to complete one. Figure 8 is the program flow chart.

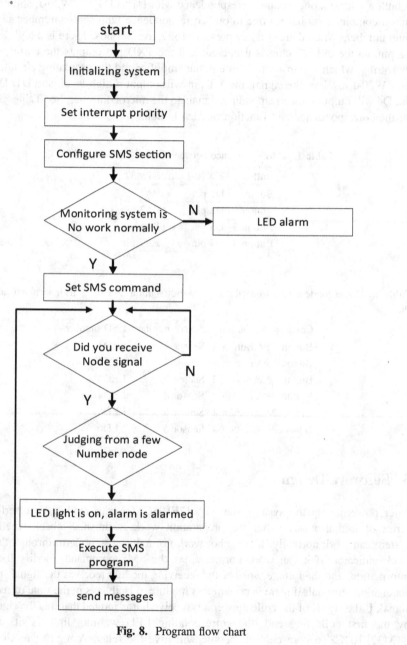

**Fig. 8.** Program flow chart

### 3.1   SMS Communication Software Part Design

The SIM900 communication module is used to send text messages to the user, and the data transmission is realized by the A10/RX and A9/TX two-pin connection of the RXD and TXD serial ports and the STM32F103C8T6 control module, and the VCC of the SIM900 communication module supports the DC 5 V voltage supply, which can be directly connected with the STM32. The VCC of the module is connected, and the communication module is controlled by the STM32F103C8T6 module to complete the operations of adding information, querying information, modifying information, and deleting information.

### 3.2   ZW20-J Wireless Receiving Module Programming

The ANT pin of the ZW20-J wireless receiving module is an antenna interface, which requires an external antenna. The four pins D1, D2, D3 and D4 are connected to the A1, A3, A5 and A7 pins of the STM32F103C8T6 core module respectively for decoding and receiving. The signal to the corresponding output starts low, the decoding is successful, the output is high, and the VT pin is an indication of the output status of the four pins D1, D2, D3, and D4, initially low. After successful decoding, the output is high, the GND pin is the ground pin, and the VDD pin is connected to the DC 5 V. According to the wiring design software, the port configuration is first used for the used I/O port, including five pins A1, A3, A5, A7, and B1, because button 1, button 2, button 3, button 4 and D1, D2 The four data transmission ports of D3 and D4 are one-to-one correspondence. They are combined with the number of the sensor end and the number of the LED alarm light, and then define the I/O port interrupt initialization function, including initial structure definition and clearing. Interrupt flag, pin selection, trigger mode selection, start interrupt, etc., as well as interrupt priority setting, call delay function, after all configuration is completed, run the program, ZW20-J module is in the state of real-time receiving signal, once When the signal is received, the data written by the pins A1, A3, A5, and A7 is compared with the initial address of the STM32 module, and it is known which sensor sends the signal, and then the corresponding LED warning light is illuminated.

## 4   System Test

According to the steps, the sensor end and the receiving end are paired and learned, and the simulation test of the landslide warning system is carried out on site. Sensors are buried in the hillside near the campus of Yanshan District, Guilin City, which are buried in different locations to monitor the landslide of the entire hillside. The main control end is placed in the dormitory and laboratory for the longest transmission distance. The shovel-activated soil was used to simulate landslides for testing. The experimental data is shown in Table 3.

It can be seen from Table 3 that the alarm trigger success rate is within 100 m, and the success rate exceeds this distance. Therefore the effective distance is within 400 m. Overall physical picture, the sensor buried field map shown in Figs. 9 and 10.

**Table 3.** The experimental data

| Test distance (unit: meter) | Test count | Successfully triggered the number of alerts |
|---|---|---|
| 200 | 20 | 20 |
| 400 | 20 | 20 |
| 500 | 20 | 16 |
| 600 | 20 | 8 |

**Fig. 9.** Overall physical picture

**Fig. 10.** The sensor buried field map

# 5   Conclusion

After the test, in terms of transmission distance, it can be seen that within 400 m, the accuracy of the system can be guaranteed. This fully meets the daily mountain monitoring needs. In terms of power consumption, if the STM32L series with lower power consumption is used, the power consumption can be controlled lower. From the sensor accuracy analysis, the tilt sensor can only distinguish whether the tilt occurs but cannot distinguish the specific angle and direction of the tilt. If the gyroscope is used, it will be more accurate, but the power consumption will also increase.

**Acknowledgments.** This work was supported by the Ministry of Science and Technology 863 Project (No. 2013AA12210504), Guangxi Education Department Guangxi University Research Project (No. 200911MS106), Guangxi Science and Technology Department Project (No. AC16380128), Qingxiu Science and Technology Department of Nanning City Project (No. 2018002) and by GuangXi key Laboratory Fund of Embedded Technology and Intelligent System.

# References

1. Wen, N.: Laboratory realization of real-time monitoring system for landslides based on sensors. College of Resources and Environmental Sciences, Chongqing University, Chongqing (2011)
2. Pei, R.: Analysis of the mechanism of landslide formation and stability study. Fujian Const. Technol. (05), 33–34+40 (2018)
3. Liu, H., Yang, X., Wang, R., Jin, W., Jia, W., Liu, W: Transmit/receive spatial smoothing with improved effective array aperture for angle and mutual coupling estimation in bistatic MIMO radar. Int. J. Antennas Propag. (2016)
4. Dong, J., Wu, Q., Wan, S., et al.: Emergency monitoring and early warning technology system for sudden landslide disasters. Sci. Technol. Eng. 18(11), 135–140 (2018)
5. Tang, P., Wang, X., Hu, L.: Upgrading STM32 MCU firmware online with IAP technology. J. Jishou Univ. (Nat. Sci. Edn.) 40(01), 21–26 (2019)

# Scheduling Method Based on Backfill Strategy for Multiple DAGs in Cloud Computing

Zhidan Hu[✉], Hengzhou Ye, and Tianmeizi Cao

Guangxi Key Laboratory of Embedded Technology and Intelligent System,
Guilin University of Technology, Guilin 541006, China
1983747709@qq.com

**Abstract.** Multiple DAGs scheduling strategy is a critical factor affecting resource utilization and operating cost in the cloud computing. To solve the problem that multiple DAG scheduling cannot meet the resource utilization and reliability when multiple DAGs arrive at different time, the multiple DAGs scheduling problem can be transformed into a single DAG scheduling problem with limited resource available time period through multiple DAGs scheduling model based on backfill. On the basis of discussing the available time period description of resources and the sorting of task scheduling when the available time period is limited, the multiple DAGs scheduling strategy is proposed based on backfill. The experimental analysis shows that this strategy can effectively shorten the makespan and improve the resources utilization when multiple DAGs arrive at different time.

**Keywords:** Cloud computing · Multiple DAGs · Backfill ·
Resource utilization · Makespan

## 1 Introduction

With the continuous advancement of modern technology, users have higher and higher requirements for the computing performance of computers. Traditional computing models can no longer meet the needs of users, which makes cloud computing with distributed processing capabilities come into being [1]. Task scheduling is a key technology in cloud computing [2]. In general, Directed Acyclic Graph (DAG) is used to represent dependent task scheduling problems [3].

In cloud computing, how to efficiently deal with the scheduling problem of multiple user and multiple DAGs sharing a group of cloud computing resources can reduce the operation cost and improve the resource utilization. In recent years, some progress has been made in multiple DAGs scheduling [4–17]. These studies mainly focus on resource utilization, fairness, scheduling length and throughput, but less on how to balance resource utilization and reliability when multiple DAGs arrives at different time. To solve this problem, this paper proposes a backfill-based multi-DAG scheduling strategy (MSBB), which can improve resource utilization and shorten makespan when considering the different arrival time of DAG.

Figure 2 shows the comparison of the makespan between the sequence scheduling based on HEFT, E-fairness, Round-Robin and MSBB algorithm by scheduling the two

© Springer Nature Singapore Pte Ltd. 2019
R. Mao et al. (Eds.): ICPCSEE 2019, CCIS 1059, pp. 278–290, 2019.
https://doi.org/10.1007/978-981-15-0121-0_21

DAGs in Fig. 1. Among them, Fig. 2(a), (b), (c), (d) respectively show the effect of scheduling these two tasks in sequence using the HEFT (Heterogeneous Earliest Finish Time) algorithm that first proposed by Topcuoglu et al. [18], using the E-Fairness and Round-Robin algorithm by Guo-Zhong et al. [12], and MSBB algorithm. And DAG-A reaches 35 time units earlier than DAG-B.

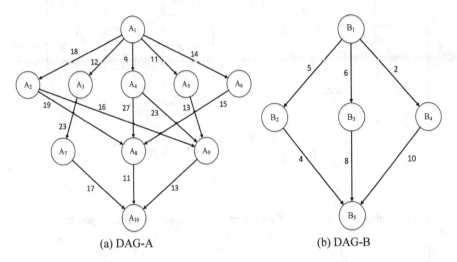

(a) DAG-A                          (b) DAG-B

**Fig. 1.** Two DAG instances

The main contributions of this paper include:

(1) A multiple DAGs scheduling strategy based on backfill is proposed. When multiple DAGs arrive at different time, higher resource utilization and shorter makepan can be obtained.
(2) A method for describing the available time period of resources is proposed, on based this, the calculation method of the earliest execution start time is given.
(3) Through simulation experiments analyze the effects of DAG number, multiple DAGs arrival time interval, number of tasks in DAG, number of resources and other factors on the algorithm.

The rest of this paper is organised as follows: Sect. 2 describes the related work. Section 3 presents the multiple DAGs scheduling strategy based on backfill. Section 4 shows the experimental results. Finally, Sect. 5 concludes this paper and discusses future work.

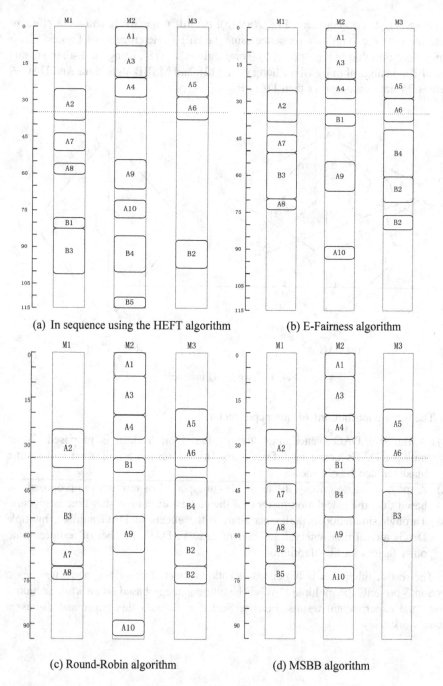

(a) In sequence using the HEFT algorithm

(b) E-Fairness algorithm

(c) Round-Robin algorithm

(d) MSBB algorithm

**Fig. 2.** Comparison of the makespan using four algorithms to schedule 2 DAGs

## 2  Related Work

Considering whether multiple DAGs can share resources, The DAG task scheduling problem can be divided into two categories, namely single DAG task scheduling and multiple DAGs task scheduling. In recent years, the research on single DAG task scheduling has gradually matured and is applied in existing computing platforms [8, 11], and the research on multiple DAGs task scheduling still needs further research.

Literature [12], in the research of multiple DAGs task scheduling, on the basis of single DAG, a scheme of cascade processing is proposed. Although multiple DAGs task scheduling is implemented, the resource utilization is extremely low and there is a large time overhead. Literature [13] proposes a multiple tasks composite, single-task scheduling scheme, which first combines multiple DAGs and then uses a single DAG scheduling method to schedule the combined DAGs to achieve multiple DAGs scheduling. The problem of task competition in such multiple DAGs scheduling schemes has not been resolved. Literature [14] proposed a multiple DAGs workflow preemptive scheduling strategy, which prioritizes DAG to achieve multiple DAGs scheduling. Literature [15], the Planner-guided multiple DAGs scheduling strategy is proposed, which also realizes the scheduling of multiple DAGs by prioritizing the DAGs. Literature [16] implements offline calculation and analysis for data-intensive scheduling problems. However, this type of scheduling strategy does not take into account the impact of the DAG task arrival time on the priority, which may result in too long task waiting time and does not guarantee the reliability of task scheduling. Literature [17], based on the mapping idea of virtual machine, the idea of segmentation execution of scheduling tasks is proposed, and the reliability is analyzed. The idea of "master copy and sub-version update" is emphasized, and all the predecessors of the task are required to be accurately acquisition, although this method can achieve a compromise between reliability and system implementation, but it ignores the start time of the minor version and the completion time of the main version node task, resulting in reduced algorithm performance.

## 3  Multiple DAGs Scheduling Strategy Based on Backfill

### 3.1  Multiple DAGs Scheduling Model Based on Backfill

As can be seen from Fig. 3, the multiple DAGs scheduling model based on backfill can be divided into four parts: the DAG receiving pool, the scheduler, the task queue on each machine, and the resource set. According to the general characteristics of the cloud computing, it is assumed that: (1) the time when the DAG arrives the DAG receiving pool is different (the $i$-th DAG is recorded as DAG-$i$); (2) the definition of the DAG is consistent with the literature [11], the execution time of node tasks in a DAG on a certain type of resource is known, and will not be described here.

The model first accepts DAGs arriving at different time through the DAG receiving pool, then waits for a short time interval $wt$ in the receiving pool, and simultaneously determines whether a new DAG arrives. If there is, merge with the arrived DAG; if not, go directly to the scheduler. After entering the scheduler, the task is assigned to the task queue on each machine in combination with the multiple DAGs backfill scheduling strategy, and finally assigned to each resource.

It is not difficult to see that through the multiple DAGs scheduling model based on backfill, the multiple DAGs task scheduling problem is finally transformed into a single DAG scheduling problem with limited resource available time.

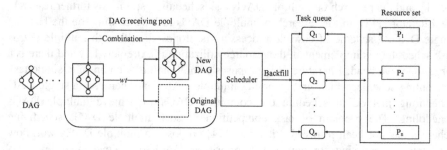

**Fig. 3.** Multiple DAGs scheduling model based on backfill

## 3.2    Task Priority When the Resource Available Time Period is Limited

In the description of the available time period of the resource, $bt_i(p)$ and $et_i(p)$ respectively represent the start time and the end time of the $i$-th idle period on the $p$ processor resource, where $i = 1, 2,..., k$ and $bt_k(p) = +\infty$. The idle period of the processor resource $p$ can be named as *Idle* $(p)$, which is defined as:

$$Idle(p) = \left\{ \bigcup_{i=1}^{k-1} [bt_i(p), et_i(p)] | bt_i(p) < et_i(p_q) < bt_{i+1}(p) \right\} \quad (1)$$

Learning form the idea of literature [19], in order to determine the earliest start time $EST(v, p)$ and the earliest completion time $EFT(v, p)$ of task $v$, it is necessary to first determine $avail(p, v)$ that is the earliest time that $p$ can perform task $v$. So the following two situations need to be considered:

(1) When $et_{i-1}(p_q) \leq EST(v,p) < bt_i(p)$, $avail(p, v)$ is defined as:

$$avail(p, v) = \min\{j | j \geq i \wedge et_i(p_q) - bt_i(p_q) > w_{v,q}\} \quad (2)$$

Where min$\{A\}$ means the smallest element in A and $w_{v,q}$ represents the time required to perform task $v$ on resource $p$.

(2) When $bt_i(p) \leq EST(v,p) < et_i(p)$, $avail(p, v)$ is defined as:

$$avail(p, v) = \begin{cases} EST(v,p), et_i(p) - EST(v,p) \geq w_{v,q} \\ \min\{j | et_{i+1}(p_q) - bt_{i+1}(p) \geq w_{v,q}\}, \ other \end{cases} \quad (3)$$

When $avail(p, v)$ is used instead of $avail[p]$ in literature [11], *Ranku* can be calculated using the method in literature [11].

### 3.3  MSBB Algorithm

Consistent with the $k$-HEFT algorithm in literature [11], the MSBB algorithm also selects tasks in the first $k$ tasks for scheduling after sorting by *Ranku*. However, unlike the $k$-HEFT algorithm, the MSBB algorithm also uses the backfill strategy to make full use of the available idle time period to improve resource utilization and shorten the makepan. The main strategy of the MSBB algorithm can be divided into two steps. First, we merge adjacent DAGs into new DAG where the interval time between adjacent DAGs is less than *wt*, and let new DAG enter the scheduling sequence. Second, the DAG is scheduled according to the description of the available time period of the resource and the idea of the $k$-HEFT algorithm.

---

Algorithm 1. MSBB

---

1: Accept DAG-$i$;
2: If (no new DAG arrives during *wt*)
3:     DAG-$i$ enters scheduling sequence;
4: Else
5:     While (new DAG arrives during *wt*) do
6:             Combine two DAGs into DAG-$s$ ;
7:             DAG-$s$ enters scheduling sequence;
8:     End while
9: While (Scheduling sequence is not empty) do
10:     Obtain an idle period of each processor;
11:     Sort all task in the first DAG by *Ranku;*
12:     Select the first $k$ tasks(less than k, take all tasks);
13:     Calculate their respective *avail(p, v)*, find the smallest one of them;
14:     Complete the scheduling in the corresponding processor;
15:     Update the idle time of each resource.
16: End while

---

The time complexity analysis of the MSBB algorithm is as follows:

The time complexity of the HEFT algorithm for processing $n$ tasks on $q$ processors is $O(q * n^2)$. According to the literature [11], we know that the $k$-HEFT algorithm has a time complexity of $k(k + 1)/2$ times that of HEFT. The MSBB algorithm also considers the DAG merging process based on the $k$-HEFT algorithm considered. Assuming the number of DAGs is $m$, the computational complexity of the MSBB algorithm should be: $O(m * q * n^2 * k^2)$.

## 4  Experiments

The experiment hardware platform is configured with Interl(R) Core(TM) i7-3630QM, 2.40 GHz, 8 GB RAM and 1 TB hard disk. And the software platform is configured with Windows 7 operating system and Java1.8.

In the section, the required parameters are described as follows. The *wt* is configured with 3, and the range of $k$ used in the $k$-HEFT is $k \leq 5$. The *avgCost* that it is average cost of tasks on processors is configured with 20. The *CCR*, the ratio of communication overhead to computation time, is set to 2.5.

The experiments compare the performance of sequential scheduling, combined scheduling, and MSBB algorithms by changing the number of DAGs, the DAG arrival time interval, the number of DAG average tasks, and the number of resources. In the comparative evaluation, we use the following three performance metrics.

- **Makespan.** The scheduling completion time of the algorithm is called makespan. The makespan value is defined as:

$$Makespan = AFT(v_{last}) - EST(v_{first}) \tag{4}$$

- **Utilization.** The value of utilization is the ratio of the sum of the running times of each task in all DAGs on the processor it schedules to the number of Makespan and the number of processors. The utilization value is defined as:

$$Utilization = \frac{\sum_{v_i \in DAGs} AFT(v_i, P_q) - AST(v_i, P_q)}{makespan \times sourceNumber} \tag{5}$$

- **RunTime.** This parameter indicates the actual running time of the program.

### 4.1 The Effect of the Number of DAG on the Performance of the Algorithm

In this subsection, the number of DAGs is configured with $\{2, 4, 6, 8, 10\}$, and the range of the average arrival time interval of each DAG is set to $[0, 2t)$, where $t$ is 125. The average number of tasks $v$ of DAGs is initialized to 15, and the number of processor resources *sourceNumber* is set to 4.

Figures 4, 5, and 6 show the fluctuations in the three strategies (sequential scheduling, merge scheduling, and MSBB) in Makespan, Utilization, and RunTime, respectively, when the number of DAGs changes. Through data analysis, we can draw the following conclusions.

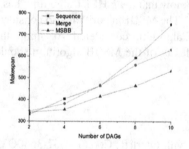 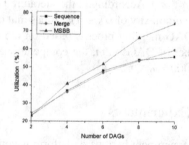

**Fig. 4.** The effect of DAG number changes on Makespan

**Fig. 5.** The effect of DAG number changes on the Utilization

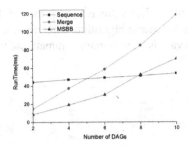

**Fig. 6.** The effect of DAG number changes on RunTime

(1) The three scheduled Makespan are not significantly different when the number of DAGs is small, but as the number of DAGs increases, the Makespan of MSBB is significantly lower than the other two algorithms. By calculation, we can learn the MSBB algorithm to be 17% lower than the sequential scheduling and 11% lower than the merge scheduling on the Makespan.

(2) The resource utilization of the three schedules is not obvious when the number of DAGs is small. For example, when the number of DAGs is 2, the three schedules are all around 23%–24%. As the number of DAGs increases, the resource utilization of sequential scheduling and merge scheduling is not much different, and MSBB is gradually superior to them. When the number of DAGs is 10, the MSBB is about 14% higher than the other two.

(4) The MSBB is significantly better than merge scheduling, but when the number of DAGs is small, it is inferior to sequential scheduling, in the comparison of RunTime.

## 4.2 The Effect of the DAGs Arrival Time Interval on the Performance of the Algorithm

In this subsection, the number of DAGs is set to 10, the range of the average arrival time interval of each DAG is set to $[0, 2t)$, where the set of values of $t$ is {25, 50, 75, 100, 125}. The average number of tasks $v$ of DAGs is initialized to 15, and the number of processor resources *sourceNumber* is set to 4.

Figures 7, 8 and 9 show the fluctuations in the three strategies (sequential scheduling, merge scheduling, and MSBB) in Makespan, Utilization, and RunTime, respectively, when the DAGs arrival time intervals changes. Through data analysis, we can draw the following conclusions.

(1) The value of $t$ has little effect on the sequential scheduling of the Makespan, but due to the sequential scheduling, there will be a lot of time fragments, which will inevitably make it the worst performer on the makepan.

(2) The value of t has a greater impact on Makespancombined scheduling and MSBB in Makespan. The Makespan curve of MSBB is relatively flat. When the value of t is small (t ≤ 65), the merge scheduling of Makespan is better than MSBB

scheduling; but when the critical value is t > 65, the MSBB scheduled Makespan starts to be lower than the merge algorithm. Therefore, it can be inferred that the DAG arrival time interval t is an important parameter affecting the performance of the algorithm.

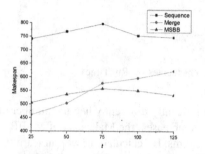

**Fig. 7.** The effect of time interval changes on Makespan

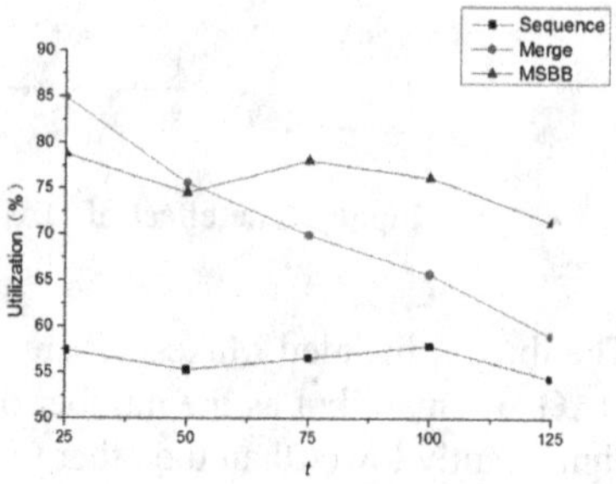

**Fig. 8.** The effect of time interval changes on Utilization

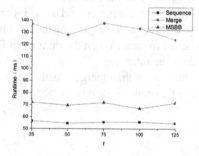

**Fig. 9.** The effect of time interval changes on RunTime

(3) When the value of t is small, the merge scheduling is higher than the MSBB scheduling. Otherwise, the MSBB's Utilization is higher than the merge scheduling.

(4) The merge schedule adapts to certain scenarios, such as when t is small. However, the RunTime of the merged schedule is very large, about 1.9 times that of the MSBB, about 2.4 times that of the sequential schedule. This shows that MSBB is more efficient.

## 4.3 The Effect of the Number of Tasks on the Performance of the Algorithm

In this subsection, the number of DAGs is set to 10, the range of the average arrival time interval of each DAG is set to [0, 2t), where t is 125. The set of the average number of tasks $v$ of DAGs is initialized to {5, 10, 15, 20, 25}, and the number of processor resources *sourceNumber* is set to 4.

Figures 10, 11 and 12 show the fluctuations in the three strategies (sequential scheduling, merge scheduling, and MSBB) in Makespan, Utilization, and RunTime, respectively, when the average number of tasks changes. Through data analysis, we can draw the following conclusions.

(1) When the number of DAG average tasks is small, the Makespan of sequential scheduling and merge scheduling is close, and the MSBB scheduling is better than the first two. As the number of tasks increases, sequence scheduling of Makespan is the highest, and MSBB and merge scheduling are getting closer. It can be predicted that the more DAG average tasks, the less obvious the advantages of MSBB.

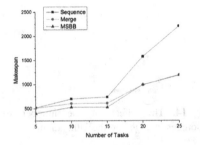

**Fig. 10.** The effect of the average number of tasks changes on Makespan

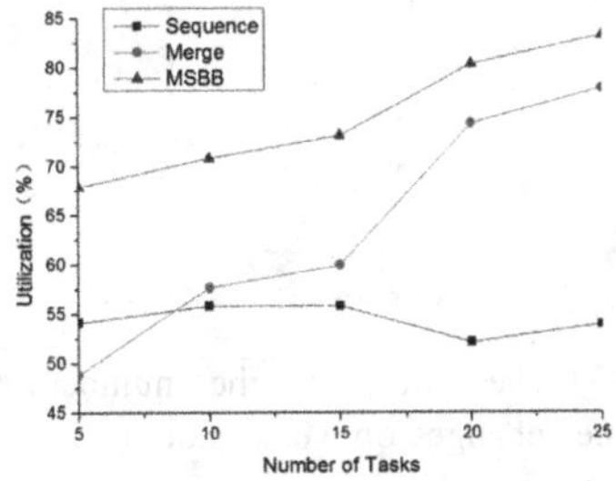

**Fig. 11.** The effect of the average number of tasks changes on the Utilization

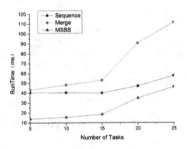

**Fig. 12.** The effect of the average number of tasks changes on RunTime

(2) Although the Utilization of MSBB scheduling is the highest compared to the other two. However, as the number of average tasks continues to increase, it can be speculated that the combined scheduling will be closer to MSBB scheduling.

(3) MSBB is optimal in RunTime compared to the other two, but as the average number of tasks continues to increase, the RunTime for merge scheduling and MSBB scheduling grows faster and faster.

### 4.4 The Effect of the Number of Resources on the Performance of the Algorithm

In this subsection, the number of DAGs is set to 10, the range of the average arrival time interval of each DAG is set to [0, 2*t*), where *t* is 125. The average number of tasks *v* of DAGs is initialized to 15, and the set of the number of processor resources *sourceNumber* is set to {2, 4, 6, 8, 10}.

Figures 13, 14, and 15 show the fluctuations in the three strategies (sequential scheduling, merge scheduling, and MSBB) in Makespan, Utilization, and RunTime, respectively, when the average number of resources changes. Through data analysis, we can draw the following conclusions.

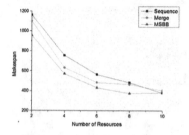

**Fig. 13.** The effect of the number of resources changes on Makespan

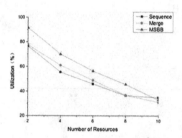

**Fig. 14.** The effect of the number of resources changes on the Utilization

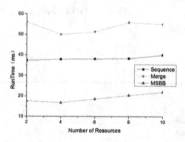

**Fig. 15.** The effect of the number of resources changes on the RunTime

(1) As the number of resources increases, the three scheduled Makespans are gradually reduced, and the Utilization is also decreasing.
(2) The sequence scheduling shows relatively good performance in the case of a large number of resources, and the Makespan and Utilization curves are gradually gradual when MSBB is used. It can be speculated that MSBB may be more suitable for situations where the number of resources is small.
(3) The RunTime of three scheduling are relatively flat, with the merge scheduling time overhead being the largest and the MSBB being the smallest.

# 5 Conclusion

Aiming at the scheduling problem of multiple DAGs arriving at different time in cloud computing, a multiple DAGs scheduling strategy based on backfill is proposed. Under the premise of not considering the priority, this strategy mixes the idea of combined scheduling and backfill scheduling strategy, transforms the multiple DAGs scheduling problem into a single DAG scheduling problem with limited resource time, and finally solves it by MSBB algorithm. The simulation experiment compares several factors such as Makepan, Utilization and RunTime with the sequence scheduling and merge scheduling.

It also analyzes the influencing factors such as the number of DAGs, the time interval of multiple DAG arrivals, the number of DAG average tasks, and the number of resources. The experimental results show that the strategy of this paper can obtain better Makepan and Utilization.

**Acknowledgment.** This work is supported by Guangxi Universities key Laboratory Director Fund of Embedded Technology and Intelligent Information Processing (Guilin University of Technology) under Grand No. 2018A-05.

# References

1. Jia-Xin, Y., et al.: Time-aware minimum area task scheduling algorithm based on backfilling algorithm. Comput. Sci. **45**(8), 100–104 (2018)
2. Huang-Ke, C., et al.: Cost-efficient reactive scheduling for real-time workflows in clouds. J. Supercomput. **74**(11), 6291–6309 (2018)
3. Jiang, X., Xiang, L.: Improved decomposition-based global EDF scheduling of DAGs. J. Circ. Syst. Comput. **27**(7), 1–23 (2018)
4. He-Jhan, J., et al.: Scheduling concurrent workflows in HPC cloud through exploiting schedule gaps. Algorithms Archit. Parallel Process. **7016**, 283–293 (2011)
5. Bittencourt, L.F., Madeira, E.R.M.: HCOC: a cost optimization algorithm for workflow scheduling in hybrid clouds. J. Internet Serv. Appl. **2**(3), 207–227 (2011)
6. Javadi, B., Tomko, M., Sinnott, R.O.: Decentralized orchestration of data-centric workflows in cloud environments. Future Gener. Comput. Syst. **29**(7), 1826–1837 (2013)
7. Casanova, F., Suter, F.: On cluster resource allocation for multiple parallel task graphs. J. Parallel Distrib. Comput. **70**(12), 1193–1203 (2010)
8. Jia-Yu, Z., Dan, X.: Path priority-based heuristic task scheduling algorithm for cloud computing. Comput. Eng. Des. **34**(10), 3511–3515 (2013)
9. Kan-Kan, L.: High performance algorithm for task scheduling in heterogeneous environment. Comput. Syst. Appl. **19**(11), 102–105 (2010)
10. Ya-Qiu, L., Hong-Run, S., Wei-Peng, J.: DAG task scheduling integrating with security and availability in cloud environment. Comput. Eng. **40**(12), 12–18 (2014)
11. Tian-Mei-Zi, C., Heng-Zhou, Y., Zhi-Dan, H.: k-HEFT: a static task scheduling algorithm in clouds. In: Proceedings of the 3rd International Conference on Intelligent Information Processing, ICIIP 2018, pp. 152–159 (2018)
12. Guo-Zhong, T., Chuang-Bai, X., Zhu-Sheng, X.: Hybrid scheduling strategy for multiple DAGs workflow in heterogeneous system. J. Softw. **23**(10), 2720–2734 (2012)

13. Yuan-Xiong, G., Yu-Guang, F.: Electricity cost saving strategy in data centers by using energy storage. IEEE Trans. Parallel Distrib. Syst **24**(6), 1149–1160 (2013)

14. Yue, S., Jiong, Y., Jian-Bo, Z.: Preemptive scheduling for multiple DAGs in cloud computing. Comput. Sci. **41**(3), 145–148 (2014)

15. Jun, Z., et al.: Efficient fault-toleran scheduling on multiprocessor systems via replication and deallocation. Int. J. Embedded Syst. **6**(2–3), 216–224 (2014)

16. Wei-Peng, J., et al.: Multiple DAGs dynamic workflow reliability scheduling algorithm in a cloud computing system. J. Xidian Univ. **43**(2), 92–97 (2016)

17. Ji, L., Long-Hua, F., Sheng-Long, F.: An greedy-based job scheduling algorithm in cloud computing. J. Softw. **9**(4), 921–925 (2014)

18. Topcuoglu, H., Hariri, S., Wu, M.Y.: Performance-effective and low-complexity task scheduling for heterogeneous computing. IEEE Trans. Parallel Distrib. Syst. **13**(3), 260–274 (2002)

19. Zhao, H., Sakellariou, R.: Scheduling multiple DAGs onto heterogeneous systems. In: Proceedings of the 20th International Conference on Parallel and Distributed Processing, IPDPS 2006 (2006)

# Education

# Study on Evaluation Method of Teaching Quality Based on Cloud Model and D-S Theory

Jianting Shi[✉], Jun Sun, Jiancai Wang, Jingping Cao,
and Chunyuan Liu

Hei Longjiang University of Science and Technology,
Harbin 150022, Hei Longjiang, China
hotmail8194@163.com

**Abstract.** The Evaluation of teaching quality is a complicated project. It's a typical problem of uncertainty. A design scheme of teaching quality evaluation method based on cloud model and D-S theory is proposed with the school teaching evaluation index. With the experiences and knowledge of experts, evaluation index was described by cloud model and judgment matrix of AHP was revised by D-S theory. The weight of evaluation indexes was determined reasonably and objectively. Applied in the actual environment, the method can make an accurate and efficient evaluation with a certain practical value.

**Keywords:** Teaching quality evaluation · Cloud model · D-S theory

## 1 Introduction

Teaching is a kind of spiritual work, but also an art. Giving value to the teaching process and teaching results can continuously promote the improvement of teaching quality. Therefore, the evaluation of teaching quality in colleges and universities is a very important part of teaching. Carrying out the quality evaluation of teachers' classroom teaching is of positive significance and important role in guiding teachers to continuously carry out teaching reform and improve the quality and level of classroom teaching [1].

Teaching quality evaluation is a complex project. The weight of the evaluation indicators is usually given directly by a few experts based on experience, and the subjectivity is very strong. The AHP method was used in the literature [2] to determine the weight, which is less objective. Do not use the objective evaluation of teaching quality. Therefore, based on the influence of the subjective factors of AHP method, this paper introduces the data fusion idea in evidence theory (D-S theory), combines the opinions of many experts to improve the judgment matrix of AHP method, and overcomes the subjective limitations of traditional AHP method to obtain weight. Students, peers and experts use natural language in the evaluation of teachers' teaching, while natural language has strong ambiguity and randomness. Most of them are qualitative descriptions of teaching quality, which is difficult to quantify. The cloud model is a qualitative and quantitative interchange model proposed by Professor Li Deyi on the basis of probability statistics and traditional fuzzy mathematics [3].

© Springer Nature Singapore Pte Ltd. 2019
R. Mao et al. (Eds.): ICPCSEE 2019, CCIS 1059, pp. 293–302, 2019.
https://doi.org/10.1007/978-981-15-0121-0_22

Therefore, this paper combines the cloud model and proposes a scientific comprehensive evaluation method of teaching quality to realize the natural conversion between qualitative language and quantitative data. It provides a practical reference for the study of teaching quality evaluation methods.

## 2  Construction of Teaching Quality Evaluation Index System

Teaching quality evaluation indicators are affected by many factors. Such as teaching content, teaching methods, teaching methods, etc. When evaluating the quality of teaching and the quality of personnel training the traditional evaluation model generally has the following problems: (1) There is lack of scientific basis for the quantification of evaluation indicators, most of which are determined according to the expert's experience value or simple calculation; (2) The evaluation index system ignores the differences between different courses; it does not reflect the teaching characteristics of the course; (3) The expression of the evaluation result is single and fails to give targeted improvement measures. In addition, teacher teaching is a complex intellectual activity that involves not only the knowledge of the courses taught, but also pedagogy, psychology, linguistics and so on [4]. Therefore, in the evaluation of teacher teaching quality, we must consider various factors. Combining with the actual teaching situation of our school and the new development of college teaching, we have established the teaching quality evaluation index system that accords with the actual situation of our college. The evaluation factor set is shown in Table 1.

**Table 1.** Teaching quality evaluation index

| Primary indicator | Secondary indicator |
|---|---|
| Teaching attitude $U_1$ | Strict and serious, well prepared, full of spirit, enthusiasm for lectures $U_{11}$ |
| | Teaching and educating people, being a model for others, caring for students, demanding strict $U_{12}$ |
| Teaching content $U_2$ | The lectures are rich in content and correct in opinion $U_{21}$ |
| | Clear ideas, accurate concepts, and innovative teaching $U_{22}$ |
| | New ideas, new concepts, and new achievements that can be reflected or related to the development of disciplines $U_{23}$ |
| | Summarize and emphasize the key points and difficulties in the lecture or discussion $U_{24}$ |
| Teaching method $U_3$ | In-depth, inspiring thoughts, and inspiring interests $U_{31}$ |
| | Communicate and interact, explain vividly, and flexibly use various teaching media to mobilize student emotions $U_{32}$ |
| | The teaching method is suitable for the characteristics of the subject and the characteristics of the students, and pays attention to the guidance of the students' learning methods $U_{33}$ |

*(continued)*

**Table 1.** (*continued*)

| Primary indicator | Secondary indicator |
|---|---|
| Teaching art $U_4$ | Language succinct, rhythm, and appealing, increasing students' interest in the subject $U_{41}$ |
| | Proficiency in lectures and teaching standards improve students' ability to analyze and solve problems $U_{42}$ |
| | The board is neat, accurate, and organized, which promotes positive thinking and inspiration $U_{43}$ |
| Teaching effect $U_5$ | Students are eager to attend classes, listen carefully, and have an active atmosphere $U_{51}$ |
| | Orderly, thoughtful, enlightened $U_{52}$ |
| | Active interaction, students reflect well $U_{53}$ |

# 3 Evaluation of Teaching Quality Based on Cloud Model

## 3.1 Cloud Model Description of the Evaluation Indicator

The cloud model has three characteristics of Ex (Expected value), En (Entropy) and He (Hyper entropy) [4]. Ex represents the central value of the qualitative concept in the domain, that is, the distribution center of the cloud; En represents the degree of dispersion of the qualitative concept and the range of values of the cloud droplets that can be accepted by the concept; He is the uncertainty measure of entropy, reflecting the uncertainty measure of the point representing the linguistic value in the number domain space, which is determined by the randomness and ambiguity of entropy. Assume the teaching evaluation attribute set to U:

Primary factor: $U_1, U_2 \ldots \ldots \ldots U_m$

a. The first-level factors are evaluated in five aspects: teaching attitude, teaching content, teaching methods, teaching art and teaching effects;
b. The secondary factors are evaluated in terms of rigorousness, adequate preparation, full spirit, and enthusiasm for lectures.

Secondary factor: $U_{11}, U_{12} \ldots \ldots . U_{1n}$

$$U_{21}, U_{22} \ldots \ldots . U_{2n}$$
$$\ldots \ldots \ldots \ldots \ldots \ldots$$
$$U_{m1}, U_{m2} \ldots \ldots . U_{mn}$$

Among of which $\{m = 1 \ldots \ldots 5, n = 1 \ldots 4\}$. Experts, peers, and students often use "very good", "good", "medium", "poor" and "very poor" to describe the effectiveness of teacher class teaching when evaluating the quality of teaching. Set the teaching evaluation result set $V = \{V_1, V_2, \ldots, V_5\}$. These linguistic values that describe the various evaluation factors of the teacher can be expressed using a cloud model. For each evaluation factor, a set of samples is established as a cloud drop according to the expert opinion, and a conceptual cloud model corresponding to the evaluation results of each level is given by the inverse cloud generator [5]. To simplify the calculation, it is assumed that the conceptual cloud models whose evaluation factors correspond to the evaluation results at all levels are normal cloud models. The cloud model identifier corresponding to each evaluation factor corresponding to the evaluation result is $C_k$ $C_k$ $(Ex_k, En_k, He_k)$ where $\{k = 1, 2, 3, 4, 5\}$ represents five evaluation results.

## 3.2    The Establishment of Teaching Evaluation Cloud Model

The process of establishing a teaching evaluation cloud model is as follows:

1. The evaluation factor $U_{ij}$ corresponds to five evaluation results. The most representative of each factor corresponding to the level $C_k$ is the intermediate value of the level, namely:

$$Ex_k = Z_k \tag{1}$$

Set the index corresponding to the evaluation level $C_k$ be the threshold values of $Z_k^1$ and $Z_k^2$, The method of determination is:

$$Z_k^2 = \begin{cases} Z_k + D^*/2, (z_{k+1} < Z_k) \\ Z_k - D^*/2, (z_{k+1} > Z_k) \end{cases}, \tag{2}$$

where $D^*$ is the modulus of the difference between the adjacent two levels of center values, namely: $D^* = |Z_{k+1} - Z_k|$. Therefore, the correspondence between rank and score is: "very good" corresponds to the score range of [90, 100], "good" corresponds to the score range of [80, 90], and "medium" corresponds to the score range of [70, 80], "bad" corresponds to a score range of [60, 70]. In general, we take 50 as the criterion for the evaluation difference, so the "worse" corresponding score range is [30, 60].

2. The threshold value is a transition from one level to one level, and belongs to the upper and lower levels, and the affiliate degrees of the two levels are equal, so $\exp\left(-\frac{(Z_k^1 - Z_k^2)^2}{8(En_k)^2}\right) \approx 0.5$, $En_k$:

$$En_k = \frac{|Z_k^1 - Z_k^2|}{2.355} \tag{3}$$

3. The super-entropy $He_k$ can be obtained from experience and repeated experiments according to the size of $En_k$, The value is 0.05.

Finally, the method which the level of each factor evaluation result is expressed in the cloud of the score field is: $D_{v1} = D(95, 10/2.355, 0.05)$, $D_{v2} = D(85, 10/2.355, 0.05)$, $D_{v3} = D(75, 10/2.355, 0.05)$, $D_{v4} = D(65, 10/2.355, 0.05)$, $D_{v5} = D(50, 10/2.355, 0.05)$.

# 4    Weight Calculation Based on Evidence Theory

The weight is to enable the comprehensive evaluation to consider the inconsistency of the overall impact of each evaluation attribute [6]. In the concentration of evaluation factors, the importance of each factor is different. Through the communication with the inspectors and the basic principles of the Analytic Hierarchy Process (AHP), the data fusion technology in D-S evidence theory is used to continuously feedback and modify. Finally determine the weight of each factor.

## 4.1    AHP Method

The AHP (Analytical Hierarchy Process) was proposed by American operations researcher Satie in the 1970s [7]. It is a combination of qualitative and quantitative decision analysis methods to determine the weight relationship between indicators. However, as a subjective weighting method, the decision-making result will be influenced by the subjective understanding of decision makers.

## 4.2    D-S Evidence Theory

Evidence theory is an artificial intelligence method of uncertainty reasoning. It can be concluded by combining multiple evidences and a reasonable information theory explanation for reasoning [8].

**Definition 1.** Evidence credibility [9]. Assume the system have N independent evidences, $\{\Theta, \Gamma_i, m_i\}$ and $\{\Theta, \Gamma_p, m_p\}$ $(i \neq p, 1 \leq i, p \leq N)$ are two different pieces of evidence.

$$SimBetP(m_i, m_p) = 1 - difBetP(m_i, m_p) \tag{4}$$

$$SupBetP(m_i) = \sum_{p=1, p \neq i}^{N} SimBetP(m_i, m_p) \tag{5}$$

$$CrdBetP(m_i) = SupBetP(m_i) / \sum_{p=1}^{N} SupBetP(m_p) \tag{6}$$

In formula, $SimBetP(m_i, m_p)$ express similarity between evidence $i$ and evidence $p$, $SupBetP(m_i)$ indicates the support of all other evidence for evidence $i$, $CrdBetP(m_i)$ demonstrate the credibility of evidence $i$ in all evidence.

Assign weight $\omega(A)$ algorithm can be described as:

$$\omega(A) = \sum_{i=1}^{N} CrdBetP(mi) \cdot m_i(A), A \subseteq \Theta \tag{7}$$

**Definition 2.** D-S evidence synthesis rules: set the system to have N independent evidence $\{\Theta, \Gamma_i, m_i\}(1 \leq i \leq N)$, Orthogonal sum [10]:

$$\begin{cases} m_\otimes(A) = \frac{1}{1-k} \sum_{A_j=A_1 \leq i \leq N} \prod m_i(A_j), A \subseteq \Theta, A \neq \emptyset \\ m_\otimes(\emptyset) = 0 \end{cases} \tag{8}$$

$m_i$ is the Probability distribution function on $\Theta$, $k$ is the coefficient of conflict

$$k = \sum_{A_j=\emptyset} \prod_{1 \leq i \leq N} m_j(A_j)(0 \leq k \leq 1) \tag{9}$$

In the AHP method, constructing a judgment matrix is the basis of the application. The judgment matrix compares the index i and the index j with each other and records $A = (a_{ij})$, represents the relative importance value of the factor $U_i$ to $U_j$, and its value is given by the expert judgment quantitative rule. Satisfy $a_{ij} > 0$, if $a_{ij} = 1/a_{ij}$, then $a_{ij} = 1$. Under normal circumstances, the expert's judgment is not completely accurate, and the overall deviation is not too large. For example, it is impossible to think of both $a_{ij} = 2$ and $a_{ij} = 1/2$. In order to avoid the influence of subjective factors, the fusion technology of D-S evidence theory is used to transform it, and the judgment matrix is modified by combining the opinions of many experts, making the evaluation result more objective.

The scale framework for D-S evidence theory is $\{\theta_1, \theta_2, \theta_3, \ldots, \theta_{20}\}$, $\theta_1 > \theta_2 > \theta_3 > \ldots\ldots > \theta_{20}$ $\theta_1 > \theta_2 > \theta_3 > \ldots > \theta_{20}$, From $\theta_1$ to $\theta_{20}$ respectively represent "good" to "poor", the symbol > indicates better than, $\theta_i$ is expert judgment comment, then make the expert give him the degree of trust in $\theta_1$, which is called $a$; If the judgment result of oneself is uncertain, it can also give the degree of trust in the uncertainty of the indicator, that is, the trust degree of the entire comment frame $\Theta$. For the different evaluation factors in the evaluation index system, the opinions of the experts 1, experts 2, ... and 20 are defined as the reliability functions Bel1, Bel2, ..., Bel20, m1, m2, ..., m20 are the basics of each expert. The probability assignment function, the expert's reliability for each focal element is shown in Table 2 (taking two experts as an example).

**Table 2.** Expert judgment reliability

| Judging level | Expert 1 confidence in judgment | Expert 2 confidence in judgment |
|---|---|---|
| $\theta_1$ | 0.1 | 0.2 |
| $\theta_2$ | 0.6 | 0.5 |
| $\theta_3$ | 0.2 | 0.3 |
| ...... | ...... | ...... |
| $\theta_{20}$ | 0 | 0 |
| Uncertain ($\Theta$) | 0.1 | 0.2 |

The values of $m(\theta)$ are calculated using Eqs. (8) and (9), and the results after data fusion are shown in Table 3. Through data fusion, the evaluation values of experts with the same opinion have been amplified. Then use formula (7) to find the distribution weight of the focal element $\omega(A)$.

**Table 3.** The result of data fusion

| $m(\theta_i)$ | Data fusion result |
|---|---|
| $m(\theta_1)$ | 0.0345 |
| $m(\theta_2)$ | 0.0576 |
| $m(\theta_3)$ | 0.1034 |
| ...... | ...... |
| $m(\theta_{20})$ | 0 |
| $m(\Theta)$ | 0.1023 |

## 4.3 Algorithm Analysis of Evaluation Factors Weight

1. Establish a hierarchical structure. Mainly divided into target layer (teaching quality evaluation), criterion level (evaluation index factor) and program layer (20 supervision experts);
2. To construct a judgment matrix A.
3. Target consistency test: consistency ratio $CR = \frac{CI}{RI}$, $CI = \frac{\lambda_{max}-n}{n-1}$ ($CI$ is the consistency indicator, $\lambda_{max}$ can be solved according to the formula $|\lambda E - A| = 0$), $RI$ is the average random consistency indicator. When $CR < 0.1$, it indicates that the weight distribution is very reasonable, otherwise it needs to be adjusted by experts. $\lambda$ vector obtained by normalizing the corresponding feature vector. The calculation results are shown in Table 4 (taking the evaluation teaching method as an example).
4. Calculation of weights: It can be proved that A has the largest eigenvalue $\lambda_{max} = m$ when matrix A satisfies the consistency requirement. The normalized feature vector of $\lambda_{max}$ is the weight vector of each factor in the criterion layer. The obtained weight vector $W = \{w_1, w_2, w_3, w_4\}$, where m represents the number of evaluation factors. Finally, the weights of each evaluation factor are obtained as $W_{U1} = \{0.5, 0.5\}$, $W_{U2} = \{0.2970, 0.2928, 0.2468, 0.1634\}$, $W_{U3} = \{0.3333, 0.3333, 0.3333\}$, $W_{U4} = \{0.2356, 0.3674, 0.3970\}$, $W_{U5} = \{0.3674, 0.2356, 0.3970\}$.

**Table 4.** Judement matrix of teaching way

| $U_3$ | $U_{31}$ | $U_{32}$ | $U_{33}$ | $W$ | Consistency test |
|---|---|---|---|---|---|
| $U_{31}$ | 1 | 2 | 2/3 | 0.3333 | $\lambda_{max} = 3$ |
| $U_{32}$ | 1/2 | 1 | 1/3 | 0.3333 | $CI = 0$ |
| $U_{13}$ | 3/2 | 3 | 1 | 0.3333 | $CR = 0 < 0.1$ |

## 4.4 Comprehensive Evaluation Result

After obtaining the weights of each evaluation factor, the hierarchical result evaluation clouds of each single factor are combined and the following formula is used:

$$E_x = \frac{E_{x_1} w_1 + E_{x_2} w_2 + \ldots + E_{x_n} w_n}{w_1 + w_2 + \ldots + w_n} \tag{10}$$

A collection of comments corresponding to each comprehensive evaluation cloud is obtained. As shown in Fig. 1.

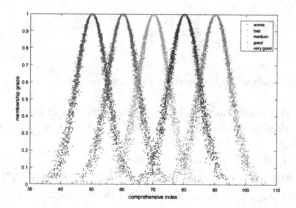

**Fig. 1.** Cloud model corresponding to each level of cloud

**Table 5.** The teacher evaluation data to be evaluated is shown in Table 4:

| Evaluation index | Evaluation factor | Comment | Quantitative value |
| --- | --- | --- | --- |
| Teaching attitude | $U_{11}$ | Very good | 93 |
| | $U_{12}$ | Very good | 96 |
| Teaching content | $U_{21}$ | Very good | 97 |
| | $U_{22}$ | Medium | 76 |
| | $U_{23}$ | Good | 84 |
| | $U_{24}$ | Medium | 73 |
| Teaching method | $U_{31}$ | Good | 89 |
| | $U_{32}$ | Good | 87 |
| | $U_{33}$ | Good | 82 |
| Teaching art | $U_{41}$ | Medium | 72 |
| | $U_{42}$ | Good | 85 |
| | $U_{43}$ | Good | 85 |
| Teaching effect | $U_{51}$ | Good | 87 |
| | $U_{52}$ | Medium | 78 |
| | $U_{53}$ | Good | 91 |

In Table 5, using the experience of experts, a cloud model description is made for each evaluation index, and the quantitative values of each evaluation are obtained, and then a comprehensive evaluation cloud is constructed according to the obtained weights, and corresponding comments are obtained. In the end, the quality of the teacher's class teaching was 92.64, but the language was not refined enough and vivid, and the enthusiasm of the students was not fully mobilized.

## 5 Conclusion

In this paper, cloud quality evaluation, AHP method and D-S evidence theory was used, a research method was proposed based on cloud model and evidence theory. The cloud model is used to implement the qualitative and quantitative conversion of evaluation index comments, and the evidence theory combined with the AHP method to correct the judgment matrix makes the evaluation result more objective. Compared with other works [11], there are many uncertain factors in teaching. The use of AHP method for quantitative analysis often ignores these uncertain factors. The results are also inaccurate. The combination of AHP and D-S evidence can add uncertainties to evaluate teaching more comprehensively.

The evaluation method has been applied to the teaching evaluation process of teachers in our school. The weight distribution of evaluation factors is reasonable and the evaluation results are reliable, indicating that the evaluation method has certain practical significance. It provides a meaningful reference value for the study of teaching quality evaluation.

**Acknowledgements.** This paper was supported by the Higher Education Teaching Reform Project of Heilongjiang named Research and Practice of Computer Basic Teaching Reform with the concept of Computing Thinking (Grant No. SJGY20170344) and Planning project of Education department of Hei Longjiang province (Grant No. GJC316157).

## References

1. Ding, J.L., Ye, J.H.: The application of analytic hierarchy process and fuzzy comprehensive evaluation in the evaluation of teachers' classroom teaching quality. J. Wuhan Univ. (Soc. Sci. Ed.) **56**(2), 241–245 (2003)
2. Shi, J.T., Hou, J.Y., Guan, F.Y.: Design and application of teaching quality evaluation system based on AHP-FUZZY. Comput. Technol. Dev. **23**(5), 188–191 (2013)
3. Li, D.Y.: Uncertain Artificial Intelligence. National Defence Industry Press, Beijing (2005)
4. Yang, X.L., Dong, C.Y., Song, X.Q.: Evaluation of urban residential land based on cloud model and evidence distance. Comput. Mod. **3**, 192–195 (2013)
5. Jia, Q., Duan, Q.C., Chen, X.N.: Cloud model for evaluation of water resources renewability in the yellow river basin. China's Popul. Resour. Environ. **20**(9), 48–52 (2010)
6. Guo, A.H., Qin, W.M.: A comprehensive evaluation method for college student party members based on system fuzzy decision theory. J. Guangdong Univ. Technol. (Soc. Sci. Ed.) **11**(1), 43–46 (2011)
7. Levine, T.: Stability and change in curriculum evaluation. Stud. Educ. Eval. **28**(1), 1–33 (2002). https://doi.org/10.1016/S0191-491X(02)00010-X

8. Zhang, Y., He, Z.Y., Lin, S.: A power system fault diagnosis method based on D-S evidence theory. In: Prognostics and Health Management Conference, pp. 5–10. IEEE Xplore (2010)
9. Sun, Q., Ye, X.Q., Gu, W.K.: A new synthetic formula based on evidence theory. Chin. J. Electron. **28**(8), 117–119 (2000)
10. Cui, W.: Application of improved D-S evidence theory in marine target fusion recognition model. Ship Sci. Technol. **8**, 89–95 (2017)
11. Wang, Z.Y., Wang, Z.Q., Chen, L.E.: Teaching quality evaluation based on analytic hierarchy process. J. Guangdong Univ. Technol. (Soc. Sci. Ed.) **10**(6), 25–27 (2010)

# Bottom-Up Teaching Reformation for the Undergraduate Course of Computer Organization and Architecture

Yanjun Shu[1][(✉)], Wei Emma Zhang[2], Yanxin Liu[1], Chunpei Wang[1],
Jian Dong[1], Zhan Zhang[1], Dongxi Wen[1], and Decheng Zuo[1]

[1] School of Computer Science and Technology,
Harbin Institute of Technology, Harbin, China
yjshu@hit.edu.cn
[2] The University of Adelaide, Adelaide, Australia

**Abstract.** According to the building requirements of "China's double first class" discipline, traditional computer organization and architecture (COA) course has new challenges including various course expectations and the limited teaching hours. Considering the hierarchy feature of COA, a bottom-up teaching mode is adopted in teaching reformation to meet the challenges. In this paper, details about COA teaching reforms were shown from aspects of teaching contents, teaching methods, hands-on assignments, and examination methods. These reform experience will benefit teachers who embark on courses related to computer hardware.

**Keywords:** Computer organization and architecture ·
Bottom-up teaching mode · Teaching reform

## 1 Introduction

Computer organization and architecture (COA) plays a significant role in undergraduate computer science programs. In most of computer science and technology curriculum, COA is the major foundation [1]. As shown in Fig. 1, computer organization and architecture course occupies a key place for jointing the low-level courses and the high-level courses. COA course not only deepens the understanding of the primary courses (e.g., digital logic and computer system), but also lays a solid foundation for subsequent courses, such as compilation technology and operation system.

In the computer organization and architecture area, there are many subjects, including Machine level arithmetic of data, Memory system organization and architecture, Interface and communication, Performance evaluation and enhancement, and so on. These subjects can be taught in either a two courses sequence or one course. Moreover, teaching the subjects of COA is a difficult task since its theories and concepts are highly abstract.

R. Mao et al. (Eds.): ICPCSEE 2019, CCIS 1059, pp. 303–312, 2019.
https://doi.org/10.1007/978-981-15-0121-0_23

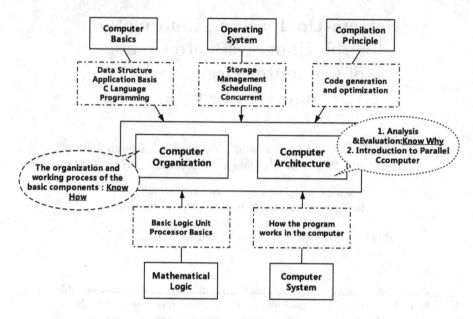

**Fig. 1.** COA course orientation

Based on the teaching content analysis of COA, we can find that most of subjects have knowledge hierarchy. For example, the subject of Memory system organization and architecture can be divided into three levels. The lowest level is the Resistor Transistor Logic (RTL) level of different memory circuit. The middle level is the component level of memory devices (i.e., cache and flash). The highest level is the system level of memory. The three levels of COA are hierarchical from the bottom to the top. Hence a bottom-up teaching mode is appropriate for COA course. Although the bottom-up teaching mode has been used [2–4], it has not been applied in the computer organization and architecture area to our knowledge. In this paper, we introduce a bottom-up teaching reform of computer organization and architecture course. We show how to decompose the teaching contents of COA and then integrate them based on the bottom-up teaching mode. The corresponding reforms for the teaching methods, hands-on assignments and examination methods are also presented.

This paper is organized as follows. Section 2 introduces the background and challenges for COA teaching reform. Section 3 shows the education objects of COA course. Section 4 depicts the details of bottom-up teaching mode reform. Section 5 offers conclusions and future works.

## 2    Background and Challenges

With the rapid development of Internet, Artificial Intelligence, and Big Data, an increasing number of related subjects have arisen in the computer science discipline. As a result, the main teaching purpose of computer science is gradually changing from the computer system principles to the computer system application. More and more computer application courses and concentrations are needed in the computer science and technology major curriculum.

COA courser consisted of two courses, Computer Organization and Computer Architecture. These two courses are taught in two continuous semesters. With two courses, students can clearly understand knowledge from Computer Organization firstly, and then they can understand content of Computer Architecture more easily with these prerequisite knowledge. However, according to requirement of faculty, these two courses are combined into one COA course since 2018. In COA course, the contents of Computer Organization and Computer Architecture are taught together. Thus, we need a reformation on teaching method to adapt to this change.

Furthermore, as a "double first class" discipline, our school of computer science and technology has finished the innovation of major curriculum. 12 professional concentrations are set up as electives. These concentrations are computer engineering, software engineering, computer science, network security, internet of technology, big data, biology information, artificial intelligence, and so on. Students in each concentration take about three courses specific to their concentration. Except for the three concentration courses, students can also take courses from other concentrations according to their interests. Thus, the undergraduate course of computer organization and architecture has students from different concentrations, and they will have various prerequisite courses. This results in a unique challenge: students may have different expectations for the course depending on their concentrations. Another challenge is the course hour reduction. As an elective course, the Computer Organization and Architecture only has 48 course hours. The teaching hours of COA course are less than the 60 core hours in the requirements of IEEE Computer Society and Association for Computer Organization and Architecture area [5].

To meet these challenges and enhance learning effectiveness of students, we decided to reform the computer organization and architecture course. The bottom-up teaching mode is applied in the course design. The traditional practice assignments are reformed into autonomous labs. Moreover, MOOC, an online courses platform, is utilized for teaching the fundamental knowledge of Computer Organization and Architecture.

## 3 Education Objectives of COA Course

Our school of computer science and technology is accredited by Chinese Engineering Education Accreditation Association (CEEAA) [6] to ensure the education quality of undergraduate students. CEEAA defines eleven Educational Objectives that are fulfilled by the sum total of all the courses. In our COA course, 6 objectives are fulfilled. The following list describes which objectives are fulfilled by this course and in what manner they are fulfilled.

(1) **an ability to apply knowledge of mathematics, science, and engineering:** Students will use mathematical and engineering concepts throughout the course to solve a series of programming assignments and on the exams.

(2) **an ability to design a system, component, or process to meet desired needs within realistic constraints such as economic, environmental, social, political, ethical, health and safety, manufacturability, and sustainability:** Students will implement a complex engineering problem related to computer system which includes a simple instruction set of CPU, memory, cache and pipeline.

(3) **an ability to function on multi-disciplinary teams:** Students will be grouped to collaborate on autonomous labs and learning reports assignments.

(4) **an ability to identify, formulate, and solve engineering problems:** Students will develop original solutions to open-end design specifications in the lab projects.

(5) **a knowledge of contemporary issues:** Students will be introduced the contemporary developments in the computer architecture field and investigate some contemporary topics on the learning report assignments.

(6) **an ability to use the techniques, skills, and modern engineering tools necessary for engineering practice:** Students will use the industry-standard development hardware board and design tools in the lab projects.

## 4  A Bottom-Up Teaching Mode Reform

The principles of "bottom-up" teaching mode can be concluded as [3,4]: (1) from simple to complex, (2) from easy to difficult, (3) from concrete to abstract. Based on these principles, we reformed the computer organization and architecture course from aspects of teaching contents, teaching methods, hands-on assignments and examination methods.

### 4.1  Reform of Teaching Contents

Teaching content should be carefully chosen and organized in accordance with teaching purposes and teaching requirements. According to the requirement for CE-Computer Architecture and Organization announced in ACM/IEEE-CS 2016 [5], there are 10 subjects in this area including (1) History of computer architecture, organization, and its role in computer engineering, (2) Standards and design tools used in computer architecture and organization, (3) Instruction set architectures, including machine and assembly level representations and assembly language programming, (4) Computer performance measurement, including performance metrics and benchmarks and their strengths and weaknesses, (5) Arithmetic algorithms for manipulating numbers in various number systems, (6) Computer processor organization and tradeoff, including data path, control unit, and performance enhancements, (7) Memory technologies and memory system design, including main memory, cache memory, and virtual memory, (8) Input/output system technologies, system interfaces, programming methods, and performance issues, (9) Multi/many-core architectures, including interconnection and control strategies, programming techniques, and performance, (10) Distributed system architectures, levels of parallelism, and distributed algorithms for various architectures. These 10 subjects need 60 course hours to complete.

Based on the above requirements, we can further formulate the teaching content of the COA course. There are lots of abstract concepts and theories in the computer organization and architecture course. The key factor in the reform of COA teaching content is how to clearly transmit the basic knowledge, train the ability of designing a system to meet desired needs within realistic constraints, make students solve engineering problems, and organize the abundant subjects in the limited hours. Considering the hierarchical feature of COA, we decide to decompose the COA teaching content and then formulate the teaching content with a bottom-up mode. Figure 2 shows the complete reformation of teaching content.

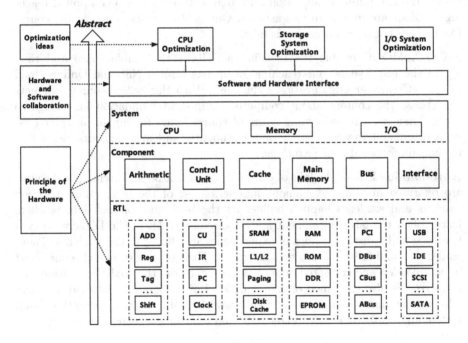

**Fig. 2.** COA teaching content decomposition

From Fig. 2, we can find that the teaching contents are organized into a hierarchy bottom-up system. The lowest level is the computer hardware principles. These hardware principles are divided into three levels from simple to complex.

Then, based on the hardware system, the hardware and software collaboration of computer is introduced. This level mainly interprets how the hardware and software collaborate deal with exceptions and interruptions. Finally, the evaluation and optimization ideas will be depicted from the view of the whole computer architecture. Considering the limited teaching hours, the basic contents of computer organization and architecture are preserved, but the subject of the distributed system is not involved. This subject will be introduced in the subsequent course "Parallel and Distributed Systems".

## 4.2   The Reform of Teaching Method

Teaching method reformation is the key issue in improving the teaching quality. A lot of modern teaching technologies have been developed in recent years. These modern teaching technologies provide us a flexible organization of teaching. In Sect. 2, we have depicted the challenges of the computer organization and architecture course in our school. Students from various concentrations with different expectations and the limited teaching hours are two main challenges.

To tackle the challenges, we have tried some modern teaching methods to reform the computer organization and architecture course. The applied teaching methods are mainly in two aspects. One is the massive open online course (MOOC). Another is autonomous labs.

**MOOC.** MOOC enable the teaching activity with flexible space and time [7]. In the last 5 years, our teaching group has built a National Online Open Course "Computer Organization Principle". With this online course, students can choose the content about computer component foundation knowledge to learn. Then, we can save some hours of course-hours for higher abstract concepts or difficult theories. Furthermore, students can learn the course according to their professional concentrations.

**Autonomous Labs.** The labs of computer organization and architecture course are reformed into an autonomous manner. Most of the interactions between teachers and students happen online. At the beginning of each lab, teachers just explain the lab environment and requirements. Then, the Digilent Nexys3 board [8] and video recordings of experiments are offered to the students. Then, students should finish lab projects in groups. Their project code and report need to be uploaded and corrected online. Moreover, before the credit evaluation, the plagiarism is also examined online. This kind of autonomous labs can improve self-learning ability of students and cultivate their abilities to develop the actual system and applying learned knowledge and technologies comprehensively.

## 4.3   The Reform of Hand-On Assignments

### 4.3.1   Lab Projects

The computer organization and architecture course is a very practical course. Many subjects of COA come from the actual computer system technologies.

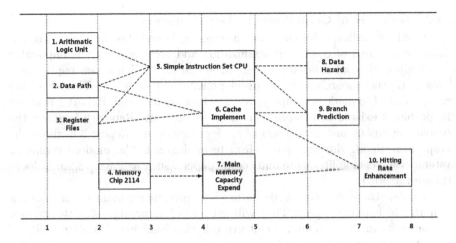

**Fig. 3.** Lab projects schedule

Thus, practical teaching is an important part of the COA course [9]. As a hardware course, COA course lab projects need the supporting hardware platform and environment. Moreover, new developments in computer technologies appear continuously. The lab projects should follow the newest hardware computer technology. Thus, the lab project design is a difficult task. Our teaching group investigates the labs of the computer system related hardware courses offered by CMU [10], Stanford University [11,12], MIT [13], Peking University [14], Toji University [15] and Nanjing University [16]. In these hardware related courses, the common lab project is to implement a simple instruction set CPU. Although the number of implemented instructions or the computer architecture (i.e., CISC or RISC) in different course labs may be diverse, the main topic is always a CPU implementation.

Moreover, based on our investigation, most labs projects just implement in a simulation environment [17], leading to the lab projects are quite different from the real hardware system development. We think the main reasons of simulation labs are high requirements of students practical ability and too many lab hours required for designing real computer hardware system.

In the reform of COA lab projects, we try to apply the real engineer development tools for training students' practical ability to use industry-standard design tools and methodologies. To implement a whole prototype computer system, we decompose it into 10 series progressive lab projects. These lab projects are demonstrated as an engineering schedule control diagram. Figure 3 shows the schedule of 10 lab projects. We can find from the figure that these lab projects are also in a "bottom-up" manner when the figure is rotated. Moreover, we give a real hardware development board (i.e., Digilent Nexys3 [8]) to each student. The video recordings and a standard design tool are also provided. As a widely used development hardware board, there are many supportive materials for Digilent Nexys3 on the internet. Student can also find them to implement the lab projects.

### 4.3.2  Learning of Contemporary Developments

As a rapid renovation field, numerous new concepts and theories appear continuously in the area of computer organization and architecture. It is impossible to introduce all contemporary developments in class. Hence, we require students to do the researches of new developments by themselves. Three learning reports related to modern computer architecture are assigned. Report 1 is about the problems existed in modern computer architecture. Report 2 is about the comparison among instruction sets of modern processors. Report 3 is about the comparison among different parallelism technologies in the modern computer system. Some classic literature and recent papers, such as Patt [18] and Moore [19] are provided.

To finish three reports, students need to investigate mainstream modern computer technology widely. They will get lots of knowledge of contemporary developments in the computer architecture field. Meanwhile, learning ability of students is also trained during the process of reports complement.

### 4.4  Reform of Examination Method

Examinations are essential to check the performance of students and the teaching effects. In order to emphasize the evaluation of study process rather than the final examination, we use various test methods, such as class testing, chapter testing, and finial examination. Moreover, based on our MOOC, students can study the fundamental knowledge of computer organization and architecture. Some online tests in the MOOC can also be used for checking students learning effectiveness. In summary, all these tests are combined with hands-on assignments for their total points. Such an examination method can stimulate students during the whole process of COA and ensure them a continuous study.

## 5  Conclusion

With the development of Information Technology, more and more subjects are emerging in computer science. Traditional computer courses need innovation to deal with new challenges. As a core course, Computer organization and architecture course always plays an important role to connect the electric hardware with the whole system. Considering the course feature, this paper presents a "bottom-up" teaching reformation for the COA course from aspects of teaching content, teaching method, hands-on assignment, and examination method. Teaching reforms of COA course will be a long process. This is our first attempt applying "bottom-up" teaching method on COA course. Taking advantage of this reformation, students are more likely to interact with teachers in both lectures and experiments, also, key and difficult points in COA course are much easier for students to understand than traditional teaching methods. As a result, we got the same teaching efficiency as two courses got with less course hours. In the future, the feedback and evaluation of COA reformation will be obtained to improve our teaching substantially.

**Acknowledgements.** This work is supported by 2016 Online Education Research Teaching Fund by Ministry of Education (General Education).

# References

1. Chen, T., Jiang, G., Hu, W., Lou, X.: The innovation and reformation of teaching method for computer organization and design course. In: The 4th International Conference on Information Engineering and Computer Science. IEEE, December 2009
2. Wang, T., Zhenqiang, W., Ren, P.: Top-down or bottom up? On the teaching mode of computer network principle course. IT Educ. **24**, 65–68 (2011)
3. Brailas, A., Koskinas, K., Alexias, G.: Teaching to emerge: toward a bottom-up pedagogy. Cogent Educ. **4**(1), 137–145 (2017)
4. Cummings, R., Phillips, Ro., Tilbrook, R., Lowe, K.: Middle-out approaches to reform of university teaching and learning: champions striding between the top-down and bottom-up approaches. Int. Rev. Res. Open Distrib. Learn. **6**(1) (2005)
5. ACM and IEEE. Computer engineering curricula 2016, December 2016. https://doi.org/10.1145/3025098. Accessed Mar 2019
6. CEEAA. China engineering education accreditation association, December 2018. http://www.ceeaa.org.cn. Accessed Mar 2019
7. Wikipedia contributors. Massive open online course – Wikipedia, the free encyclopedia (2019). https://en.wikipedia.org/w/index.php?title=Massive_open_online_course&oldid=884223275. Accessed Mar 2019
8. Digilent. Nexys 3 spartan-6 fpga trainer board (2019). https://store.digilentinc.com/nexys-3-spartan-6-fpga-trainer-board-limited-time-see-nexys4-ddr/. Accessed Mar 2019
9. Wang, Q., Chen, H., Liu, H., Xu, L., Zhang, Y.: Application of project management in undergraduates' innovation experiment teaching. In: Zhou, Q., Miao, Q., Wang, H., Xie, W., Wang, Y., Lu, Z. (eds.) ICPCSEE 2018. CCIS, vol. 902, pp. 564–572. Springer, Singapore (2018). https://doi.org/10.1007/978-981-13-2206-8_47
10. Carnegie Mellon University. Introduction to computer architecture[eb/ol] (2011). http://www.ece.cmu.edu/~ece447/. Accessed Mar 2019
11. Stanford University. Digital systems i[eb/ol] (2011). https://ccnet.stanford.edu/ee108a/. Accessed Mar 2019
12. Stanford University. Digital systems ii[eb/ol] (2010). http://www.stanford.edu/class/ee108b/. Accessed Mar 2019
13. Massachusetts Institute of Technology. Computation structures [eb/ol] (2011). http://6004.csail.mit.edu/. Accessed Mar 2019
14. Peking University. Computer organization and architecture course (2018). http://mprc.pku.edu.cn/courses/organization/autumn2018/. Accessed Mar 2019
15. Qin, G., Hu, Y., Huang, L., Guo Y.: Design and performance analysis on static and dynamic pipelined CPU in course experiment of computer architecture. In: The 13th International Conference on Computer Science and Education. IEEE, August 2018
16. Nanjing University. Computer organization and architecture course. http://media.njude.com.cn/course/jsjzcyl/index.htm. Accessed Mar 2019

17. Liang, X.: A survey of hands-on assignments and projects in undergraduate computer architecture courses. In: Sobh, T. (ed.) Advances in Computer and Information Sciences and Engineering, pp. 566–570. Springer, Dordrecht (2008). https://doi.org/10.1007/978-1-4020-8741-7_101
18. Patt, Y.: Requirements, bottlenecks, and good fortune: agents for microprocessor evolution. Proc. IEEE **89**(11), 1553–1559 (2001)
19. Moore, G.E., et al.: Cramming more components onto integrated circuits (1965)

# Education Platform of Congenital Heart Disease Based on Mixed Reality Technology

Yuwei Ji[1], Xiangjun Zhang[1], Hanze Tang[1], Hao Luo[1],
Shengwei Zhao[2], Zhaowen Qiu[2], Qinghua Zhang[3], Kun Wang[3],
and Liwei Diao[4(✉)]

[1] Harbin Medical University, Harbin 150080, Heilongjiang, China
[2] Northeast Forestry University, Harbin 150040, Heilongjiang, China
[3] The Second Affiliated Hospital of Harbin Medical University,
Harbin 150086, Heilongjiang, China
[4] Department of Cardiac Surgery, Second Affiliated Hospital of Harbin Medical
University, Harbin, China
13613667413@126.com

**Abstract.** Recently the Mixed Reality (MR) technology has a rapid development and strong application prospect in various fields. This technology has been successfully applied in clinical work by means of surgical navigation and puncture positioning. However, there is few reports about MR technology's applications in China. Therefore, based on the MR technology, a medical education platform was designed for related professions through the reconstruction of 3D heart model. Because of the various types of congenital heart diseases and the comprehensive medical knowledge system involved, it will start from congenital heart disease and expand to other organs later. The teaching mode was enriched and basic and clinical teaching materials were provided. And students' interests were motivated and learning efficiency was strengthened. An equal communication mechanism between teachers and students was constructed. Experimental results show that both teachers and students benefit from this proposed platform.

**Keywords:** Education · Congenital heart disease · Mixed reality

## 1 Research Background

### 1.1 Education Background

**Traditional Teaching Mode**

The education of clinical medicine mainly includes the mastery of basic knowledge, the understanding of anatomical structure and the operation of clinical treatment.

In the traditional education mode, the basic knowledge is mainly learned by engrafting the knowledge points in textbooks, such as concept definition, classification, etiology, and students often learn by rote.

Anatomical structure understanding includes classroom two-dimensional multimedia pictures, small class of anatomy practice, after-class independent learning. Human

© Springer Nature Singapore Pte Ltd. 2019
R. Mao et al. (Eds.): ICPCSEE 2019, CCIS 1059, pp. 313–334, 2019.
https://doi.org/10.1007/978-981-15-0121-0_24

anatomy is an important part of clinical teaching and the most complex and difficult to master. In class, teaching by pictures on power point requires high 3D consciousness of students, and it is very difficult for most students to understand the actual structure of the human body and the adjacent relationship. In the small class of anatomy practice, students observed the cadaver sample made by themselves, but the samples were limited and affected by time and place in the actual use process. Moreover, many cadavers often fail to meet the requirements of anatomy teaching due to the variation of processing time and fine structure in the actual autopsy. Autonomous learning after class mainly includes anatomy atlas, teaching video, 3D Body application. These methods are more delicate and flexible than picture teaching in class, but they are still limited to the two-dimensional plane.

Clinical treatment operation mainly includes viewing video and actual operation. Many teaching video is time-consuming and labor-intensive, so the teaching video is used for a long time, which does not conform to the more advanced operating specifications in real time. For actual operation, because of the consumption of the material, students have few opportunities to operate, and no review courses are arranged to consolidate the operation standards.

**New Forms of Teaching**

With the rapid development of science and technology, many new contents have been added to medical teaching forms, such as 3D printed models, AR and VR.

3D printing can deepen the understanding of complex anatomical structures and shorten the learning time. However, the most important limitation of 3D printing in medical teaching application is its high cost rather than its practicality. The available software programs, printers and printing materials are very expensive [1].

The application of AR and VR technology provides more detailed and richer knowledge for medical students' anatomy learning. But there is a problem that cannot be ignored. The design of virtual reality and augmented reality teaching platforms and resources emphasizes form over content. Some knowledge points provided by virtual reality education platforms are still in the real world. The monotony and dullness of textbook contents are not alleviated by the existence of software, and the explanation of knowledge points is not made more vivid, interesting and targeted [2].

## 1.2   Disease Background

There are many kinds of diseases related to human body, and we choose the representative congenital heart disease as the breakthrough point to conduct in-depth research. The teaching content of congenital heart disease is comprehensive, including early embryonic development, cardiac structure, pathological changes, surgical treatment and other aspects. Therefore, starting from the congenital heart disease, it will be more clear and rigorous to expand to other organs and systems in later.

Moreover, congenital heart disease is a birth defects disease with the first high incidence and high mortality in China. There will be 180,000 to 200,000 babies was born with congenital heart disease every year, according to the monitoring results of birth defects published by the national health and safety commission. The incidence of congenital heart disease in Chinese fetuses is 7.4%, according to a report on the status of congenital heart disease in Chinese fetuses released by health news on September 15, 2017. A total of 2452249 pregnant women were investigated, and 18171 cases of congenital heart disease were detected. For complex and severe congenital heart disease in neonatal period, if the heart surgery is not carried out within 1 month after birth, the case fatality rate is as high as 50% [3]. However, the number of surgical treatment of congenital heart disease in China is only about 80,000 every year, less than half of the number of new cases every year. It is a very difficult and risky to perform heart surgery in neonatal period, which is a great challenge for surgeons [4].

## 1.3 Technical Background

Mixed reality (MR) technology is a further development of virtual reality (AR) technology, which builds an interactive feedback information loop between the virtual world, the real world and users by introducing the real scene information into the virtual environment to enhance the sense of reality of user experience.

MR technology is at the stage of exploration and development at home and abroad. At present, it only involves the surgical treatment of individual cases, and there is still a gap in teaching.

"Compared with our counterparts who do not use VR (virtual reality), students who use these VR simulators in surgery courses complete their learning tasks in less time. The surgeon's understanding is also enhanced by showing views that are not possible in VR, such as proximity to "invisible" peripheral structures. In addition, the adjustable cutting plane allows visualization of how do visual surgical tools interact with anatomical structures to help prepare for surgery" [5]. MR (mixed reality) has a stronger sense of interaction and authenticity than VR (virtual reality), and may achieve better results in teaching.

## 1.4 Questionnaire Survey

### Questionnaire Object

The second affiliated hospital of Harbin medical university in Harbin, heilongjiang province has 30 undergraduate students majoring in clinical medicine, 10 teachers and 10 cardiac surgeons.

### Questionnaire Design

50 questionnaires were designed for undergraduate clinical medicine students, teaching teachers and cardiac surgeons, aiming at the practicality and feasibility of MR teaching platform, and the survey questions were scientifically designed. (see annex 1 for the questionnaire content – questionnaire survey based on the MR congenital heart disease teaching platform).

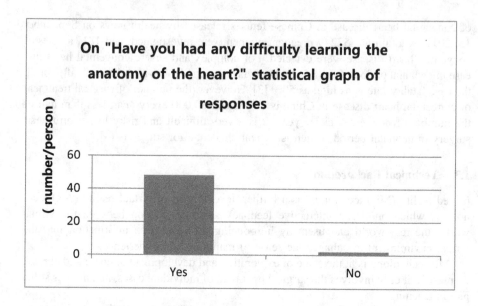

Fig. 1.

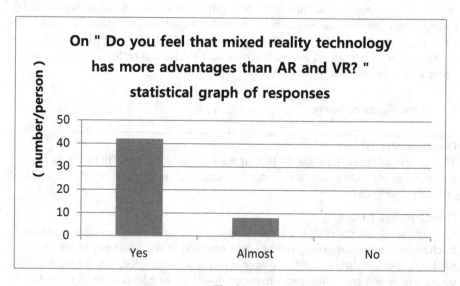

Fig. 2.

## Questionnaire Survey Results

According to the statistical analysis of the questionnaires filled by teachers and students, we can draw a conclusion that the application of MR technology to medical teaching has practical significance, advantages and feasibility.

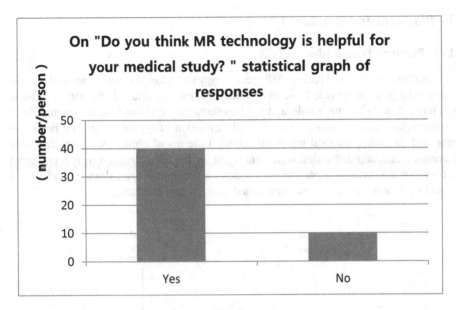

Fig. 3.

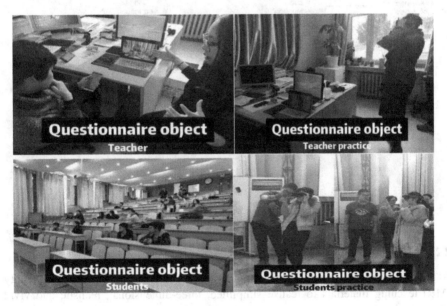

Fig. 4.

## 2  Platform Is Introduced

### 2.1  Platform Design Idea

This platform uses mixed reality (MR) technology and takes congenital heart disease as the starting point to provide basic and clinical teaching materials. In the aspect of basic teaching, it includes the whole and split explanation of normal heart structure, animation restoration of embryo evolution, classification of typical congenital heart disease and simplified surgical treatment model. In terms of clinical teaching, real case teaching and group discussion were conducted for the reconstruction of real enhanced CT in patients with multiple complex types of congenital heart disease, and virtual model operation evaluation was conducted through this platform.

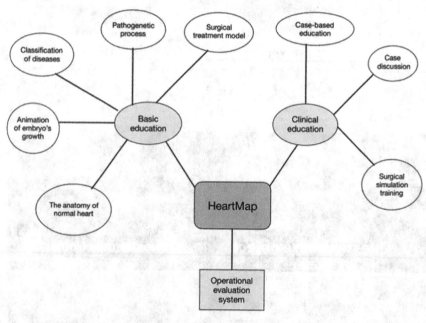

**Fig. 5.**

### 2.2  Main Functions of the Platform

At the same time, it also serves teachers and medical students, and applies mixed reality (MR) teaching materials to realize simplified, three-dimensional, realistic and vivid medical education.

For teachers, our platform can provide a variety of teaching materials, and teachers can freely choose the materials in the platform to conduct teaching stitching and design electronic teaching plans to achieve mixed reality (MR) teaching.

For medical students, our platform can simplify the writing teaching, by MR technical implementation class 3D experience, improve students' interest in learning, enhance the student to the understanding and the understanding of disease anatomical structure, structure of close observation of disease, interact with holographic images, fundamentally improve the study effect, provide more communication opportunities, more in line with the education concept of time now.

## 2.3   Main Contents of the Platform

### Basic Teaching Materials

(1)  Normal anatomical structure

This module is based on the normal anatomical structure. According to the anatomical map, 3D reconstruction image is designed to provide teachers and students with a real, fine, beautiful and powerful 3D model of the standard heart. In the traditional teaching mode, especially when it comes to the explanation of anatomy learning, teachers will look for teaching materials before class, such as various kinds of beautifully made pictures, video, solid or 3D models, so that students can observe the heart from all angles and deepen their understanding. Now, a virtual hologram can replace all the pictures and video by using the 3D model in our material library and combining with the MR equipment. Students can watch from all angles and do not need to make a physical model. The whole model is the overall structure of the heart, while the split structure is a separate model of various structures of the heart cavity, blood vessel and valve. Other parts can also be hidden in the overall structure for viewing. For teachers, MR equipment can be used for demonstration in the process of teaching, and students can watch the teacher's demonstration while listening to the teacher's explanation in the glasses, so that the teacher's explanation can be delivered to every student accurately, truly achieving a seamless connection of information between teachers and students. For students, they can use their MR devices to watch and operate the 3D

**Fig. 6.**

images of the normal heart in the material library, so as to have a correct perceptual understanding of the heart structure before class. After class, teachers can further understand and digest the knowledge taught by them through holographic images, so that all parts of the heart can be separated, such as the myocardium removed or transparent, and the internal structure, such as the valve and the heart cavity, can be viewed to form a rational understanding of the heart structure.

(2)  Animation restoration of embryos

This module mainly focuses on various congenital diseases. Taking congenital heart disease as an example, mixed reality (MR) animation is presented here. We condense the process of embryonic development into a small animation, which can be watched repeatedly or carefully in a certain structure. The most critical animation frame can be viewed from multiple perspectives. The animation starts from the fertilized egg, and gradually evolves into the original embryonic structure, and then forms various embryonic tissues, and then further develops into various organ systems. In the teaching of congenital heart disease, we will first focus on the embryonic development process of the heart, see how the heart starts from the primitive embryo tissue development, explore the structure and functional characteristics of the heart during the embryonic period, and finally get the mature heart. Later, we will focus on specific parts of the lesion and gradually differentiate into specific pathological models, highlighting the evolution of the lesion structure. Students can clearly understand the specific evolution process of the whole congenital heart disease, which is conducive to the vivid teaching of teachers and the overall understanding of the pathogenesis of students. Before class, you can also learn about embryo development by watching video in advance. In class, you can choose classic ones such as tetralogy of fallot to display, and after class, students can watch and review again. In the future, we will expand the platform to other organs and systems to improve students' vertical understanding of organ systems.

(3)  Disease classification

This module is the highlight of our platform, and we will focus on the anatomical structure model of each disease classification. Congenital heart disease can be divided into many types, each type has its own characteristics, the pathological anatomical structure is also very different, the anatomical structure of the abnormal almost involves all the heart structure, accurate understanding and grasp is a very difficult thing. Therefore, we made typical models for each species according to the atlas, and accurately restored the anatomical abnormalities of this disease. In order to show clearly, we have specially processed the main anatomical lesion site. Teachers or students can highlight the lesion site or hide other normal cardiac structures to view the lesion site. Moreover, they can play with this abnormal structure in the palm of their hand. In this way, teachers and students can follow a "whole – part – whole" approach, which is conducive to flexible teaching of teachers, more conducive to the formation of students' comprehensive understanding of the heart structure of congenital heart disease, and improve the teaching effect. For some of the more complex clinical congenital heart disease, such as some of the combination of a variety of structural abnormalities or secondary to a variety of structural changes in the case, we will use different colors to distinguish, to facilitate your understanding. If you think there are too many colors, we can show them in the same order according to the primary or secondary relationship of anatomical abnormalities. For example, in the case of right ventricular outflow tract stenosis, we will first show the stenosis of right ventricular outflow tract, and then in turn show secondary changes such as pulmonary artery enlargement and right ventricular hypertrophy. We will be supplemented by a variety of tips, such as each case corresponding to the etiology, pathogenesis, pathological change and imaging examination performance, etc., so that the teacher is a kind of choice in the lecture, they can use the traditional teaching mode in combination with holographic images to explain, and can be directly in MR equipment combined with holographic images directly to complete the whole teaching process of relevant basic knowledge. Another great advantage of this design is that it is convenient for students to learn after class. In this way, it not only provides students with a source of learning materials, but also frees teachers to focus all their efforts on the teaching of key cases, so that students can fully understand a knowledge and master corresponding learning methods.

(4)  Pathogenesis

In this module, we will focus on presenting the pathophysiological process of a disease, which leads to the pathological changes and clinical manifestations of various diseases. Before we have provided the embryonic development, anatomical structure, this part mainly explains the structural abnormalities caused by the change of hemodynamics and organ motion state, help you to understand all kinds of clinical symptoms or signs of the congenital heart disease changes, such as the general condition of patients with abnormal, the characteristics of physical examination, specific imaging performance, etc. We plan to show the hemodynamic effect with three-dimensional animation, color distinction of moving and venous blood, and then show the abnormal shunt, such as left to right or right to left shunt of atrial septal defect, so that the viewers can know the hemodynamic changes and their consequences. For some typical cases, we will also reveal the abnormal heart beat from a three-dimensional perspective, and this effect will also be realized through three-dimensional animation, so that users can understand the heart movement state. The clinical manifestations of CHD are systemic, not only limited to the heart, but also to other organ systems. In the animation of the pathogenesis, we will also show the manifestations of some other typical organs in the case of congenital heart disease, such as the effects on the lungs. The idea is basically to follow the practice of the heart above. We also hope to extend similar techniques and ideas to other organ systems.

(5)  Surgical treatment model

After learning the anatomical abnormalities and relevant basic knowledge of congenital heart disease, teachers and students need to understand the treatment process. This module is designed to present this process. The traditional teaching mode is most limited to this part of teaching. The book can only use limited pictures and large paragraphs of text for explanation, and the teaching video demonstration is also limited. If it is too short, it cannot explain the problem, and if it is too long, it will take a lot of time. In this platform, teachers can choose several classic surgical methods for the treatment of congenital heart disease for 3D demonstration in class. By simplifying various surgical schemes, key steps can be extracted, simplified 3D models can be made, and dynamic demonstration or operation demonstration can be carried out to demonstrate the operation principle and procedure in an all-round and multi-angle way. This is conducive to the authenticity of the teacher's explanation, which saves the time and energy for students to deal with a lot of text and two-dimensional pictures video, and improves the teaching quality of teachers and students' learning interest. Teachers and students can interact with the 3D model by hand, so that students can experience the implementation of the operation, enhance the immersion of the teaching process, and increase the interest and interactivity of the class. Students can learn the key steps of the surgical scheme by themselves through the platform after class. They can watch the parts they don't understand repeatedly to deepen their specific understanding of the surgical procedure.

## Clinical Teaching Materials

(1)  Case teaching

This module is designed for the purpose of closely combining clinical teaching. A large number of cases are saved in the database in the platform, and the imaging data in the cases are generated into mr-based 3D images for the convenience of teachers' selection and access. This platform provides the image data of common diseases and even rare diseases from the department of cardiac surgery of the second hospital of Harbin medical university. The advantageous teaching position of teaching affairs department and surgical department in the second hospital of Harbin medical university provides us with advanced teaching materials and mode selection. According to the large number of congenital heart disease cases collected in the database, the basic information, medical history, diagnosis and treatment process of the cases are recorded truthfully. The imaging data are first input into the original data, which will be generated into a 3D model with the corresponding technology. In this way, in the teaching process, the teacher can make a detailed explanation of the case based on the 3D model, and analyze each pathological change and its corresponding clinical manifestations and treatment measures. Students can combine 3D models to deepen their understanding of cases and develop clinical thinking.

A case of right ventricular outflow tract stenosis followed by other changes is described briefly. Patients with related inspection diagnosis of right ventricular outflow tract stenosis, pulmonary stenosis, cardiac function III level. Electrocardiogram examination indicated sinus rhythm, complete right bundle branch block, and st-t changes. Figure 7 is one of the imaging manifestations of this case.

This case if it is hard to imagine just see traditional imaging findings of its overall performance, but if I can have a 3D holographic image, as shown in Fig. 8, and can in MR equipment and its interaction, the teacher can will clearly right ventricular outflow tract stenosis, secondary thickening of right ventricular hypertrophy, pulmonary valve and pulmonary artery trunk enlargement performance easily presented to everybody, makes everybody can efficiently and accurately understand and avoid the traditional teaching mode between students in the teaching effect is uneven. Moreover, the

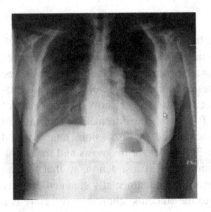

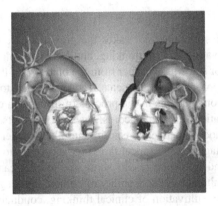

Fig. 7.

Fig. 8.

advantages of such teaching are not limited to the short term, but also benefit students in the long run, because students can understand the traditional imaging performance more easily after they have learned the near-real performance through holographic images. This is very important, because in the future, the traditional imaging examination methods will still dominate, and medical survivors need to master their film reading methods to achieve the diagnosis of most diseases in the first time. At this stage, the process will take at least a decade, during which countless films will be watched and summaries made. After adopting the 3D image assistant method combined with MR technology, this process is bound to be greatly shortened, which is beneficial to the students of clinical department, especially the students of imaging department.

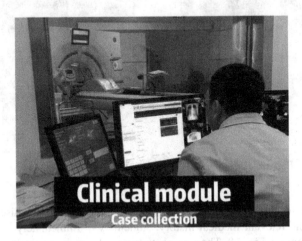

Fig. 9.

(2) Case study

This module undertakes the previous case teaching module, with the purpose of cultivating clinical thinking, and USES the platform database, but this module is prepared for students. Teacher from the platform of the database and pick out a few typical cases, appropriate adjustments, and distributed to the classmate of everyone, group discussion, students in the process of discussion content platform can show the classmate of patient medical records, inspection results, CT images, the information such as 3 d models, students can imitate the teacher teaching cases discussion, analysis and diagnosis of clinical thinking, according to the 3D model for diagnosis and treatment, and for group results show, the teacher can from the next instruction, so that we can cultivate students ability of clinical thinking, analysis, fundamentally changed the way the cultivation of clinical thinking, conducive to the students' clinical study. Here is a case of left ventricular outflow tract stenosis. Wang mou, male, 12 years old. He was admitted to hospital at 7:00 am on March 2, 2019 due to "chest tightness and shortness of breath after the activity for 2 years". The patient showed symptoms of chest tightness and shortness of breath after the activity, accompanied by palpitation, which could be relieved after rest, and was diagnosed as left ventricular outflow tract stenosis and cardiac function by relevant examination Can III level. Electrocardiogram examination indicated sinus rhythm, left ecg deviation, complete right bundle block, and first-degree atrioventricular block. Figure 10 is one of the imaging manifestations of this patient. Figure 11 is a reconstructed 3D model.

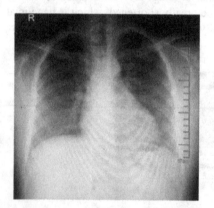

Fig. 10.

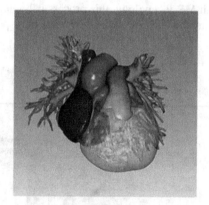

Fig. 11.

Suppose the teacher divides the students into several groups to discuss the case. After reading the relevant medical history, students will immediately study the chest radiography shown in Fig. 10. Then, most students will see the aortic node bulging and left ventricle enlargement in a very short time, presenting an aortic heart on the whole, which is the accumulation of a lot of previous learning. However, there will definitely be students who want to know the overall shape of the heart in this case, who will want to know the specific shape and location of this stenosis, and then form a comprehensive understanding of this case, which is basically impossible for students to get from the existing information. However, if there is a 3D image of this case, and it can be viewed and interacted through the MR device, the anatomical abnormalities of this case can be fully mastered. For a structural heart disease, congenital heart disease, abnormal anatomical structure is the key of the key, if the anatomical structures can be in the discussion of the students to master, then at the back of the clinical manifestation, diagnosis, the performance of the various examination and treatment knowledge can be on the basis of the knowledge of physiology, pathology, physiology and so on have many students' communication and the master, to achieve a better teaching effect.

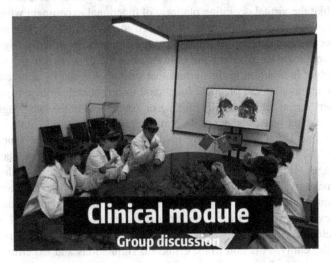

**Fig. 12.**

(3)  Surgical simulation training

The surgical simulation training module belongs to the clinical teaching category of the platform, but it is actually derived from the simplified model module of surgical treatment in the field of basic teaching. On this basis, a variety of operation options are developed to comprehensively review clinical knowledge and increase the fun of the platform. This module is more like a small game program, which aims to enable interns to test what students have learned and consolidate their learning results in an easy and efficient way through a small game after the previous class or discussion. In this small program, we extracted the main operation steps, interns can put on MR glasses, with gesture to replace various operations, such as cutting, sewing, take equipment, etc., to a model of congenital heart disease treatment, pathogenesy, for example, students can pick up the patch, to repair the defect and stitching, immersed in the operation process, experience the feeling of the surgeon to careful, by the way can also review the anatomic knowledge. If you do not know how to start the situation, the system will be in the form of knowledge points to prompt. Students can follow these tips to eventually cure the congenital heart disease model, and get some rewards, like a video game, let students learn knowledge in the process of playing. We will perfect the simulation training program, add some atypical cases to the platform, and let them automatically generate corresponding disease manifestations, and then allow students with higher requirements to conduct simulated surgical training on these models. It will be further developed and fully integrated into other organ systems.

## Operating Evaluation System

After a period of courses, teachers can use the operating evaluation system to test students, and students can also use MR equipment to operate and learn in their spare time. The operating evaluation system is still based on 3D images. Students can operate virtual images in the MR device, which is similar to a game, but each step will be scored, just like the virtual operating platform of functional experiment which has been realized now. This process can not only be used to assess students, but more importantly, students can review the corresponding anatomical structure again in this immersive game experience, which not only expands their vision, but also consolidates their knowledge. In addition to the system that can be assessed by simulated surgery, we will also add anatomical practice, functional experimental simulation, animal surgery, and even various clinical skills training or assessment in the later stage to further enrich your learning materials and replace many training that requires physical objects or in-depth clinical practice. Anatomy field mainly depends on the function to the ascent of the MR equipment technology and the scale of the image, we plan to further optimize images, simulated human body structure, the cascade relationship as far as possible to make the layers of connective tissue, and let every inch of organization can be found from a little in MR equipment removed, just like the real anatomical specimens, help everyone closer to the real understanding of the anatomical structure, a more complete and comprehensive, at the same time, the largest extent, instead of the anatomy of the entity. Function to experimental simulations and experiments on animals, and these two functions now machine can learn the experimental teaching platform of similar functions in the operating process on the same, we mainly focused on the characteristics of realistic 3D images to replace their regular 2D image, and combined with the corresponding dynamic effects, such as the reaction of animals, such as blood flow effect, on the process of experiment can maximize the close to the real experiment, and can avoid unnecessary waste of resources. Clinical skills training. In this function, we also plan to apply the technology in animal experiment simulation, and combine big data and artificial intelligence to build a three-dimensional virtual dummy platform. Based on this process system, users can be graded and their skills can be improved, laying a solid foundation for the real clinical practice.

## 3  Platform Analysis

According to the statistical analysis of the questionnaire survey, we summarized the evaluation of teachers and students on this platform.

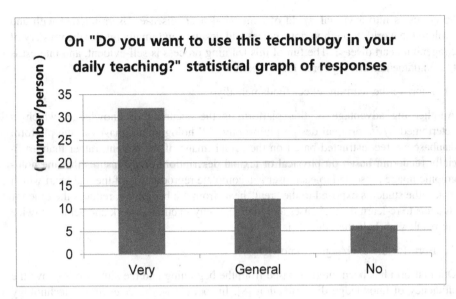

**Fig. 13.**

## 3.1   Advantages of Platform

(1)  Enhance the interaction, experience, fun of the learning process

HoloLens glasses on our platform can throw away computers and mobile phones without using keyboards, mice or display screens. Hands can operate in the air by gestures or Bluetooth remote control [2]. The heart can be zoomed in and out and rotated in real time. Fix the heart in the space, the user can view the heart from multiple angles and can also approach the heart, analyze the internal structure and unobstructed the internal structure of the heart. Not only that, but our platform is also displayed to the platform users from the aspects of sight, hearing and touch. After using the glasses, they can feel the immersive feeling and strongly stimulate the motivation of the learners. According to the statistics of our team, ordinary visitors only need 10 min of simple training, they can master the glasses gesture operation, and truly realize the pure gesture that can be realized in science fiction movies. If the gesture is too complicated, the voice control (in both Chinese and English) can meet the needs of the learner. As long as the corresponding instructions are issued, a required model will be displayed in front of the user. With this teaching platform based on MR technology, students don't have to look at anatomical maps and a dense description of the text. Pick up the MR glasses and open the teaching platform. A corresponding three-dimensional hologram can solve any obstacles in understanding the anatomical structure. This way helps students to observe and experiences themselves, increases the opportunities for students to exchange and cooperate, and enables students to build a clearer knowledge framework and deepens their understanding of cumbersome medical knowledge. This platform is not so much a teaching material library, but rather a "mobile library on a network platform." Compared to traditional teaching materials, it is revolutionary, brings users into a virtual world of congenital heart disease. Users interact with and explore the world through various voices and gestures and finally master the mystery of congenital heart disease. The fun of this learning process is self-evident, and interest is the fundamental driving force for all learning.

(2)  Improve the learning effect fundamentally

Another big advantage of this platform is the accuracy of knowledge, which is determined by its original design philosophy. All holographic images in the platform database are reconstructed based on the actual image of the patient, rather than artificially rendering based on physical or textual descriptions. This ensures that the holographic image presented to each user is a complete reproduction of the real heart which makes the students exposed to the "real" heart from the beginning, reduces the time for students to re-learn the real cases and their anatomy in clinical work and achieves twice the result with half the effort [7].

(3)  Born for the exchange of knowledge

Our platform has been internet-based from the beginning and the aim is to improve the efficiency of knowledge dissemination [8]. In today's world, science and technology are developing rapidly, the speed of knowledge dissemination is accelerating, and the speed of updating medical knowledge is getting faster and faster. More and more teachers and students need to exchange platforms and make progress together. The central database of our teaching platform is deployed in the form of a server. Connect the platform to a local area network. People can share the images in their glasses with others, use their glasses to express their opinions, and the teacher can teach them together. If we can, different doctors can even discuss cases on this platform, exchange experiences, and let knowledge be sent to everyone in need.

(4)  Meet the development concept of the information age

With the rapid development of economy and science and technology, the rise of the third largest industry of Internet and upgrading of information technology, the emerging of augmented reality gradually comes into the public view. The research of hybrid reality technology is no longer just on paper or in science fiction movies, but more applied to various fields such as education and industry through practice. Our teaching platform not only uses Mixed Reality (MR) technology, but also uses technologies such as the Internet and Big Data to create a central database of teaching platforms. At present, the central database of our teaching platform is deployed by server, uses a single server WIFI connection, plans to switch to 5G connection in recent years, server adopts cloud platform mode and is deployed in Harbin Yungu.

(5)  Provide favorable conditions for independent learning

With the changes of the times, the ability of independent learning has become increasingly important, and more and more colleges and students pay attention to the ability of independent learning. The mixed reality teaching materials that our platform can provide can be stored in network operation platforms, desktop devices, mobile devices and paper books. Students can autonomously learn by using mixed display technology teaching materials in different devices for anytime and anywhere [10].

Knowledge points that are not available in the classroom can also be re-learned to increase the consolidation and understanding of knowledge. Sometimes students fail to study in the classroom for special reasons, and can also make up after class [2].

(6) Promote the equalization of quality resources

Due to China's vast territory, numerous colleges and universities, and uneven regional economic development, it has created a huge teaching gap between major universities. Such a teaching gap has caused the inevitable result of uneven distribution of educational resources. High-quality colleges and universities have sufficient funds and various forms of education to enjoy high quality educational resources. Other colleges and universities have insufficient forms of education and relatively scarce educational resources. Our platform is a medical teaching platform for colleges and universities across the country. The teaching resources and materials on the platform are high-quality data that have been carefully designed and proofread many times. It can alleviate the uneven differentiation of educational resources and expand the scope of enjoyment of high-quality resources. It also reduces the teaching gap between colleges and universities and optimizes the allocation of educational resources. It is the best way to achieve fairness in medical education.

### 3.2 Shortcomings of Platform

(1) High data collection requirements

It requires high quality and thickness of the collected data. And the 3D reconstruction of the data is the basis, while the data quality has a high impact on the 3d reconstruction, such as motion artifacts, scanning time, layer thickness and contrast development will affect the quality of the 3D reconstruction model.

(2) Hardware still needs to be optimized
① System compatibility needs to be improved.
② Spatial localization algorithm needs to be optimized.
③ Some people are easy to produce dizzy feeling after wearing for a long time.
④ Endurance and weight need to be optimized.
⑤ Hardware costs are so high that a consumer version of the device with lower price needs to be introduced to further popularization.
(3) Interaction technology and interface paradigm

Registration tracking technology is still not stable, there are still errors and delays between the real physical world and the holographic virtual world, and the error rate is high.

(4) Equipment adaptability problem

Most people use traditional teaching mode to study all the year round. It is difficult to accept new learning methods psychologically. Some people may have difficulty in focusing attention, it is difficult to adapt to equipment use, and even some people may have dizziness problems. Mixed-reality (MR) devices have high requirements on the network, and some functions can only be realized through 5G networks. When the network speed is insufficient, it often causes delay, resulting in a decrease in user experience.

(5)  Research resource shortage

Due to the current low profit in the education market, inadequate equipment, lack of technical skills of developers, and many college teachers have not been exposed to mixed reality technology, the development technology resources and teaching resources of the teaching platform are relatively short, but today the mixed reality technology is developing rapidly. To be applicated is a trend that is imperative today. We believe that mixed reality technology will bring subversive changes to traditional medical education.

(6)  High equipment prices

Most companies that study hybrid display technology have high costs in the early stage, the number of companies studied is small, and the sales volume of equipment is low, which make Most of the mixed reality (MR) equipment prices high. Moreover, most universities have insufficient funds and limited purchasing power. Colleges and universities have certain difficulties, and the concentration of one or two colleges and universities will make the allocation of educational resources more unequal and the education gap will further increase.

(7)  Technical conditions limit platform development

At present, mixed reality equipment needs further technical improvement. System problems need to be optimized. The technical requirements for animation playback in HoloLens are extremely high, and the research funding is huge. And the hemodynamic research problem is not properly solved. Big data and artificial intelligence technology can't keep up with current platform development requirements. Therefore, the requirements of the technical conditions cannot be fully met at present, resulting in the platform being limited by technical conditions.

### 3.3  Economic Benefit

(1)  The teaching of congenital heart disease is very complicated, and it requires more multimedia teaching materials such as pictures and videos. The 3D holographic image teaching material has great advantages compared with traditional teaching materials, and there is no platform for providing such teaching materials [9]. The teaching materials provided in this platform directly attack the pain points in the teacher's teaching process. The teacher has such a 3D holographic image that can replace many traditional teaching materials such as pictures and videos, optimize the course of the class, improve the quality of teaching, and there is great attraction for the teachers of the medical college. If the price is reasonable, the teachers must be very happy to download and use. The only thing that will cause concern for teachers or schools is the high price of the MR glasses, but with the popularity of this technology and advances in technology, the price of the device will definitely fall within a range that the relevant users can afford. But even before the price has

dropped to a very low level, schools can still buy a few small-scale applications. As long as the corresponding teaching process and form are arranged, students and teachers can still appreciate the charm of this revolutionary teaching method. In this way, the platform can create a lot of economic benefits by relying on a large number of downloads and visits by teachers.

(2) The knowledge of congenital heart disease, especially the knowledge about anatomical structure, is difficult to understand. It is difficult for students to ensure that they can fully understand the class. It is often necessary to find a large amount of learning materials for review. The cardiac anatomical model of 3D holographic image can display the anatomy of the heart accurately, multi-angle, all-round and intuitively. It can replace a lot of traditional teaching materials such as text, pictures and videos, and achieve the goal of once and for all. If our cost control and price setting are reasonable, the cost of using our materials can be lower than the sum of other materials, or the learning effect can be better than the use of general teaching materials. Then this platform will have very good market prospects. Considering that students who are related to the medical profession, including those in clinical medicine, basic medicine, preventive medicine, and stomatology, can basically use these materials, we can launch a personal version, let students register first, and then provide the corresponding students free and high-end fee materials so that we can create a lot of profit.

### 3.4   Social Benefit

The platform applies the most advanced (MR) mixed reality technology in the field of information technology to the teaching of congenital heart disease. It is an innovation of teaching mode, which greatly improves teaching efficiency. If this teaching platform can be used in large-scale medical colleges and universities, it will train a large number of qualified cardiology and cardiologists to serve the society. Nowadays, with the advancement of inspection technology, the detection rate of congenital heart disease has increased, leading to an increase in the incidence rate. The impact of congenital heart disease on the patient and his family is enormous and can have many negative effects on their body, mind and quality of life. However, many congenital heart disease can be cured or relieved in the early intervention, but the surgery related to congenital heart disease is technically demanding. According to the data we have checked and the interview records of relevant doctors, a qualified cardiologist needs a large number of practice in order to be proficient in the relevant technology, it is expected that there will still be a large demand for this professional cardiologist in the future. Our teaching platform actually promotes the training process of cardiac surgeons in the basic stage, fundamentally reduces the risk of surgery, improves the success rate of surgery, and thus improves the cure rate of congenital heart disease, and finally benefits the majority of patients with congenital heart disease and improves the qualities of patients' and their families' lifes. It is in line with the "Healthy China" concept.

## 4 Platform Outlook

(1) Begin to solve the hemodynamics and dynamic effects in 3D holographic images. Combine the two-dimensional ultrasound, M-mode ultrasound, color Doppler ultrasound, spectral ultrasound and other real-time detection methods of the heart to reconstruct the hemodynamics and real motion effects of the heart in three dimensions, instead of making virtual animation effects based on experience alone. For the study of the normal anatomy of the heart, the addition of dynamic effects is more conducive to the user to fully understand the heart from the perspective of structure and function. For the learning of the heart model under pathological conditions, the addition of dynamic effects allows the user to directly see the changes in anatomical structure, such as exercise state, hemodynamics and other cardiac functions, and deepen the understanding of the corresponding pathology and pathophysiology.

(2) The ultimate goal of the simulated surgical training design is to combine the dynamic effects of hemodynamics and the comprehensive application of MR technology, from the general game to the full simulation of the real surgical operating environment. We will perfect the entire surgical procedure based on the surgical simulation training. For example, when the operation starts, the user needs to first perform an anesthesia step for the operation object. The anesthesia method, anesthetic medicine, anesthetic dose, etc. require the user to select within the system, and then operate according to certain prompts, so that the virtual operation object enters anesthesia induction status. Then the operator can perform disinfection, drape, open the skin, open the chest, establish extracorporeal circulation, etc. All procedures are the same as the real ones. Then there are those classic steps, such as completing the defect of the atrioventricular septum, replacing the abnormally developed valve, and forming the heart chamber. After these steps are completed, the operator needs to step through the heart, pericardium, pull out the extracorporeal circulation, and finally close the chest, seam, etc., to complete the entire surgical procedure [6]. This idea goes beyond the game, but more like the training system that has been used in aviation pilot training, aircraft maintenance, etc., in order to improve the user's relevant operational level.

(3) We will use congenital heart disease as a starting point to expand into various organs and systems such as the digestive system, respiratory system, and exercise system, and gradually develop models for various systemic diseases. And we will make efforts to restore each real case, gradually expand from a disease to each disease model, save it to the database, establish an independent and complete teaching database, constantly improve and maintain the teaching platform, and gradually push the teaching platform to the entire medical teaching system.

(4) The remote guidance education platform is based on HoloLens's remote communication and support sharing and virtual instrumentation and Internet of Things functions. With the support of 5G network, it is designed to guide the remote surgery simulation training and the guiding education platform during surgery. Teachers and students from different regions can wear HoloLens glasses to learn and communicate with the same knowledge points or cases. Teachers can

demonstrate the relevant operations in HoloLens. Students can also express their opinions while operating the model, so that there is no limit to the dissemination of knowledge, and ultimately the equalization of medical and medical education resources [6].

(5) Our teaching platform does not stop there. In the later stage, we will build a virtual standardized patient platform based on big data and artificial intelligence, so that each case can be applied to this platform as much as possible [8]. After the case information is input, the textual case information is virtualized. The performance of the person is presented. At this time, the students seem to have seen a real case with various clinical manifestations and signs in front of them. With this virtual standardized patient platform, the case teaching, case study, surgical treatment model, and the function of the operation evaluation system module in the clinical teaching module of this platform can be applied to the platform to further expand their functions. For example, in case teaching and case study, teachers and students can perform diagnostic operations such as consultation and physical examination on virtual patients, and virtual patients will respond accordingly. The case can then be treated, and the virtual patient can respond to the treatment, such as for some medical conditions, if the correct medication is entered, the symptoms of the virtual patient will gradually ease. For some surgical diseases, students can heal them through surgical procedures. In the surgical treatment model, because of the addition of artificial intelligence, the operation of the surgical case will be more diverse and completely controlled by the case information you input. In the operation evaluation system, the above operations can be split and assigned according to the steps, and then the assessment can be performed. In this way, this virtual standardized patient platform integrates the important functions of the teaching platform to make them more closely connected.

**Acknowledgments.** Our project was supported by Educational Scientific Research Project of Harbin Medical University-The Practice of MR in the teach of Cardiovascular Disease.

# References

1. Yoo, S.-J., Thabit, O., Kim, E.K., et al.: 3D printing in medicine of congenital heart diseases. 3D Printing Med. **2**, 3 (2016)
2. Wang, T.: Application and prospect of virtual and augmented reality (VR/AR) technology in teaching. Digit. Educ. 1–10 (2017)
3. Wang, S., Su, Z., Xu, Z., et al.: Surgical strategies for neonatal and small infant heart disease. Chin. J. Pediatr. Surg. **27**(4), 177–181 (2006)
4. Fu, Y., Liu, X., Qu, Y., et al.: Neonatal congenital heart disease 2001 to 2014 single-center surgical treatment status Lingnan. J. Cardiovasc. Dis. **22**(6), 678–683 (2016)
5. Ong, C.S., Krishnan, A., Chen Yu, H., et al.: Role of virtual reality in congenital heart disease. Congenital Heart Dis. 1–5 (2018)
6. Zhu, Q., Huang, X., Zhou, Z., et al.: Application of virtual reality technology (VRT) in surgical training skills. Educ. Teach. BBS **5**, 41–42 (2019)

7. Zhang, Y., Chen, P.: Virtual simulation: opening a new vision of medical teaching. J. Inner Mongolia Med. Univ. **40**(S1), 16–18 (2018)
8. Ning, R.: Thoughts on clinical medicine teaching in the era of big data. Popular Sci. Technol. **20**(233), 82–83 (2018)
9. Liu, Q., Liu, X.: Informatization and medical teaching reform. PLA Hospital Manag. J. **26** (22), 183–187 (2019)
10. Li, Y., Wang, C., Wang, R.: Application research of multimedia in medical classroom teaching in colleges and universities. Intell. – Innovative Educ. (2018)

# Construction and Practice of Higher Vocational Colleges' Quality Assurance System Based on Evaluation

Pingping Song[1] and Binwen Huang[2(✉)]

[1] Information Technology Center,
Wuhu Institute of Technology, Wuhu 241006, Anhui, China
songping@whit.edu.cn
[2] Information Technology Center, Hainan Vocational College of Political
Science and Law, Haikou 571100, Hainan, China
64471362@qq.com

**Abstract.** Higher vocational colleges pay more attention to the establishment and improvement of the internal quality assurance system. An objective and reasonable evaluation system was established by analyzing the law of the development of higher education and vocational education in the world. Through the integration of the evaluation system and the information platform, it showed the advantages of the efficient flat structure of the information platform and the zero-distance advantages of the participants. And teachers and students were served efficiently. By analyzing all kinds of data in real time, the problems were discovered in each link in time. A normal diagnosis and improvement working mechanism were gradually formed, and internal quality assurance system of sustainable development was constructed for school governance services based on objective data, realizing the continuous improvement of the quality of higher vocational talents training.

**Keywords:** Evaluation · Indicator · Intelligent campus · Data center

## 1 Introduction

The Ministry of Education, in accordance with the general requirements of the Central Committee of the Party and the State Council of "Streamlining government and delegating authorities, combining decentralization with management and optimizing service", and in accordance with the "Decision of the State Council on Accelerating the Development of Modern Vocational Education" (National Development [2014] No. 19), issued in May 2015 "Some Suggestions on Further Promoting the Separation of Educational Management Running and Evaluation, and Transformation of Government Functions". To deepen the comprehensive reform in the field of education in an all-round way and focus on the overall objective of "perfecting and developing the socialist education system with Chinese characteristics, promoting the modernization of the educational governance system and governance capacity", it is clearly put forward that the core task is to implement the principal position of schools and stimulate their vitality in running schools, so as to speed up and improve the operation

R. Mao et al. (Eds.): ICPCSEE 2019, CCIS 1059, pp. 335–351, 2019.
https://doi.org/10.1007/978-981-15-0121-0_25

mechanism of independent-development and self-discipline. "In order to further promote the relevant work of vocational colleges, the Ministry of Education has issued "The notice on the Establishment of a System for Diagnosis and Improvement of Teaching Work in Vocational Colleges" ([2015] No. 2), "The action Plan for Innovation and Development of Higher Vocational Education (2015–2018)" ([2015] No. 9), and "The Guidance Plan of Higher Vocational Colleges for Diagnosis and Improvement of Institutional Quality Assurance System (Trial)" ([2015] No. 168). The promulgation of a series of documents indicates that the teaching diagnosis and improvement of internal quality assurance system will be carried out nationwide.

After several rounds of demonstration college construction and backbone college construction, the development of vocational colleges has entered a new stage of development. How to set up modern quality culture, guide higher vocational colleges to improve quality awareness, establish and improve quality standard system and evaluation system, constantly improve standard connotation, and promote all-process all-round education for all staff; How to enable higher vocational colleges to fulfill the main responsibility of quality assurance of talents cultivation work, combine college governance with informatization, and reflect problems with data, quality with data, and evaluation with data; These new demands of the new era of higher vocational development indicate that the teaching diagnosis and improvement of internal quality assurance system is also the inner demand and inevitable choice of the long-term development of higher vocational colleges.

## 2  Establishment of Evaluation System

Internal quality assurance system of higher vocational colleges should follow the concept of quality sustainable development, which centers on improving the quality of talents cultivation, grasps the improvement of teachers' comprehensive level, takes specialty construction and curriculum construction as the foundation, and guarantees high-level college management services, college should do the top-level design, build indicator system of students growth, teacher development, specialty construction, and curriculum construction quality; Give full play to the main role of students and teachers, independent diagnosis and improvement. Through the evaluation indicator system, it guides the value orientation of students and teachers, stimulates the internal development motivation of teachers and students, and provides support for scientific assessment and college governance. Therefore, we select five key goals of student, teacher, major, course and college to build their evaluation system.

### 2.1  Growth of Students

Talents cultivation is the central work of the college. The growth indicators of students include moral quality (D), academic level (Z), sports level (T) and ability quality (N). The comprehensive score is S = D × 30% + Z × 30% + T × 10% + N × 30%.

(1) Moral quality D = base score + extra points − minus points. Bonus indicators: national honorary title, provincial honorary title, college-level awards (excellent student cadre, merit student, excellent league cadre, excellent league member), school-level awards (excellent student cadre, merit student, excellent league cadre, excellent league member); College-level individual awards, social practice, advanced individual for young volunteers, excellent cadets for military training; Advanced individuals such as party school, college newspaper, advanced individual of college level social practice, young volunteer, college-level civilized dormitory, college-level civilized dormitory, college-level circular commendation, school-level circular commendation, class level commendation, voluntary blood donation, participation in social practice, youth volunteer and other public welfare activities, counselor evaluation etc. Reductions indicators: be kept in school but placed under surveillance, demerit, serious warning, warning, criticism, classroom hygiene reported to the person responsible or on duty, dormitory hygiene reported to criticism, dormitory due to lock doors and other reasons for refusal of inspection, night out, be (including experimental class, teaching, production practice etc.) late, early leave phenomenon, absenteeism etc.

(2) Academic level $Z = \dfrac{\sum(\text{course grade} \times \text{course credit})}{\sum \text{course credit}}$

Courses such as examination and internship are graded according to the five-grade system, and the scores are 95 for excellent, 85 for good, 75 for medium, 65 for passing and 45 for failing.

(3) Sports level T = sports benchmark score T1 + bonus points
Sports benchmark score: in the academic year with physical education, T1 = physical education score × 50% + standardized score × 50%; for an academic year without physical education, T1 = standardized score (physical fitness test score). Sports score indicators: provincial competition, Jiangnan area and municipal competition, college-level competition, and school-level competition.

(4) Ability quality N = benchmark score + extra points.
Bonus points: take part in the national, provincial and college all kinds of competition, innovation, entrepreneurship, intellectual property, pass the College English Test Band 4 (425 points) in this school year, second level of the national computer rank examination, obtained the intermediate above professional qualification certificate, national junior or industry professional qualification certificate (college recognition), this year through the higher education self-study exam course, post points, literature and art joint performance and other race, papers and articles. The evaluation score of students is shown in Table 1.

**Table 1.** Student evaluation addition and subtraction scale table

| Indicators category | Add and subtract items | Specific indicator | Score | Instructions |
|---|---|---|---|---|
| Moral quality | Bonus | National honorary title | 20 | Honor commendation |
| | | College-level excellent student cadre | 10 | Honor commendation |
| | | Advanced individual of College -level young volunteer | 6 | Honor commendation |
| | | Voluntary blood donation | 3 | Public welfare practice |
| | | Participate in social practice | 0.5 | Public welfare practice |
| | Reductions | Academic probation | −15 | Punishment |
| | | Demerit | −12 | Punishment |
| | | Late arrival | −0.5 | Class attendance |
| | | Truancy | −1 | Class attendance |
| Sports level | Bonus | Provincial first place in sports competition | 15 | Sports competition |
| | | Provincial eighth place in sports competition | 6 | Sports competition |
| | | Provincial level participants in sports competitions | 3 | Sports competition |
| | | Sports competition Jiangnan area and the municipal first prize | 10 | Sports competition |
| | | Sports competition Jiangnan area and the municipal eighth prize | 4 | Sports competition |
| Ability quality | Bonus | Develop a career plan or annual plan | 10 | Career planning |
| | | First prize in the national competition | 15 | Knowledge skills competition |
| | | CET-4 (above 425) | 4 | Skills certificate |
| | | National computer rank examination level 2 | 4 | Skills certificate |
| | | President of student union | 12 | ACTS as a post |
| | | Provincial first prize in art competition | 15 | Artistic competition |
| | | Publications above provincial level | 15 | The article published |
| | | Intellectual property | 15 | Intellectual property rights |
| | | Project leader (registered company or individual business) | 15 | Entrepreneurship practice |

## 2.2    Teacher Development

The teacher information contains annual data, and has clear professional attributes and course attributes, as well as specific professional attributes and course attributes for various projects that the teacher participates in. The evaluation indexes of teachers include teacher's morality, teaching ability, teaching research ability, scientific research ability and practice ability.

(1)  Teacher's morality bonus: the national famous teachers, exemplary teachers, excellent teachers, excellent education workers, excellent party member, outstanding party workers and model workers, provincial teaching teachers, exemplary teachers, excellent teachers, excellent education workers, excellent party member, outstanding party workers and model workers and so on the title, won the field of professional leaders, curriculum chief lecturer, exemplary teachers, excellent teachers, excellent education workers, excellent party member, outstanding party workers, etc. Reductions: Academic integrity problem, teaching accidents, disciplinary action, integrity and self-discipline.

(2)  The indicators of teaching ability include leader evaluation, student evaluation, supervision evaluation, "double qualification" teachers certificate, demonstration teaching, academic lecture, overseas (frontier) visiting study training, domestic visiting study training, enterprise (administrative) industry practice training, and whether it is lower than the minimum workload.

(3)  The index of teaching and research ability includes teaching and research project, teaching material construction, teaching achievement award, professional practice achievement, guiding competition result.

(4)  The indicators of scientific research and practical ability include papers, scientific research projects, scientific research awards and achievements promotion. The score of teacher development bonus and reductions is shown in Table 2.

**Table 2.** Teachers' evaluation score table

| Indicators category | Bonus and reductions items | Specific indicator | Bonus points | Instructions |
|---|---|---|---|---|
| Teacher's morality | Bonus | National honorary titles | 10 | Title of honor |
| | | College-level honorary title | 3 | Title of honor |
| | Reductions | Level 1 teaching accident | −15 | Teaching accident |
| | | A demerit | −20 | Punishment situation |
| | | Academic integrity Problem | −30 | Punishment situation |
| Teaching ability | Bonus | Advanced certificate of "double qualification" teachers | 6 | Visiting training |
| | | Demonstration teaching, academic lecture | 5 | Visiting training |
| | Reductions | The number of class hours is less than the minimum school workload | −10 | Punishment situation |

(*continued*)

**Table 2.** (*continued*)

| Indicators category | Bonus and reductions items | Specific indicator | Bonus points | Instructions |
|---|---|---|---|---|
| Teaching and research ability | Bonus | Participate in Category I education teaching and research project (top 3) | 10 | The teaching and research project |
| | | Chief editor or deputy editor of national planning textbooks | 10 | Curriculum reform and textbook construction |
| | | National special award for teaching achievements | 20 | Teaching achievements |
| | | Category I specialty practice performance | 12 | Specialty practice performance |
| Scientific research and practical ability | Bonus | Category I scientific research projects | 20 | Scientific research project |
| | | Category (SCI,EI), the first author | 15 | Papers published |
| | | National scientific research award (top 10) | 20 | Scientific research reward |
| | | Provincial first prize of scientific research award (top 8) | 15 | Scientific research reward |
| | | Category I achievement promotion | 10 | Achievement promotion |

## 2.3  Specialty Construction

The specialty construction information contains annual data, including specialty construction plan, talents cultivation program, annual specialty research report, specialty construction steering committee and other materials. The evaluation indexes of specialty construction include five aspects: talents cultivation, resource construction, social service, exchange and cooperation, and specialty reputation.

(1) Talents cultivation indicators: teacher development level, teaching achievement award, skills contest awards (such as multiple professional sharing project, score should be divided by the share number), teaching research, teaching reform and teaching basic construction situation (such as multiple professional sharing project, score should be divided by the share number), third party evaluation, MyCOS report or large sample data acquisition quality such as employment rate, employment quality and employment satisfaction rate, career expectation fit, turnover rate, alumni evaluation (Recommendation degree).

(2) Resources construction indicators: the experimental practice base construction, college-enterprise cooperation, named class and order class situation (on the basis of unified management in college), participate in the curriculum reform and construction of teaching material (such as multiple specialty sharing project, score should be divided by the share number), research and practice achievements (such as multiple specialty sharing project, score should be divided by the share number), research projects, scientific research reward.

(3) Social service indicators: specialty practice achievement (such as multiple specialty sharing project, score should be divided by the share number), achievement promotion (such as multiple specialty sharing project, score should be divided by the share number), specialty social service ability, innovation and entrepreneurship.

(4) Exchange and cooperation indicators: credit mutual recognition in cooperation with related majors of foreign college, overseas (frontier) visit training, students' overseas exchange and study, foreign teachers' visit to our college, overseas students' visit to our college, domestic study and further training, enterprise (business) practice training, demonstration teaching, academic lectures.

(5) Specialty reputation indicators: students' recognition degree, graduates' recognition degree, employers' recognition degree, graduates' status in industrial enterprises and teachers' recognition degree. The bonus table of specialty construction is shown in Table 3.

**Table 3.** The bonus table of specialty construction

| Indicators category | Specific indicator | Bonus points | Instructions |
|---|---|---|---|
| Talents cultivation | Professor | 5 | Teacher development level |
| | Associate professor | 3 | |
| | Doctor | 3 | |
| | National team | 10 | |
| | Provincial team | 5 | |
| | National famous teacher | 10 | |
| | Provincial famous teacher | 5 | |
| | Senior "double qualification" teachers | 6 | |
| | National special award for teaching achievements | 20 | Teaching achievement award |
| | First prize of national teaching achievements | 15 | |
| | Second prize of national teaching achievements | 10 | |
| | Category I education teaching and research project | 10 | Teaching program |
| | Category II education teaching and research project | 6 | |
| | Category III education teaching and research project | 3 | |
| Resources construction | Chief editor or deputy editor of national planning textbooks | 10 | Curriculum reform and teaching materials |
| | Participate in compiling national planning textbooks | 5 | |
| | Chief editor or deputy chief editor at provincial level or above | 6 | |

(continued)

**Table 3.** (*continued*)

| Indicators category | Specific indicator | Bonus points | Instructions |
|---|---|---|---|
| | Participate in compiling provincial and above planning textbook | 3 | |
| | Other monograph editor | 4 | |
| | Other monograph participants | 2 | |
| | First author of Category 1 | 10 | Papers published |
| | Second author and later of Category 2 | 5 | |
| | First author of Category 2 | 6 | |
| | Second class author and after of Category 2 | 3 | |
| | Category 1 project (top 10) | 20 | Scientific research project |
| | Category 2 project (top 8) | 15 | |
| | Category 3 project (top 5) | 10 | |
| Social service | Category I specialty practice achievements | 12 | Specialty practice achievements |
| | Category II specialty practice achievements | 8 | |
| | Category III specialty practice achievements | 4 | |
| | National practical training base | 10 | Practical training base construction |
| | Provincial practical t training bases | 5 | |
| | National teacher training | 10 | Specialty service society |
| | Provincial teacher training | 5 | |
| | Municipal social training | 1 | |
| | Cooperation among colleges (with cooperation achievements) | 5 | |
| | To support and guide the development of local economy and industries | 5 | |
| | Host national competition | 10 | |
| | Host provincial competition | 5 | |
| Exchange and cooperation | Mutual credit recognition of related majors cooperation with foreign colleges | 15 | Overseas exchange |
| | Students go abroad for exchange and study | 4 | |
| | Foreign teachers visit our college | 6 | |
| | Foreign students study in our college | 4 | |
| | Overseas visit training | 8 | Visiting training |
| | Domestic visiting and learning training | 5 | |
| | Enterprises and institutions practice training | 5 | |
| | Demonstration teaching, academic lecture | 5 | |

## 2.4    Curriculum Construction

The curriculum construction information contains annual data, including course construction plan, course teaching plan, annual course survey report, etc., and the material has a total benchmark score of 60 points. The course construction indicators include the teacher construction level, course influence, teaching process record and course evaluation.

(1) Teacher construction level indicators: course leader, professor, associate professor, doctor, national famous teacher, provincial famous teacher, senior "double qualification" teachers, intermediate "double qualification" teachers, "double qualification" teachers, professor proportion of course team, associate professor above proportion.
(2) Course influence indicators: teaching and research situation, teaching material construction, etc.
(3) Teaching process record indicators: students' attendance, process assessment, students' online learning, classroom activities, etc.
(4) Course evaluation indicators: weighted average of the evaluation of all teachers in the course. Course construction bonus table is shown in Table 4.

**Table 4.** Course construction bonus table

| Indicators category | Specific indicator | Bonus points | Instructions |
|---|---|---|---|
| Teacher construction level | Professor | 5 | Teacher construction level |
| | Associate professor | 3 | |
| | Doctor | 3 | |
| | The course leader is a national teacher | 10 | |
| | The course leader is a provincial teacher | 5 | |
| | Senior "double qualification" teacher | 6 | |
| | Intermediate "double qualification" teacher | 4 | |
| | "Double qualification" teachers | 2 | |
| | Professor proportion > 20% | 5 | |
| | Associate professor proportion > 50% | 5 | |
| | Category 2 teaching research project | 6 | The teaching program |
| | Category 3 teaching research project | 3 | |
| Course influence | Chief editor or deputy editor of national planning textbooks | 20 | Curriculum reform and teaching materials |
| | Participate in compiling national planning textbooks | 12 | |

*(continued)*

**Table 4.** (*continued*)

| Indicators category | Specific indicator | Bonus points | Instructions |
|---|---|---|---|
| | Chief editor or deputy chief editor at provincial level or above | 12 | |
| | Participate in compiling provincial and above planning textbook | 9 | |
| | Other monographs editor | 9 | |
| | Other monographs participant | 5 | |
| | Category1 teaching research project | 20 | |
| | Category 2 teaching research project | 15 | |
| | Category 3 teaching research project | 10 | |

## 2.5   College Observation Indicator

(1) Freshman quality: Arts/science average admission scores, first voluntary reporting rate, independent enrollment rate, registration rate, special student rate, physical health status, and mental health status.

(2) Statistical analysis of students, teachers, majors and courses.

(3) Campus culture: campus activity, academic forum, number of student associations, number of entries of student association.

(4) Media influence: print media, broadcast media, network media, We-Media.

(5) Annual budget and expenditure.

(6) Online service statistics: process quantity, processing quantity, maintenance quantity and feedback statistics.

(7) Teacher-student satisfaction: teaching environment satisfaction, management level satisfaction, career development satisfaction, welfare level satisfaction, logistics service satisfaction, specialty satisfaction, learning environment satisfaction and life environment satisfaction.

# 3   Information Platform Implementation

## 3.1   System Framework

Business systems will be integrated into the online service hall, and the data generated by each business system will be entered into the data center. The evaluation system will extract the information data through the data center and make tabular or graphical diagnosis and analysis. See Fig. 1 for the system framework.

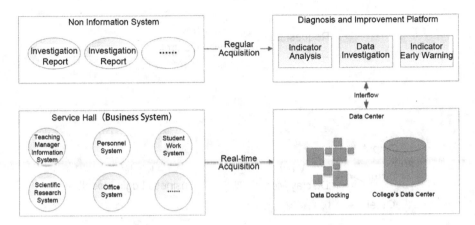

**Fig. 1.** Informatization platform framework of Wuhu Institute of Technology

## 3.2 Technical Route

The system development is based on the B/S architecture and the MVC application development model, which gathers the business logic into a component. While improving and customizing the interface and user interaction, it does not need to rewrite the business logic. The logic block diagram is shown in Fig. 2 and the business flow diagram in Fig. 3.

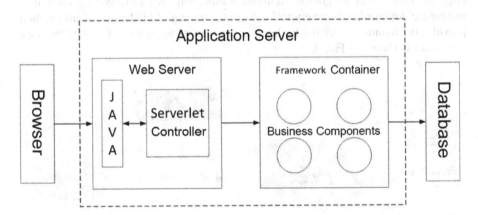

**Fig. 2.** Logical diagram of MVC development on the application server

## 3.3 Data Interface Design

Evaluation system should be seamlessly integrated into the intelligent campus big platform, it should follow the general principle of once data acquisition and data sharing, try to call the data of each application system, it can extend the data in the related application systems if the evaluation needed, data interface integration is to solve the data format of the description of the definition, rules, data sorting and rework. Data integration needs to solve the problem of consistent view of different data sources

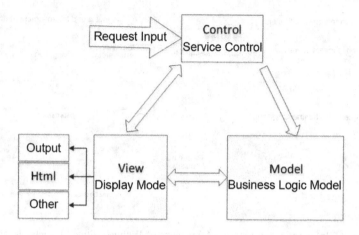

**Fig. 3.** MVC business process diagram

and organize the content of different database systems as a whole. The data must be easily queried and uniquely identified across the entire integration application. The construction of the evaluation data warehouse can plan, balance, coordinate and edit a large number of scattered and independent databases, identify and catalogue the data, determine the metadata model, and enable the data to be distributed and shared in the integrated system. The advantage of data warehouse is that it can still adapt to the upgraded system after integration. At the same time, with the rapid development of data mining and knowledge discovery technology, the mining of hidden useful information provides the foundation for the further development of the school. The data interface flow chart is shown in Fig. 4.

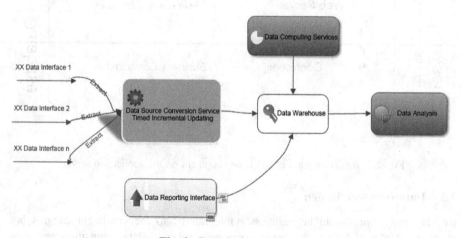

**Fig. 4.** Data docking process

Data interface is the key of this system. The following is an excerpt of the core code:

```java
public void syncScourseInfo() {
List<Map<String, Object>> list = T_BZKS_KSCJService.selectAllScourseInfo();
if (ValidateTools.isValid(list)) {
    for (Map<String, Object> map : list) {
            try {
                    T_BZKS_KSCJService.updateScourseInfo(map);
            } catch (Exception e) {
                    e.printStackTrace();
                    continue;
            }
    }
}
}
@Transactional
public void updateScourseInfo(Map<String, Object> map) {
    if(ValidateTools.valid(map)){
        if (map.get("XN") != null && map.get("XQDM") != null) {
                String year = (String) map.get("XN");
                if (map.get("XQDM").equals("1")) {
                        year = year.substring(0, 4);
                } else {
                        year = year.substring(year.length() - 4, year.length());
                }
                map.put("year", year);
        }
        /* map.get("s_number") and map.get("t_number") */
        map.remove("XN");
        map.remove("XQDM");
        List<Map<String, Object>> list = scourseMapper.selectScourse(map);
        Map<String, Object> course = scourseMapper.getCourse(map);//get course
information
        if(ValidateTools.valid(course)){
                scourseMapper.updateCourse(map);
        }
        if (ValidateTools.isValid(list)) {
                Integer id = (Integer) list.get(0).get("id");
                Map<String, Object> stMap = list.get(0);
                stMap.remove("id");
                DecimalFormat df = new DecimalFormat("0.##");
                if (stMap.get("score") != null) {
                        stMap.put("score", df.format(stMap.get("score")));
                }
                if (stMap.get("credit") != null) {
                        stMap.put("credit", df.format(stMap.get("credit")));
                }
                if (stMap.get("level_score") != null) {
                        stMap.put("level_score", map.get("level_score"));
                }
                if (stMap.get("point") != null) {
                        stMap.put("point", df.format(stMap.get("point")));
                }
                if (!MapUtil.compareMap(stMap, map)) {
                        map.put("id", id);
                        scourseMapper.updateScourse(map);
                }
        } else {
                scourseMapper.insertScourse(map);
        }
    }
}
```

## 3.4   Growth Track of Students

In addition to the evaluation of schools, majors, courses, teachers and students, the system also records major events of students over the years, forms a time record axis, improves students' awareness of self-value realization and realizes motivation education guidance. Here are some of the core codes in the student growth system:

```java
①control layer
@RequestMapping(value = "/loadCredit")
    @ResponseBody
    public Map<String,Object> loadCredit(HttpServletRequest request,Integer year){
            Map<String, Object> map = new HashMap<String, Object>();
            Student s = (Student) request.getSession().getAttribute("student");
            String              schoolNumber           =              (String)
request.getSession().getAttribute("schoolNumber");
            // My growth system score
            Map<String, Object> myCredit = studentCreditService.myCredit(schoolNumber,
s.getNumber(), year);

            List<Integer> years = DateTools.myYears(s.getYear(), s.getEduLength(), 0);
            List<Map<String, Object>> trend =
                        studentCreditService.getCreditByYear(schoolNumber,
s.getNumber(), s.getmNumber(), s.getcNumber(), year, years);

            map.put("myCredit",myCredit);
            map.put("years", years);
            map.put("trend", trend);
            return map;
    }
②Business layer
@Override    }
    public List<Map<String, Object>>getCreditByYear(String schoolNumber,
                    String s_number, String mnumber, String cnumber, Integer year,
List<Integer> years) {
            List<Map<String, Object>> temp = new ArrayList<Map<String,Object>>();
            if(ValidateTools.isValid(years)){
                    for(Integer y : years){
                            Map<String, Object> maptemp = new HashMap<>();
                            maptemp.put("year", y);
                            maptemp.put("pe_integral", "--");
                            maptemp.put("total", "--");
                            maptemp.put("ability_integral", "--");
                            maptemp.put("sch_integral", "--");
                            maptemp.put("moral_integral", "--");
                            temp.add(maptemp);
                    }
            }
            Map<String, Object> map = new HashMap<String, Object>();
            /* map.put schoolNumber, s_number, m_number, c_number, year */
            List<Map<String, Object>> result = studentCreditMapper.getCreditByYear(map);
            if(ValidateTools.isValid(result)){
                    for(Map<String, Object> r:result){
                            for(Map<String, Object> t : temp){
                                    if(r.get("year").equals(t.get("year"))){
                    /* if r.get pe_integral, total, ability_integral, sch_integral, moral_integral is not
null, t.put it */

                                    }
                            }
                    }
            }
            return temp;
    }
```

## 3.5   Autonomous Diagnosis and Improvement

The system combines the characteristics of "PDCA" circulation in quality management with big data, and designs the quality improvement spiral of autonomous diagnosis and improvement. The two subjects, students and teachers, can set their own goals for multiple indicators. The system will automatically track and give regular warning to realize continuous improvement and improvement of quality spirals. See Fig. 5 for the logical block diagram.

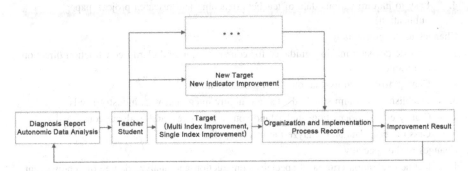

**Fig. 5.** Quality spiral of autonomous diagnosis and Improvement

**Table 5.** Changes brought by the evaluation platform to various roles

| Changes to college governance | |
|---|---|
| 1. | Look at college governance through a single entry, and promote the application of diagnosis and improvement |
| 2. | Timely find and locate various problems in college governance, provide basis for decision-making, and do well in coping strategies |
| 3. | To open up the data island and improve the informatization level |
| 4. | Comprehensively master the real and objective data in college governance, improve the transparent management, and support the construction of the college quality evaluation system |
| 5. | Based on the data analysis platform, improve the overall efficiency |
| 6. | Real-time early warning analysis and targeted management of abnormal data |
| 7. | Define the division of responsibilities and quickly lock the responsible person |
| Changes for students | |
| 1. | Visualization of various process data during student learning (courses, credits, grades, awards, practice) |
| 2. | The student growth path is shown directly, and the traditional single evaluation method for students is changed |
| 3. | Through horizontal and vertical comparison, we can realize the analysis of self-advantages and self-disadvantages |
| 4. | Real-time early-warning data not only accurately locates student problems, but also gives improvement suggestions based on real data |
| 5. | Based on process data, students can realize personalized and hierarchical education |

*(continued)*

**Table 5.** (*continued*)

| Changes to college governance |
|---|

| Changes for teachers | |
|---|---|
| 1. | Through horizontal and vertical comparison, we can realize the analysis of self-advantages and self-disadvantages |
| 2. | Reducing teachers' burden and reducing the preparation for teachers' Professional Title Selection |
| 3. | Locate problems in teaching timely and make self-adjustment |
| 4. | Easy to inquire various data of teachers (teaching and research project, paper publication) |

| Changes to the curriculum | |
|---|---|
| 1. | Provide decision-making guidance for course design, and clarify construction direction and scope |
| 2. | Change from "end evaluation" to "process evaluation" |
| 3. | Establish curriculum standards that are highly integrated with industry needs |
| 4. | Clarify course orientation, understand course resources, and evaluate course value |
| 5. | Grasp course data and analyze the teaching effect |

| Changes to the specialty | |
|---|---|
| 1. | Define evaluation criteria of specialty construction and analyze degree of achievement |
| 2. | Comprehensive analysis of specialty construction |
| 3. | Improve the work efficiency of specialty responsible person |
| 4. | Through horizontal and vertical comparison, understand the advantages and disadvantages of the specialty |

## 4  Conclusion and Significance

Table 5 is the positive significance of the evaluation platform, through the evaluation platform, the college leaders, management departments, teachers and students can dynamically monitor and diagnose at anytime and anywhere, to achieve comprehensive assessment of the key objectives, to provide a good start for continuous exploration and improvement of indicators and KPI, to lay the foundation for the reform of scientific assessment, incentive and distribution system.

**Acknowledgements.** This work was financially supported by the 2018 Scientific Research Major Project of Higher Educational Institutes of Anhui Province (KJ2018ZD059) and 2012 Education and Teaching Research General Project of Anhui Provincial Department of Education (2012jyxm638) and 2018 Education and Teaching Research Major Project of Hainan Provincial Department of Education (Hnjg2018ZD-17).

# References

1. The state department's decision to accelerate development of the modern profession education, 02 May 2014
2. Ministry of education. Higher vocational education action plan for innovation and development (2015–2018 year), 19 October 2015
3. The ministry of education. Notice concerning the establishment of a vocational college teaching diagnosis and improvement of the work system, 23 June 2015
4. Ministry of education. Guidance program for diagnosis and improvement of internal quality assurance system in higher vocational colleges (trial), 30 December 2015
5. Yang, Y.: The treatment is not a "straitjacket" for the school. China Education, 5 July 2016
6. Wang, X., Ding, J.: Role orientation and professional quality of teachers in higher vocational colleges. Res. High. Eng. Educ. (4) (2014)
7. Li, X.: The role transformation of higher vocational teachers from the perspective of "clinical reform". China High. Educ. Res. (06), 100–103 (2017)
8. Wang, D., Qian, X., Xiao, Y.: Research and implementation of educational administration management system based on diagnosis and improvement work. China Occup. Technol. Educ. (11), 76–80 (2017)
9. Chen, X., Yuan, H.: Research on student development diagnosis and improvement index system in higher vocational colleges. China Occup. Technol. Educ. 35(24), 27–30 (2016)
10. Jiang, M., Liu, J.: "Evaluation" to "treatment reform": the change of education quality view in higher professions. J. wuxi Inst. Vocat. Technol., (1), 1–4 (2017)
11. Zhao, H., Zhao, H.: Exploration and practice of diagnosis and improvement of internal quality assurance system in higher vocational colleges. China Training (10), 296–297 (2017)
12. https://baike.baidu.com/item/MVC%E6%A1%86%E6%9E%B6/9241230?fr=aladdin&fromid=85990&fromtitle=MVC

# Research on Teachers' Development from International Perspective

Tiejun Zhu[✉], Jun Liu, Yue Zheng, Min Ou, and Chengxing Jin

Anhui Polytechnic University, Wuhu 241000, Anhui, China
ztj@ahpu.edu.cn

**Abstract.** Competition of excellent teachers and talents has increasingly become a key element of global higher education competition at present. The sustainable and efficient international education development of university teachers and the expansion and promotion of their international vision and ability are the key areas and important contents for the intensive development of China's higher education. Compared with the economically developed provinces, the current international development of Anhui university teachers is not optimistic, and the existing problems are obvious. Aiming at removing bottleneck in the development, this paper put forward specific practical strategies in the development orientation, management system and mechanism, space expansion, language enhancement and financial guarantee, based on the comparative research and analysis between Anhui and other provinces, it effectively promotes the development process of Anhui university teachers from international perspective.

**Keywords:** Higher education · International vision · Teachers' development

## 1 Introduction

The internationalization of higher education has not only become a recognized development trend in the international community and the educational circles, but also brought about profound change, which is an important driving force for the continuous improvement of higher education. As a key link in the internationalization of higher education, the cultivation of internationalized teachers is very important. The cultivation of teachers' internationalization concept, the expansion of their internationalization vision and the enhancement of their internationalization ability can not only make teachers have better responsibilities, but also play an important role in personnel training, teaching and scientific research, as well as in promoting and serving local economy, culture and social development.

In the day of 31st Oct 2017, there are 46 undergraduate colleges and universities in Anhui Province, and the number of university teachers is also growing. Based on different development histories, different levels of running schools, the development positioning of Anhui ordinary undergraduate colleges and Universities is bound to form differences and hierarchies, but due to the strong impact of internationalization of higher education, the development goals of opening up and internationalization should become the current and future development direction of Anhui undergraduate colleges and universities without exception. In addition, at the national level, both in the

R. Mao et al. (Eds.): ICPCSEE 2019, CCIS 1059, pp. 352–363, 2019.
https://doi.org/10.1007/978-981-15-0121-0_26

economic and education fields, the current development strategy focuses on the connotation construction of quality. How to promote and realize the continuous improvement of the quality and level of university teachers in depth, and then form the talent-driven traction effect and healthy development of universities should be an important target and an important subject of reform for Anhui undergraduate colleges and universities, as well as for Anhui Province to speed up the construction of a strong province of higher education. Therefore, under the background of the internationalization of higher education, it is particularly important to plan and set the scientific path for the education and development of teachers in general undergraduate colleges and universities in Anhui Province.

## 2  Background

In a long term perspective, the major developed countries in the world have attached great importance to the internationalization of higher education and internationalization education and development of university teachers. They have not only started early, but also made great strides. Among them, the measures and experience of promoting the internationalization of teachers' education and development are very mature. Now, only Germany and the United States, which have more cooperation among undergraduate universities in Anhui Province, are taken as examples.

The internationalization of German higher education and teacher education have been enjoying a high reputation all over the world. On the one hand, German higher education has a long history. The birthplace of modern universities is Germany, and the situation of European integration of education also provides a fertile soil and establishes a broad platform for its development. On the other hand, the German government fully recognizes the country of teachers. Internationalized training plays an important role in enhancing the international competitiveness of German higher education, and even in expanding the international influence of Germany. The main approaches are as follows: Firstly, to carry out substantive measures around the internationalization of teacher education, which mainly involves the international academic exchanges of teachers, the internationalization requirements of teachers' academic visit, academic degrees or related qualifications, the development of internationalized courses, teachers' bilingual education and so on. Since the 21st century, Germany has attached great importance to incorporating the relevant factors of internationalization of higher education into school development plan and curriculum setting. Secondly, it has adequate financial support. Whether it is teachers' subjective continuing education and learning needs or objective requirements, the accommodation and transportation costs during their study period can be solved, thus the teachers' enthusiasm and willingness to continue learning are strong. Thirdly, a large number of overseas education projects, distance education projects, international cooperation projects in education and research, as well as multinational enterprise cooperation projects implemented by

German higher education institutions have greatly promoted the need for university teachers to have high quality and internationalization ability to cope with and adapt to various international environments, and also provides a multi-platform for their international education and development.

Undoubtedly, the internationalization level of USA universities has become a benchmark in the world. It has already realized not only the international diversity and two-way flow at the student level, but also the two-way flow at the teacher level. On the one hand, a large number of university teachers have exchanged and cooperated around the world. On the other hand, it has also attracted a large influx of excellent teachers from other countries. The United States is one of the countries that made the international exchange of university teachers as the strategic elements to realize the internationalization of higher education at the national level. "In 1862, the United States Congress passed the Morrill Land-Grant Act, and a large number of experts, scholars and students were sent to Europe for further study." [1] One of the greatest characteristics of the United States in the development of teacher education in colleges and universities is that, apart from the U.S. government departments, a large number of foundations such as the very famous Carnegie Foundation, Rockefeller Foundation also fully subsidize all kinds of activities carried out by USA university teachers in international exchanges. In addition, the U.S. government attaches great importance to the acquisition of international language and culture of university teachers. A large number of foreign language teaching centers, international cultural research centers and regional research centers have been established successively. These institutions not only provide a good communication platform for university teachers, experts and scholars, but also broaden their horizons, enhance their knowledge reserves and capabilities, and provides a strong support for the international development of colleges and universities.

# 3   Current Situation and Existing Problems of Internationalized Education for Teachers in Undergraduate Colleges and Universities in Anhui Province

## 3.1   Development Status

In recent years, although the undergraduate colleges and universities in Anhui Province pay more attention to the internationalization and development of teachers' education, and the funds for teachers' training are also increasing year by year, the overall development is not balanced. Compared with the surrounding provinces and cities, such as Jiangsu and Shanghai, there is a big gap in scale. Through many investigations, the number of teachers' overseas visits and exchanges can be inferred. Table 1 lists some universities in Anhui, Jiangsu and Shanghai:

**Table 1.** A list of the numbers of teachers visiting and exchanging overseas in Colleges and Universities in Anhui, Jiangsu and Shanghai in 2017.

| No. | Name of University | The number of teachers visiting and exchanging overseas in 2017 |
|-----|--------------------|-----------------------------------------------------------------|
| 1 | Anhui University | 348 |
| 2 | Anhui University of Technology | 20 |
| 3 | Anhui Normal University | 65 |
| 4 | Anhui University of Science and Technology | 98 |
| 5 | Anhui Polytechnic University | 71 |
| 6 | West Anhui University | 26 |
| 7 | Hefei University | 59 |
| 8 | Huangshan University | 19 |
| 9 | Nanjing Normal University | 350 |
| 10 | East China Normal University | 1678 |

## 3.2    Current Problems

### 3.2.1    Emphasizing Indicators but Lightening Quality

Through investigation and research, it is found that many undergraduate colleges and universities in Anhui have clearly formulated international education and training plans for teachers, and some universities have included overseas research and learning experience into the necessary conditions of professional title evaluation, such as Anhui University, Anhui Normal University, Anhui Polytechnic University, etc. Therefore, they have implemented better in dispatching strength and allocation of indicators, but the common problem is that the performance of teachers' overseas study is not satisfactory and the learning quality is not high.

### 3.2.2    Insufficient Understanding of Ideas

Although colleges and universities in Anhui Province have reached a consensus on the direction and situation of the internationalization development of higher education, their ideas and understanding are insufficient when it comes to how to realize the benign interaction between the internationalization development and the development of teacher education. Some cooperation channels and projects are still at the exploratory stage, and there are many problems to be solved urgently at the level of understanding, some even lack of confidence.

### 3.2.3    Limitations of Institutional Mechanisms

Compared with the developed provinces, Anhui undergraduate colleges and universities have greater limitations in school-running system, management mechanism and financing channels. In addition, the flexibility, characteristics of discipline setting and the profundity of education and teaching reform deserve to be greatly improved.

### 3.2.4    Inadequate Funding

Lack of funds is an important obstacle to restrict the development of international education for teachers in undergraduate colleges and universities in Anhui Province. The main manifestation is that the sources of funds mainly depend on government financial allocation, almost no donations such as foundations and alumni associations are involved, and the government financial allocation is insufficient. In addition, teachers' funds for scientific research, communication and cooperation are relatively low, and teachers are generally unwilling to use these funds to carry out over- seas research.

### 3.2.5    Weak English Proficiency

In addition to foreign language teachers, other professional teachers have poor English language competence and lack of international exchange experience and ability. Although their English reading and writing ability is still acceptable, it is still weak to use English to carry out professional knowledge study and grasp international research hotspots.

### 3.2.6    Regional Environmental Impact

Compared with some economically developed provinces, the level of internationalization of colleges and universities in Anhui Province is generally weak because of the constraints of geographical location, regional environment and policy mechanism, so the guarantee for teachers' internationalization education development is relatively limited.

## 4    Innovative Development Path of Teachers in Anhui General Undergraduate Colleges and Universities Based on the International Perspective

### 4.1    Reasonable Orientation to Realize Differentiated Hierarchical Development

The history and development basis of 46 undergraduate colleges and universities in Anhui are different, and the orientation of running schools and the structure of subjects and specialties are also different. Therefore, the convergence of orientation should be avoided in the setting of internationalization path and the direction of teacher education development. For example, 14 undergraduate colleges and universities in Anhui named by universities, of which many have similar development paths in this field, in fact, even if they are all named universities, the strength of running a school is obviously different. As development orientation of many independent colleges is close to that of the parent universities, we must orientate and construct them according to the actual level and ability of the schools. We can neither be satisfied with the status quo, lack of foresight and planning and form blind areas and misunderstandings, nor blindly compare with each other, far-reaching and blindly strive for higher education, so as to create higher education. Thus causes the resources waste and the limits high quality resources to compete fiercely, it is necessary to consider comprehensively and balanced, "insist on the spirit of doing certain things and refraining from doing other things", [2] rationally orientate itself to realize its own characteristic, differentiation and hierarchical development. For example, the key universities with the right to grant doctoral degrees should rely on their core strength, dock with the world-class universities and institutions, highlight the high level, pay attention to innovation, realize the synchronous leap of teachers' education quality and quantity; other key construction colleges and Application-oriented Undergraduate Colleges in Anhui Province should focus on advantageous disciplines, while relying on industry characteristics, building a teacher education and training system with distinctive orientation. For example, Anhui Polytechnic University has innovatively and demonstratively carried out the construction of the International College of Engineers in the province, and put forward the idea of "internationalization, engineering, entrepreneurship and diversification", which is bound to put forward higher and more specific requirements for teachers' internationalization education and development. Schools that have just entered the ranks of undergraduate colleges and universities should fully understand their family background and act according to their abilities, find out the key factors and necessary conditions for the internationalization of teachers' education, and then comprehensively evaluate and analyze to choose the path of the internationalization of teachers' education in line with the orientation of schools and the actual development of teachers' education, in order to achieve steady development.

## 4.2 Actively Expand and Make Full Use of International Space and Educational Resources

Due to the influence of regional environment, the space and channels for the internationalization development of universities and teachers' internationalization education in Anhui Province are limited. Therefore, for the undergraduate colleges and universities in Anhui Province, mobilizing and integrating all forces as far as possible to actively expand and make full use of international space and resources is the fundamental guarantee for the implementation of teachers' education development, and The primary task is to build a multi-channel, multi-level and innovative teacher education and development platform. Firstly, the government platforms and resources such as the construction and exchange of friendly provinces and prefectures in Anhui Province, Anhui Overseas Intellectual Introduction Workstation and Anhui Province's "100-person Program for Foreign Specialized Education" can be used to establish contacts and open up channels. Secondly, it can combine with the school's profession- al structure, integrate school resources, accurately dock and match overseas universities and institutions. Through active exploration and innovative design to create a number of distinctive teacher education brand cooperation projects, "Establish a mechanism for experts and scholars to send and visit each other among universities, enterprises and research institutes", [3] and form a long-term effect. Third, vigorously expand and implement a large number of Sino-foreign cooperative school-running projects, Sino-foreign joint training projects, exchange student projects, cultural tourism camps and other short-term student exchange projects, international students program and other student programs to achieve frequent communication and interaction between teachers, and at the same time, an international atmosphere in the campus can be formed, so that the majority of teachers can feel the advanced ideas and teaching methods of foreign countries. Fourthly, it introduces the network resources reasonably, especially the platform of admiring courses, micro-courses and distance education in overseas universities. On the one hand, it provides students with diversified international education resources. The omni-directional mode of breaking through the boundaries of time and space can effectively enhance students' international thinking ability and knowledge acquisition, thus forcing teachers to carry out international education and learning in the process of teaching. On the other hand, the introduction of unlimited network education resources can also enable teachers to timely understand the latest international frontier knowledge information of the subjects they teach and track the latest developments of the subjects. Building a broad platform for communication and teachers' self-development, and forming a strong guarantee and eternal power for teachers' sustainable development.

### 4.3    Creating a Management System of Internationalization of Education

"Open and competitive management system is a powerful driving force for the internationalization of Universities" [4]. Therefore, it is of great significance to establish and gradually improve the management system of internationalization of education. On the one hand, undergraduate colleges and universities in Anhui Province need to sort out, summarize and analyze the information of teachers' educational background, age, academic background and international education background comprehensively. On the other hand, they need to solicit and listen to the opinions, suggestions and reasonable demands of teachers extensively, so as to draw up a special plan and institutional measures for the internationalization of teachers from bottom to top. At the same time, they also need to conduct situation research and judgment, specially, the top-level design and the top-down introduction of a more mature and perfect management system are carried out for teachers' internationalization education and the focus of the annual administrative work to be specified and task decomposition. Anhui Polytechnic University's key administrative work in 2018 as an example, which clearly pointed out: "Accelerate the opening up of schools, create a new plateau for cooperation and exchange. Supporting teachers to study abroad, participating in international cooperation, research and academic exchanges, and promoting managers' overseas training in an orderly manner" [5]. On the other hand, through the reform of management mechanism, we can guide and promote the innovation of personnel training, discipline, specialty, curriculum and teaching methods. At present, most of the undergraduate colleges in Anhui Province are still in the period of expanding student numbers. A large number of enrollment expansion and the high standards of teachers' employment and introduction cause the phenomenon that the number of students in colleges and universities is relatively high, teachers in schools have to put a lot of energy into dealing with daily teaching, so that there is no more time to selfcharge, and it's more difficult to achieve international education and learning. Therefore, we should start with the details of teachers' daily teaching and scientific research, draw lessons from successful foreign management experience, and arrange teachers' workload rationally by flexible adjustment of training programs, disciplines and curriculum system, and encourage them to formulate appropriate study abroad plans and put them into practice.

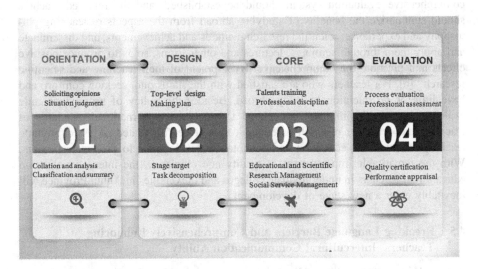

**Fig. 1.** Steps to create a management system of internationalization of education

## 4.4 Perfecting Process Evaluation Mechanism and Performance Appraisal Mechanism

In view of the phenomenon that teachers' overseas learning performance and efficiency are not high in Anhui undergraduate colleges and universities, reasonable setting of process evaluation mechanism and performance evaluation mechanism is an effective way to solve the problem. Universities that started and implemented the teacher's overseas education learning plan earlier should mobilize teachers who have obtained overseas learning qualifications and returned to school to complete their learning tasks according to their own learning experience, discuss their experience and insufficiency, and the relevant functional departments of schools should summarize their opinions, especially the problems reflected, and work with overseas universities and institutions to provide the best learning conditions and environment for the teachers who are going to travel in the later period. We should strengthen the process management and evaluation for the teachers who still have not completed their studies abroad. We can learn from the management model of the State Overseas Studies Fund Committee and according to the period of overseas study, the process assessment of overseas study teachers should be carried out regularly. For those teachers who have strong willingness to study abroad and have actively prepared and signed up for the relevant teachers' overseas study and research projects, on the one hand, they are required to draw up a study plan before going abroad. The goal setting and feasibility of the individual study plan will be an important criterion for qualification selection. According to this, after determining their eligibility, signing agreements and commitments with them, specify the tasks and objectives they need to accomplish, reward and punishment conditions, carrying out process management according to the provisions of the plan and the agreement during their study abroad. After returning home, "a follow-up tracking and comprehensive evaluation system should be established, and the returned teachers should summarize the benefits of studying abroad from the aspects of teaching philosophy and experience, scientific research methods and achievements, and disseminate and apply them through seminars, lectures and other forms, so as to produce positive effects and promote the comprehensive development of their teaching and scientific research" [6]. For teachers who are satisfied with the status quo, seeking comfort, and have little or no intention to study abroad. Besides the policy of professional title evaluation, more flexible policies related to salary and international standards of teaching and scientific research assessment should be formulated so as to promote teachers' international education and connotative development to the greatest extent. With the remarkable improvement of quality and efficiency, the internationalization education of teachers in undergraduate colleges and universities in Anhui Province will inevitably enter a new stage of development.

## 4.5 Breaking Language Barriers and Comprehensively Enhancing Teachers' Intercultural Communication Ability

The weak application ability of English is a common problem that lingers in most non-English majors' in colleges and universities. Because of this, many teachers are not willing to study abroad. "Modern university teachers should have the ability of bilingual teaching and have strong international communication ability." [7] It can be said that bilingual teaching is a good medicine for curing the crux of teachers' English language problems. Practice and attempt in localized campus can effectively enhance teachers' English application ability and self-confidence. For undergraduate colleges in Anhui Province, it is an inevitable trend to introduce and set up a large number of international courses. International courses can not only improve students' international quality and attract more foreign students, but also improve teachers' international teaching level and ability through curriculum practice, which is also an important way for teachers' international education and development. Firstly, we can learn from the high-quality curriculum system and talent training program of foreign universities, and firstly start the curriculum setting and the use of foreign original textbooks, then gradually follow up and integrate the supporting teaching management system, such as teaching evaluation system and teaching evaluation mechanism; secondly, we need to combine the characteristics of schools, teachers and students to carry out the international mode of school-based curriculum. In the process of developing internationalized curriculum, teachers must deeply understand and learn language culture and teaching methods, so it is undoubtedly the process of their internationalization education and development. Thirdly, international understanding education need to be carried out, schools can penetrate and integrate international understanding education according to the characteristics of foreign curriculum systems, For example, foreign humanities and arts can help teachers understand the differences between eastern and western cultures, while natural sciences can help teachers understand foreign scientific thinking and spirit [8]. Through the implementation of the above measures, university teachers will be confident in offering bilingual teaching courses, their English language application ability and the construction of the school's international curriculum system will make significant progress. In addition, we can draw lessons from the practice of USA colleges and universities, set up language and culture research centers, regional and cultural research centers, Chinese and foreign cultural research centers, and increase support for international cooperation projects in education and research, guide and encourage teachers to make continuous progress in the development of language communication, cultural exchanges, educational and research exchanges and ideological blending.

## 4.6    Financing by Various Means to Form a Strong Financial Guarantee for Teachers' Internationalized Education

According to the survey, the funds for international education and training of teachers in Anhui undergraduate colleges and universities mainly come from three aspects: the Ministry of Education and the State Overseas Study Fund Committee, the Education Department of Anhui Province and the school itself. The proportion of teachers funded by educational research projects and social support is relatively low. Taking Anhui University of Engineering as an example, in 2017, the number of its overseas teachers was 71, among which 64 were funded by the education department of Anhui province or the school. The remaining seven people were sent through educational research projects or other channels, accounting for only 9.85%, and none of them were supported by social funds. These mainstream channels are not only limited in quota and competitiveness, but also limited in funds. They are difficult to fully meet the needs of teachers. It also affects the diversity of communication and the richness of communication objects. Therefore, it is necessary to strengthen social propaganda, society recognize and understand the internationalization of teachers' education and development. "The knowledge, experience, methods and level of internationalization of teachers," [9] will directly have an important impact on personnel training, teaching quality, scientific research innovation, academic dissemination and social services in Colleges and universities, so as to form a consensus among various social forces and multi-funds, and then to support teachers to carry out high-level continuing education. Secondly, it is necessary to give full play to the role of alumni associations, cooperative units and school-run enterprises in colleges and universities. It is encouraged to set up various kinds of foundations and special projects for teacher education development. Thirdly, in view of the government's financial allocation funds, it is necessary to establish a fund allocation system that integrates the cost and performance of teacher education organically to promote the rational use of such funds and maximize their benefits.

# 5 Conclusion

"In the view of internationalization of higher education, the development of teachers' professionalization and the cultivation of teachers' professional qualities require that university teachers have the consciousness of international cooperation and communication" [10]. The background of internationalization of higher education also makes the international exchanges and cooperation of university teachers richer and more frequent. However, the current situation of Anhui undergraduate colleges, the first front of higher education in Anhui Province, is not optimistic in the field of internationalization education and training of teachers, and they are facing many bottlenecks. Therefore, it is necessary to rationally orientate and realize differentiated hierarchical development, actively expand and fully utilize international space and educational resources, create a management system for internationalization of education, and improve the process evaluation mechanism and performance evaluation mechanism, then open a smooth channel for the development of internationalization of teachers' education at macro level and institutional mechanism, Supported by the establishment of school-based international language and culture application platform and the guarantee of diversified special funds, it will promote internationalization education and development of Anhui undergraduate teachers sustainably, healthily and efficiently.

**Acknowledgements.** This research is supported by the Sino-foreign Cooperation Training Project of Provincial Quality Engineering in 2018 "Financial Engineering Major of Sino-foreign Cooperation in Running School of Anhui Polytechnic University" (2018zwpy001); Anhui Education Science Research Project "Research on the Creation and Application of Online Teaching and Maker Practice Platform of Art Theory Course under the Background of Intelligence Classroom"; The Key Project of Higher Education Research of Anhui Polytechnic University in 2019 "Research on Connotation Construction of International Engineering Institute of Anhui Polytechnic University" (2019gjzd002); Anhui Polytechnic University Creation Group Project "Local Culture and Creative Design Research Group"; Design Principle Funds of Anhui Polytechnic University; Anhui Province philosophy and social science planning project "Anhui province higher education internationalization strategy research" (AHSKY2016D29); Anhui province quality engineering teaching and research key project "School running characteristic exploration and innovation practice of International Engineering Institute of Anhui Polytechnic University" (2016jyxm0091).

# References

1. Liu, J.: The present situation and case study of the internationalization of American College Teachers. Master's degree thesis of Hunan University (2015)
2. Fang, Q., Sun, D., Ping, P.: Analysis of the orientation of local undergraduate Universities in Anhui. High. Archit. Educ. **13**(6), 1–6 (2013)
3. Education Department of Anhui Province. The 13th five-year plan for education development of Anhui Province, 15 February 2017. http://www.ahedu.gov.cn/168/view/19908. Accessed 05 Feb 2018
4. Li, S.: Research on the development of internationalization of universities in Anhui Province. Sci. Technol. Inf. **33**(12), 698–699, 706 (2012)

5. Zhao, X.: Key administrative work of Anhui Polytechnic University in 2018. http://www.ahpu.edu.cn/2018/0202/c5a99379/page.html. Accessed 05 February 2018
6. Ying, Z.: Investigation and analysis on the short-term overseas travel of college teachers in anhui province - taking a university of science as an example. J. Fuyang Inst. Technol. **28**(2), 88–91 (2017)
7. Gao, Y., Deng, F.: Research on the quality of college teachers in the context of internationalization of higher education. Educ. Occup. **25**(5), 42–44 (2011)
8. Sun, G.: Reflections on teachers' professional development in the context of internationalization of education. Teach. Monthly **31**(10), 37–39 (2015)
9. Xia, W.: Teachers' participation in the internationalization of university curriculum. High. Educ. Res. **20**(3), 64–70 (2010)
10. Liu, H.: Research on the professional development of college teachers from the perspective of internationalization of higher education. High. Agric. Educ. **36**(1), 47–49 (2013)

# Research on the Teaching Model
# of Experimental Virtualization in Digital Logic
# and Digital System Design Course

Pan Liu[1(✉)], Qing Wang[2], Huipeng Chen[2], Lei Xu[2],
and Yanhang Zhang[2]

[1] Harbin Institute of Technology (Shenzhen), Shenzhen 518055, China
liupan@hit.edu.cn
[2] Harbin Institute of Technology, Harbin 150001, China

**Abstract.** By stating the problems faced by the experiment of digital logic
design course in traditional laboratory, the necessity of hardware experiment
virtualization is analyzed, and then two virtual experiment methods are introduced. The pilot implementation of remote virtual experimental platform and
virtual component library shows that the virtualization of hardware experiment
can effectively break the time and space limitation of traditional hardware
experiment, and improves the learning enthusiasm and autonomy of students,
which is worth further promoted.

**Keywords:** Virtual experiment · Digital logic · FPGA platform ·
Virtual component library

## 1 Difficulties Faced by Traditional Experimental Models

The course "Digital Logic and Digital System Design" (hereinafter referred to as
"Digital Logic") is a core curriculum for computer major, whose main theoretical
knowledge is presented on the basis of hardware circuit. Therefore, the experimental
part of the course is a significant portion of the teaching process and occupies a very
important position. However, the Digital Logic Experimental Boxes that are utilized
during the traditional hardware experiments are highly customized, which could merely
support the designated experiments. As those boxes are designed for some given
experiments, it is difficult to carry out innovative experiments and individuation
experiments on them, and this situation limits the development of high-level new
experimental projects, and subtracts the utilization rate.

Currently, a majority of colleges and universities conduct the experiment of Digital
Logic course in two different fashions, namely wire connection experiment and simulation experiment. However, differing from the software series experiments, both
modes require students to accomplish a complete system on actual experimental
platforms or development boards except for computers, so that, the teaching laboratories should provide more equipment, and students should be familiar with more
devices. In recent years, "online + offline", "classroom + extracurricular", "MOOC +
SPOCs" and other hybrid teaching modes have achieved good results and been widely

promoted in college teaching practice. However, the difficulty of networking of hardware experiments hinders the networking process of experimental teaching. Traditional experimental models are facing some problems that need to be solved urgently.

## 1.1 Hardware Laboratory Investment Is Huge, but It Is Difficult to Maintain the Advancement of Hardware Resources for a Long Time

Experiments of Digital Logic are divided into wire connection mode and simulation mode, students have to utility real chips, wires, voltmeter, etc. during the former and FPGA boards during the latter. Since wastage occurs in all devices at any time, it is necessary for laboratories to deploy at least 20% extra equipment to deal with unexpected instances in class, and hence more purchase funds, more storage space and longer maintenance time are demanded.

In addition, the well-known Moore's law reveals that when the price remains unchanged, the number of components that can be accommodated in integrated circuits will double every 18–24 months, and the performance will also double. With the rapid development of programmable devices in recent years, the FPGA chips purchased more than 3 years will soon be unable to adapt to the new hardware market. In order to maintain the advancement of experimental devices, laboratories have to update the FPGA development boards constantly. The update speed of the hardware equipment on the market is much faster than the construction speed of the laboratory hardware resources, and the cost of updating the development board is enormous.

## 1.2 Hardware Laboratory Equipment Can Not Be Moved, and It Is Difficult to Support After-Class Experiments

At present, the length of experimental teaching hours in class is limited by the curriculum settings. Students can only design and operate experimental projects in class, and the learning time is restricted. Therefore, the mixed "classroom + after-class" teaching mode is more and more popular among students.

However, the real experimental environment required by hardware experiment can not be shifted freely in physical space. One solution is opening the laboratory after class, but there are many factors to consider, such as school management institutions, laboratory safety management, experimental course arrangement and so on, and the implementation process is complex.

Moreover, a large number of students require the opening of laboratories in order to improve their experimental results, so the number of students will increase dramatically during several weeks in the duration of the course, and the equipment is obviously inadequate, but after the course ends, no one is interested and the open laboratory becomes idle again.

### 1.3  Errors in Hardware Experiments Are Hard to Reproduce, and It Is Difficult to Debug

The realization of hardware experiment is affected not only by the correctness of design logic, but also by the completeness of experimental chips and wires. When students gain incorrect experimental results, they generally demolish the whole connections violently and reconnect again, the solution of "violent demolition" makes it difficult for experimental instructors to reproduce the errors and to figure out the point of errors.

Besides, unlike the automatic debug function in software development environment, the hardware experiment can only be implemented by layer-by-layer analysis and manual measurement. It takes a certain amount of time for students to correct errors and teachers to guide, and there are a great deal of rough elements.

### 1.4  The Cyberization Degree of Hardware Experiment Is Low, and It Is Difficult to Popularize on Internet

Under the background of "Internet + education", promoting the transformation of special educational resources to general educational resources has become the important target of the current education and teaching reform. Nevertheless, the realization of hardware experiment is closely related to the actual chip and experimental platform. This feature has obvious geographical limitations, which seriously restricts the Web-based dissemination and promotion of high-quality experimental teaching resources.

Subject to all these limitations, the traditional experimental teaching mode of digital logic has been unable to fully meet the new demands of modern talents training. Virtual experiments can meet the requirements of learners with different careers in different places and at different times. Virtual experiment has strong reusability and high reproducibility, and it can effectively share data, equipment and resources, overcome the drawbacks of traditional hardware experiments, and provide a learning platform for learners to collaborate and share.

## 2  Research Status of Virtualization Experiment

As early as the 1990s, the researches and applications of virtual experiment system for teaching have been carried into effect, and then the developments of teaching-oriented virtual experiment system have been greatly launched with the popularization of network, the comprehensive application of multimedia technology, computer technology and virtual reality technology. Nowadays, numerous universities and research institutes both domestic and international have initiated to exploit virtualization experiment system.

### 2.1  International Research Status

Foreign universities attach great importance to the deep integration of experiment and education, they pay much attention to support academic activities, and many institutions have launched the researches on virtual experiment system from early days and acquired abundant achievements.

The electronic circuit virtual experiment system developed by University of Victoria in Canada is an earlier virtualization experiment system [1]. For the sake of making the virtual experiment system more similar to the real experimental scene, some of the virtual chips used by students in the system are not available with uncertain probability, which is consistent with the actual experimental situation.

Learn Anytime Anywhere Physics Lab [2] developed by the research team from the University of North Carolina is programmed for giving service to physics experiments. The system is developed in Java language, it utilizes browsers as user interfaces, and uses VRML to display a 3-D virtual environment.

Moreover, a 3D chemical virtual experiment system [3] developed by Multimedia Systems Laboratory of Mari State Technical University is provided for simple chemical experiments.

In the field of data science, Fang, Zhai, etc. have created a virtual lab for information retrieval, one subdomain of data science [4]. And then, Chase Geigle, Ismini Lourentzou, etc. of University of Illinois at Urbana-Champaign have generalized this to address creating a virtual lab for any data science domain or application [5].

## 2.2    Domestic Research Status

In China, as early as 2008, the Ministry of Science and Technology has launched the "Research and Application Demonstration of Virtual Experimental Teaching Environment" as "Eleventh Five-Year Plan" support project, which involves the development, evaluation and management of virtual experiments, as well as the key technology of visual design of virtual experimental teaching. In the Outline of the Ten-Year Development Plan of Education Informatization (2010–2020), it is distinctly pointed out that "virtual laboratory should be built", especially in the field of vocational education, "practice bases for virtual simulation training should be built, the number of virtual simulation training software should be increased, application satisfaction and professional coverage should be enhanced". Thus we can see that, the state attaches great importance to virtual experiment and its application.

So far, some colleges and universities in China have established virtual experiment system, some of which have been put into normal operation.

Tsinghua University adopts large-scale and ultra-large-scale united circuits and programmable devices, combines the design ideas and methods both of software and hardware, and exploits a integrate hardware experimental platform [6]. The platform unifies all backbone experiments of computer hardware curriculum into one platform, which makes the experiments originally completed on the decentralized platforms get unified support, reduces the teaching costs brought by various platforms, and effectively improves the learning efficiency of students.

The Computer Department of Peking University has developed a virtual experiment system [7], which supports a large amount of computation and has powerful interaction functions. In the interface of the virtual experiment system, it provides specific experimental instruments and experimental requirements for each experiment. After the user completes the experiment, they could save the data, and the system would then return the processed results to them.

The virtual experiment lab developed by Central South University [8] is a virtual experiment platform related to digital signal processing. It is realized by taking advantage of JAVA and MATLAB mixed programming technology to support students to build experiment process smoothly, which is conducive to experimental innovation.

EVLab, a three-dimensional electronic circuit experiment system based on virtual space environment manufactured by Institute of Modern Educational Technology of Beijing Normal University, is a three-dimensional electronic circuit experiment environment based on virtual space. Through the EVLab system, students can master the common methods of instrument operation in electronic circuit experiments, and own a deeper understanding of the basic circuit experiments.

# 3 Methods of Experimental Virtualization Teaching

At present, the experiment of Digital Logic course in the experimental center involves wire connection and simulation. Therefore, it is divided into two portions in the process of experiment virtualization, and different models are used to achieve the experiment virtualization. At last, the two models are combined to unify the usage mode, which is convenient for students to use.

## 3.1    Remote Virtual Experimental Platform of Hardware Courses

Aiming at the phenomenon of "one lesson, one platform" existing in the current computer hardware experiment course, the experimental center proposes to establish a remote virtual experiment platform for hardware experiment course based on FPGA, which can effectively settle the shortcomings of hardware experiment course limited by time and space, and actualize the objectives of multi-purpose and multi-course integration platform [9]. The platform adopts the design concept of "local computer virtual simulation software + real hardware platform with remote access", and achieves the construction goal of "combining virtual with real". The system physical photo and experiment running picture are shown in Figs. 1 and 2 respectively.

The experimental platform establishes an entity consisting of server clusters and FPGA development boards placed in the laboratory. When using the platform, visitors do not need to install any plug-ins and clients, nor even simulation integrated development environment. They only need to visit the website of the experimental platform in the browser with the local computer under their own account. Although they are carrying on simulation experiments of Digital Logic Design course, they can observe the real appearance of real components running on the actual FPGA development boards. At the same time, the platform also offer personal storage space, exchange and discussion forum and so on, which provides great convenience for the users.

In the traditional laboratory environment, in order to complete the digital logic experiment of simulation type, the laboratory should provide experimental sites, computers installed large Integrated Development Environment software, and FPGA development boards for every student. However, these physical devices are centralized in the remote virtual experiment platform, and students need only one computer to complete the virtual experiments to obtain the equivalent teaching effect to the

    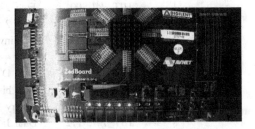

**Fig. 1.** Remote virtual experiment platform          **Fig. 2.** Experiment running picture

traditional experiments. In addition, the employ of virtual platform breaks the time and space constraints in traditional experimental model, and achieves the goal of experiment at anytime and anywhere veritably.

The remote virtual experiment platform for hardware courses based on FPGA has been applied in the undergraduate teaching of Computer College and Software College of Harbin University of Technology, and is being gradually promoted. Practice has proved that the platform not only saves the time of occupying computers in laboratory, improves the efficiency of the experiment, but also greatly reduces the workload of the experiment management.

### 3.2  Virtual Component Library

For the virtualization of wire connection type experiment teaching of Digital Logic, the concrete realization method of the experiment center is to utilize Virtual Component Library. The virtual component library takes the actual components as the template, and restores the characteristics and functions of the actual devices as real as possible in the aspects of device function, device shape and operation method, so as to reduce the distance between virtual and real experiments. Furthermore, considering that the virtual component library needs to be consistent with the remote virtual platform in terms of usage method and environment, and it should support the dual usage both online and offline, we abandoned the graphical interface and C/S architecture. Instead, we designed and implemented a platform-independent virtual component library which can be applied in general environment including more than 30 common chip types, and Table 1 has exhibited partially of them. Virtual component library and remote virtual experiment platform adopt a unified IDE environment, and run with the same tool. This design combines wire connection experiment with simulation experiment, saves the time for students to learn different development environments, and can effectively improve the efficiency of experimental learning.

**Table 1.** Part chips of virtual component library

| Number | Virtual component | Main function |
|--------|-------------------|---------------|
| 1 | HIT_74LS00 | Four 2-input nand gates |
| 2 | HIT_74LS02 | Four 2-input nor gates |
| 3 | HIT_74LS04 | Hex inverter |
| 4 | HIT_74LS20 | Dual 4-input nand gates |
| 5 | HIT_74LS74 | Dual D flip flop |
| 6 | HIT_74LS83 | 4-bit binary adder with fast carry |
| 7 | HIT_74LS90 | Binary-Pental Counter |
| 8 | HIT_74LS112 | Dual JK flip flop |

With usage of the chips in the virtual component library, learners can observe the experiment results in two ways, simulating with waveform and programming on development board. The procedure of waveform simulation is divided into three steps, compiling simulation files, running simulation operations, and observing simulation waveforms. The advantage of this method is that it can record the whole operation process through simulation files, which is convenient for guidance and network dissemination. The disadvantage is that the phenomenon is not intuitive enough, which is different from the observation of the real experiment.

However, if the experimental design content is implemented logically and downloaded to the FPGA development board, the experimental operation means are completely consistent with the real experiment.

In the real experiment, the experimental operation means is to observe the final phenomenon through the monochrome lamp or digital tube by changing the high/low level of the given input port of the dial switch, and the experimental operation means on the development board are exactly the same.

Figure 3 shows the operation and observation interface of electronic system design using virtual component library and real chips. The two methods are highly consistent in both operation and observation, which can effectively shorten the distance between virtual experiment and real experiment, and greatly increase awareness of students on experimental content.

In addition, virtual component library can also support flexible expansion of user-defined components. In practical use, if it is found that there is a lack of certain components, the experimental instructor can increase the types of components to meet their own needs according to the general process of IP core design under IDE environment. The design process is consistent with the general design process and facilitate to operate.

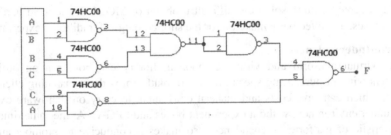

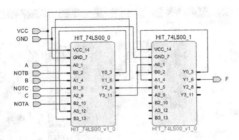

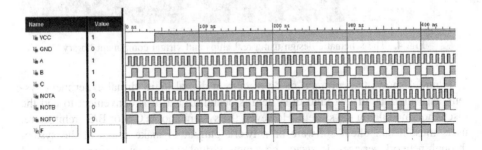

**(a)** Logic schematic diagram with real chips

**(b)** Connection schematic with virtual component library

**(c)** Waveform of experiment result

**Fig. 3.** The schematic design using real chips and virtual component library

## 3.3 Characteristics and Innovation of Virtual Experiments

### Making up for the Shortcomings of Traditional Experiments

By making use of network technology and virtualization technology, remote virtual platform and virtual component library supply a gap of traditional teaching methods. In virtual experimental teaching, we can get rid of the limitation of hardware resources, break through the limitation of time and space in traditional laboratory, adjust the difficulty degree flexibly, solve the difficult problem of MOOC learning in hardware series courses, and effectively make up for the shortcomings of traditional experiments.

**Platform-Independent**

Remote virtual platform and virtual component library can both run on ordinary computers without purchasing special hardware platform, without installing clients or plugins, which can save costs and make it convenient to carry out hardware experiments under environment without experiment boxes and FPGAs. At the same time, the Characteristic of platform independence also makes it conducive to promote among universities and social learners outside universities for lifelong learning.

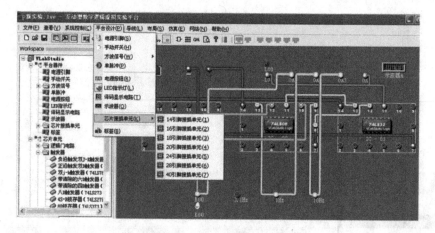

**Fig. 4.** The schematic design using real chips and virtual component library

Figure 4 shows the common mode of current digital logic virtual experiment system. The interface of this kind of system is beautiful, and it is convenient to add the discussion module on the system. However, the system adopts C/S or B/S architecture, users must install client software or network plug-in to make usage of the system through network services. In stead, the remote virtual experiment platform and virtual component library studied in this paper are both available in Vivado universal environments, and it is unnecessary to install any client or plug-in. The platform-independent features have obvious advantages in communication, discussion and remote guidance.

**High Authenticity**

The ultimate carrier of the remote virtual platform is actual FPGA development boards, and the original intention of the virtual component library design is to accord with the real components as far as possible. Despite the convenience of virtualization, both of them are very close to the real experimental means, which facilitates students to switch between virtual experiment and real experiment seamlessly, and enhances their understanding of the experiment.

# 4 Conclusion

On account of the limitations of traditional experiments in time, space, equipment and other aspects, virtual experiment has become a necessary experimental measure in many universities. The remote virtual simulation experiment platform based on FPGA established by our experimental center adopts the design idea of combining local computer virtual simulation software with remote accessible real hardware platform, which has the characteristics of high expansibility and flexibility. It breaks through the time and space limitations of traditional experiments and enables students to design virtual simulation experiments at anytime and anywhere. On the basis of be accordance with to real devices, virtual component library solves the time, apace and equipment constraints of wire connection experiment, and provides strong support for MOOC teaching and mixed teaching of digital logic experiment.

Based on the large-scale programmable logic device technology and Internet of Things technology, the two technologies are highly consistent in the use environment and methods, which truly realizes the full virtualization of digital logic experiments. Hardware experiment virtualization can not only achieve ubiquitous hardware course learning and hardware experiment operation, but also further develop students' self-learning consciousness, and help students improve their ability to design, implement and analyze digital systems for specific problems.

**Acknowledgements.** This research is supported by 2018 Education and Teaching Reform Project of Heilongjiang Province (Exploration and Research on Virtual Experimental Teaching Model of Digital Logic and Digital System Design Course from the Perspective of "Double-first class" Construction) and the first batch of Cooperative Education Projects between Industry and University in 2018 (Reform on Experimental Teaching Content of Digital Logic and Digital System Design Course).

# References

1. Serra, M., Wang, E., Muzio, J.C.: A multimedia virtual lab for digtial logic design. In: MSE 1999 IEEE International Conference on Microelectronic Systems Education (1999)
2. Jacobson, A.: Virtual physics lab close to reality. Comput. Sci. Eng. **5**(4), 3–5 (2003)
3. Morozov, M., Tanakov, A., Gerasimov, A., Bystrov, D., Cvirco, E.: Virtual chemistry laboratory for school education. In: Proceedings of IEEE International Conference on Advanced Learning Technologies, 30 August–1 September 2004, pp. 605–608 (2004)
4. Fang, H., Wu, H., Yang, P., Zhai, C.X.: VIRLab: a web-based virtual lab for learning and studying information retrieval models. In: Proceedings of SIGIR 2014, pp. 1249–1250. ACM (2014)
5. Geigle, C., Lourentzou, I., Sundaram, H., Zhai, C.X.: CLaDS: a cloud-based virtual lab for the delivery of scalable hands-on assignments for practical data science education. In: Proceedings of the 23rd Annual ACM Conference on Innovation and Technology in Computer Science Education, ITiCSE 2018, 2–4 July 2018, Larnaca, Cyprus. pp. 176–181 (2018)
6. Li, S.S., Liu, J.H., Quan, C.B.: Explorations in circuit logic experiment teaching for computer specialty. Exp. Sci. Technol. **14**(2), 115–118 (2016)

7. Jiang, G.C.: Virtual laboratory technology. Beijing Institute of Technology Press, Beijing, pp. 8–16 (2011)
8. Peng, Y.: Research and implementation of digital signal processing virtual lab based on web. Central South University, Hunan (2012)
9. Zhang, Y., Liu, H., Chen, H., Su, X., Lou, J.: Remote virtual experimental platform of hardware courses based on FPGA. Exp. Technol. Manag. **34**(1), 16–20 (2017)

# Teaching Reform of Ideological and Political Education in the Course of Electromagnetic Field and Electromagnetic Wave

Aili Wang[✉], Bo Wang, and Lanfei Zhao

Higher Education Key Lab for Measuring & Control Technology and Instrumentations of Heilongjiang, Harbin University of Science and Technology, Harbin 150001, China
aili925@hrbust.edu.cn

**Abstract.** To strengthen moral education, improve the quality of talent culti- vation, our school carries out "the aggregation action of the ideological and political education in all the courses" to comprehensively improve the quality of personnel training. It strongly encourages the reform of education and teaching of "ideological and political education in the courses" throughout the school, and requires all course teachers to raise their awareness of teaching and edu- cating to explore and build a big ideological education system of the whole staff and all the courses. Taking the teaching of the course "electromagnetic field and electromagnetic wave" as an example, this paper analyzes the necessity and feasibility of carrying out the political education in the courses when profes- sional teachers in colleges teach students' specialty knowledge and train their specialty quality. Simultaneously, some concrete measures were given and the reform of the course teaching and practice were carried out to make the political education run through the course learning.

**Keywords:** Electromagnetic field and electromagnetic wave · Ideological and political education · Reform in education

## 1 Introduction

Under the background of General Secretary Xi Jinping's statement at the National Conference on Ideological and Political Work in Colleges and Universities that "we should persist in taking morality as the central link and make political education run through the whole process of education and teaching", we should promote the organic integration of all kinds of courses with ideological and political education, and tap and enrich the resources of Ideological and political education in all kinds of courses. "Course Ideology and Politics" fully reflects the educational function of each course, the educational responsibility of each teacher, and improves the ability and con- sciousness of moral education of all teachers. It is helpful to change the phenomenon that professional teachers only teach infertile morality and ideological and political education teachers fight alone, and to construct the whole course personality education bureau [1].

R. Mao et al. (Eds.): ICPCSEE 2019, CCIS 1059, pp. 375–381, 2019.
https://doi.org/10.1007/978-981-15-0121-0_28

In view of the teaching practice of the course "Electromagnetic Field and Electromagnetic Wave", while consolidating the students' theoretical knowledge, this paper explores and practices the path and specific measures of Ideological and political education in the course of teaching professional courses from the different dimensions of introduction, detailed explanation of engineering examples and national defense applications, full utilization of 10 min between classes, etc. It also discusses how to carry out ideological and political education in the course of teaching professional courses. Ways and means to let political education run through the whole process of students' learning. In the construction of student-centered political education of electromagnetic field and electromagnetic wave course, teachers must clarify their sense of responsibility and mission, firmly establish the concept of student-centered, highly responsible to students, constantly improve their ideological and moral quality, enrich their knowledge structure, so as to impart theoretical knowledge to students [2]. Through various educational links and teaching processes, students are guided to establish correct outlook on life, world outlook, values and norms of life, to guide students how to behave, do things and learn [3].

"Electromagnetic field and electromagnetic wave" is a compulsory basic professional course for engineering electrical specialty. The basic theory of the course is the necessary knowledge for scientific and technological workers engaged in the field of electrical specialty. Based on the knowledge of electromagnetics in College Physics, and with Maxwell's equations as the core, this course further elaborates the basic concepts, principles, laws and basic analysis methods of electromagnetic field and wave and their application in engineering practice. This course is abstract in concept, with many formulas and strong in theory. It is also recognized as one of the most difficult courses for engineering students. At the same time, the theory involved in this course plays an important role in the frontier disciplines such as optical fiber communication, bioelectromagnetics, electromagnetic compatibility, guided wave optics, microwave technology, and it is also the basis of other frontier disciplines and interdisciplinary disciplines. Through the study of this course, we can master the basic macro-electromagnetic theory and have the ability to analyze and solve the basic electromagnetic field engineering problems. The development of modern science shows that the basic theory of electromagnetic field is the growth point of some interdisciplinary disciplines and the basis of the development of new frontier disciplines. It plays a long-term role in improving students' quality, enhancing their adaptability and creativity.

However, this course is always reflected as a "teacher is difficult to teach, students are difficult to learn" course. Through analysis and summary, we think that there are several reasons for the difficulty of the course: first, the concept of "electromagnetic field and electromagnetic wave" course itself is abstract, there are many formulas, the content of the course involves too many basic theories of mathematics and physics, and it is difficult for students to understand in depth because of the difficulty of introducing and learning; second, schools attach importance to training "applied" talents, and the theoretical basis of class hours is one. It is obviously not an easy thing to be compressed and explained thoroughly in a limited period of time. It is also very difficult for students to fully understand the content of teaching in such a short time. To learn this course, we should not only grasp the basic theory of the course and the methods of solving problems, but also make clear the great role of this course in engineering practice and national defense industry. While teaching curriculum knowledge, teachers must give some examples to guide. Therefore, the selection of engineering practice and national defense application examples is particularly important. In the classroom, teachers should always spread positive energy with a positive attitude, impart correct moral values to students, infect students with their enthusiasm, and influence students' outlook on life through these examples.

In recent years, many domestic colleges and universities have done a lot of research work in the teaching reform of this series of courses, such as Ye Yuhuang's reform of electromagnetic field and microwave technology series courses. Bian Li re-studied the idea of reconstructing and optimizing the teaching contents of electromagnetic field and microwave technology. Zhao Chunhui put forward a preliminary plan for the construction and reform of electromagnetic field and Microwave Technology Series Courses Based on the reform of course system and the optimization of teaching content according to the actual situation of Communication Engineering Specialty. Chen Yu put forward the idea of teaching reform of mobile communication, data communication principle and radar technology for students majoring in communication engineering, with circuit, electric field and radio wave as the main lines.

Taking the teaching of "Electromagnetic Field Theory" as an example, Tian analyzed the necessity and feasibility of carrying out ideological and political education while teaching professional knowledge in class and cultivating students 'professional quality, gave specific measures, and carried out curriculum reform and practical exploration, so as to make political education run through the whole process of curriculum learning [4]. Xi takes electrician and electronic technology course as an example to analyze the necessity of carrying out political education reform in the basic course of technology. From the perspective of front-line teachers in teaching practice, she explores and practices the construction of student-centered "curriculum ideological and political education" from the aspects of teachers' sense of responsibility, teaching design and teaching reform [5].

## 2  Combining and Reconstructing Curriculum Knowledge Points and Integrate Political Content into Curriculum

Based on the application, the actual teaching content should be determined. We should adjust and revise the syllabus and teaching plan properly, weaken the formula deduction in theoretical explanation in class, and focus on practical application, so as to improve students' ability of integrating theory with practice. Based on the principle of "learning for application", each concept is explained in conjunction with common applications, which can not only enable most students to grasp the main content of the course, but also enable students interest in the course to learn and improve in depth. In this process, a very important content is to explain the outstanding contributions and achievements made by Chinese scientists in the field of electromagnetics and their position in the research of electromagnetics in the world, to point out the mission of contemporary college students, to stimulate students' patriotic enthusiasm, and to stimulate students' thirst for knowledge and enthusiasm for learning this course.

The electromagnetic phenomena are simulated and analyzed by using MATLAB, Mathematics and finite element simulation software. The distribution of electromagnetic field in time and space and its changing law are simulated and analyzed, and the parameters are adjusted and modified flexibly so that students can see the electromagnetic field and wave intuitively, and have an intuitive understanding of its physical significance and propagation characteristics, and achieve real-time interaction between teachers and students, which is helpful for students to understand and master the theoretical knowledge of electromagnetic field and wave.

## 3  Introducing Case Teaching Method Combined with Micro-Lesson and Inserting Scientific Research Examples

The case teaching method combined with micro-lessons refers to the use of video or audio production software by teachers to record each case within 10 min. Teachers will systematically explain knowledge points and related cases in class, and then pass the micro-lessons to students. Students can watch the micro-lessons through mobile phones and other terminals at any time in their spare time. For example, in the introduction class, the video is about Lin Weigan, the father of China's microwave, who has persevered in the field of microwave for more than 50 years and made outstanding contributions to the development of China's electromagnetic science. In the lecture on the transmission and polarization of electromagnetic waves, the video gives Beidou satellite navigation system, the most important instrument in China, which shows that China has been in the leading position in the world in rocket technology and space technology. Micro-class has the advantages of short time-consuming and watching at any time. In addition, students are encouraged to find their own application cases of electromagnetic waves in life, and then make micro lessons to share with their classmates.

In the classroom, teachers should always spread positive energy with a positive attitude, impart correct moral values to students, infect students with their enthusiasm, and influence students' outlook on life through these examples. For example, we choose Beidou Satellite Navigation System as an example when we teach the polarization part of electromagnetic wave. Our goal is to build an independent, open, compatible, technologically advanced, stable and reliable satellite navigation system covering the whole world, to promote the formation of the satellite navigation industry chain and improve the national satellite navigation application industry support, promotion and guarantee system to promote the wide application of satellite navigation in all sectors of the national economy and society. This year, we should cover all the countries along the belt and road, and achieve global service capacity around 2020. We should emphasize the important role of Beidou Satellite Navigation System in Wenchuan Earthquake and Beijing Olympic Games. We should also tell students that the interruption of GPS signals in the United States had a significant impact on us. In this way, students can understand why our country has to build the Beidou system at a huge cost, rather than relying entirely on the GPS system of the United States.

For example, the 500-m spherical radio telescope (FAST) in Dawodan depression, Pingtang County, Guizhou Province, built under the auspices of the National Observatory of the Chinese Academy of Sciences, is the largest and most powerful single-antenna radio telescope in the world. FAST extends China's space measure and control capability from geosynchronous orbit to the outer edge of the solar system. It is the most accurate pulsar timing array in the world and can conduct high-resolution microwave patrols. This is the specific application of antenna technology in "electromagnetic field theory". The establishment of FAST shows that our country is in the leading position in many high-tech fields such as antenna design and manufacture, high-precision positioning and measurement, high-quality radio receiver and so on.

Use examples close to life to explain concepts in a general way, so as to improve students' understanding of ideological and political work. For example, in the stage of classical electromagnetic field theory, the law of charge conservation, Coulomb's law, Gauss's law, Faraday's law of electromagnetic induction and Maxwell's equations should be introduced. The representative scientists in this period include Faraday, Maxwell, Hertz and so on. The background at that time and personal life are briefly introduced. In the theoretical stage of computational electromagnetic field, the development background of computational electromagnetics is introduced, which is mainly due to the improvement of the complexity of problem solving and the development of computer technology.

Computational electromagnetics can be divided into forward and inverse problems. The forward problems include high-frequency method and low-frequency method. The inverse problems include linear reconstruction and non-linear reconstruction. The representative scientists include Harrington, Born, Yee and so on. Through this explanation, students will be clear about the development of electromagnetic field theory and the current development situation. In this process, a very important content is to explain the outstanding contributions and achievements made by Chinese scientists and Chinese scientists in the field of electromagnetics and their position in the research of electromagnetics in the world, to point out the mission of contemporary college students, to stimulate their patriotic enthusiasm, and to stimulate their thirst for knowledge and enthusiasm for learning this course.

Explain a practical application scenario of a concept. For example, when explaining "electrostatic field", cases such as electrostatic precipitation, electrostatic spraying, electrostatic copying can be selected; when explaining "Faraday's law of induction", teaching cases such as electronic induction accelerator, motor, generator, maglev train and other application cases can be selected; when explaining "dynamic field", practical application fields such as impulse radar, eddy current phenomenon in electromagnetic oven can be explained. Scene; When explaining "electromagnetic shielding", we can explain the application of electromagnetic shielding in EMC technology; when explaining "electromagnetic wave propagation", we can combine stealth and anti-stealth technology and target recognition technology; when explaining "electromagnetic radiation problem", we can explain it with the development course and principle of phased array radar.

## 4 Diversified Experimental Teaching Reform Combining Simulation Software and Hardware

The simulation software HFSS can solve electromagnetic field and wave quickly and accurately. The content of simulation experiment can be selected according to the requirements of teaching content, and the propagation and spatial distribution of electromagnetic field and wave can be explained with examples. It can make students have a deeper understanding of relevant theoretical knowledge and improve their interest in learning.

The electromagnetic wave comprehensive experimental instrument can complete many experimental projects, including displacement current verification, design and fabrication of electromagnetic wave induction devices, electromagnetic wave propagation characteristics, electromagnetic wave polarization, passive interference and electromagnetic shielding verification. According to the requirement of the syllabus, different projects can be designed. The experiment content can be designed flexibly, which makes the experiment close to the key and difficult points of the course.

The experiment types include demonstration experiment, confirmation experiment, comprehensive experiment and design experiment. The experiment teaching of electromagnetic field and electromagnetic wave with electromagnetic wave comprehensive experimental instrument can well fulfill the requirements of experiment task, enable students to deeply understand and appreciate some important knowledge points, establish image thinking mode, strengthen students' understanding of the process of electromagnetic wave generation, transmission and reception, cultivate students' innovative ability, and help them learn electromagnetic field better. In every routine experiment, we can make full use of the existing experimental instruments to carry out diversified experiments, and then use the theoretical knowledge taught and learned to reasonably explain the experimental phenomena, so that we can turn the simple confirmatory experiment into a heuristic experiment.

## 5 Influencing Students by Teachers' Thoughts and Behaviors

"Virtue is a teacher and integrity is a model". Teachers in Colleges and universities have received good scientific and cultural education and ideological and moral education, and have higher cultural quality and better moral quality. Most teachers have good learning habits, diligent learning attitude and earnest working attitude. If teachers can use a few minutes in class to show students their academic transcripts, narrate their hard work experience and examples of excellent students in past and current students of our school, they can influence and inspire students by telling stories. It can effectively communicate the feelings between teachers and students, and enable students to receive education in happy emotional experience, so as to receive the effect of "kissing their teachers and believing in their ways".

If we ask students not to play with mobile phones, actively exercise, think positively, study hard, play their role in their own jobs in the future, contribute to the construction of the motherland, establish a correct outlook on life, world outlook and values, then teachers should also ask themselves to play less mobile phones in life, exercise actively every day, and care about the country. With the development of military and science and technology, we should strengthen the study of political theory, abide by discipline and laws, standardize our own behavior, and speak positive words in class, and actively disseminate positive energy. Teachers can make some PPTs based on the national flag according to the teaching content, influence students with visual impact, stimulate students' national pride, strengthen patriotism sentiment, and strengthen the belief of the Chinese dream.

# 6 Conclusions

This reform of political education for Electromagnetic Field and Electromagnetic Wave combined teaching knowledge points and improved the connection between theoretical courses and practical teaching throughout. The experiment teaching of electromagnetic field and wave with simulation software and instrument can be abstracted into concrete, dull into vivid, students can change from passive acceptance of knowledge to active exploration and research, and students' learning enthusiasm can be mobilized. Diversification of teaching methods, the introduction of case-based teaching method combined with micro-lessons, naturally interspersed with some of the truth about life, so that students can feel the same and arouse resonance.

# References

1. Yang, H.: From "course of ideological and political education" to "courses for ideological and political education". J. Yangzhou Univ. (High. Educ. Study Ed.) **22**(2), 98–104 (2018)
2. Gu, Y.: Discussion on ideological education reform of aviation medical courses based on OBE concept. Sci. Technol. Vis. 34–35 (2017)
3. Ke, X., Lu, J.: The science education article collects, the path of college English teachers' career development under the background of "ideological and political theory teaching in all courses. **12**(444), 19–21 (2018)
4. Tian, Y., Li, F.: Thinking and discuss of political education in the course of electromagnetic field theory. Electron. Educ. China **3**, 10–14 (2018)
5. Xi, C., Wang, M.: How to do the construction of the student-centered ideological and political education in the course of "electrical and electronic technology". Idel. Polit. Educ. **32**(11), 84–86 (2018)

# Application of Grey Correlation Analysis in Physical Fitness Assessment of Higher Vocational Students

Lei Wang, Yinan Chen, Shijie Cai, and Xia Liu[✉]

Sanya Aviation and Tourism College, Sanya 572000, Hainan, China
252009507@qq.com, 305780161@qq.com, paolo_lx@qq.com

**Abstract.** Taking students from Sanya Aviation and Tourism College as research objects, the relationship between physical course and physical health is discussed with the method of grey relational analysis. The index with the highest degree of correlation between female students and teaching assessment was the index related to the physical core competence exercise, while the indexes with the highest degree of correlation between male students and teaching assessment were the indexes related to the maintenance of strong and healthy body shape and physical core competence. Thus, in the physical education course setting of higher vocational colleges and in the process of the integrated development between their teaching and students' physical health, their teaching mode should aim at students' exercise needs and promoting the quality improvement of their physical education teaching.

**Keywords:** Grey system theory · Physical education teaching · Physical health · Assessment

## 1 Introduction

With the rapid development of modern society, the demands of various industries on talents are getting increasingly higher, especially that the physical and mental health of graduates in higher vocational colleges has attracted more and more attention from employers. At present, physical health has become one of the core competitiveness for the employment of students from vocational college while the physical education teaching of higher vocational colleges is particularly closely related to the development of their students' physical and mental health [1]. The teaching objectives of physical education course in higher vocational colleges are mainly to improve students' physical health, supplemented by mastering sports skills and shaping their physique ability for their future employment. The improvement of PE teaching quality is directly related to the improvement of students' basic physical ability and health level. Therefore, the relationship between physical education teaching and students' physical health is mutually related and influenced.

In recent years, with the development of the national economy and the gradual improvement of people's living conditions, the physical quality of students in higher vocational colleges in China has generally declined significantly in an obvious way [2].

© Springer Nature Singapore Pte Ltd. 2019
R. Mao et al. (Eds.): ICPCSEE 2019, CCIS 1059, pp. 382–389, 2019.
https://doi.org/10.1007/978-981-15-0121-0_29

Yet, With the attention of education departments at all levels, the endurance and aerobic ability of students in higher vocational colleges has been significantly improved, but the strength quality has shown a downward trend year by year. In view of this phenomenon, some higher vocational colleges have taken the assessment of students' physical health as an important part of their physical education teaching, and meanwhile, in order to change the declining trend of students' physical health year by year, Sanya Aviation and Tourism College has carried out targeted and effective reform measures on its physical education teaching.

Grey relational analysis is mainly to study the degree of similarity or difference among the index factors, and then to measure the relational degree among the index factors [3]. Since there are randomness and uncertainty between the physical health and physical education teaching of students in higher vocational colleges, their physical health and physical education teaching are constantly changing on the time axis. Combined with the lack of data and dynamic data of grey theory and due to the characteristics of the academic system in higher vocational colleges, Because of the characteristics of higher vocational education system, this study uses nearly four years of small sample data, and has grey nature. At the same time, it uses grey correlation analysis method to calculate the correlation degree between each index in the system and each index in another system, which is irreplaceable by other methods [4].

## 2    Construction of Assessment Index System

### 2.1    Assessment Indexes of Physical Education Teaching

Through Establishment of Assessment System of Physical Education Teaching Needs multiple indicators. And at the same time, teaching indicators should be reasonable and scientific. Therefore, the establishment of teaching assessment index is an important factor in testing the quality of teaching [5]. This paper selected "pre-teaching (X1)" to assess the pre-teaching preparation and teaching documents, "in-teaching (X2)" to assess the teaching process, teaching organization, teaching method and exercise load arrangement, and "after teaching (X3)" to assess the improvement of students' skills and physical quality.

### 2.2    Evaluation Indexes of Students' Physical Fitness and Health

According to the national standard of students' physical fitness and health, the unified physical fitness test for higher vocational students was carried out. The test items included body shape, body function and physical fitness, specifically covering: height (Y1), weight (Y2), vital capacity (Y3), sit-and-reach stretch (Y4), 50 m run (Y5), standing-broad jump (Y6), girls' 800 m run (Y7), boys' 1000 m run(Y8), girls' sit-ups (Y9), and boys' pull-up (Y10).

## 2.3    Sample Selection

The students' physical health test data from the year of 2014 to the year of 2017 in Sanya Aviation and Tourism College were used as the study sample, including 1500 male students and 1500 female students. The teaching assessment data was derived from the questionnaires for the assessment of physical education teaching quality issued by some higher vocational colleges. The design of the questionnaire in this paper adopted the questionnaire for the assessment of physical education teaching quality in common colleges and universities presented in the Reference No.5. The data was shown in Tables 1 and 2.

**Table 1.** Statistics on the indicators of female students from 2014 to 2017

| Year | 2014 | 2015 | 2016 | 2017 |
|---|---|---|---|---|
| Score $X_1$ | 8.29 | 8.01 | 8.47 | 8.09 |
| Score $X_2$ | 8.61 | 8.56 | 9.17 | 8.11 |
| Score $X_3$ | 7.99 | 8.12 | 8.13 | 8.42 |
| Height $Y_1$ | 164.15 | 164.50 | 165.50 | 164.10 |
| Weight $Y_2$ | 52.30 | 54.25 | 54.30 | 53.81 |
| Vital capacity $Y_3$ | 2887.35 | 2404.00 | 2713.65 | 2995.50 |
| Standing-broad jump $Y_4$ | 167.96 | 165.20 | 166.75 | 174.49 |
| 50 m run $Y_5$ | 9.15 | 9.25 | 9.35 | 9.76 |
| 800 m run $Y_6$ | 269.00 | 273.00 | 272.00 | 254.00 |
| Sit-and-reach stretch $Y_9$ | 15.30 | 16.70 | 18.30 | 12.94 |
| Sit-up for one minute $Y_{10}$ | 30.10 | 28.85 | 31.40 | 29.37 |

**Table 2.** Statistics on the indicators of male students from 2014 to 2017

| Year | 2014 | 2015 | 2016 | 2017 |
|---|---|---|---|---|
| Score $X_1$ | 8.07 | 7.98 | 8.01 | 7.79 |
| Score $X_2$ | 9.12 | 9.12 | 7.99 | 8.73 |
| Score $X_3$ | 8.11 | 8.46 | 9.01 | 8.19 |
| Height $Y_1$ | 173.80 | 174.45 | 174.90 | 174.10 |
| Weight $Y_2$ | 65.20 | 65.90 | 65.20 | 64.26 |
| Vital capacity $Y_3$ | 4207.45 | 3518.90 | 3997.60 | 3854.32 |
| Standing-broad jump $Y_4$ | 225.52 | 218.17 | 218.05 | 225.47 |
| 50 m run $Y_5$ | 7.55 | 7.80 | 7.85 | 7.72 |
| 1000 m run $Y_7$ | 277.00 | 283.00 | 282.00 | 250.00 |
| Pull-up $Y_8$ | 6.65 | 6.10 | 5.50 | 8.19 |
| Sit-and-reach stretch $Y_9$ | 11.45 | 11.55 | 13.60 | 11.55 |

# 3   Grey Relational Analysis

## 3.1   Dimensionless Processing of Assessment Indexes

The method of grey relational analysis was applied to carry out the analysis on the physical education quality and students' physique health in colleges [6]. Due to the different units of each assessment index and to ensure the comparability of each factor, the data in Tables 1 and 2 needed to be dimensionlessly processed. As the data indexes in this paper were divided into forward indexes and reverse indexes, the interval relative value was used to process the original data dimensionlessly. The specific algorithms were as follows.

The positive indexes are:

$$X_{nm1} = \frac{x_{nm} - minx_m}{max\,x_m - min\,x_m}$$

The reverse indexes are:

$$X_{nm2} = \frac{maxx_m - x_{nm}}{max\,x_m - min\,x_m}$$

Among them, $x_{nm}$ denotes the original value of the m index in the n year while $X_{nm}$ refers to the dimensionless value corresponding to the original data. The concrete values are shown in Tables 3 and 4.

**Table 3.** The non-dimensionality of indexes for female students from 2014 to 2017

| Year | 2014 | 2015 | 2016 | 2017 |
|---|---|---|---|---|
| Score $X_1$ | 0.609 | 0.000 | 1.000 | 0.174 |
| Score $X_2$ | 0.472 | 0.425 | 1.000 | 0.000 |
| Score $X_3$ | 0.000 | 0.302 | 0.326 | 1.000 |
| Height $Y_1$ | 0.036 | 0.286 | 1.000 | 0.000 |
| Weight $Y_2$ | 0.000 | 0.975 | 1.000 | 0.755 |
| Vital capacity $Y_3$ | 0.817 | 0.000 | 0.523 | 1.000 |
| Standing-broad jump $Y_4$ | 0.297 | 0.000 | 0.167 | 1.000 |
| 50 m run $Y_5$ | 1.000 | 0.836 | 0.672 | 0.000 |
| 800 m run $Y_6$ | 0.211 | 0.000 | 0.053 | 1.000 |
| Sit-and-reach stretch $Y_9$ | 0.440 | 0.701 | 1.000 | 0.000 |
| Sit-up for one minute $Y_{10}$ | 0.490 | 0.000 | 1.000 | 0.204 |

**Table 4.** The non-dimensionality of indexes for male students from 2014 to 2017

| Year | 2014 | 2015 | 2016 | 2017 |
|---|---|---|---|---|
| Score $X_1$ | 1.000 | 0.679 | 0.786 | 0.000 |
| Score $X_2$ | 1.000 | 1.000 | 0.000 | 0.655 |
| Score $X_3$ | 0.000 | 0.389 | 1.000 | 0.089 |
| Height $Y_1$ | 0.000 | 0.591 | 1.000 | 0.273 |
| Weight $Y_2$ | 0.573 | 1.000 | 0.573 | 0.000 |
| Vital capacity $Y_3$ | 1.000 | 0.000 | 0.695 | 0.487 |
| Standing-broad jump $Y_4$ | 1.000 | 0.016 | 0.000 | 0.993 |
| 50 m run $Y_5$ | 1.000 | 0.167 | 0.000 | 0.433 |
| 1000 m run $Y_7$ | 0.182 | 0.000 | 0.030 | 1.000 |
| Pull-up $Y_8$ | 0.428 | 0.223 | 0.000 | 1.000 |
| Sit-and-reach stretch $Y_9$ | 0.000 | 0.047 | 1.000 | 0.047 |

## 3.2   The Difference Sequence and the Maximum and Minimum Values

The reference sequence should be selected in the comparison of the indexes. This paper carried out a study on the relationship between the teaching quality of physical education in higher vocational colleges and the physical health of students, which meant that the reference sequence was $X_i(k)$. By using the reference sequence to get the difference sequence, then

$$\Delta_i(k) = \left| X_i(k) - Y_j(k) \right|$$

Among them, k represents the year, and I stands for the teaching quality assessment index; $i = 1, 2, 3$; j; represents the test item for students' physical health.
Maximum value:

$$\Delta max = max_i max_i \Delta_i(k)$$

Minimum value:

$$\Delta max = max_i max_j \Delta_i(k)$$

After calculations, $\Delta max = 1$, and $\Delta max = 0$.

## 3.3   Calculation of Correlation Coefficients

The so-called degree of correlation is, in essence, the degree of difference in geometric shapes between curves. Therefore, the value of the difference between curves can be used as a measure of the degree of correlation. The correlation coefficient $\xi(K_i)$ of the comparison sequence and the reference sequence at all times (that is, every point in the curve) is calculated by the following formula.

$$\xi_{ij}(K) = \frac{\Delta min + \rho\Delta max}{\Delta_i(K) + \rho\Delta max}$$

Among them, the resolution coefficient p = 0.5.

## 3.4    Correlation Calculation

The formula for calculating the correlation degree is as follows:

$$R_{ij} = \frac{1}{n}\sum_{k=1}^{n}\xi_{ij}(K)$$

Among them, $i = 1, 2, 3; j = 1, 2, 3\ldots8; n = 1, 2, 3, 4, 5$.

The correlation degree between physical education teaching quality and students' physical health can be obtained by formula, as shown in Tables 5 and 6.

**Table 5.** Correlation degree between teaching quality and physical health of female students

| Item | $Y_1$ | $Y_2$ | $Y_3$ | $Y_4$ | $Y_5$ | $Y_6$ | $Y_9$ | $Y_{10}$ |
|------|-------|-------|-------|-------|-------|-------|-------|----------|
| $R_{x1}$ | 0.707 | 0.559 | 0.644 | 0.588 | 0.565 | 0.565 | 0.723 | 0.937 |
| $R_{x2}$ | 0.829 | 0.597 | 0.494 | 0.498 | 0.660 | 0.469 | 0.896 | 0.804 |
| $R_{x2}$ | 0.665 | 0.631 | 0.680 | 0.752 | 0.435 | 0.743 | 0.462 | 0.485 |
| R | 2.201 | 1.787 | 1.819 | 1.838 | 1.659 | 1.778 | 2.080 | 2.225 |

**Table 6.** Correlation degree between teaching quality and the physical health of male students

| Item | $Y_1$ | $Y_2$ | $Y_3$ | $Y_4$ | $Y_5$ | $Y_7$ | $Y_8$ | $Y_9$ |
|------|-------|-------|-------|-------|-------|-------|-------|-------|
| $R_{x1}$ | 0.633 | 0.712 | 0.694 | 0.538 | 0.605 | 0.384 | 0.428 | 0.597 |
| $R_{x2}$ | 0.446 | 0.610 | 0.625 | 0.733 | 0.767 | 0.562 | 0.612 | 0.365 |
| $R_{x2}$ | 0.861 | 0.576 | 0.518 | 0.399 | 0.488 | 0.498 | 0.494 | 0.879 |
| R | 1.940 | 1.898 | 1.838 | 1.671 | 1.859 | 1.443 | 1.535 | 1.842 |

It can be seen from Table 5 that, in the relationship between physical education quality and female students' physical health in colleges, the correlation degree is as follows: $R_{y10} > R_{y1} > R_{y9} > R_{y4} > R_{y3} > R_{y2} > R_{y6} > R_{y5}$. Therefore, the female students' sit-ups, heights and sit-and-reach stretches are the important factors that affect the assessment of physical education quality of female college students among all the testing indexes of the physical health, while the correlation degree of the indexes of 50 m run and 800 m run ranks in the last two places. According to the analysis of correlation degree, the main purpose of girls' participation in physical exercises is to keep a good body shape, and they shown less favor for the running exercises, while the in-situ exercises are popular with female students because of its small amount of exercise. In Hainan area, due to the hot weather, the ultraviolet radiation is strong. Therefore, the female student's resistance to the running exercise is more obvious.

From Table 6, it can be seen that in the relationship between physical education quality and male students' physical health in colleges, the sequence of the correlation degree is as follows: $R_{y1} > R_{y2} > R_{y5} > R_{y9} > R_{y3} > R_{y4} > R_{y8} > R_{y7}$. Therefore, the male students' height, weight and sit-and-reach stretch are the important factors that affect the assessment of physical education quality of male college students among all the test indexes of the physical health, while standing-broad jump, pull-up and 1000 m run have the least impact on the assessment of physical education quality in colleges. According to the grey relational analysis, compared with girls, boys pay more attention on bodybuilding and the main purpose of participation in physical exercises is to develop their speed and power qualities.

### 3.5  Result Analysis

From the view of grey correlation degree, most of the correlation degree is higher than 0.5, which shows that the physical health test indexes selected in this paper have a high correlation degree with the assessment of physical education quality in colleges and that the physical health of students has a significant impact on the quality of physical education teaching in colleges. Therefore, the statistical data index of the research object selected in this paper is reasonable and scientific.

In the process of reform and innovation of PE teaching and physical fitness future employment, Sanya Aviation and Tourism College pays attention not only to the promotion of students' physical fitness reserve for their future employment through physical education, but also to the promotion of their' physical health level. According to the above analysis, The index of core physical competence which male and female students focus on has the highest correlation with the assessment of teaching quality and the improvement of physical education quality have high correlation with the needs of students' exercises and bodybuilding. In other words, the physical education in higher vocational colleges, based on the improvement of students' physical fitness for their future employment, can not only improve the quality of teaching, but also meet the needs of students' physical exercises.

## 4  Conclusion and Recommendations

### 4.1  Conclusion

The indicators of male and female students emphasizing physical core competence development have the highest correlation with the evaluation of teaching quality. Male students pay more attention to maintain their body shape through bodybuilding, while female students are more resistant to the speed exercise. In addition, in the process of the integrated development of physical education teaching and students' physical fitness, the teaching mode aimed at satisfying students' exercise needs is an important way to improve the quality of physical education teaching.

## 4.2   Recommendations

By using grey relational analysis, this paper analyzed the relationship between physical education teaching and students' physical health in Sanya Aviation and Tourism College, and found that the teaching quality of public physical education still has the following problems: First, although it has relatively complete sports hardware facilities, the use of the sites and the mining of students' interest in sports are not sufficient; second, the public physical education course setting are not in consistent with students' needs for fitness and exercise.

Therefore, in view of the above problems existing between the physical education teaching and students' physical health, it is suggested to solve these problems from the following two aspects:

(1) Enrich campus sports culture. Students should be offered with guidance to take more exercises in their daily life and integrate physical exercises from class into life. Also, they should be encouraged to participate actively in extracurricular or out-of-school physical exercises and competitions as part of the assessment of their performance in physical education. Meanwhile, those who receive awards in the extracurricular and out-of-school competitions can take such achievements or awards as their corresponding physical education achievements.

(2) The public physical education curriculum should be set up to meet the needs of students' physical training, and the teaching mode of public physical education should be reformed actively. In the teaching of public physical education, "modular" teaching should be adopted for the purpose of students' fitness exercise and diversified and personalized teaching methods should be formulated. The teachers should be aimed at improving the students' skill level and practical operation ability in the course of class and ultimately realize those goals.

**Acknowledgment.** Project supported by the Education Department of Hainan Province, project number: Hnjg2019-129.

# References

1. Fang, D.: Research on the development of physical health education in colleges and universities: a review on the research of curriculum of physical education and health under quality education. Contemp. Sci. Educ. **15**, 26 (2015)
2. Lu, H.: Exploration on the reform of physical education work in colleges and universities. J. Sports Cult. **11**, 86–89 (2011)
3. Liu, S., Dang, Y., Fang, Z., et al.: Grey System Theory and Its Application. Beijing Science Press (2010)
4. Zhang, E., Wang, C.: Correlation analysis on the integration of culture industry and tourism industry in Shaanxi Province. J. Xi'an Technol. Univ. **34**, 53–56 (2014)
5. Ye, L., Tang, L.: Research on the construction of physical education assessment system in colleges and universities based on cloud theory under the background of big data. J. Yunnan Normal Univ. (Nat. Sci. Ed.) **5**(3), 73–78 (2016)
6. Ye, X., Xia, Z., Zhou, M., Hu, W., Jiang, Z.: A study on the correlation assessment between the teaching quality of physical education and students' physical health. J. Kunming Univ. **40** (03), 115–119+2 (2018)

# Empirical Research on the Status Quo of Teachers of Common Required Courses in Vocational Higher Education Institutions

Zhipeng Ou[1], Dan Ren[1], Yinan Chen[1], and Mingrui Chen[2(✉)]

[1] Sanya Aviation and Tourism College, Sanya 572000, Hainan, China
hainanozp@126.com, 340459345@qq.com, 305780161@qq.com
[2] Information and Technology School, Hainan University,
Haikou 570228, Hainan, China
1607885098@qq.com

**Abstract.** In order to build a high-quality team of teachers in vocational higher education institutions, and investigate the current situation of teachers of common required courses in vocational higher education institutions of Hainan Province, an empirical research is conducted. A questionnaire survey was carried out in aspects of satisfaction of teaching and scientific research, professional quality, work saturation and teacher's job satisfaction. The research results shows that the professional quality of teachers of common required courses cannot significantly influence the satisfaction of teaching and scientific research; Job saturation can positively influence teaching and research satisfaction; teaching and research satisfaction can positively influence teachers' job satisfaction.

**Keywords:** Vocational higher education institutions ·
Common required courses · Teacher · Empirical research

As a significant part of higher education, higher vocational education aims to cultivate high-skilled talents for the first-line positions of production, construction, management and service. The talent cultivation quality in higher vocational college is not only related to the existence and development of a school, but also influences the social and economic development of a country. It has become an instant historical mission for higher vocational education to speed up the cultivation of high-quality and high-skilled talents. Strengthening the construction of teaching staff is not only a special requirement for higher vocational education, but also an inevitable demand for the times as well as social and economic development. "Some Advice on Deepening the Educational Reform in Vocational Education and Improving the Quality of Talent Cultivation (exposure draft)" conducted by Ministry of Education has clearly mentioned "Reinforce and improve the teaching of basic socializing courses. Strengthen the cultural and basic education while guaranteeing the cultivation quality of students' technical skills. On the basis of teaching programme (curriculum standards) issued by Ministry of Education, vocational schools should provide abundant, complete and first-class basic socializing courses for students, including moral education, Chinese, mathematics, English, history, sports and health, art, basics and professional theories of computer application." Thus, the status of public courses in higher vocational colleges has been

© Springer Nature Singapore Pte Ltd. 2019
R. Mao et al. (Eds.): ICPCSEE 2019, CCIS 1059, pp. 390–404, 2019.
https://doi.org/10.1007/978-981-15-0121-0_30

intensified by the state. Thus, the construction of teaching staff for public courses is especially important, and the professional development of teaching staff for basic socializing courses is the key to improve the teaching quality and cultivate talents' social adaptation ability in higher vocational education.

At present, the systematic study on teaching staff construction of public courses in higher education institutions is far from many, and these researches are mainly conducted from the following aspects, including the structure of teaching staff [1, 2], teachers' academic competence [3], teachers' basic quality [4], job satisfaction [5–8], professional development [9–11], job burnout [12, 13], etc. A questionnaire survey was conducted on teachers of public courses in higher vocational colleges in Haikou, Sanya and Qionghai, in order to know the current situation of these teachers in Hainan.

# 1 Descriptive Analysis of Data Statistics

As it is shown in Table 1, among the 258 test samples, 158 participants are females, accounting for 61.2%. 149 participants are aging 31 to 40, accounting for 57.8%. About 139 participants have undergraduate education background, while 109 participants have postgraduate education background, accounting for 96.1% in total; 147 participants have master's degree, accounting for 57%. 121 teachers get lectureship as professional qualification, accounting for 46.9%. The participants are mainly responsible for the teaching of college English, other specialized courses, ideological and political course, computer, and the number of teachers for these courses are respectively 59, 51, 51 and 40, accounting for 78% in total. 211 participants are full-time teachers, accounting for 81.8%; 102 participants are double-professionally-titled teachers, accounting for 39.5%. In recent three years, 133 participants published 1–3 papers, accounting for 51.6%; 164 participants are not the leading researcher of any teaching and scientific projects, accounting for 63.6%; 137 teachers participate in 1 or 2 research projects, accounting for 53.1%. In the recent year, the average weekly class hours for 82 participants are between 8 and 12, and 73 participants are between 13 and 16, accounting for 28.3% in total. 208 participants adopt information-based teaching, accounting for 80.6%; 163 participants think it is difficult for public courses teachers to get a raise in liberal wages and benefits, as well as get promotion, training and learning opportunities, accounting for 63.2%.

The frequency statistics of multiple choices are shown in Table 2. Among the 208 samples adopting information-based teaching, the main information-based teaching method used in class is mixed teaching method, about 71.60% of the teachers choosing this option, followed by on-line and off-line teaching, about 42.30% of the teachers choosing this option. The rest options are also chosen by some teachers. Thus, the teaching methods used by teachers of public courses in Hainan province have transformed from traditional to information-based and diversified. The information-based teaching methods are also diversified, and this may have something to do with the age of teachers. As more than half of the teachers for public courses are young and middle-aged, and it is relatively easy for them to master information technology, especially mixed teaching method, which is the main part of information-based teaching method in higher vocational colleges.

**Table 1.** Frequency statistics (N = 40)

| Variables | Attribute | Frequency | Percentage (%) |
|---|---|---|---|
| Gender | Male | 100 | 38.8 |
| | Female | 158 | 61.2 |
| Age | <30 | 40 | 15.5 |
| | 31–40 | 149 | 57.8 |
| | 41–50 | 53 | 20.5 |
| | 51–60 | 13 | 5 |
| | >60 | 3 | 1.2 |
| Education background | Undergraduate | 139 | 53.9 |
| | Postgraduate | 109 | 42.2 |
| | Doctoral candidate | 10 | 3.9 |
| Degree | Bachelor | 101 | 39.1 |
| | Master | 147 | 57 |
| | Doctor | 10 | 3.9 |
| Professional qualifications | Professor | 11 | 4.3 |
| | Associate professor | 54 | 20.9 |
| | Lecturer | 121 | 46.9 |
| | Teaching assistant | 41 | 15.9 |
| | No qualification | 31 | 12 |
| The course teachers are responsible for | Computer basic course | 40 | 15.5 |
| | Ideological and political course | 51 | 19.8 |
| | Advanced mathematics | 6 | 2.3 |
| | College English | 59 | 22.9 |
| | Physical education | 24 | 9.3 |
| | Other basic courses | 27 | 10.5 |
| | Specialized course | 51 | 19.8 |
| Full-time teacher | Yes | 211 | 81.8 |
| | No | 47 | 18.2 |
| Double-professionally-titled teacher | Yes | 102 | 39.5 |
| | No | 156 | 60.5 |
| Number of papers published in recent 3 years | None | 57 | 22.1 |
| | 1–3 | 133 | 51.6 |
| | 4–5 | 42 | 16.3 |
| | >5 | 26 | 10.1 |

*(continued)*

**Table 1.** (*continued*)

| Variables | Attribute | Frequency | Percentage (%) |
|---|---|---|---|
| Number of teaching and scientific research projects presided in recent 3 years | None | 164 | 63.6 |
| | 1–2 | 87 | 33.7 |
| | 3–4 | 3 | 1.2 |
| | >4 | 4 | 1.6 |
| The number of teaching and scientific research projects participated in recent 3 years | None | 79 | 30.6 |
| | 1–2 | 137 | 53.1 |
| | 3–4 | 31 | 12 |
| | >4 | 11 | 4.3 |
| Average weekly teaching hours in the recent year | <8 h | 65 | 25.2 |
| | 8–12 h | 82 | 31.8 |
| | 13–16 h | 73 | 28.3 |
| | >16 h | 38 | 14.7 |
| Using information-based teaching | Yes | 208 | 80.6 |
| | No | 50 | 19.4 |
| Liberal wages and benefits, opportunities for promotion, training and further-study | Relatively difficult | 163 | 63.2 |
| | Average | 87 | 33.7 |
| | Relatively easy | 8 | 3.1 |

**Table 2.** Frequency statistics of multiple choices (N = 208)

| Variables | Attribute | Frequency | Percentage |
|---|---|---|---|
| Main information-based teaching method | MOOC | 40 | 19.20% |
| | SPOC | 5 | 2.40% |
| | Micro-lecture | 53 | 25.50% |
| | Flipped class | 51 | 24.50% |
| | On-line and off-line teaching | 88 | 42.30% |
| | Mixed teaching | 149 | 71.60% |

The description statistics of the current situation of teaching staff is shown in Table 3. All the items are rated on a 5-point format, with 3 points the median. As it is shown in the table, participants think that the number of teachers as well as the teaching and scientific research funds for public courses are relatively in short supply; teachers aging 30 to 45 are relatively in large quantity; the working environment for teachers of public courses is relatively good, but the policy environment for them to publish a paper is not so satisfactory; the further-study and training opportunities for teachers of public courses are relatively few, and their salary and social benefits are relatively low; the possibility for teachers to apply for research project related to public courses is relatively low; teachers are not satisfied with the reward system for publishing a researching paper, no matter it is a paper related to educational reform or scientific research.

**Table 3.** Description statistics of the current situations of public courses teaching group

| Items | Option for 1 point | Option for 5 points | Average | Standard deviation |
|---|---|---|---|---|
| Number of teachers | Very few | Very many | 2.86 | 0.777 |
| Working environment | Very bad | Very good | 3.07 | 0.674 |
| Frequency of further-study | Very low | Very high | 2.55 | 0.841 |
| Salary and social benefits | Very low | Very high | 2.66 | 0.711 |
| Number of teacher aging 30 to 45 | Very few | Very many | 3.5 | 0.8 |
| Expenditure for teaching and scientific research of public courses | Very little | Very much | 2.5 | 0.795 |
| Difficulty in applying for a research project related to public courses | Very difficult | Very easy | 2.39 | 0.778 |
| Policy environment to publish a paper | Very bad | Very good | 2.58 | 0.82 |
| Reward system for publishing a paper related to educational reform | Very unsatisfied | Very satisfied | 2.72 | 0.737 |
| Reward system for publishing a paper related to scientific research | Very unsatisfied | Very satisfied | 2.74 | 0.726 |

# 2 Reliability and Validity Test

Factor analysis is used to respectively analyse teaching and scientific research satisfaction, professional quality, working saturation, and teachers' job satisfaction. By using single factor test, Tables 4 and 5 report the first factor analysis results of each variable. Only the explained variance of teachers' job satisfaction reaches 57.762%, while that of the rest variables is less than 50%. The factor loading value for teaching and scientific research satisfaction 2 and 4, professional quality 3 and 6, working saturation 1 and 6 are all less than 0.5. So, these items are excluded and factor analysis is used on the rest items again. In the second factor analysis, teaching and scientific research satisfaction 1 and 3 are also excluded due to their low explained variance.

**Table 4.** Result of the first factor analysis

| Variables | KMO value | Bartlett's test | | | Characteristic root | Explained variance (%) |
|---|---|---|---|---|---|---|
| | | Approximate chi-square | df | p | | |
| Teaching and scientific research satisfaction | 0.599 | 143.819 | 15 | 0.000 | 1.892 | 31.528 |
| Professional quality | 0.714 | 238.163 | 15 | 0.000 | 2.282 | 38.036 |
| Working saturation | 0.751 | 297.245 | 15 | 0.000 | 2.468 | 41.140 |
| Job satisfaction | 0.798 | 420.356 | 10 | 0.000 | 2.888 | 57.762 |

Tables 6 and 7 report the final results of factor analysis. All the items of teaching and scientific research satisfaction are excluded except Item 5 and 6, thus factor analysis will not be used on this variable again. As it is shown in Table 6, the KMO value of professional quality, working saturation and teachers' job satisfaction are all above 0.6, and Bartlett's test reaches 0.001, suggesting factor analysis can be used on these samples. The number of factors with the characteristic root above 1 is 1, and the explained variance of each single factor are all above 50%. As for factor loading, each index is above 0.5, suggesting that the single factor structure of professional quality, working saturation and teachers' job satisfaction is good, that is, the construct validity is good.

**Table 5.** Factor loading of first factor analysis

| Index | Factor loading | Index | Factor loading | Index | Factor loading | Index | Factor loading |
|---|---|---|---|---|---|---|---|
| Teaching and scientific research satisfaction 6 | 0.689 | Professional quality 2 | 0.806 | Working saturation 2 | 0.841 | Teachers' job satisfaction 4 | 0.819 |
| Teaching and scientific research satisfaction 5 | 0.665 | Professional quality 1 | 0.792 | Working saturation 3 | 0.757 | Teachers' job satisfaction 5 | 0.789 |
| Teaching and scientific research satisfaction 3 | 0.628 | Professional quality 5 | 0.704 | Working saturation 4 | 0.753 | Teachers' job satisfaction 2 | 0.759 |
| Teaching and scientific research satisfaction 1 | 0.611 | Professional quality 4 | 0.586 | Working saturation 5 | 0.698 | Teachers' job satisfaction 3 | 0.718 |
| Teaching and scientific research satisfaction 2 | 0.453 | Professional quality 3 | 0.386 | Working saturation 6 | 0.358 | Teachers' job satisfaction 1 | 0.711 |
| Teaching and scientific research satisfaction 4 | 0.058 | Professional quality 6 | −0.13 | Working saturation 1 | −0.067 | | |

Reliability test is used on teaching and scientific research satisfaction, professional quality, working saturation and teachers' job satisfaction, and Table 8 reports the result. As it is shown in Table 8, the Cronbach's $\alpha$ coefficient of teaching and scientific research satisfaction, professional quality, working saturation and teachers' job satisfaction are all above 0.6, suggesting that the reliability is good.

**Table 6.** Result of the final factor analysis

| Variables | KMO value | Bartlett's test | | | Characteristic root | Explained variance (%) |
|---|---|---|---|---|---|---|
| | | Approximate chi-square | df | p | | |
| Professional quality | 0.706 | 213.882 | 6 | 0.000 | 2.182 | 54.549 |
| Working saturation | 0.759 | 268.820 | 6 | 0.000 | 2.383 | 59.564 |
| Job satisfaction | 0.798 | 420.356 | 10 | 0.000 | 2.888 | 57.762 |

**Table 7.** Factor loading of final factor analysis

| Index | Factor loading | Index | Factor loading | Index | Factor loading |
|---|---|---|---|---|---|
| Professional quality 2 | 0.811 | Working saturation 2 | 0.843 | Teachers' job satisfaction 4 | 0.819 |
| Professional quality 1 | 0.791 | Working saturation 3 | 0.775 | Teachers' job satisfaction 5 | 0.789 |
| Professional quality 5 | 0.732 | Working saturation 4 | 0.763 | Teachers' job satisfaction 2 | 0.759 |
| Professional quality 4 | 0.602 | Working saturation 5 | 0.699 | Teachers' job satisfaction 3 | 0.718 |
| | | | | Teachers' job satisfaction 1 | 0.711 |

**Table 8.** Results of reliability test

| Variables | Cronbach's α coefficient | Index number |
|---|---|---|
| Teaching and scientific research satisfaction | 0.601 | 2 |
| Professional quality | 0.717 | 4 |
| Working saturation | 0.771 | 4 |
| Job satisfaction | 0.816 | 5 |

# 3  Correlation Analysis

Description statistics and correlation analysis are used to analyse the four variables including teaching and scientific research satisfaction, professional quality, working saturation and teachers' job satisfaction, and Table 9 reports the result. The result of description statistics shows that the average value of the second and the third variables is above 3, suggesting that the participants' professional quality, and job satisfaction are relatively high. The average value of the first variable is lower than 3, suggesting that the participants' teaching and scientific research satisfaction is relatively lower.

The average value of working saturation is quite close to 3, suggesting that the working saturation of the participants is in medium level. The result of correlation analysis shows that in the three variables including professional quality, working saturation and job satisfaction, there are significant correlation between each two variables ($p < 0.001$). Teaching and scientific research satisfaction has significant correlation respectively with working saturation and job satisfaction ($p < 0.001$). But there is no significant correlation between teaching and scientific research satisfaction and professional quality ($p > 0.05$).

**Table 9.** Description statistics and correlation analysis

|  | Teaching and scientific research satisfaction | Professional quality | Working saturation | Job satisfaction |
|---|---|---|---|---|
| Teaching and scientific research satisfaction | 1 |  |  |  |
| Professional quality | 0.088 | 1 |  |  |
| Working saturation | 0.321*** | 0.372*** | 1 |  |
| Job satisfaction | 0.366*** | 0.351*** | 0.586*** | 1 |
| Mean value (M) | 2.548 | 3.450 | 3.001 | 3.063 |
| Standard deviation (SD) | 0.738 | 0.518 | 0.620 | 0.583 |

Note: *stands for $p < 0.05$, **stands for $p < 0.01$, ***stands for $p < 0.001$.

## 4 Structural Equation Model

Structural equation model is built to test the relationship among professional quality, working saturation, teaching and scientific research satisfaction, job satisfaction. As the modified index (MI) of professional quality 4 and 5, teaching and scientific research satisfaction 5 and 6, teachers' job satisfaction 3 and 4 are relatively high, correlation is set on each residual respectively to modify the model, and the final structural equation model is shown as Fig. 1.

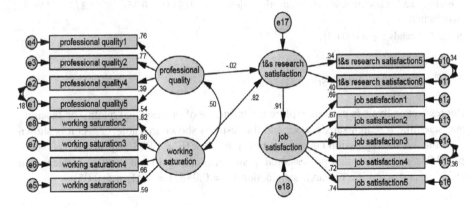

**Fig. 1.** Structural equation model

The main fit index of each item in this structural equation model is shown in Table 10. X2/df = 2.473 < 3, RMSEA = 0.076 < 0.08, RMR = 0.039 < 0.05, GFI = 0.909 > 0.9, IFI = 0.905 > 0.9, CFI = 0.902 > 0.9, all of them are inside the fitting range, suggesting that this structural equation model is acceptable.

**Table 10.** Global fitting result of structural equation model

| Model fitting index | Critical value | Research model | Fitting judgement |
|---|---|---|---|
| $X^2$/df | <3 | 2.473 | Yes |
| RMSEA | <0.08 | 0.076 | Yes |
| RMR | <0.05 | 0.039 | Yes |
| GFI | >0.9 | 0.909 | Yes |
| IFI | >0.9 | 0.904 | Yes |
| CFI | >0.9 | 0.902 | Yes |

Table 11 reports the path coefficients of this structural equation model. As it is shown in the result, professional quality has no significant positive influence on teaching and scientific research satisfaction ($\beta$ = −0.015, t = −0.179, p > 0.05), but working saturation has a significant positive influence on teaching and scientific research satisfaction ($\beta$ = 0.821, t = 3.999, p < 0.001); teaching and scientific research satisfaction has a significant positive influence on teachers' job satisfaction ($\beta$ = 0.913, t = 4.273, p < 0.001).

**Table 11.** The Path coefficient of structural equation model

| Influence path | $\beta$ | S.E. | C.R. | P |
|---|---|---|---|---|
| Professional quality→teaching and scientific research satisfaction | −0.015 | 0.069 | −0.179 | 0.858 |
| Working saturation→teaching and scientific research satisfaction | 0.821 | 0.152 | 3.999 | *** |
| Teaching and scientific research satisfaction→job satisfaction | 0.913 | 0.351 | 4.273 | *** |

Note: ***stands for p < 0.001.

## 5    Regression Analysis

Regression analysis is used to explore the influence of professional quality on teaching and scientific research satisfaction, and the result is shown in Table 12. As it is shown in Table 12, the regression equation is not significant (F = 1.988, adjusted R2 = 0.004, p > 0.05), suggesting that professional quality has no significant positive influence on teaching and scientific research satisfaction ($\beta$ = 0.088, t = 1.41, p > 0.05).

**Table 12.** Regression analysis of professional quality to teaching and scientific research satisfaction

|  | Nonstandard coefficient B | Standard error (S.E.) | Standard coefficient $\beta$ | t | p |
|---|---|---|---|---|---|
| (Constant quantity) | 2.117 | 0.31 |  | 6.838*** | 0.000 |
| Professional quality | 0.125 | 0.089 | 0.088 | 1.41 | 0.16 |
| R2 | 0.008 |  |  |  |  |
| Adjusted R2 | 0.004 |  |  |  |  |
| F | 1.988 |  |  |  |  |

Regression analysis is used to explore the influence of working saturation on teaching and scientific research satisfaction, and the result is shown in Table 13. As it is shown in Table 13, the regression equation is significant (F = 29.387, adjusted R2 = 0.099, p < 0.001), suggesting that working saturation has a significant positive influence on teaching and scientific research satisfaction ($\beta$ = 0.321, t = 5.421, p < 0.001).

**Table 13.** Regression analysis of working saturation to teaching and scientific research satisfaction

|  | Nonstandard coefficient B | Standard error (S.E.) | Standard coefficient $\beta$ | t | p |
|---|---|---|---|---|---|
| (Constant quantity) | 1.401 | 0.216 |  | 6.481*** | 0.000 |
| Working saturation | 0.382 | 0.071 | 0.321 | 5.421*** | 0.000 |
| $R^2$ | 0.103 |  |  |  |  |
| Adjusted $R^2$ | 0.099 |  |  |  |  |
| F | 29.387*** |  |  |  |  |

Regression analysis is used to explore the influence of teaching and scientific research satisfaction on teachers' job satisfaction, and the result is shown in Table 14. As it is shown in Table 14, the regression equation is significant (F = 39.646, adjusted R2 = 0.131, p < 0.001), suggesting that teaching and scientific research satisfaction has a significant positive influence on teachers' job satisfaction ($\beta$ = 0.366, t = 6.297, p < 0.001).

**Table 14.** Regression analysis of teaching and scientific research satisfaction to teachers' job satisfaction

|  | Nonstandard coefficient B | Standard error (S. E.) | Standard coefficient $\beta$ | t | p |
|---|---|---|---|---|---|
| (Constant quantity) | 2.326 | 0.122 |  | 19.097 | 0.000 |
| Teaching and scientific research satisfaction | 0.289 | 0.046 | 0.366 | 6.297 | 0.000 |
| $R^2$ | 0.134 |  |  |  |  |
| Adjusted $R^2$ | 0.131 |  |  |  |  |
| F | 39.646*** |  |  |  |  |

## 6 Check Analysis of Mesomeric Effect

Taking infrastructure, teaching situation and public course as independent variables, reward for scientific research as mediator, and job satisfaction as dependent variable, regression analysis is conducted to test the mesomeric effect of reward for scientific research between the independent variables and dependent variable. The result is shown in Table 15.

**Table 15.** Check analysis of mesomeric effect

| Variables | Model 1 (job satisfaction) | | Model 2 (reward for scientific research) | |
|---|---|---|---|---|
|  | $\beta$ | T | $\beta$ | T |
| Infrastructure | .281*** | 4.965 | .516*** | 8.087 |
| Teaching situation | .159** | 2.994 | −.070 | −1.049 |
| Public course | .307*** | 5.903 | .231*** | 3.619 |
| Reward for scientific research | .155** | 3.130 |  |  |

As it is shown in Table 15, in model 1, regression analysis is conducted with infrastructure, teaching situation, public course and reward for scientific research as independent variables, and job satisfaction as dependent variable, and the result suggests that significant difference exists between each two variables. As it is shown in the result, infrastructure can significantly influence job satisfaction ($\beta = .281^{***}$), that is, the better the infrastructure is, the higher the job satisfaction will be; teaching situation can significantly influence job satisfaction ($\beta = .159^{**}$), that is, better teaching situation can obviously lead to higher job satisfaction; public course can significantly influence job satisfaction ($\beta = .307^{***}$), that is, the more abundant the public course is, the higher the job satisfaction will be; reward for scientific research can significantly influence job satisfaction ($\beta = .155^{**}$), that is, more reward for scientific research can obviously lead to higher job satisfaction.

In model 2, regression analysis is conducted with infrastructure, teaching situation, public course as independent variables, and reward for scientific research as dependent variable. As it is shown in the result, infrastructure has a positive significant influence on reward for scientific research ($\beta = .516^{***}$), that is, the better infrastructure can obviously lead to better reward for scientific research; public course can significantly and positively predict reward for scientific research ($\beta = .231^{**}$), that is, more abundant public course can significantly lead to better reward for scientific research.

As it is shown in the result, scientific research experience has no mesomeric effect between teaching situation and job satisfaction, while it has mesomeric effect among infrastructure, public course and job satisfaction. Figure 2 reports the result.

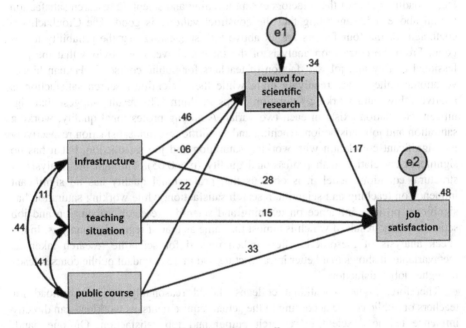

**Fig. 2.** Mesomeric effect

## 7  Conclusion

The quality of teaching staff is a direct decisive factor for talent cultivation quality, and the teaching staff of public courses plays a significant role in improving students' artistic appreciation and moral quality, and preparing students a solid foundation for the study of specialized courses. As it is shown in the description of current situation, females take up more than a half of all the participants, and about half participants are undergraduates with medium-grade professional title; 39.5% of the participants are double-professionally-titled teachers, and this percentage is a little bit low; the percentage of the teachers presiding over or participating in teaching and scientific research projects is low. All the data suggests that the teaching staff of public courses in

Hainan province is relatively weak, and teachers of public courses need to improve their education background and work hard to get double-professionally-titled. Teachers should be encouraged to preside over or participate in all kinds of scientific research projects, thus to improve their scientific research ability, to integrate teaching with scientific research. Moreover, higher vocational colleges should take public courses teachers' workload into consideration, provide opportunities for them to improve their education background, and improve their salary and social benefits according to the actual requirement, thus to motivate teachers to cherish their posts and devote wholeheartedly to work.

Factor analysis is conducted on the four factors including teaching and scientific research satisfaction, professional quality, working saturation, job satisfaction, and the factor loading value of these factors except teaching and scientific research satisfaction are all above 0.5, suggesting that the construct validity is good. The Cronbach's $\alpha$ coefficient of the four factors are all above 0.6, suggesting that the reliability test is good. From the correlation analysis of the factors above, it is obvious that the professional quality and job satisfaction of teachers for public courses in Hainan higher vocational colleges are relatively high, while their scientific research satisfaction is relatively low, and working saturation keeps medium. The results suggest that significant correlation exists in each two variables among professional quality, working saturation and job satisfaction; teaching and scientific research satisfaction respectively has significant correlation with working saturation and job satisfaction, but it has no significant correlation with professional quality ($p > 0.05$). Through the analysis of structural equation model, it is obvious that professional quality has no significant influence on teaching and scientific research satisfaction, while working saturation has significant positive influence on teaching and scientific research satisfaction and job satisfaction, the result of which is almost the same as that of regression analysis. In the check analysis of mesomeric effect, when reward for scientific research taken as metavariable, it shows that better infrastructure and more abundant public courses leads to higher job satisfaction.

Therefore, higher vocational colleges should reasonably assign workload for teachers of public course according to the actual requirements, as workload can directly influence teachers' scientific research output and job satisfaction. On one hand, workload can influence teachers' income; on the other hand, overload makes teachers have no or little time to do scientific research. Moreover, overload can also affect teachers' daily life. Working saturation, job satisfaction, scientific research satisfaction can greatly influence public courses teachers' professional development, so the relationship among the three factors should be attached importance to. As scientific research has great influence on teachers' professional title evaluation and promotion, offering young and middle-aged teachers scientific research guidance and help is especially important, and proper reward for scientific research is helpful to motivate teachers' enthusiasm to take part in scientific research. In order to cultivate a group of high-quality teaching staff in higher vocational college, the administrators should fully consider teachers' professional development, reasonably use the factors including professional quality, teachers' workload, and scientific research workload, create nice teaching environment, provide good teaching facilities, exploit different kinds of public courses, take teachers' need into full consideration. Only by doing these can higher vocational colleges build "double-professionally-titled" teaching staff in a better and more stable way, and improve their level of school management.

Under the background of "Notice of the State Council on Issuing the Implementation Plan of Reforming Vocational Education", the government attaches great importance to vocational education, which makes a higher request on the teaching staff in vocational education institutions. Each vocational education institutions should take different measures to build a group of "double-professionally-titled" teachers, explore how to build a group of creative and high-quality structural teaching staff, organize teachers to cooperate and make their own contribution in modular teaching [14]. Meanwhile, higher vocational colleges should make full use of this historic opportunity, clearly understand the particularity of teaching staff training, preferably promote teachers' professional development, meet teachers' needs, complete both the contents and procedure design of teaching staff training, supply further-study and training opportunities for teachers of public courses, create nice working environment, and formulate policies which is helpful to teachers' development. Moreover, teachers of public courses in higher vocational colleges have a strong desire in professional title promotion, and their teaching and scientific research satisfaction not only influence their job satisfaction, but also influence their development of professional career and job burnout, as well as the build and development of high-quality teaching staff.

# References

1. Yang, C., Sun, Z., Bi, H.: Strengthening the construction of teaching staff for basic socializing courses in order to adapt to higher education development—a research report on the construction of teaching staff for basic socializing courses in Liaoning higher education institutions. Liaoning Educ. Res. **04**, 29–31 (2002)
2. Li, Y.: A survey on the current situation of teaching staff for basic socializing courses in Sichuan higher education institutions and the research of its countermeasures. Acad. J. Sichuan Normal Univ. (Soc. Sci.) **30**(1), 59–65 (2003)
3. Liu, X.: An reflection on academic development and teaching competence of teachers in Guangxi higher vocational colleges. Educ. Vocat. **10**, 83–86 (2017)
4. Gu, M.: The current difficulty and realization route of developing young teachers' teaching and academic competence in higher education institutions. Contemp. Sci. Educ. **11**, 65–68 (2018)
5. Wang, Q.: Quality requirements of teachers in basic socializing courses in vocational-technique schools. China Adult Educ. **24**, 79–80 (2012)
6. Meng, H., Yang, L.: The construction of teaching staff for basic courses in higher vocational colleges from the perspective of constructivism. Educ. Vocat. **12**, 72–73 (2013)
7. Liu, B.: An Theoretical and Empirical Research on Teachers' Working Hours in Higher Education Institutions. Capital University of Economics and Business, Peking (2014)
8. Meng, J., Liu, S., Song, T.: Acceleration or obstruction: an empirical research of the influence of a school organization structure on teachers' job satisfaction. Contemp. Educ. Manag. **12**, 79–84 (2018)
9. Lin, J., Li, J.: The influence of college teachers' scientific research pressure on scientific research performance—the intermediary and regulating effect of job satisfaction and emotional intelligence. J. Fujian Normal Univ. (Philos. Soc. Sci.) **06**, 72–78 (2018)

404    Z. Ou et al.

10. Tang, Y.: The professional development of basic courses teachers in higher vocational colleges from the perspective of NYP Model. Educ. Vocat. **10**, 85–86 (2013)
11. Zhu, S., Ning, S., Ren, S.: An reflection of teachers' professional development in private higher education institution from the perspective of ecology of education. J. Zhejiang Shuren Univ. (Hum. Soc. Sci.) **19**(01), 20–24 (2019)
12. Huang, Y.: A Research on the Hardship of the Demonstration Construction of Teaching Staff in Higher Vocational Colleges—Take College D in Shanghai As an Example. East China Normal University, Shanghai (2014)
13. Ji, H.: A diagnosis of teachers' job burnout of basic socializing courses in Jiangsu. J. Nanjing Univ. Posts Telecommun. (Soc. Sci.) **13**(4), 104–108 (2011)
14. The state council: Notice of the State Council on Issuing Implementation Plan of Reforming Vocational Education [EB/OL]. http://www.gov.cn/zhengce/content/2019-02/13/content_5365341.htm. Accessed 14 Feb 2019

# Evaluation Model of Teaching Quality Based on Cloud Computing

Xiaofeng Li[1(✉)] and Qiushi Wang[2]

[1] Department of Information Engineering, Heilongjiang International University,
Harbin 150025, China
mberse@126.com
[2] Development and Planning Office, Heilongjiang International University,
Harbin 150025, China

**Abstract.** In the teaching quality evaluation, due to the influence of various human factors, the teaching quality evaluation system is presented with some grey characteristics. Traditional evaluation methods cannot solve the problem effectively. Through gray relational analysis, the grey system of partial information known and unknown can be analyzed. It is difficult for the qualitative description language evaluation index to use traditional mathematical model for accurate representation. In this paper, cloud model is combined with grey relational analysis, and an evaluation method is proposed based on the cloud-grey relational analysis, in order to realize the mutual transformation between the qualitative information and quantitative numerical. Three university teachers' information was collected by questionnaire survey, and the teaching quality was evaluated through the evaluation model based on grey relational analysis built in MATLAB. The results were compared with those of fuzzy comprehensive evaluation and the BP neural network evaluation. The results show that the teaching quality evaluation model based on cloud grey relational analysis has certain rationality and feasibility.

**Keywords:** Cloud model · Teaching quality · Gray relational analysis

## 1 Introduction

As the reform of educational system deepens, the teaching quality evaluation has become an important part of improving teaching quality. Its evaluation methods are diversified, including student symposium, after-school interview, online evaluation and questionnaire. Each evaluation method has its own characteristics. Among them, the most common ones are online evaluation and questionnaire method [3, 8]. These two methods are carried out based on the corresponding evaluation questionnaire. In the teaching evaluation system, the factors that affect the teaching quality do not have a definite quantitative relation. Therefore, in the questionnaire, it is difficult to describe the precisely numerical evaluation contents, so the qualitative language is adopted. Besides, gray relational analysis [7] needs to be based on the accurate survey data, and the results are obtained by the relevant mathematical calculation [10, 11].

© Springer Nature Singapore Pte Ltd. 2019
R. Mao et al. (Eds.): ICPCSEE 2019, CCIS 1059, pp. 405–416, 2019.
https://doi.org/10.1007/978-981-15-0121-0_31

On this basis, this paper combines the qualitative quantitative interchanging model-cloud model with gray relational analysis, and thus builds the teaching evaluation model based on cloud-gray relational analysis. The model integrates many of the same type of language values into a more generalized language value by taking advantage of the comprehensive cloud theory, complete the comprehensive evaluation of teaching quality by using grey relation theory as reference, and realizes the interchanging between qualitative information of language value and the quantitative quantity of precise value, which has solved difficulty of using the traditional mathematical model to accurate express through the qualitative language description evaluation index.

## 2  Teaching Quality Evaluation Model Based on Cloud - Grey Relational Analysis

### 2.1  Cloud Theory

In the mathematics field space, the cloud is a scalable, no-edge and one-to-many mathematical map. Supposing U is the quantitative domain and C is the qualitative concept in the domain U, if $x \in U$ is a stochastic implementation of C, and the degree of x determination of C $\mu(x) \in [0, 1]$ satisfies $\mu : U \to [0, 1]$, $\forall x \in U, x \to \mu(x)$ then the distribution of x in the domain U is called the cloud, denoted as cloud C(X) [2, 6, 9]. The overall property of the qualitative concept can be characterized by the digital feature $(Ex, En, He)$ of the cloud. Where, $Ex$ (Expected value) is the expected value of qualitative concept in the domain, which is entirely under the concept of this concept. $En$ (Entropy) is the measurement of the qualitative concept's fuzziness, which is called entropy. It indicates the range of values that can be accepted by this concept. The larger the $En$ is, the more fuzzy the concept is. $He$ (Hyper entropy) is the uncertainty measurement of $En$, which is called super entropy. The larger the $He$ is, the larger the dispersion degree is, and the greater the thickness of the cloud is.

In the cloud model, the interchanging between the qualitative concept and the quantitative value is realized through the cloud generator. Normal cloud generator achieves the interchanging from the qualitative concept to quantitative value, whereas the reverse cloud generator achieves the interchanging from the quantitative value to qualitative concept.

---

**Algorithm1** The concrete algorithm of forward cloud generator

---

**Input:** $Ex, En, He$ and the number of cloud drops to generate n

**Output:** n cloud drops $x_i$ and its determination $\mu_i$.

1. The expected value is $En$ , the variance is a normal random number $En_i' = NORM(En, He^2)$ for $He^2$.

2. The expected value is $Ex$ , the variance is a normal random number $x_i = NORM(En, He^2)$ for $En_i'^2$.

3. Equation (1) gives the determination $\mu_i$ of $x_i$ .

$$\mu_i = e^{\frac{(x_i - Ex)^2}{2En_i'^2}} \tag{1}$$

Where, the determination $\mu_i$ of $x_i$ as a cloud droplet, repeat steps 1-3 until n cloud drops are generated.

---

---

**Algorithm2** The concrete algorithm of reverse cloud generator

---

**Input:** The quantitative value $x_i$ of n cloud drops.

**Output:** A qualitative concept $Ex, En, He$ expressed by n cloud drops

1. Calculated cloud drops $x_i$ the sample mean $\overline{X}$ and sample variance $S^2$ , it is shown in Equation (2).

$$\overline{X} = \frac{1}{n}\sum_{i=1}^{n} x_i, S^2 = \frac{1}{n-1}\sum_{i=1}^{n}(x_i - \overline{X})^2 \tag{2}$$

2. Calculated expectation value $Ex$ , it is shown in Equation (3).

$$Ex = \overline{X} \tag{3}$$

3. Calculated entropy $En$ , it is shown in Equation (4).

$$En = \sqrt{\frac{\pi}{2}}\frac{1}{n}\sum_{i=1}^{n}|x_i - Ex| \tag{4}$$

4. Calculated hyper entropy $He$ , it is shown in Equation (5).

$$He = \sqrt{S^2 - En^2} \tag{5}$$

---

## 2.2 Teaching Quality's Comprehensive Evaluation Based on Cloud-Gray Relational Analysis

Cloud-Relational Analysis is a new evaluation method that combines the advantages of Gray Relational Analysis and the cloud model. The implementation steps are as follows:

### 2.2.1 Determine the Cloud Model of Teaching Evaluation Level

According to the domain classification of each evaluation level in Table 1, the numerical characteristics $Ex, En, He$ corresponding to five levels are calculated according to the method in Table 2. Table 2 calculates the numerical characteristics of the teaching evaluation level cloud, as shown in Table 3.

**Table 1.** Teacher's teaching quality evaluation grade corresponding table.

| Evaluation grade | Range of value |
|---|---|
| Excellent | 90–100 |
| Good | 80–90 |
| Medium | 70–80 |
| Qualified | 60–70 |
| Unqualified | 0–60 |

**Table 2.** Digital characteristic and range relation of the cloud model

| Division of domain | $U_1 \in [a_1, b_1]$ | $U_i \in [a_i, b_i]$ | $U_m \in [a_m, b_m]$ |
|---|---|---|---|
| $Ex$ | $a_1$ | $(a_i + b_i)/2$ | $b_m$ |
| $En$ | $(b_1 - a_1)/3$ | $(b_i - a_i)/6$ | $(b_m - a_m)/3$ |
| $He$ | constant | constant | constant |

**Table 3.** Teaching evaluation grade cloud model

| Grade of teaching evaluation | Division of domain | Digital feature |
|---|---|---|
| Excellent | [90, 100] | (100, 10/3, 0.05) |
| Good | [80, 90] | (85, 10/6, 0.05) |
| Medium | [70, 80] | (75, 10/6, 0.05) |
| Qualified | [60, 70] | (65, 10/6, 0.05) |
| Unqualified | [0, 60] | (0, 20, 0.05) |

### 2.2.2   Determine the Weight of All Indexes in the Secondary Index Layer

First of all, according to the teaching quality evaluation index system, carry out the qualitative evaluation over all indexes in the secondary index layer as per five-level scoring criterion; quantify the qualitative description information of $n$ score givers over $m$ secondary indexes through the normal cloud generator, and get the corresponding quantitative value $x_{ij}$, in [4], determine the weight of secondary index $w_j$ based on the obtained quantitative value $x_{ij}$.

### 2.2.3   Determine the Cloud Model of All Indexes in the Primary Index Layer

According to cloud model of the secondary index and its corresponding weights, this paper uses the integrated cloud technology [5], and gets cloud model of all primary indexes according to Eq. (6). Suppose that in the teaching quality evaluation index system, the "teaching method" is one of the primary indexes, the "teaching method" is divided into $q$ secondary indexes, and $q$ secondary indexes correspond to the cloud model $C_1(Ex_1, En_1, He_1), \ldots, C_q(Ex_q, En_q, He_q)$.

The corresponding weight is $w_j$. A cloud model generated by a cloud model of the q secondary index, is a cloud model for the first level index. Calculation formula of digital feature $Ex, En, He$ is shown in Eq. (6).

$$
\left|
\begin{aligned}
Ex &= \frac{Ex_1 \times En_1 \times w_1 + Ex_2 \times En_2 \times w_2 + \ldots + Ex_q \times En_q \times w_q}{En_1 \times w_1 + En_2 \times w_2 + \ldots + En_q \times w_q} \\
En &= En_1 \times w_1 + En_2 \times w_2 + \ldots + En_k \times w_k \\
He &= \frac{He_1 \times En_1 \times w_1 + He_2 \times En_2 \times w_2 + \ldots + He_q \times En_q \times w_q}{En_1 \times w_1 + En_2 \times w_2 + \ldots + En_q \times w_q}
\end{aligned}
\right.
\tag{6}
$$

### 2.2.4 Determine the Cloud Model of the Target Layer's "Teaching Quality" to Be Evaluated

According to the determination method of primary index cloud model, the cloud model of the target layer's "teaching quality" be evaluated could be obtained in a similar way.

### 2.2.5 Solve Relation Coefficient and Correlation Between the Target Layer's "Teaching Quality" Be Evaluated and the Teaching Evaluation Level

1. According to the numerical features of the target layer's "teaching quality", produce n cloud drops $x_i$ and their corresponding deterministic $\mu_i$ through the normal cloud generator. The calculation results are recorded as $\mu_i = [\mu_1, \mu_2, \ldots, \mu_n]$.
2. By taking n cloud drops ix produced based on Step 1 as the input of five teaching evaluation level cloud models of Sect. 2.2.1, this paper calculates the determination degree of n cloud drops $x_i$ relative to five teaching evaluation levels through the normal cloud generator. The calculation results are recorded as

$$
\mu_{ki} =
\begin{bmatrix}
\mu_{11} & \mu_{12} & \cdots & \mu_{1n} \\
\mu_{21} & \mu_{22} & \cdots & \mu_{2n} \\
\mu_{31} & \mu_{32} & \cdots & \mu_{3n} \\
\mu_{41} & \mu_{42} & \cdots & \mu_{4n} \\
\mu_{51} & \mu_{52} & \cdots & \mu_{5n}
\end{bmatrix}.
$$

3. Select the determination degree $\mu_i$ of the target layer's "teaching quality" as the reference sequence, select five determination degrees $\mu_{ki}$ of the known teaching evaluation level as the comparison sequence, and solve the gray relation coefficient $\xi_{ki}$ of $\mu_i$ to be evaluated for $\mu_{ki}$ according to Eq. (7).

$$
\xi_{ij} = \frac{\min_i \min_j |x'_{0j} - x'_{ij}| + \rho(j) \max_i \max_j |x'_{0j} - x'_{ij}|}{|x'_{0j} - x'_{ij}| + \rho(j) \max_i \max_j |x'_{0j} - x'_{ij}|}.
\tag{7}
$$

4. According to the obtained correlation coefficient $\xi_{ki}$, get the relation $r_k$ between the target layer's "teaching quality" of the evaluation level and the five known teaching evaluation levels according to Eq. (8).

$$r_k = \frac{1}{m} \sum_{j=1}^{m} \xi_{ij} \tag{8}$$

### 2.2.6 Determine the Evaluation Levels of Teaching Quality

The greater the gray correlation $r_k$ is, the closer the target layer's "teaching quality" to be evaluated to the kth level among five known levels. Therefore, based on the principle of maximum selection, solve maximum value $r_t = \max\{r_k\}$ of the correlation $r_k$, so the teaching evaluation level $t$ corresponded to the maximum correlation $r_i$ is the final teaching quality level.

## 3 Application of Cloud - Grey Relation Analysis Method in Teaching Quality Evaluation

To facilitate the comparison, this paper selects specific example from Literature [1] for analysis. Its evaluation index system is divided into 4 primary indexes and 18 secondary indexes, and four-level scoring criterion is selected. Based on the value range of the four levels, the numerical characteristics corresponding to each evaluation level calculated from Table 1 are: "Excellent" = A1(95, 10/3, 0.05), "Good" = A2(85, 10/3, 0.05), "Medium" = A3(75, 10/3, 0.05), "Poor" = A4(65, 10/3, 0.05).

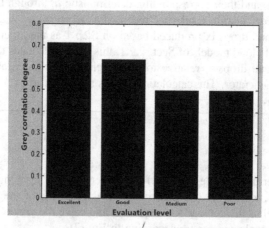

**Fig. 1.** The relationship between the teaching quality of the target layer and the level of teaching evaluation

According to the evaluation of the teaching quality evaluation questionnaire in this literature, the teaching evaluation model based on cloud- grey relational analysis proposed in the present study is adopted to evaluate the classroom teaching quality. The evaluation results are shown in Fig. 1. In Fig. 1, the final teaching evaluation level is "excellent". This evaluation result is consistent with the result in the original literature.

Therefore, it is feasible to apply the cloud-gray relational analysis model to the teaching quality evaluation for obtaining more accurate evaluation result.

# 4 Experiment and Analysis of Teaching Quality Evaluation

## 4.1 Data Selection

In the process of teaching quality evaluation, in order to get more highly reliable final evaluation results, it is essential to guarantee the authenticity of the original data. Due to various limitations such as conditions, time and other factors, this paper fails to get a large number of existing student evaluation data. Therefore, the information related to actual teaching situations can only be obtained by investigating students. The questionnaire method is an investigation means most frequently adopted in investigation, because it is an investigation method based on certain purposes to formulate the corresponding questionnaires and then collect the relevant information of respondents through the questionnaire.

### 4.1.1 The Preparation of Questionnaires

In the process of questionnaire formulation, the questions can be divided into three types - the structural type (there are fixed answer options to questions), the open type (there are no optional answers, and the questions will be freely answered by respondents), and the semi-structural type (between the structural type and the open type). The questionnaire used in this study is semi-structural type, that is, according to select five-level scoring standard, the respondents are required to select the level corresponding to each secondary evaluation index from multiple scoring levels based on actual teaching situations of evaluation objects (teachers), and give the corresponding evaluation scores as per the scoring interval of the level; the evaluation score reflects respondents' satisfaction with teachers.

### 4.1.2 The Implementation of Questionnaire Survey

The questionnaire survey was conducted in September 2016, which selected three literature and history, science and engineering teachers (numbered A, B and C) from a normal university as evaluation objects and respectively selected students of such three teachers as respondents. Besides, 150 questionnaires were distributed and collected on the site, with the recovery rate of 100%. In addition to few invalid questionnaires that were not filled or filled in an unsatisfactory manner, there are 138 valid questionnaires, with the efficiency up to 92%.

### 4.1.3 Survey Data Processing

Due to different dimensions, numerical levels and manifestation forms of all evaluation indexes in the teaching quality evaluation index system, it is hard to compare and analyze them. Therefore, these original data must be normalized. In the linear dimensionless method, the threshold method has no strict requirements on the number of index data and its distribution, and it has relatively little raw data in the process of dimensionless processing. As a result, this paper selects the minimization dimensionless method in the threshold method to normalize the required data, and the results are shown in Fig. 2. In the Fig. 2, 19 secondary indexes of the teaching evaluation index system are in the horizontal coordinate, and the values obtained by data dimensionless processing are in the longitudinal coordinate. The three histograms corresponding to each secondary index are the results of dimensionless processing of the average of teacher A, teacher B and teacher C on the basis of the index scoring.

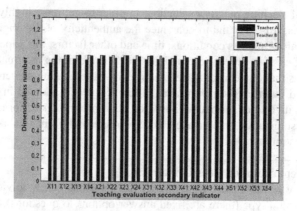

**Fig. 2.** Results of dimensionless processing

## 4.2    Determination of Weight

By combining with the dimensionless processed data and following the reference sequence selection principle in the given cloud-grey relational analysis in Sect. 2.2.2, this paper determines the reference sequence $x_0 = \{1, 1, 1, 1, 1, 1, 1, 1, 1, 1, 1, 1, 1, 1, 1, 1, 1, 1, 1\}$

and the comparison sequence of $x_{ij} = \begin{bmatrix} 0.9366 & 0.9635 & 0.9702 & 0.9623 \\ 0.9643 & 0.9660 & 0.9932 & 1.0000 \\ 1.0000 & 1.0000 & 1.0000 & 0.9954 \end{bmatrix}$ and inputs

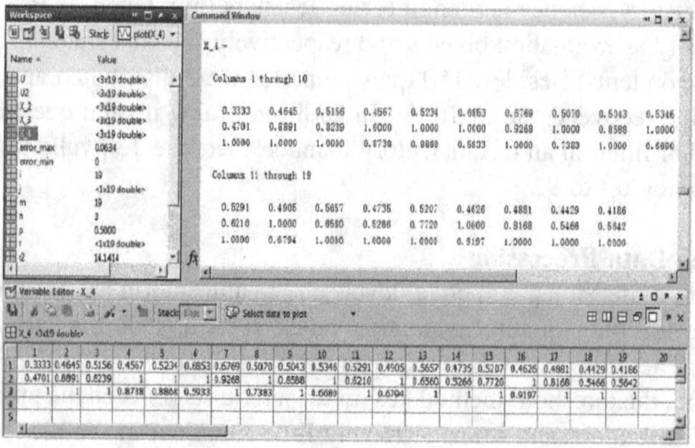

**Fig. 3.** The relation coefficient between comparison sequence and reference sequence

the reference sequence and comparison sequence into MATLAB. Afterwards, according to the established teaching quality evaluation model based on cloud-grey relational analysis, this paper obtains the relation coefficient between the reference sequence and comparison sequence, correlation and weights of all evaluation indexes. It is shown in Fig. 3.

## 4.3 Evaluation Result Analysis

In the example of present study, the semi-structural questionnaire method is adopted, the corresponding evaluation scores are given by respondents according to the score interval of evaluation level, and it is unnecessary to conduct quantitative conversion by using the cloud model, therefore, this example is analyzed by selecting the teaching evaluation model based on grey relational analysis. In MATLAB, the main algorithm for the teaching quality evaluation model based cloud - grey relational analysis is as follows:

---

**Algorithm3** Teaching quality evaluation model based cloud - grey relational analysis

**Input:** $Ex, En, He$ and the number of cloud drops to generate n

**Output:** n cloud drops $x_i$ and its determination $\mu_i$.

1. For j=1: m;  a=0;
2. For i=1: n;  a=X_3(i, j)+a;
3. X_33 (j)=a/n;
4. For j=1: m; p1 (j) = X_33(j)/error_max;
5. For j=1: m;
6. If (error_max>0 && error_max<=2*X_33(j)) p(j)=0.9;
7. Else if (error_max>2*X_33 (j) && error_max<=3*X_33 (j)) p(j)=2*p1(j);
8. Else error_max>3*X_33 (j) p(j)=1.5*p1(j);
9. For i=1: m
10. For j=1: n
11. X_4 (i, j) = (error_min+p(j)*error_max)/(X_3(i,j)+p(j)*error_max);
12. r =X_4* $w$ ;
13. End.

---

According to the evaluation model of teaching quality based on cloud - grey relational analysis, the teaching evaluation results of A, B and C are shown in Figs. 4, 5 and 6. $r$ is the relation between teacher's score and evaluation level.

The correlation between teachers' A scores and each evaluation grade is (0.9367, 0.9443, 0.7828, 0.6686, 0.4426), where, the maximum correlation degree $r\_max$ is 0.9443, and the corresponding evaluation level of the maximum correlation $k$ is 2. Therefore, the evaluation level of teacher A is good. It is shown in Fig. 4. The correlation between teachers' B scores and each evaluation grade is (0.9637, 0.8883, 0.7483, 0.6465, 0.4380), where, the maximum correlation degree $r\_max$ is 0. 9637, and the corresponding evaluation level of the maximum correlation $k$ is 1. Therefore, the evaluation level of teacher B is excellent. It is shown in Fig. 5. The correlation between teachers' C scores and each evaluation grade is (0.9795, 0.8568, 0.7285,

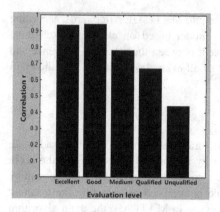

**Fig. 4.** The evaluation results of teacher A    **Fig. 5.** The evaluation results of teacher B

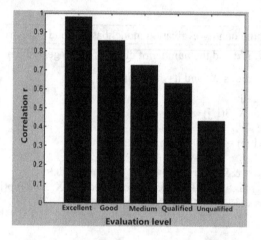

**Fig. 6.** The evaluation results of teacher C

0.6337, 0.4353), where, the maximum correlation degree $r\_max$ is 0.9795, and the corresponding evaluation level of the maximum correlation $k$ is 1. Therefore, the evaluation level of teacher C is excellent. It is shown in Fig. 6.

In the same data environment, the teaching quality of three teachers was analyzed by fuzzy comprehensive evaluation method and BP neural network evaluation method respectively. The evaluation results are shown in Table 4.

Table 4 shows that based on the five levels of "excellent, good, medium, qualified and unqualified", this paper obtains the fuzzy comprehensive evaluation vector (0.7002) of teacher A by the fuzzy comprehensive evaluation method, and the evaluation level of teacher A is obtained to be "excellent" according to the principle of maximum membership principle, and his evaluation result is 89.9541 by using BP neural network evaluation method; according to the values corresponding to all evaluation levels in Table 4, the evaluation level of teacher A is "excellent"; the evaluation level of teacher A is "good" through the Grey Relational Analysis. In combination with

**Table 4.** Comparison of teaching quality evaluation result

| Evaluation object | Evaluation method | Evaluation result | Evaluation level |
|---|---|---|---|
| Teacher A | BP neural network evaluation | 89.9541 | Good |
| | Fuzzy comprehensive evaluation | (0.7002 0.2456 0.0531 0.0009 0) . | Excellent |
| | Cloud-grey relational analysis | (0.9367 0.9443 0.7828 0.6686 0.4426) | Good |
| Teacher B | BP neural network evaluation | 91.9902 | Excellent |
| | Fuzzy comprehensive evaluation | (0.8573 0.1314 0.0111 0 0) | Excellent |
| | Cloud-grey relational analysis | (0.9637 0.8883 0.7483 0.6465 0.4380) | Excellent |
| Teacher C | BP neural network evaluation | 92.3811 | Excellent |
| | Fuzzy comprehensive evaluation | (0.8835 0.0977 0.0187 0 0) | Excellent |
| | Cloud-grey relational analysis | (0.9795 0.8568 0.7285 0.6337 0.4353) | Excellent |

the questionnaire, the score obtained by teacher A is almost within the range of (87, 91), so the evaluation results obtained by BP neural network evaluation method and cloud-gray relational analysis are more consistent with the actual situation. However, the cloud-gray relational analysis is characterized by simple structure and no strict requirements over the number of data and distribution rules, which has overcome the deficiencies of BP neural network evaluation method requiring a lot of data for the network training, as well as network learning or memory instability. In addition, cloud-gray relational analysis possesses the unique advantages of dealing with cloud-gray system, so that it can analyze and process the teaching quality evaluation system in a better way.

## 5 Conclusion

In this paper, on the basis of the assessment information for qualitative description language, for teaching quality evaluation system of gray feature, combined with cloud model to qualitative concept and quantitative uncertainty between numerical conversions, a teaching quality evaluation model based on cloud - grey relational analysis is proposed. The application of this method in teaching evaluation is analyzed and verified by concrete examples in MATLAB. The final analysis results show that the teaching quality evaluation method based on cloud-grey relational analysis has certain rationality and is a new feasible method for teaching evaluation.

# References

1. Hu, S., Jiang, X.: Evaluation method of cloud model of teacher's classroom teaching quality. J. Wuhan Univ. (Philos. Soc. Sci. Ed.) **60**(3), 455–460 (2010)
2. Li, D., Du, Yi.: Uncertainty in Artificial Intelligence, pp. 143–147. National Defence Industry Press, Beijing (2015)
3. Liao, M., Yuan, J.L.: Construction of computer laboratory system based on cloud platform. Comput. Educ. **12**, 150–154 (2017)
4. Liu, S.: Grey System Theory and Its Application, pp. 1–4. Science Press, Beijing (2010)
5. Luo, S., Zhang, B.M.: Comprehensive evaluation of image map quality based on cloud model. Surv. Mapp. Sci. **33**(3), 44–46 (2008)
6. Ren, H., Yan, Y.G.: Evaluation and research on the cooperative partner of giant project organization based on cloud model and grey relational degree method. J. Civil Eng. **44**(8), 147–152 (2011)
7. Sun, X.: Several Decision-Making Methods and Applications Based on Grey Relational Analysis, pp. 2–5. Qingdao University (2016)
8. Wang, L.: Research on application of cloud classroom network assisted teaching platform based on cloud computing technology. Technol. Commun. **9**(22), 137–138 (2011)
9. Zhou, Y., Li, B.: An early warning method for online public opinion based on cloud model. J. Intell. **31**(8), 861–874 (2011)
10. Xuesheng, G., Cheng, H., Yuping, Z.: Analysis of mixed teaching quality evaluation model. J. Hubei Correspondence Univ. **31**(7), 29–131 (2018)
11. Yan, F., Liping, M.: An evaluation model of teaching quality in colleges and universities based on optimized BP neural network. Stat. Decis.-Making **34**(2), 80–82 (2018)

# Study on Digital Exchange Platform of University Information Resources

Jiancai Wang[✉], Jianting Shi, Yanjing Cui, and Jiuyang Hou

Hei Longjiang University of Science and Technology,
Harbin 150022, Hei Longjiang, China
154539860@qq.com

**Abstract.** Digital exchange platform is a comprehensive service platform. In this paper, the construction scheme of the platform is proposed. By constructing the data exchange platform, a people-oriented information environment was built to meet demands of faculty, staff and students on information resources. And thus a digital campus was constructed to provide efficient online services and information resources. The experimental results show that the platform is user-friendly, runs orderly, integrates the service, and provides a good reference value for the data integration of enterprises and universities in the future.

**Keywords:** Digital exchange platform · Data exchange · Heterogeneous business systems

## 1 Introduction

Educational informatization is the new requirement of information society for education and the inevitable trend of the development of educational modernization [1]. With the advancement of digital campus construction and the wide application of information system in colleges and universities, the dependence of teaching, scientific research and management on information system is constantly enhanced, and the quality requirements of information service are greatly improved. But under the condition of the absence of unified data standards and construction norms, the existing application system can only solve the business processing and information sharing of the department within the college, which has not yet been done on information sharing and integration across departments, business and applications [2]. Also, between different business systems, there is data inconsistency and the sharing degree is not high very much, so it is difficult to bring into playing real-time and accurate features of informatization, which seriously hinders the further construction of digital campus.

Data exchange platform is a comprehensive service platform proposed to solve various data exchange and sharing needs in colleges and universities [3]. It can realize the information interaction between its heterogeneous business systems, providing a practical solution for data sharing among existing internal systems, and can effectively guide the establishment of new systems. Meanwhile, it also provides a sample of the available data exchange platforms for other industries. Therefore, the construction of data exchange platform project has very important practical significance.

R. Mao et al. (Eds.): ICPCSEE 2019, CCIS 1059, pp. 417–421, 2019.
https://doi.org/10.1007/978-981-15-0121-0_32

## 2  Construction Project

Data center is the integration of multiple application systems of colleges and universities, which can guarantee the uniqueness and authority of college information and provide accurate data basis for college decision-making [4]. The data center construction of our college has foundation. For example, the business system of our university has considered the problem of system integration in the future when purchasing, and takes the national university information standard as a condition for the selection of information system to standardize the system interface. In addition, the existing business system of our college has been used for many years and has accumulated a large number of business data. In the construction of data exchange platform, therefore, we need to make our school information standard, and build the data exchange and sharing platform under the guidance of the standard, and then we combed and analyzed the existing data resources of our college to form the data center, formed the theme database of each functional domain, and finally built a data exchange platform for information resources comprehensive services.

### 2.1  Construction Goals

Taking Hei Longjiang University of science and technology as an example, according to the relevant national standards and the standards of the ministry of education, combined with the characteristics of our college and the status quo of informationization, the formation of information standards is in line with the school, so as to achieve the standardization of information. At the same time, based on the integration and exchange of various information resources in the college, construct a supporting environment of open and collaborative operation which meets the requirements of the overall management and service of the school to provide a digital exchange platform for the management of the university.

### 2.2  Specific Implementation Steps

#### Develop Basic Standards for Information Resource Management
The basic standard of information resource management is the foundation of information resource development and utilization and information system construction [5]. For the planning of information resources in colleges and universities and the establishment of shared database, it is difficult to make effective work without establishing basic standards of information resource management [6]. The ministry of education has issued 'the standards for educational management informatization' [7]. Taking our university as an example, our own data standards have been established by referring to the national, ministry of education and relevant industry standards, which can provide basis and standard for related work. The basic standards of information resource management include information standard, authority standard, exchange standard and public data standard [8].

**Establishment of a Unified Data Exchange Platform**

The construction and implementation of the unified data exchange platform will be able to solve the problem of sharing and integration of heterogeneous data between application systems (including Oracle, SQL server, mysql, db2, etc.) [9]. providing global data view, global data access view and improving data exchange service for the application system of our college, which can effectively solve the problem of information island [10].

The specific operation steps are as follows:

1. based on the specific business database (such as personnel database), taking the needs of other departments into account, unify basic information in the data format and coding rules, and develop data dictionary and unify data exchange rules;
2. Determine the authoritative source of data;
3. Establish a central database based on the authoritative source of data and shared information required by other departments;
4. According to the development of the data dictionary and unified data exchange rules to build a unified data exchange process between the source database and the target database.

According to the actual situation and requirements of our school, the input data in the unified data exchange platform are Shared data generated by the information system of various business departments. The source of a certain type of Shared data is unique, that is, such data can only be generated from a certain management information system, and the corresponding data in other management information systems must be consistent with the authoritative data. For example, both the personnel system and the scientific research system have the basic information of teachers, but only the information in the personnel system is authoritative information. The information in the scientific research system must be consistent with the personnel system.

**Theme Database Construction**

Information standards have been formulated and a data exchange and sharing platform has been established. On this basis, subject databases in various functional areas conforming to the standards have been formed. Divide the subject database according to the functional area, store and manage the data according to the subject, and realize the integration and orderliness of the data. According to information resource planning, the database of business topics can be divided into eight categories: students (XS), teaching (JX), human resources (RS), financial library (CW), scientific research (KY), equipment assets (SB), alumni (XY), and others. In addition, considering the requirements of data sharing and exchange between functional areas, the database is logically divided into 10 large databases.

As shown in Table 1:

**Table 1.** Topic database partitioning

| The serial number | Subject library | Number of topics |
|---|---|---|
| 1 | Students subject | Multiple |
| 2 | Teaching topic | Multiple |
| 3 | Human resources theme | Multiple |
| 4 | The financial theme | Multiple |
| 5 | Scientific research topic | Multiple |
| 6 | Equipment asset subject | Multiple |
| 7 | Alumni | Single |
| 8 | Archives, logistics, others | Multiple |
| 9 | Common base database | |
| 10 | Exchange database | |

**Information Resources Integrated Service Data Exchange Platform**

The construction of data exchange platform for integrated service of information resources includes five aspects.

First, information resource center construction should be carried out to establish a unified, safe, cross-platform and cross-application data exchange platform to realize information resource sharing.

Second, a comprehensive platform for digital teaching should be established to make full use of information technology to facilitate teaching reform and provide schools with a fully functional digital teaching environment.

The third, to establish a comprehensive business system that covers the main business scope of the management department with the core of educational administration, students, personnel, scientific research and asset management, and can provide real-time information service for the whole school.

Fourth, an information security and operation guarantee mechanism should be established to guarantee the reliable operation of information system and resource center.

Fifth, the establishment of digital resources, public service platform, hardware and system software etc.

## 3   Anticipated Benefits

The construction of the data exchange platform for the integrated service of information resources can help improve the information service for students and staff, promote the development of teaching and scientific research, realize the information sharing, data exchange and transparent access of all departments, and greatly improve the response speed of transaction processing. Especially for some jobs that consume a lot of labor, the introduction of information means will save the working time and labor cost of the management department. At the same time, it can standardize the information management of our college, strengthen education supervision, and improve the scientific decision-making ability and management level of all functional departments and decision-making levels in our college. The construction of information resources comprehensive service data exchange platform will greatly improve the popularity of our college, establish the university brand, and bring considerable social benefits to our college.

# 4  Conclusion

By the construction of colleges and universities information resources data exchange platform with comprehensive services, a people-oriented information environment was built which meets demand, leads the demand, is easy to use, run orderly, and integrate the services provided by these application systems in the form of a campus information portal to provide efficient online services and information resources for the college leaders, administrators at all levels, workers, teacher and students. This data exchange platform has been applied in the construction of digital campus of our university, which has certain universality and expansibility, and provides a good reference value for the data integration of enterprises and universities in the future.

**Acknowledgements.** This paper was supported by the Higher Education Teaching Reform Project of Heilongjiang named Research and Practice of Computer Basic Teaching Reform with the concept of Computing Thinking (Grant No. SJGY20170344) and Planning project of Education department of Hei Longjiang province (Grant No. GJC316157).

# References

1. He, K.: Meeting the challenge of the new stage of educational informatization development. CET China Educ. Technol. **8**, 5–11 (2006)
2. Meng, Y.: Research on interdepartmental sharing mechanism of government information resources under e-government. China Manag. Inf. **22**, 149–151 (2017)
3. Deng, Z.: Research and Design of SOA-Based Data Exchange Platform. Guangzhou University of Technology, Guangzhou (2009)
4. Clements, P., Northrop, L.: Software Product Lines: Practices and Patterns, pp. 550–554. Addison-Wesley Longman Publishing Co., Inc., Boston (2001)
5. Wang, L., Li, L., Zhu, Y.: Analysis on data center construction in universities. Heilongjiang Educ. (High. Educ. Eval.) (9), 69–71 (2006)
6. Wei, X.: Cloud computing in information resource management. Comput. Knowl. Technol. **13**(17), 6–7 (2017)
7. Lu, H.: Design and Implementation of Service-Oriented Digital Campus Data Exchange Platform. Lanzhou University, Lanzhou (2009)
8. Zhang, X., Chen, D., Deng, F.: Research on top-level design of standardization system of government information resources management. Inf. Stud.: Theory Appl. **4**, 10–15 (2017)
9. Ministry of Education of the People's Republic of China. Management Information Standards for Institutions of Higher Learning. People's Posts and Telecommunications Press, Beijing (2008)
10. Share data exchange platform solutions [EB/OL] (2014). http://wenku.baidu.com

# Integration of University-Enterprise in Applied Universities

Jingping Cao[✉] and Jiancai Wang

Hei Longjiang University of Science and Technology,
Harbin 150022, Hei Longjiang, China
154539860@qq.com

**Abstract.** Because higher education in China attaches more importance to theory rather than to practice, there is a gap between academic research and technical practice. With transformation and development of universities, to cultivate a large number of technical talents who can make scientific and technological achievements into practice, an educating strategy of integrating enterprises into higher education is proposed. From universities' perspective, the cooperation with enterprises was further strengthened by linking the Industry-University-Research cooperation. With university-enterprise cooperation to in-depth integration, the talent supply was further improved. And an in-depth ecological mechanism of government-led, market-driven and common development between universities and enterprises was established. The research result show that with the proposed strategy "win-win" cooperation between universities and enterprises is achieved.

**Keywords:** University-enterprise · Integration · Ecological mechanism

## 1 Introduction

China has the largest system of higher education in the world. With the economic development gradually entering a new environment and a new situation, the relationship between talent supply and market demand has undergone profound changes [1]. The structural contradictions in higher education have become increasingly prominent, and the homogenization tendency has become increasingly serious. The structure and quality of talent training in colleges and universities are still far from being fully adapted to the changes and requirements of economic and social development and industrial upgrading. For the first time in the report of the 19th National Congress, the major conclusions of "deepening the integration of production and education, university-enterprise cooperation" were put forward. The development of local undergraduate universities is both an opportunity and a mission of the new era. As an application-oriented undergraduate university with industry background and distinctive mining characteristics, how to deepen the training mechanism of university-enterprise integration and collaborative education in the face of the opportunity to comprehensively promote the transformation and upgrading of new engineering? How to deepen the construction of connotation and improve the quality of running universities? How

© Springer Nature Singapore Pte Ltd. 2019
R. Mao et al. (Eds.): ICPCSEE 2019, CCIS 1059, pp. 422–429, 2019.
https://doi.org/10.1007/978-981-15-0121-0_33

to build a new, practical, and application-oriented talent training model? The topic has important practical significance and research value.

## 2    The Status and Characteristics of University-Enterprise Cooperation

Cooperation between universities and enterprises has been existed for a long time. Especially since the state promulgated the "Guiding Opinions on introducing Some Local Undergraduate Universities to Applied Types" in 2015, most undergraduate universities have begun to transform into applications and began to post tabs called " application-oriented", the "university-enterprise cooperation" has become the hottest buzzword in colleges and universities [2]. The current form and content of cooperation between universities and enterprises can be summarized as five "simplification".

The first is the simplification of the form of cooperation. The so-called cooperation is mostly in the form of students' internships with enterprises, and the company's personnel help the students' graduation design (thesis).

The second is the simplification of the cooperation model. The so-called cooperation is mostly a framework agreement between the universities and enterprises to sign the practice base, mostly to complete the teaching links in the training program, and the enterprises are passive and helpless.

The third is the simplification of cooperation content. The so-called cooperation is mostly short time, many forms, and less content. It can only be seen and can't be touched, and turning around is not a deep, systematic, and real way.

The fourth is the simplification of cooperation personnel. The so-called cooperative universities are mostly professional teachers who lack rich practical work experience; most of them are professional and technical personnel, who are not familiar with the knowledge system of university personnel training.

The fifth is the simplification of the cooperation subject. The so-called cooperation is mostly active in schools, active in schools, and hot in schools, while enterprises are mostly passive, do not want to participate, and are not active.

## 3    "University-Enterprise Cooperation" and In Depth Integration

The so-called "cooperation" refers to the joint mode of cooperating with each other for a certain goal, which is often phased; the so-called "integration" refers to the deep cooperation between the two sides, mutual support, mutual benefit and win-win, which is generally persistent. In the process of sustainable development of colleges and universities, with the socialization of the goal of talent training, the university-enterprise cooperation is flourishing in different forms, and it is in the process of talent cultivation in universities in different degrees, but it is limited to the special nature of personnel training, the interests of development of enterprise and the lack of social mechanisms make the relationship between most universities and enterprises superficial, shallow, temporary, single, and unstable. Most of them can only be called "cooperation". There is a big gap in "convergence" [3].

The so-called university-enterprise integration is a two-way integration. The two sides are the main body of mutual benefit and win-win, and the two are coexisting. The two are proactive, bound to some cause and long-term strategic cooperation, working together for long-term stability and far-reaching development, as shown in Fig. 1. On the contrary, in the cooperation between university and enterprise, the former is the main body, and they are the unilateralism who actively close to the enterprises which often have no motivation, no interests, no strategic prospects for cooperation, and no more experience after cooperation. Instead, they bear so much security responsibilities and social responsibilities that they are afraid to avoid or deal with things.

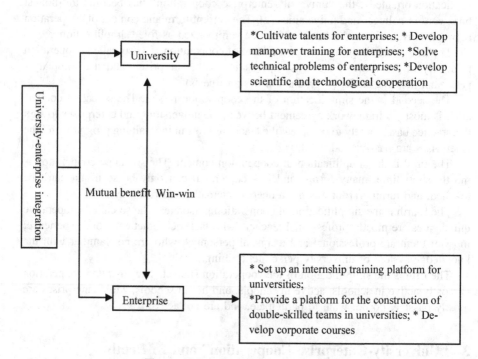

**Fig. 1.** University-enterprise integration

## 4  Deep Integration and Ecological Mechanism

The term "ecology" belongs to the biological category in a narrow sense. In a broad sense, it involves all aspects of society. Any interaction with environmental factors such as the social environment and the formation of a benign symbiotic system belong to the category of ecological development. As a university personnel training is its fundamental task, the development of colleges and universities is essentially to establish a virtuous circle system that serves the quality of talents and social development. The purpose of deep integration of university-enterprise is to establish an ecological education mechanism for symbiotic development through substantial participation and cooperation between universities and enterprises, and ultimately realize the return of enterprise benefits and the improvement of talent quality in universities. As shown in Fig. 2.

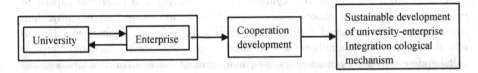

**Fig. 2.** Sustainable development of university-enterprise integration

Integrity: The two sides are a fusion development. The university is based on industrial needs and social development. The company participates in the whole process of talent cultivation and realizes technical support, intellectual security and social recognition.

Intercommunity: The two sides are the interactive body of development. Colleges and universities should pay close attention to the industrial situation and social needs. Enterprises must participate in the process of talent cultivation in colleges and universities; constantly adjust the curriculum and teaching links and content, continuous communication, adjustment, and optimization.

Development: The ecological mechanism of deep integration of schools and enterprises is dynamic balance. Colleges and universities can constantly reflect on self-examination according to social evaluation. Enterprises can absorb talents that meet the needs and further feed back their own development, thus achieving symbiosis and common prosperity.

## 5 Constructing the Ecological Mechanism of Deep Integration of University-Enterprise in Applied Universities

To establish a solid bridge between universities and enterprises and build a platform for sound development, we need to build a deep-integrated ecological mechanism of "government-led, market-driven, school-enterprise, and common development" to collaboratively cultivate high-quality applied talents. Inject technology and talent reserves for enterprise development and promote industry, regional and social economic development.

### Efficient and Complete Deep Integration Management Mechanism
Establish a university-enterprise collaborative expert advisory committee composed of government agencies, universities, industry, and business circles, and hold regular meetings to conduct research and guidance on school development planning, discipline construction, teaching plans, personnel training programs, and management mechanisms. Provide consultation, conduct demonstration, discussion and analysis on enterprise

technical problems, industry development situation, talent demand structure, school-enterprise cooperation channels, etc., smoothly integrate schools into local economic development channels, broaden the social competitive advantages of enterprises, and promote the integration of production and education and the in-depth development of school-enterprise collaboration.

Established a mechanism for regularly inviting industry and enterprise experts to participate in school governance, guiding school professional construction, curriculum construction, and personnel training, and formulated relevant systems to adapt to school-enterprise collaborative talent training, and established a school-enterprise collaborative two-tier evaluation teaching management mechanism. It guarantees the realization of the goal of school-enterprise collaborative training of applied talents. Established a curriculum replacement mechanism, introduced the enterprise curriculum system, paid attention to the advanced and practical content of the teaching content, and expanded the professional courses and elective courses in the traditional sense to the business expertise, professional ability and professional quality, and gradually realized academic courses, technical courses and occupations. The qualification courses are combined.

1. To improve flexible and deep integration incentives
    a. Give full play to the leading role of the government. Mobilize the enthusiasm of both universities and enterprises and avoid "one-hot" so that improve the government's working mechanism, and give certain preferential and incentive measures to enterprises that support university-enterprise cooperation in terms of systems, management methods and taxation, so that enterprises can take up education. Social responsibility allows companies to take the initiative to enter the campus, take root in the campus, and serve the campus.
    b. Through the university-enterprise collaboration platform, on the one hand, the university practice training project through the teacher, planned professional teachers, especially young teachers to the cooperative enterprise to carry out professional practice, industry-university-research cooperation, job-hanging training. Teachers have passed "practice", thus improve the teachers engineering practice ability and enriching practical experience. The teachers bring the knowledge gained in practice back to the classroom, to supplement and update the teaching content, conduct case teaching with practical problems and to promote the combination of teaching and practice. Finally good teaching results are gained. On the other hand, by hiring experts and engineering technicians with rich practical experience in the cooperative enterprise as part-time teachers to combine curriculum with production practices and new demands for talent to the classroom. A team of "double-type" teachers with strong practical ability and high teaching level has been established.
    c. Universities and enterprises focus on the target of training applied talents, capacity development as mainline, strengthening engineering practice ability, engineering design ability and engineering innovation ability, reconstructing curriculum system and teaching content, and jointly develop talent training programs. Conduct basic theory, experiment and training in the university and carry out internships, production internships, graduation internships, post-workouts, corporate courses,

graduation design (thesis), etc. outside the university. Universities and enterprises jointly develop talent training programs, which can ensure the cultivation of talents according to the needs of enterprises, which enhances the fit of talent training and enterprise needs, and realize the seamless connection between talent training and job requirements.

d. According to the first-line technical backbone and management personnel of the enterprise, they need to improve their quality, actively organizing teachers to vigorously carry out service-to-door activities, and train key management personnel, safety engineering technicians, safety supervisors, engineers, etc. through short- and medium-term technical training and academic classes. At the same time, through project cooperation, enterprise settlement, co-construction of laboratories and joint technical problems, we will share the benefits of transformation.

2. Deep integration and restraint mechanism

At present, the bottleneck that plagues the development of colleges and universities is the phenomenon of "one hot". Enterprises are not actively involved in it. The most critical factor is the lack of social supervision and regulation mechanism for university-enterprise cooperation. The main position of enterprises has not been manifested. The obligation of both sides is not clear.

a. The government's macroeconomic regulation and control, improve the legislative system. The government should introduce laws and regulations related to school-enterprise cooperation, such as tax reduction and exemption policies, corporate loan subsidies, independent enrollment authority, and tuition fee standard control measures, and encourage and restrict enterprises' participation in education and teaching from the aspects of social responsibility and business behavior; Undertaking education missions and national development to encourage and constrain universities to actively improve the quality of education. Governments at all levels should clarify the methods, contents, rights and obligations of university-enterprise cooperation at the legislative level, restraining the actions of both parties, guarding their respective interests, enhancing their consciousness, strengthening the legalization and standardization, and provide an excellent social environment for both parties.

b. Social coordination and supervision, to improve the evaluation system. The evaluation expert group will be set up to include the university-enterprise cooperation status as the content of the enterprise qualification assessment, and the university-enterprise cooperation effectiveness will be included in the special supervision work of the university, and the annual assessment will be conducted every year, and a special inspection will be conducted every three years. We will focus on supervising the implementation of policies, the depth of cooperation, and the effectiveness of the policies to ensure the sustainable development of the deep integration of schools and enterprises.

c. Both parties take their responsibilities and improve their responsibilities system. Government dominant position: It is mainly responsible for formulating and improving relevant laws, regulations and supervision systems, providing environmental, financial and social support, establishing communication and coordination mechanisms between universities and enterprises, and building bridges and ties.

The main body status of colleges and universities: First, based on the goal of talent training, taking social development and serving the regional economy as its own responsibility, taking the Lideshu people as its fundamental task, giving full play to the role of "think tanks" in colleges and universities, closely matching the needs of industry and social development, and doing strategic planning and finding Quasi-conformity to achieve deep integration; Second, proactive and close contact. Sign strategic cooperation agreements with enterprises, establish a large number of stable cooperation bases, jointly develop training programs, jointly develop curriculum systems, jointly cultivate teachers, and jointly research, Co-constructed laboratories and other forms to form a close-knit cooperation pattern of "you with me contain each other".

The main position of the enterprise: The development of the enterprise requires a large number of core technical talents. The research of enterprise technology is also in urgent need of high-level professional talents to support it. Education and economic development itself are an inseparable whole, and the two promote each other and develop together. Therefore, it is also the social responsibility of the company to cultivate high-quality qualified personnel for the society. First, we must actively participate in the formulation of university talent training programs, practical teaching, curriculum development, quality evaluation, etc. Second, we must actively seek strategic opportunities, establishing alliance mechanisms, jointing research development, practice, and promotion. Training channels, let students really learn to do, learning the true skills, doing real things, providing a real and practical experience learning platform.

3. Establish a comprehensive deep integration guarantee mechanism

These words such as "university-enterprise cooperation", "university-enterprise integration", "university-enterprise cooperation" and so on appear in the development strategies and work results of various universities, how to make them stronger and stronger, to maintain stability that is necessary to establish a supporting mechanism.

a. System guarantee. First, the government and the competent authorities should introduce relevant documents and management systems to promote the integration of colleges and enterprises. Second, increase financial input and supporting, and lead enterprises to actively participate in the training of talents in universities. Third, Government reform the enrollment system, increase school autonomy, and support set up a university-enterprise joint construction professional, clearing the university-enterprise professional tuition standards; Fourth, introduce a tax reduction policy for cooperative enterprises.

b. Funding guarantee. First, the government should set up special funds to support university-enterprise cooperation to pilot key universities and university-enterprise projects; second, universities should set up special funds to support cooperation, personnel training, team building and evaluation awards; Special funds should be set up to support technology promotion, platform establishment and base establishment.

c. Personnel protection. The first step is to set up a working group set up by experts from both colleges and enterprises to guide the implementation of the work of university-enterprise cooperation and strengthen the input of intelligence; the second is to strengthen deep integration, establish " university workstations" in enterprises, and establish "enterprise stations" in universities. It not only integrates the resources and advantages of scientific research in universities, but also grasps the development of technology and the needs of enterprises, and builds a scientific research platform, a platform for cooperation between industry, universities and research institutes, and an innovative technology platform to promote the co-construction and integration of universities and enterprises.

d. The establishment of the university-enterprise integration ecological mechanism is an important opportunity for the development of colleges and universities. It is also an important strategy and broad prospect for future education development. It is also an important guarantee to support local economic development and industry demand. Today, in the continuous replacement of high-tech industries, universities only have Continuous innovation, professional only continuous upgrading, colleges and universities only continue to deepen the integration of enterprises; enterprises only pay attention to personnel training, the company's personnel only continue to deepen undergraduate teaching, so the university and enterprise deep integration can bring the commonality of colleges and universities development and high prosperity. Therefore, all colleges and universities must establish a stable and long-term ecological mechanism for university-enterprise integration, and constantly explore new educational models and reform measures to better achieve talent supply and technological innovation.

**Acknowledgements.** This paper was supported by the Key authorized Project of Higher Education Teaching Reform in Heilongjiang named Research and Practice on the Cooperative Education Mechanism between Colleges and Enterprises in Local Applied Undergraduate Universities (Grant No. SJGZ20180054) and the Higher Education Teaching Reform Project of Heilongjiang named Research and Practice of Computer Basic Teaching Reform with the concept of Computing Thinking (Grant No. SJGY20170344).

# References

1. Hao, C.B.: The "Four Integration" applied talents training system in colleges and universities from the perspective of collaborative innovation. Heilongjiang High. Educ. Res. (4), 139–142 (2016)
2. Lin, Q., He, D.P., Chen, M.X.: Research on the construction of deep cooperation ecosystem and mechanism of local undergraduate universities and enterprises. J. Hengyang Normal Univ. **10**(6), 128–132 (2017)
3. Zhu, H.J., Chen, L.L.: Construction of ecosystem of industry-university collaborative innovation in applied universities. Educ. Obs. **9**(17), 56–58 (2018)

# Application

# Adaptive Flower Pollination Algorithm Based on Chaotic Map

Yu Li[1,2] ⓘ, Juan Zheng[2](✉), and Yi-ran Zhao[2]

[1] Institute of Management Science and Engineering, Henan University,
Kaifeng 475004, China
[2] Business School, Henan University, Kaifeng 475004, China
2625847482@qq.com

**Abstract.** Flower pollination algorithm (FPA) is one of the well-known evolutionary techniques used extensively to solve optimization problems. Despite its efficiency and wide use, the identical search behaviors may lead the algorithm to converge to local optima. In this paper, an adaptive FPA based on chaotic map (CAFPA) is proposed. The proposed algorithm first used the ergodicity of the logistic chaos mechanism, and chaotic mapping of the initial population to make the initial iterative population more evenly distributed in the solution space. Then at the self-pollination stage, the over-random condition of the gamete renewal was improved, the traction force of contemporary optimal position was given, and adaptive logarithmic inertia weight was introduced to adjust the proportion between the contemporary pollen position and disturbance to improve the performance of the algorithm. By comparing the new algorithm with three famous optimization algorithms, the accuracy and performance of the proposed approach are evaluated by 14 well-known benchmark functions. Statistical comparisons of experimental results show that CAFPA is superior to FPA, PSO, and BOA in terms of convergence speed and robustness.

**Keywords:** Flower pollination algorithm · Chaotic map · Traction force · Adaptive

## 1 Introduction

The Flower Pollination Algorithm (FPA) [1] is a new swarm intelligent optimization algorithm proposed by Yang in 2012. The algorithm simulates the pollination process of flowering plants in nature, and put the pollination activities of flowering plants evolved into the solving process of the optimization problem. Combining the advantages of Cuckoo Search (CS) [2] and Bat Algorithm (BA) [3, 4], FPA is simple to implement and adjust, and has few parameters, a novel structure, good search ability and strong robustness. The algorithm has been already applied fields including function optimization [5], wireless sensor network [6], power system [7], feature selection [8], shape matching [9], electromagnetic problems [10], parameter estimation [11], combination optimization [12] and knapsack problem [13].

The FPA, like other intelligent algorithms [14], is prone to fall into local optimum, with slow convergence speed at a later stage and the optimization accuracy is low,

© Springer Nature Singapore Pte Ltd. 2019
R. Mao et al. (Eds.): ICPCSEE 2019, CCIS 1059, pp. 433–444, 2019.
https://doi.org/10.1007/978-981-15-0121-0_34

therefore many scholars have carried out improvement studies on it. Some of these improvements combine the algorithm with other intelligent optimization algorithms, for instances: FPPS [15] combines the standard flower pollination algorithm (FP) with the particle swarm optimization (PSO) algorithm to improve the quality of FPA initial solution, thus improving the searching accuracy and convergence rate; HFPCHS [16] integrates the standard Flower Pollination algorithm (FP) with the chaotic Harmony Search (HS) algorithm to improve the searching accuracy and the quality of solution; BFP [17] combines FPA with the Bat Algorithm, which effectively avoids local minima and is successfully applied to the synthesis of non-equidistant linear antenna arrays; MFPA [18] hybridizes the Clonal Selection Algorithm (CSA) with the FPA and achieves a higher accuracy. In addition, to avoid the declining of convergence speed and the quality of the algorithm that is caused by inter-dimensional interference, DDIFPA [19] is proposed, which helps improve the algorithm's optimization speed and exploration performance; GSFPA [20] which is based on the gravitational search mechanism greatly improves the algorithm convergence speed, accuracy and robustness; ALFPA [21] which is based on mutation operators has also achieved good improvement.

This paper proposes an Adaptive Flower Pollination Algorithm based on Chaotic Map (CAFPA), which applies chaotic strategy to map the initial solution to improve the quality of the initial iterative population, and introduces the adaptive logarithmic inertia weight in the local update phase to adjust the ratio between contemporary pollen position and the disturbance to improve the convergence speed and accuracy of the algorithm. The study selects 14 test functions to do simulation tests and the results show that CAFPA has greatly improved the convergence speed and accuracy compared with the original FPA, and the solution results are stable, which proves that the improvement is feasible and effective.

## 2   The Flower Pollination Algorithm

Pollination can be achieved by self-pollination or cross-pollination. Cross-pollination, or allogamy, means pollination can occur from the pollen of a flower of a different plant, generally requires pollinators, and can occur in random places far away. In FPA, this kind of pollination is considered as global pollination (global optimization). While self-pollination is the fertilization of one mature flower from pollen of the same flower or different flowers of the same plant, which often occurs when there is no reliable pollinator available. FPA refers to such pollination as local pollination (local optimization). FPA simulates the pollination process of flowers of flowering plants in nature, and the following ideal conditions are set:

1. Biotic cross-pollination can be considered as a process of global pollination, and pollen-carrying pollinators move in a way that obeys Lévy flights (Rule 1). This rule can be represented mathematically as:

$$x_i^{t+1} = x_i^t + L(g_* - x_i^t) \tag{1}$$

where $x_i^t$ is the pollen $i$ or solution vector $x_i$ at iteration $t$, and $g_*$ is the current best solution found among all solutions at the current generation. The parameter $L$ is the strength of the pollination, which essentially is a step size, and its calculation formula is:

$$L \sim \frac{\lambda \Gamma(\lambda) sin\left(\frac{\pi\lambda}{2}\right)}{\pi} \frac{1}{s^{1+\lambda}} (s \gg s_0 > 0) \tag{2}$$

Here $\Gamma(\lambda)$ is the standard gamma function, and this distribution is valid for large steps $s > 0$. In all our simulations below, we have used $\lambda = 1.5$.

2. Abiotic and self-pollination are considered as local pollination (Rule 2), and it can be represented as:

$$x_i^{t+1} = x_i^t + \varepsilon\left(x_j^t - x_l^t\right) \tag{3}$$

where $x_j^t$ and $x_l^t$ are pollens from the different flowers of the same plant species, which is equivalent to two random solutions of the population. $\varepsilon$ is a random number subject to uniform distribution on $[0, 1]$.

3. Flower constancy, namely the probability of reproduction, is proportional to the similarity of two flowers (Rule 3).

4. Controlled by the conversion probability $p \in [0, 1]$, FPA can realize the dynamic conversion of local pollination and global pollination. Due to physical proximity and other natural factors such as wind, local pollination accounts for a larger proportion in the whole pollination activity (Rule 4). A large number of experiments have been carried out in the literature [22] to prove that the selection of 0.2 is most suitable value for $p$.

# 3 Improved FPA

## 3.1 Chaos-Based Initialization

Haupt et al. pointed out that for the intelligent optimization algorithm which based on swarm iteration, the quality of the initial population affects its solution accuracy and global convergence speed [23]. A initial population with a higher diversity can improve the optimization performance of the algorithm. Before the update iteration, the FPA uses random method to generate the initial solution, which cannot guarantee the uniform distribution of the search agent in the search space, resulting in a low algorithm search efficiency and solution accuracy.

Chaos is characterized by randomness, ergodicity, and extreme sensitivity to initial conditions. It can traverse all feasible solutions non-repeatedly based on its own law in the target space, so that gametes are more evenly distributed in the solution space [24, 25]. The use of chaotic motion for population initialization can fully extract useful information in the search space, and at the meantime maintain the diversity of the population and improve the distribution quality of the initial gametes, thus laying a good foundation for the iterative update of the algorithm and adding the probability of finding a higher quality solution. (logistic map, also known as parabolic map).

In this study, the classical chaotic dynamics model Logistic is used to nonlinearly map the initial solution generated by the algorithm. The steps are as follows:

1. Assign $k = 0$, map the decision variable $x_{i,j}^k, (i = 1, 2, \ldots, n; j = 1, 2, \ldots, D)$ as a chaos variable which between 0 to 1;

2. Calculate the chaos variable $s_{i,j}^{k+1} = 4s_{i,j}^k \left(1 - s_{i,j}^k\right)$ of the next iteration;

3. Convert chaotic variable $s_{i,j}^{k+1}$ into decision variable $x_{i,j}^{k+1}$, $x_{i,j}^{k+1} = x_{min,j} + s_{i,j}^{k+1}\left(x_{max,j} - x_{min,j}\right)$, among which, $x_{max,j}$ and $x_{min,j}$ represent the upper and lower limits of the variable search in dimension $j$ respectively;

4. Evaluate the new solution based on the decision variable. If the new solution is better than the original one $X_i^{(0)} = \left[X_{i,1}^{(0)}, X_{i,2}^{(0)}, \ldots, X_{i,D}^{(0)}\right]$ or if the chaotic search has reached the maximum iteration steps, use the new solution as the search result, otherwise assign $k = k + 1$ and turn to step 2.

## 3.2   Local Update of Adaptive Weight Adjust the Ratio

In the global search, the FPA adopts the Lévy flight mechanism which generate large jumps and uneven random moving steps can help the flower individual avoid being attracted by the local extremum to some extent, and give the algorithm an excellent global exploration ability. However, in the self-pollination stage, FPA has a rather insufficient development ability for the following two reasons:

First, the gamete update is guided by no factor and too random. Although it can enhance the diversity of the population, but it is difficult to meet the requirements of local accuracy development and update, and also limits the convergence speed of the algorithm.

Second, the local update process is too flat, lacks jumps, and is easy to fall into local optimum, which is mainly due to its update mechanism. The dynamic transformation for global search or local development is only controlled by the $p$ value, which is easy to understand and implement. It can also perform local search with high probability by setting the value $p$ reasonably, which makes it closer to reality. However, the smaller the value of $p$, the smaller the probability of performing the global search. Even if the global search is performed in a certain iteration, the Lévy flight mechanism cannot guarantee long-distance jumps, therefore, it is possible for the gamete position after each iteration may still lingers around the initial position, which affecting the accuracy of the solution and the speed of convergence.

In view of the above two deficiencies, this paper considers replacing the original disturbance with the difference vector: $g_* - x_i^t$, which is obtained via processing $x_i^t$ (random information) and the current optimal solution $g_*$ (deterministic information). This helps maintain the diversity while exerting the traction force of the contemporary optimal position, and avoid the problems of low search efficiency and slow convergence speed caused by large randomness. To avoid the second situations, this work proposes to introduce weight in the original local update formula, which could adaptively adjust the ratio between the contemporary pollen position and the disturbance as the number of iterations, and guarantee the activity of pollen position updates at the beginning of the iteration. The final proposed local position update formula is:

$$x_i^{t+1} = K \cdot x_i^t + (2 - K) \cdot \varepsilon \left(g_* - x_l^t\right) \tag{4}$$

Among which,

$$K = 1 - m \tag{5}$$

$$m = w_{max} - (w_{max} - w_{min}) * \frac{log(t)}{log(N_{iter})} \left(w_{max} = 0.3, w_{min} = 0\right) \tag{6}$$

When considering the weight that adjusts the ratio between contemporary pollen position and the disturbance, it is better that the weight is slightly less than 1 at the beginning of the iteration and as close as possible to 1, which can not only ensure the inheritance of contemporary pollen location, but also make local renewal more affected by disturbance. Because the population difference is slightly larger at the early stage of the iteration, the disturbance part can search a larger range under the traction of the optimal position. The introduction of weight expands the influence of the disturbance, make the algorithm have certain jumping ability at the early iteration even without global search, so as to improve the iteration efficiency of the algorithm. As the number of iterations increases, the weight can quickly return to the vicinity of the value of 1, and perform local accuracy update is on the basis of the previous iteration. As the value of $t$ increases, the $\frac{log(t)}{log(N_{iter})}$ is logarithmically increasing in range of $[0, 1]$, and the $m$ has logarithmic decrease from 0.3 to 0. Correspondingly, the value $K$ of the inertia weight has logarithmic increases rapidly from 0.7 to 1, meeting the requirements of the improvement. The change trend is shown in Fig. 1.

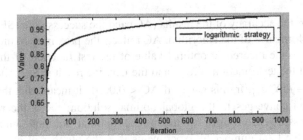

**Fig. 1.** Change trend of the value K of the weight ratio

### 3.3 Implementation Steps of CAFPA

To sum up, the steps of CAFPA proposed in this paper are as follows:

Step 1: Parameter initialization Settings: flower population number $n$, switch probability $p$ and maximum steps of chaotic initialization iteration $C_{max}$.

Step 2: Initialize a population of $n$ flowers/pollen gametes with random solutions. For each initial solution, a sequence of chaotic solutions is generated according to the chaotic initialization steps in Sect. 3.1.

Step 3: If $p > rand$, according to formula (1), the chaotic solution sequence is updated and transacted.

Step 4: If $p < rand$, according to formula (6), (5) and (4), the chaotic solution sequence is updated and transboundary processing is carried out.

Step 5: For $n$ population individuals, the fitness value of the new solution obtained by Step 3 or Step 4 is calculated. If the new fitness value is better, the current solution and the current fitness value will be replaced by the new solution and its fitness value respectively; otherwise, the current solution and the current fitness value are retained.

Step 6: If the new fitness value is better than the global optimal value, the global optimal solution and the global optimal value will be updated.

Step 7: If the end condition is met, exit the program and output the best solution found; otherwise, Step 3 is executed.

# 4  Numerical Experiment and Analysis

## 4.1  Test Functions and Performance Indicators

In order to evaluate the optimization performance of CAFPA comprehensively, this paper selects 14 different types of test functions for simulation experiments. f1–f7 are high-dimensional unimodal functions, which are mainly used to test the optimization accuracy of the algorithm. f4 is a classical non-convex ill-conditioned function that is extremely difficult to find the global optimal solution; f8–f12 are high-dimensional multimodal functions, which have large number of local minima distributed in the solution space, easy to fall into local optimum, and difficult to find the global optimal solution. Among them function f10 is a typical representative, f13 and f14 are low-dimensional functions. The specific characteristics of these test functions are shown in Table 1.

In this paper, the accuracy of the results (AC) and the success ratio (SR) are used to evaluate the performance of the algorithm. AC reflects the proximity of the algorithm's iterative results to the theoretical optimal value of the test function. If the theoretical optimal value of the test function is $X_{opt}$, and the iterative result is $S_{best}$, then the AC is $AC = |f(S_{best}) - f(X_{opt})|$. In this paper, if AC < 0.0001, it means that the optimizing is successful and converges to the global optimal solution; SR is the ratio that the algorithm successfully converges to the global optimal solution among multiple experiments. If the total number of experiments is $z$, the number of experiments that converge to the global optimal solution is $z'$, then $SR = \frac{z'}{z} \times 100\%$.

This experiment is carried out on a laptop with an operating system of Windows10, a CPU of Intel Core i5-8520, a clock speed of 1.6 GHz and a memory of 8 GB. The programming language is Matlab R2014a.

**Table 1.** Benchmark test functions

| Function | Dim | Formula | Range | Optimum | Optimal point |
|---|---|---|---|---|---|
| Sphere | 30 | $f_1(x) = \sum_{i=1}^{d} x_i^2$ | [−100, 100] | 0 | (0, …, 0) |
| Step | 30 | $f_2(x) = \sum_{i=1}^{d} (\lfloor x_i + 0.5 \rfloor)^2$ | [−100, 100] | 0 | ([−0.5, 0.5), …, [−0.5, 0.5)) |
| Schwefel 1.2 | 30 | $f_3(x) = \sum_{i=1}^{d} \left(\sum_{j=1}^{i} x_j\right)^2$ | [−100, 100] | 0 | (0, …, 0) |
| Rosenbrock | 30 | $f_4(x) = \sum_{i=1}^{d-1} [100(x_{i+1} - x_i^2)^2 + (x_i - 1)^2]$ | [−30, 30] | 0 | (1, …, 1) |
| Schwefel 2.22 | 30 | $f_5(x) = \sum_{i=1}^{d} |x_i| + \prod_{i=1}^{d} |x_i|$ | [−10, 10] | 0 | (0, …, 0) |
| Sum squares | 30 | $f_6(x) = \sum_{i=1}^{d} i x_i^2$ | [−10, 10] | 0 | (0, …, 0) |
| Quartic | 30 | $f_7(x) = \sum_{i=1}^{d} i x_i^4 + random()$ | [−1.28, 1.28] | 0 | (0, …, 0) |
| Griewangk | 30 | $f_8(x) = \frac{1}{4000} \sum_{i=1}^{d} x_i^2 - \prod_{i=1}^{d} \cos\left(\frac{x_i}{\sqrt{i}}\right) + 1$ | [−600, 600] | 0 | (0, …, 0) |
| Alpine | 30 | $f_9(x) = \sum_{i=1}^{d} |x_i \cdot \sin(x_i) + 0.1 \cdot x_i|$ | [−10, 10] | 0 | (0, …, 0) |
| Rastrigin | 30 | $f_{10}(x) = \sum_{i=1}^{d} [x_i^2 - 10\cos(2\pi x_i) + 10]$ | [−5.12, 5.12] | 0 | (0, …, 0) |
| Ackely | 30 | $f_{11}(x) = -20\exp\left(-0.2\sqrt{\frac{1}{d}\sum_{i=1}^{d} x_i^2}\right) - \exp\left(\frac{1}{d}\sum_{i=1}^{d} \cos(2\pi x_i)\right) + 20 + e$ | [−32, 32] | 0 | (0, …, 0) |
| Zakharov | 30 | $f_{12}(x) = \sum_{i=1}^{d} x_i^2 + (\frac{1}{2}\sum_{i=1}^{d} i x_i)^2 + (\frac{1}{2}\sum_{i=1}^{d} i x_i)^4$ | [−5, 5] | 0 | (0, …, 0) |
| Schaffer F6 | 2 | $f_{13}(x) = \frac{\sin^2\sqrt{x_1^2 + x_2^2} - 0.5}{[1 + 0.001(x_1^2 + x_2^2)]^2} - 0.5$ | [−100, 100] | −1 | (0, …, 0) |
| Six-Hump | 2 | $f_{14}(x) = 4x_1^2 - 2.1x_1^4 + \frac{1}{3}x_1^6 + x_1 x_2 - 4x_2^2 + 4x_2^4$ | [−5, 5] | −1.0316285 | (−0.0898, 0.7126) & (−0.0898, 0.7126) |

## 4.2 Results of the Experiments

To evaluate the performance of the CAFPA, it is compared to those algorithms which are widely employed to compare the performance of algorithms. The classical Particle Swarm Optimization (PSO) [26, 27], FPA and newly proposed Butterfly Optimization Algorithm (BOA) [28] are selected in this paper. In each run, we use of 1000 iterations as the termination criterion of these algorithms. In order to reduce statistical errors and generate statistically significant results, each function is repeated for 30 times and all the experiments are conducted under the same conditions. The mean, standard deviation, best, and worst results of these algorithms are recorded. The population number is uniformly set to 5 and other parameters are set as follows (Table 2):

**Table 2.** Parameter settings of each algorithm

| Algorithm | Parameter settings |
|---|---|
| CAFPA | $p = 0.2, C_{max} = 10$ |
| FPA | $p = 0.2$ |
| PSO | $c1 = c2 = 2, w = 0.9 - (0.9 - 0.4) * \left(\frac{t}{N\_iter}\right), \text{Vmax} = 1, \text{Vmin} = -1$ |
| BOA | $p = 0.8, a = 0.1, c = 0.01$ |

The experimental results are shown in Table 3 (the best results of the four algorithms are shown in bold):

**Table 3.** Simulation results of benchmark functions for different algorithms

| Function | Algorithm | Best | Mean | Worst | Std. | SR |
|---|---|---|---|---|---|---|
| $f_1(x)$ | CAFPA | **1.130245E−47** | **1.494507E−36** | **2.543323E−35** | **5.192518E−36** | **100** |
| | FPA | 8.014025E+03 | 1.513736E+04 | 3.169144E+04 | 6.279109E+03 | 0 |
| | PSO | 8.429342 | 1.820178E+02 | 1.211013E+03 | 2.504888E+02 | 0 |
| | BOA | 1.952883E−14 | 2.321364E−14 | 2.626873E−14 | 1.538759E−15 | 100 |
| $f_2(x)$ | CAFPA | **0** | **0** | **0** | **0** | **100** |
| | FPA | 7103 | 1.779603E+04 | 28847 | 5.897634E+03 | 0 |
| | PSO | 101 | 1.383700E+03 | 10692 | 2.588205E+03 | 0 |
| | BOA | **0** | **0** | **0** | **0** | **100** |
| $f_3(x)$ | CAFPA | **4.196681E−43** | **6.818941E−35** | **1.851407E−33** | **3.371600E−34** | **100** |
| | FPA | 7.111491E+03 | 2.388804E+04 | 8.913362E+04 | 1.582758E+04 | 0 |
| | PSO | 2.630702E+03 | 8.414531E+03 | 4.411498E+04 | 7.811326E+03 | 0 |
| | BOA | 2.019611E−14 | 2.323020E−14 | 2.561121E−14 | 1.254444E−15 | 100 |
| $f_4(x)$ | CAFPA | **28.909834** | **28.954560** | **28.988243** | 0.022782 | **0** |
| | FPA | 2.246452E+06 | 2.137835E+07 | 5.796906E+07 | 1.558962E+07 | 0 |
| | PSO | 8.598202E+02 | 7.244505E+03 | 4.269489E+04 | 8.103718E+03 | 0 |
| | BOA | 28.937676 | 28.984347 | 29.000000 | **0.018369** | 0 |
| $f_5(x)$ | CAFPA | **2.547986E−23** | **3.845174E−19** | **8.766171E−18** | **1.612985E−18** | **100** |
| | FPA | 30.301129 | 49.937731 | 74.273834 | 11.015127 | 0 |
| | PSO | 3.465601 | 19.000223 | 61.961196 | 11.971155 | 0 |
| | BOA | 1.143470E−11 | 1.364504E−11 | 1.554693E−11 | 9.586633E−13 | 100 |
| $f_6(x)$ | CAFPA | **6.016118E−46** | **2.369623E−36** | **7.094325E−35** | **1.295150E−35** | **100** |
| | FPA | 8.812280E+02 | 2.222573E+03 | 4.244824E+03 | 8.678802E+02 | 0 |
| | PSO | 18.883642 | 3.838387E+02 | 2.677386E+03 | 6.033490E+02 | 0 |
| | BOA | 1.943172E−14 | 2.195519E−14 | 2.479959E−14 | 1.379682E−15 | 100 |
| $f_7(x)$ | CAFPA | **9.916713E−05** | **0.001102** | **0.003288** | 8.154502E−04 | **3.3** |
| | FPA | 1.845188 | 10.983406 | 42.796231 | 8.808240 | 0 |
| | PSO | 2.050119 | 22.554092 | 1.146269E+02 | 22.756956 | 0 |
| | BOA | 4.280353E−04 | 0.001772 | 0.003346 | **7.885036E−04** | 0 |
| $f_8(x)$ | CAFPA | **0** | **0** | **0** | **0** | **100** |
| | FPA | 46.834412 | 1.617819E+02 | 2.870986E+02 | 63.692688 | 0 |
| | PSO | 2.691691 | 48.585018 | 2.682568E+02 | 59.462119 | 0 |
| | BOA | 0 | 1.088019E−14 | 2.853273E−14 | 1.266155E−14 | 100 |
| $f_9(x)$ | CAFPA | **1.527790E−26** | **3.523633E−21** | **5.316262E−20** | **1.031698E−20** | **100** |
| | FPA | 12.745252 | 21.451801 | 27.815056 | 4.119630 | 0 |
| | PSO | 3.334188 | 11.891386 | 24.624057 | 5.485091 | 0 |
| | BOA | 2.518392E−14 | 5.102062E−12 | 6.204387E−11 | 1.292128E−11 | 100 |
| $f_{10}(x)$ | CAFPA | **0** | **0** | **0** | **0** | **100** |
| | FPA | 1.151195E+02 | 1.519949E+02 | 2.435122E+02 | 26.628055 | 0 |
| | PSO | 71.116289 | 1.510790E+02 | 2.524735E+02 | 45.928878 | 0 |

(*continued*)

**Table 3.** (continued)

| Function | Algorithm | Best | Mean | Worst | Std. | SR |
|---|---|---|---|---|---|---|
| | BOA | **0** | 1.107418E+02 | 2.342855E+02 | 1.063628E+02 | 36.7 |
| $f_{11}(x)$ | CAFPA | **8.881784E−16** | **1.006602E−15** | **4.440892E−15** | **6.486338E−16** | **100** |
| | FPA | 9.551700 | 15.046453 | 18.134831 | 2.211245 | 0 |
| | PSO | 5.078551 | 9.617750 | 19.966769 | 4.809484 | 0 |
| | BOA | 1.599609E−12 | 1.556722E−11 | 2.253930E−11 | 3.858133E−12 | 100 |
| $f_{12}(x)$ | CAFPA | **8.154877E−47** | **6.133627E−37** | **1.471499E−35** | **2.711932E−36** | **100** |
| | FPA | 9.329856 | 48.591733 | 1.783537E+02 | 38.767985 | 0 |
| | PSO | 31.532024 | 1.212364E+02 | 2.297592E+02 | 46.302665 | 0 |
| | BOA | 1.579707E−14 | 1.930631E−14 | 2.268601E−14 | 1.564195E−15 | 100 |
| $f_{13}(x)$ | CAFPA | **−1** | **−0.99384659** | **−0.99028409** | **0.00476208** | **37** |
| | FPA | −0.99999874 | −0.97020984 | −0.58533181 | 0.09177025 | 13 |
| | PSO | −1 | −0.96911031 | −0.50112819 | 0.09023480 | 3.3 |
| | BOA | −0.99028329 | −0.95975744 | −0.92141436 | 0.01910638 | 0 |
| $f_{14}(x)$ | CAFPA | −1.0316284386 | −1.0049775545 | −0.3453969591 | 0.1262168088 | 16.7 |
| | FPA | −1.0316284535 | −1.0044229455 | −0.2154638244 | 0.1490105888 | 96.7 |
| | PSO | −1.0316284535 | −1.0316284535 | −1.0316284535 | 5.9751784E−16 | 100 |
| | BOA | −1.0315947591 | −1.0295956184 | −1.0122502899 | 0.0036370226 | 20 |

According to the statistical results in Table 3, the minimum standard deviation of 11 functions was obtained by CAFPA, and the worst standard deviation obtained by CAFPA was only 0.126217. It can be seen that CAFPA has strong robustness and high stability.

Under the fixed accuracy, CAFPA can converge to the global optimal solution of the other 13 functions except f4, and can converge to the theoretical optimal value of the three functions f2, f8 and f10 every time, which is more stable than the BOA with the highest accuracy in the comparison algorithms. Although CAFPA cannot converge 100% to the theoretical optimal value of f13, it is still better than the solution result of the comparison algorithms and has a higher success rate of optimization.

The accuracy of CAFPA in solving the four functions f1, f3, f6 and f12 is about 40 orders of magnitude higher than that of FPA, and 20 orders of magnitude higher than that of BOA. The CAFPA improves the solution accuracy of the three functions f5, f9 and f11 by 16+ orders of magnitude compared to the original flower pollination algorithm, and 4–9 orders of magnitude higher than the highest accuracy of the three comparison algorithms. CAFPA has greatly improved the accuracy of the solution.

The function f4 is a non-convex ill-conditioned classical high-dimensional uni-modal test function, whose global optimal value is extremely difficult to find. The solution of it using CAFPA is modest, but the CAFPA manages to obtain its optimal result. Although the solution result of the function f7 by CAFPA is only several orders of magnitude higher than comparison algorithms, only CAFPA can converge to its global optimum under the set accuracy.

Generally speaking, CAFPA's optimization accuracy is greatly improved, with high stability, strong robustness. The improvement is obvious.

Given the limited space, the fitness convergence curves of six of the functions are shown as follows (except f10 and f13, the objective function values of other functions are taken the base 10 logarithm):

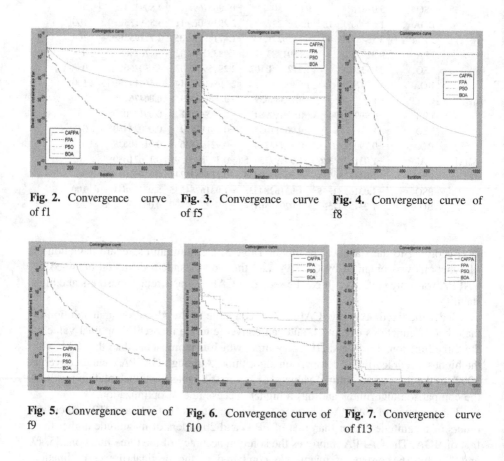

**Fig. 2.** Convergence curve of f1     **Fig. 3.** Convergence curve of f5     **Fig. 4.** Convergence curve of f8

**Fig. 5.** Convergence curve of f9     **Fig. 6.** Convergence curve of f10     **Fig. 7.** Convergence curve of f13

The convergence curves of Figs. 2, 3, 4, 5, 6 and 7 clearly show that compared with the other three algorithms, CAFPA can approach the better solution at a faster speed in the initial stage of the iteration, with a faster convergence speed, a higher optimization accuracy, and a better ability to jump out of the local optimal solution. For example, for the functions f10 and f13, the three comparison algorithms fall into local optimum several times in the solution process, while CAFPA keeps approaching the global optimal solution in the optimization process, with an extremely fast convergence speed, and a high solution accuracy.

# 5 Conclusion

To tackle the problems that the basic Flower Pollination Algorithm (FPA) is easy to fall into local optimum, the late convergence rate is slow, and the optimization accuracy is low, this work proposes to improve the algorithm in two aspects—the generation of the initial iterative generation and the local search. This work uses the ergodicity of chaos to improve the diversity and quality of the initial iterative population, exerts the optimal positional traction force in the self-pollination stage, and introduces the adaptive logarithmic inertia weight to adjust the ratio between contemporary pollen position and disturbance and ensure the activity of the pollen in the initial iteration. The simulation results show that the CAFPA greatly improves the optimization precision and the convergence speed. The improved algorithm is effective and has certain advantages. At present, the theory and application research of FPA is still in its infancy, and many problems need further study. For the future research work, the theoretical basis of research parameter setting and the application of the improved algorithm in practical engineering is recommended to work on.

**Acknowledgments.** This study is supported by the National Natural Science Foundation of China (No. 71601071), the Science & Technology Program of Henan Province, China (No. 182102310886 and 162102110109), and an MOE Youth Foundation Project of Humanities and Social Sciences (No. 15YJC630079). We are particularly grateful to the suggestions of the editor and the anonymous reviewers which is greatly improved the quality of the paper.

# References

1. Yang, X.-S.: Flower pollination algorithm for global optimization. In: Durand-Lose, J., Jonoska, N. (eds.) UCNC 2012. LNCS, vol. 7445, pp. 240–249. Springer, Heidelberg (2012). https://doi.org/10.1007/978-3-642-32894-7_27
2. Yang, X.S., Deb, S.: Cuckoo search via Lévy flights. In: 2009 World Congress on Nature & Biologically Inspired Computing(NaBIC), pp. 210–214. IEEE, Coimbatore (2010)
3. Yang, X.S.: A new metaheuristic bat-inspired algorithm. Comput. Knowl. Technol. **284**, 65–74 (2010)
4. Li, Y., Guo, Q., Liu, J.: Improved bat algorithm for vehicle routing problem. Int. J. Perform. Eng. **15**(1), 317–325 (2019)
5. Yang, X.S., Karamanoglu, M., He, X.: Multi-objective flower algorithm for optimization. Int. Conf. Comput. Sci. **18**(1), 861–868 (2013)
6. Binh, H.T.T., Hanh, N.T., Van Quan, L., Dey, N.: Improved Cuckoo search and chaotic flower pollination optimization algorithm for maximizing area coverage in Wireless Sensor Networks. Neural Comput. Appl. **30**(7), 2305–2317 (2018)
7. Peesapati, R., Yadav, V.K., Kumar, N.: Flower pollination algorithm based multi-objective congestion management considering optimal capacities of distributed generations. Energy **147**, 980–994 (2018)
8. Sayed, A.E.F., Nabil, E., Badr, A.: A binary clonal flower pollination algorithm for feature selection. Pattern Recogn. Lett. **77**, 21–27 (2016)
9. Zhou, Y.Q., Zhang, S., Luo, Q.F., Wen, C.M.: Using flower pollination algorithm and atomic potential function for shape matching. Neural Comput. Appl. **29**(6), 21–40 (2018)

10. Singh, U., Salgotra, R.: Synthesis of linear antenna array using flower pollination algorithm. Neural Comput. Appl. **29**(2), 435–445 (2018)

11. Xu, S., Wang, Y., Liu, X.: Parameter estimation for chaotic systems via a hybrid flower pollination algorithm. Neural Comput. Appl. **30**(8), 2607–2623 (2018)

12. Zhang, W., Yang, Y., Shuai, Z., Yu, D., Li, Y.: Correlation-aware manufacturing service composition model using an extended flower pollination algorithm. Int. J. Prod. Res. **56**(14), 4676–4691 (2018)

13. Abdel-Basset, M., El-Shahat, D., El-Henawy, I.: Solving 0–1 knapsack problem by binary flower pollination algorithm. Neural Comput. Appl. 1–19 (2018). https://doi.org/10.1007/s00521-018-3375-7

14. Li, Y., Pei, Y., Liu, J.: Bat optimal algorithm combined uniform mutation with Gaussian mutation. Control Decis. **32**(10), 1775–1781 (2017)

15. Abdel-Raouf, O., Abdel-Baset, M., El-henawy, I.: A new hybrid flower pollination algorithm for solving constrained global optimization problems. Int. J. Appl. Oper. Res. **4**(2), 1–13 (2014)

16. Lenin, K., Reddy, B.R., Kalavathi, M.S.: Shrinkage of active power loss by hybridization of flower pollination algorithm with chaotic harmony search algorithm. Control Theory Inform. **4**(8), 31–38 (2014)

17. Salgotra, R., Singh, U.: A novel bat flower pollination algorithm for synthesis of linear antenna arrays. Neural Comput. Appl. **30**(7), 2269–2282 (2018)

18. Nabil, E.: A modified flower pollination algorithm for global optimization. Expert Syst. Appl. **57**, 192–203 (2016)

19. Wang, R., Zhou, Y.Q.: Flower pollination algorithm with dimension by dimension improvement. Math. Prob. Eng. **4**, 1–9 (2014)

20. Xiao, H.H., Wan, C.X., Duan, Y.M., Tan, Q.L.: Flower pollination algorithm based on gravity search mechanism. Acta Automatica Sinica **43**(04), 576–592+491+493+594 (2017)

21. Salgotra, R., Singh, U.: Application of mutation operators to flower pollination algorithm. Expert Syst. Appl. **79**, 112–129 (2017)

22. Draa, A.: On the performances of the flower pollination algorithm – qualitative and quantitative analyses. Appl. Soft Comput. **34**, 349–371 (2015)

23. Haupt, R.L., Haupt, S.E.: Practical Genetic Algorithms, 2nd edn. Discrete Applied Mathematics, New York (2004)

24. Li, R., Liu, Q., Liu, L.: Novel image encryption algorithm based on improved logistic map. IET Image Proc. **13**, 125–134 (2018)

25. Fan, J.L., Zhang, X.F.: Piecewise logistic chaotic map and its performance analysis. Acta Electronica Sinica **37**(04), 720–725 (2009)

26. Eberhart, R., Kennedy, J.: A new optimizer using particle swarm theory. In: Proceedings of the 6th Int Symposium on Micro Machine and Human Science, pp. 39–43. IEEE, Piscataway (1995)

27. Kennedy, J., Eberhart, R.: Particle swarm optimization. In: Proceedings of the 1995 IEEE International Conference Neural Networks, vol. 4, no. 8, pp. 1942–1948 (2011)

28. Arora, S., Singh, S.: Butterfly algorithm with Lèvy flights for global optimization. In: International Conference on Signal Processing, Computing and Control, pp. 220–224. IEEE, Waknaghat (2015)

# Application of Data Analysis in Trend Prediction of Different Crime Types in London

Dianting Liu[1,2], Jingwen Tang[2,3(✉)], and Chenguang Zhang[1]

[1] College of Mechanical and Control Engineering,
Guilin University of Technology, Guilin 541004, Guangxi, China
[2] College of Information Science and Engineering,
Guilin University of Technology, Guilin 541004, Guangxi, China
564414779@qq.com
[3] Information School, University of Sheffield,
Western Bank, Sheffield S10 2TN, UK

**Abstract.** This paper aims to analyze the data of different crime types in the London area for the past three years. The data is from data.police.uk provided by City of London Police. The R language and ARIMA model were applied to forecast different crime types in London. Seven sets of stable non-white sequences were obtained by preprocessing the original data, stationary test, and non-stationary sequence differential processing. The time series analysis method was used to realize the ARIMA model modelling. After comparison, the prediction was performed with the optimal ARIMA model. In terms of anti-social behaviour (i1), public order (i9), robbery (i10), shoplifting (i11), theft from the person (i12), the predicted results show a small fluctuation. And it shows a large fluctuation trend in bicycle theft (i2), but the overall number is lower than the previous months. Violence and sexual offences (i14) show a downtrend.

**Keywords:** Data prediction · ARIMA model · Data analysis

## 1 Introduction

This article mainly analyzes and predicts the crime data provided by The UK Police Dataset. The author selected London, the capital of England and the largest city in England, as a research target. Through data cleansing, the author divides the crime data recorded by the police in the past three years into 14 different datasets according to the crime type, then analyzes and predicts them by using the ARIMA model. The specific process and results will be explained in the second and third chapters and presented in visual graphics.

With the rapid growth of global data volume and data analysis needs, massive data has been used more and more in journalism, government and financial institutions. It has also brought opportunities for the rapid development of predictive police technology. The source of crime prediction can be traced back to the United States in the 1980s. New York City Traffic Police Captain JackMaple tried to create a "Charts of the Future" on the wall (Dussault 1999). It helped the police reduce the number of subway robberies by 27% between 1990 and 1992 (Dussault 1999). Since then, the concept of

© Springer Nature Singapore Pte Ltd. 2019
R. Mao et al. (Eds.): ICPCSEE 2019, CCIS 1059, pp. 445–467, 2019.
https://doi.org/10.1007/978-981-15-0121-0_35

"predictive policing" has received more and more attention. Many countries have analyzed and studied crime trends and proposed strategies for crime prevention. Analyze crime trends by studying the changing patterns of specific case locations, time, means, and types. For example, crime map technology is used to analyze the time and place of night theft, and to predict the change of time and place of crime (Bowers et al. 2004); By analyzing the basic data of national statistics in the past five years, it is possible to predict the victimization rate of people of different age groups (Chu and Kraus 2008); Using multivariate discrete analysis method to qualitatively and quantitatively analyze massive crime data, and predict the change of case types by analyzing the inherent relationship of cases and the correlation between cases and cases (Methe et al. 2008).

By using data-based predictive policing techniques, police officers can extract useful data from massive case information and social resource information. In addition, analyze the characteristics of violent crimes, terrorist attacks, vicious group events and other types of cases, then conduct data-driven crime prediction. Law enforcement agencies can adjust and deploy police power based on early warning information to resolve crimes in the bud. Because the analysis of the criminal trend of data will become a very effective way to combat illegal crime, the author believes that studying crime prediction is a topic worthy of study and discussion.

## 2  Methodology

### 2.1  Data Source, Data Preparation and Cleaning

First, the data used in this article is derived from the British police data set. This is an open data, and each month, each police unit generates crime and ASB documents and police outcome documents in a fixed format (Data.police.uk n.d.). The author downloaded data of the London City Police Department from January 2016 to November 2018. These 35 CSV files are used as the original data source of this article. Take the original data of November 2018 as an example, as shown in Fig. 1, the data contains 12 variable types such as Crime ID, Month, etc. and there are missing values in it. The following will explain the data cleanup in detail.

```
> head(police)
# A tibble: 6 x 12
  `Crime ID` Month `Reported by` `Falls within` Longitude Latitude Location `LSOA code` `LSOA name`
  <chr>      <chr> <chr>         <chr>              <dbl>    <dbl> <chr>    <chr>       <chr>
1 NA         2018~ City of Lond~ City of Londo~    -0.115     51.5 On or n~ E01000914   Camden 028B
2 9fa951f19~ 2018~ City of Lond~ City of Londo~    -0.112     51.5 On or n~ E01000914   Camden 028B
3 NA         2018~ City of Lond~ City of Londo~    -0.0974    51.5 On or n~ E01000001   City of Lo~
4 NA         2018~ City of Lond~ City of Londo~    -0.0976    51.5 On or n~ E01000001   City of Lo~
5 NA         2018~ City of Lond~ City of Londo~    -0.0974    51.5 On or n~ E01000001   City of Lo~
6 NA         2018~ City of Lond~ City of Londo~    -0.0976    51.5 On or n~ E01000001   City of Lo~
# ... with 3 more variables: `Crime type` <chr>, `Last outcome category` <chr>, Context <lgl>
```

**Fig. 1.** The first six items of raw data

In the data preparation phase, based on the goal of predicting different types of crimes in this paper, author extracted the data of the crime type from the data which from January 2016 to November 2018 and used this data to create a new dataset. The packages used include: **tidyverse, rvest, dplyr**. The specific process: through the **group_by** function, the column of "Crime type" in 35 original data sets is extracted separately; In the unit of the year (2016, 2017, 2018), the extracted columns are sequentially merged by **dplyr::left_join** to form a new crime type data set, taking the 2016 crime type data set as an example, as shown in Fig. 2; Finally, combine three-year crime type datasets by 14 types of crimes with the **rbind** function, taking the anti-social behaviour of a certain crime type as an example, as shown in Fig. 3. In the process, the system detected that the crime type anti-social behaviour and public order contained missing values, and deleted (**na.omit**) the number to solve the problem of missing values due to the small amount (Dowle n.d.). Through the above steps, data cleaning is basically completed.

```
> head(crimedata16)
  month Anti-social behaviour Bicycle theft Burglary Criminal damage and arson Drugs Other crime Other theft
1   01                     65            15       14                          29    36          11          84
2   02                     92            14       15                          24    29          18         109
3   03                     57            16       16                          28    34          20         144
4   04                     79            34       17                          21    29           9         114
5   05                     50            26       21                          14    27          17         104
6   06                     64            38       11                          13    27          13         138
  Possession of weapons Public order Robbery Shoplifting Theft from the person Vehicle crime Violence and sexual offences
1                     5           22       2          54                    33            13                           63
2                     2           25       5          55                    33            14                           74
3                     3           25       1          37                    35             9                           74
4                     4           16       3          64                    25            19                           78
5                     3           17       1          49                    36            20                           73
6                     1           18       1          44                    39            15                           78
```

**Fig. 2.** Data sets for different crime types in 2016

```
> head(itemi1)
# A tibble: 6 x 1
  `Anti-social behaviour`
                    <dbl>
1                      65
2                      92
3                      57
4                      79
5                      50
6                      64
```

**Fig. 3.** The first type of crime type "anti-social behaviour" three-year data set

## 2.2 Time Series Test and ARIMA Model

Time series analysis is an important part of statistical analysis. It is a column of data in fixed time intervals. The main purpose of time series analysis is to predict the future based on existing data (Prabhakaran 2016). This article applies a powerful algorithm ARIMA in time series. ARIMA is an abbreviation for AutoRegressive Integrated

Moving Average. The autoregressive (AR) term refers to the hysteresis of the series of differences, the moving average (MA) term refers to the hysteresis of the error, and "I" is the number of differences used to make the time series stationary (Chatterjee 2018). ARIMA was jointly proposed by Box and Batholomew in 1970 and is a typical time series analysis method (Bartholomew 1971). The basic idea is to observe the time series, find the internal law from the perspective of sequence autocorrelation, and use its change law to predict the future situation.

The specific method: First, plot the data and observe its stationarity. If the data is a non-stationary time series, a d-order difference operation is performed to convert it into a stationary time series. Here "d" is the "d" in the ARIMA (p, d, q) model (University of Washington [UW] 2005). Calculate the autocorrelation function (ACF) and the partial autocorrelation function (PACF) for the obtained stationary time series (Sangarshanan 2018). Through the analysis of the autocorrelation graph and the partial autocorrelation graph, the values of the highest order of the autocorrelation function and the partial correlation function in the short-term correlation model are roughly judged. At the same time, the R language prediction package is used to automatically determine the order and obtain the best level "p" and order "q". Finally, "d", "q", and "p" obtained by the above method constitute an ARIMA model (Kang 2017). The flexibility of the ARIMA model is that it is possible to select several sets of models for comparison by repeatedly trying to select a relatively optimal model. The entire forecasting process and the results of each step are analyzed and explained in the next section.

# 3   Results and Discussion

## 3.1   Results

**Stationarity Test**
This section will show the results of the stationarity check for 14 crime types in the London area from January 2016 to November 2018. The t-series package needs to be loaded before verification (Trapletti et al. 2018).
According to historical data, time series diagram of 14 crime type data is drawn, as follows (Fig. 4):

**i1 <- ts(itemi1,start = c(2016), frequency = 12)** #Use the first crime type code as an example to convert data into time series format
   **Plot.ts(i1)**

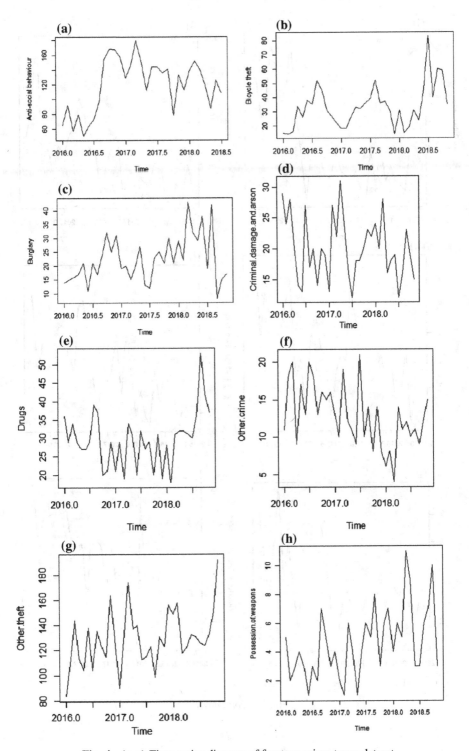

**Fig. 4.** (a–n) Time series diagram of fourteen crime types dataset

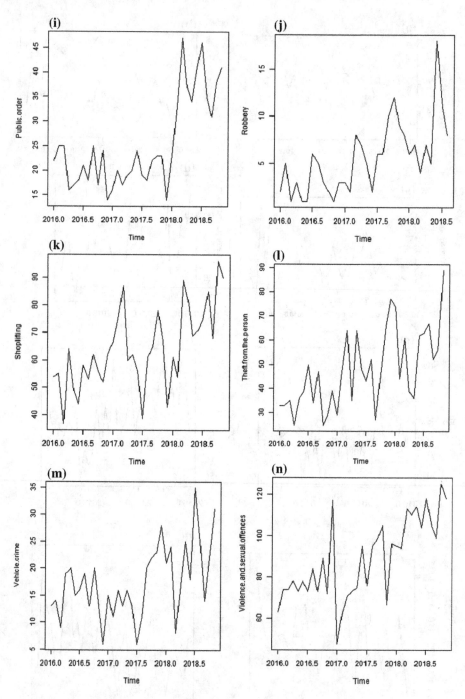

**Fig. 4.** (*continued*)

As can be seen from the figure, there are no obvious periodic and seasonal trends in the time series of the 14 crime types.

The results of the Unit root test results are as follows (Fig. 5):

**Adf.test(i1)** #Use the first crime type code as an example to perform unit root test

| Type of crime | data | Dickey-Fuller | Lag order | p-value |
|---|---|---|---|---|
| Anti-social behaviour | i1 | -1.7436 | 3 | 0.6715 |
| Bicycle theft | i2 | -3.6051 | 3 | 0.04691 |
| Burglary | i3 | -2.0793 | 3 | 0.5423 |
| Criminal damage and arson | i4 | -3.6879 | 3 | 0.04067 |
| Drugs | i5 | -1.0329 | 3 | 0.9188 |
| Other crime | i6 | -2.1354 | 3 | 0.5205 |
| Other theft | i7 | -3.0643 | 3 | 0.1598 |
| Possession of weapons | i8 | -4.1023 | 3 | 0.01729 |
| Public order | i9 | -1.6684 | 3 | 0.7018 |
| Robbery | i10 | -3.5867 | 3 | 0.04921 |
| Shoplifting | i11 | -3.4814 | 3 | 0.06195 |
| Theft from the person | i12 | -4.6259 | 3 | 0.01 |
| Vehicle crime | i13 | -3.0882 | 3 | 0.1506 |
| Violence and sexual offences | i14 | -2.2483 | 3 | 0.4767 |

**Fig. 5.** Unit root test

The crime types of unit root test results $p > 0.05$ are Anti-social behaviour, Burglary, Drugs, Other crime, Other theft, Public order, Shoplifting, Vehicle crime, Violence and sexual offences, to determine that it is a non-stationary sequence.

The data were tested for pure randomness and the results were as follows (Fig. 6):

**Box.test(i1)** #Take the first crime type code as an example and conduct a pure randomness test

| Type of crime | data | X-squared | df | p-value |
|---|---|---|---|---|
| Anti-social behaviour | i1 | 12.852 | 1 | 0.0003371 |
| Bicycle theft | i2 | 8.2351 | 1 | 0.004109 |
| Burglary | i3 | 0.77537 | 1 | 0.3786 |
| Criminal damage and arson | i4 | 0.8281 | 1 | 0.3628 |
| Drugs | i5 | 3.2103 | 1 | 0.07318 |
| Other crime | i6 | 0.40035 | 1 | 0.5269 |
| Other theft | i7 | 0.72529 | 1 | 0.3944 |
| Possession of weapons | i8 | 2.1241 | 1 | 0.145 |
| Public order | i9 | 18.901 | 1 | 1.38E-05 |
| Robbery | i10 | 8.2678 | 1 | 0.004035 |
| Shoplifting | i11 | 6.7578 | 1 | 0.009334 |
| Theft from the person | i12 | 4.614 | 1 | 0.03171 |
| Vehicle crime | i13 | 3.7138 | 1 | 0.05397 |
| Violence and sexual offences | i14 | 6.3684 | 1 | 0.01162 |

**Fig. 6.** Pure randomness test (Color figure online)

The items with P > 0.05 after the test are shown in red in the figure above, that is, these items are pure random sequences, also known as white noise time series. There is no correlation between the values of the sequence, and the sequence is subjected to a completely disordered random fluctuation, which can terminate the analysis of the sequence (QuantStart Team n.d.). The white noise time series is a stationary sequence with no information to extract. The remaining items are not white noise sequences and can be modelled.

Through the above-mentioned stationarity check, "Anti-social behaviour", "Public order", "Shoplifting", "Violence and sexual offences" of the 14 crime types are non-stationary sequences. Firstly, the logarithmic transformation of non-stationary sequences. The logarithmic transformation is mainly to reduce the vibration amplitude of the data, making its linearity law more obvious. The logarithmic transformation is equivalent to adding a penalty mechanism (Ryu 2018). The larger the data, the bigger the penalty, the smaller the data, the smaller the penalty. Then differentiate the transformed data. The first-order difference is the value of the previous item minus each item of the original number. The second-order difference is the first difference based on the first-order difference. Continuously make the difference straight to get a smooth sequence. R uses the **diff()** function to perform differential operations on time series.

**i1log <- log(i1)** # takes the first crime type code as an example and performs logarithmic transformation

**i1diff1 <- diff(i1log, differences=1)** # Take the first crime type code as an example and make a first-order difference

After the first-order difference, perform a stationarity test again. The unit root test and the pure randomness test result are as shown in the following table (Fig. 7):

| Type of crime | data | p-value(adf-diff1) | p-value(Box-diff1) |
|---|---|---|---|
| Anti-social behaviour | i1 | 0.01381 | 0.02559 |
| Public order | i9 | 0.03849 | 0.07458 |
| Shoplifting | i11 | 0.01 | 0.009467 |
| Violence and sexual offences | i14 | 0.07845 | 0.0002642 |

**Fig. 7.** The unit root test and the pure randomness test result after the first-order difference

It can be seen that only "Anti-social behaviour", "Shoplifting" two types are smooth non-white noise sequences. Therefore, the second order difference is applied to the other two items.

**i9diff2 <- diff(i9log, differences=2)** # Take the ninth crime type code as an example for second-order difference

After the second-order difference, perform a stationarity test again. The unit root test and the pure randomness test result are as shown in the following table (Fig. 8):

| Type of crime | data | p-value(adf-diff1) | p-value(Box-diff1) |
|---|---|---|---|
| Public order | i9 | 0.01 | 0.0005965 |
| Violence and sexual offences | i14 | 0.01 | 1.22E-05 |

**Fig. 8.** The unit root test and the pure randomness test result after the second-order difference

It can be seen that after the second order difference, the other two items "Public orders" and "Violence and sexual offences" also become smooth non-white noise sequences.

Through the above stationarity check and differential retest, it can be concluded that the stationary non-white noise sequence is: "Bicycle theft" (i2), "Robbery" (i10), "Theft from the person" (i12); The stationary non-white noise sequence obtained after the first order difference are "Anti-social behaviour" (i1), "Shoplifting" (i11); smooth non-white noise sequences obtained by second-order difference are "Public order" (i9), "Violence and sexual offences" (i14). The above sequence can be modelled. ARIMA model testing and parameter estimation will be analyzed and explained in the next section.

## ARIMA Model Test and Parameter Estimation

Next is to choose the appropriate ARIMA model, that is, determine the appropriate "p" and "q" values in ARIMA (p, d, q). This article use the "**acf()**" and "**pacf()**" functions in R to make judgments (Dalinina 2017). "**acf()**" and "**pacf()**" set "plot=FALSE" to get the true values of autocorrelation and partial correlation. Take the first crime type Anti-social behaviour as an example, and see the appendix for the autocorrelation plots and partial autocorrelation plots for the remaining crime types.

Anti-social behaviour auto-correlation diagram is as follows (Fig. 9):

**Acf(i1diff1, lag.max = 100)**
**Acf(i1diff1, lag.max = 100, plot = F)**

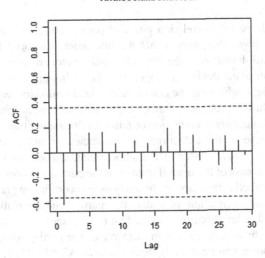

**Anti.social.behaviour**

**Fig. 9.** Anti-social behaviour auto-correlation diagram

Anti-social behaviour partial auto-correlation diagram is related to the following (Fig. 10):

**Pacf(i1diff1, lag.max = 100)**
**Pacf(i1diff1, lag.max = 100, plot = F)**

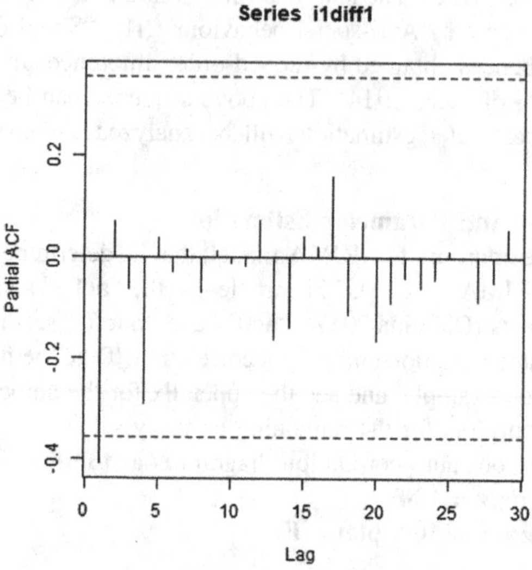

**Fig. 10.** Anti-social behaviour partial auto-correlation diagram

Analyze the above autocorrelation plot and partial autocorrelation plot: the auto-correlation graph shows that, except that the first-order autocorrelation coefficient is outside the standard deviation range, the other order autocorrelation coefficients fluctuate within the standard deviation range. This is an important feature of the auto-correlation coefficient, which can be determined to be a stationary sequence with short-term correlation. In the process of further examining the overall change of the auto-correlation coefficient, there is no obvious sinusoidal fluctuation trajectory. The process of the overall change of the autocorrelation coefficient is not a continuous gradual process, and the first-order censor of the autocorrelation coefficient can be obtained. Then analyze the process of the overall change of the partial autocorrelation coefficient, and the analysis method is the same as the analysis autocorrelation graph. It can be seen from the partial autocorrelation plot that the partial autocorrelation coefficient is truncated at the first order. According to the first-order truncation of the autocorrelation coefficient and the first-order truncation property of the partial autocorrelation coeffi-cient, the fitting model can be initially determined as ARIMA (1, 1, 1) (Adhikari and Agrawal 2013). In order to avoid the problem of inaccurate model recognition caused by insufficient personal experience, R provides the auto.arima function (Dowle (n.d.)).

The function automatically recognizes the model order based on the principle of minimum information, and gives the parameter estimates of the model. This process requires loading the zoo and forecast packages first (Bar 2016). After analysis according to the above method, the preliminary fitting model of the remaining crime types and the ARIMA model obtained after automatically recognizing the model order are shown in the following table (Fig. 11):

| Type of crime | data | arima | arimaauto |
|---|---|---|---|
| Anti-social behaviour | i1diff1 | (1,1,1) | (1,0,0) |
| Bicycle theft | i2 | (0,0,0) (0,0,0)[12] | (0,0,0) (0,1,0)[12] |
| Public order | i9diff2 | (0,2,1) | (1,0,0) |
| Robbery | i10 | (1,0,1) | (0,1,0) |
| Shoplifting | i11diff1 | (0,1,1) | (0,0,1) |
| Theft from the person | i12 | (0,0,0) | (0,1,1) |
| Violence and sexual offences | i14diff2 | (0,2,1) (0,2,1)[12] | (3,0,0) (0,1,0)[12] |

**Fig. 11.** ARIMA model

The AIC (Akaike's Information Criterion) value is an estimate of the comparison model. The lower the value, the better the model (Stephanie 2015). After comparing the above ARIMA models, the following model can be derived as the best model for each sequence.

**Arimaauto i1diff1 (1,0,0)**
**Arimaauto i2(0,0,0)(0,1,0)[12]**
**Arimaauto i9diff2 (1,0,0)**
**Arimaauto i10 (0,1,0)**
**Arimaauto i11diff1 (0,0,1)**
**Arimaauto i12 (0,1,1)**
**Arimaauto i14diff2(3,0,0)(0,1,0)[12]**

**Model Prediction**

Next, the crime types are predicted according to the best model. The results are as follows (Fig. 12):

**i10forecast <- forecast(i10arimaauto,h=5,level=c(99.5))** #Use the tenth crime type code as an example
**i10forecast**
**Autoplot(i10forecast)**

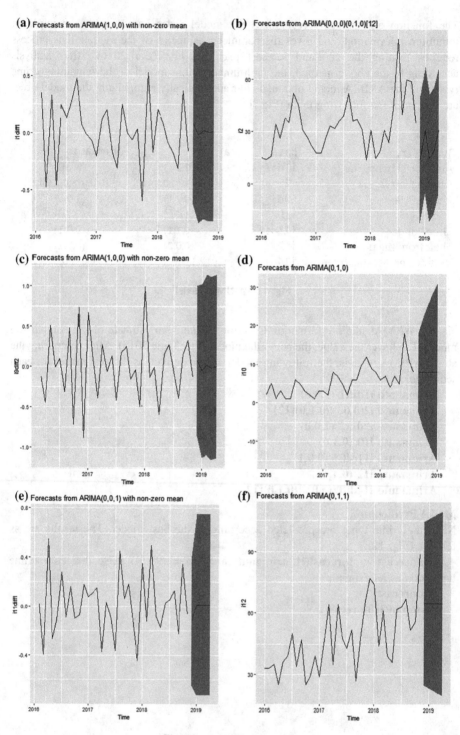

**Fig. 12.** (a–g) Forecast map

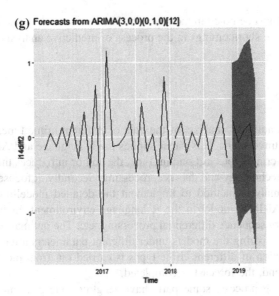

**(g)** Forecasts from ARIMA(3,0,0)(0,1,0)[12]

**Fig. 12.** (*continued*)

From the above prediction chart, it can be analyzed: Anti-social behaviour (i1), Public order (i9), Shoplifting (i11), these types of crimes show a steady small fluctuation; Bicycle theft (i2) has a large fluctuation trend, but the overall number is lower than the previous months; Violence and sexual offences (i14) show a trend of decreasing volatility; In addition, there is no obvious fluctuation in Robbery (i10) and Theft from the person (i12) two items, and the prediction model can be improved or the historical data can be added to make the prediction result more accurate.

## 3.2 Discussion

The development of crime prediction based on big data is in the ascendant and very vital. Its advantages are very significant (Balakrishnan 2018). First, crime prediction based on big data helps the police understand the risks of present and foreseeing the future. Second, many methods of crime prediction can be combined, mutually validated, and searched for optimal solutions, which can make the predicted results more accurate, so as to better guide practice. Finally, this kind of crime prediction has the characteristics of convenience and high efficiency. Using big data technology can improve the accuracy of crime prediction (Clark 2017). Comprehensive consideration of the relevance of various factors and multi-angle prediction of crime. Proactive intervention to achieve crime prevention.

However, through the crime prediction of big data, it is impossible to directly judge the causal relationship. If the forecast requires finding a causal relationship, it still requires a lot of manpower for analysis, which is inevitably subject to the influence of self-prejudice. Second, crime prediction based on big data threatens personal privacy and personal freedom to some extent, which requires more efforts on data sources and data protection. Finally, there is some information that is not easily quantifiable, and that information must be noted in the research. The next section will summarize the full text and explain the shortcomings in the process of predictive analysis and the areas to be improved.

## 4   Conclusion

The above is the author's analysis of the data of different crime types in the London area for the past three years. This paper applies the R language and ARIMA model to forecast different crime types in London. First, the author introduces the background of predictive crime technology and data preprocessing. Secondly, it focuses on how to use the time series analysis method to implement the detailed modelling steps and test methods of the ARIMA model in the R language environment, including stationary test, non-stationary sequence differential processing, etc. The optimal ARIMA model is determined by comparing the models under different parameters after detection. Next, the annual prediction of different crime types is carried out from the monthly dimension, and in the end, the forecast is completed.

In the encoding process, some parts have lengthy code phenomena. The author hopes to achieve the expected functionality and results in the most simplified code by further learning the R language. For example, when reading a file, the author implements batch import of files in a few lines of code; however, in the process of data cleansing and prediction, due to the same processing method, multiple lines appear in the same type of code set. Based on this situation, the code part of this analysis study has yet to be succinct and improved. For the time series analysis method, the ARIMA model pricing part needs to manually determine the "acf" and "pacf" diagrams to determine the order, which has the influencing factors of human error and human bias. In addition, the UK area crime data provided by UK Police Dataset is only the last three years of data. After data cleansing, the data of each crime type is only 35. At the same time, there are some cases where the type data is missing. After the phenomenon of data loss is solved, the data becomes less. For the time series analysis, more historical data can increase the accuracy of the prediction, so the amount of historical data has yet to be expanded.

Predicting policing technology has unprecedented room for development, and law enforcement agencies are becoming more dependent on data. By using big data-based predictive policing techniques, law enforcement agencies are able to make the most of the people and the information resources they have. In addition, monitor, measure and predict future criminal activity and crime trends in order to effectively deploy resources and quickly handle cases.

## Appendix

## A1. Autocorrelation Graph and Partial Autocorrelation Graph

(See Figs. 13, 14, 15, 16, 17, 18, 19, 20, 21 and 22).

**Public.order**

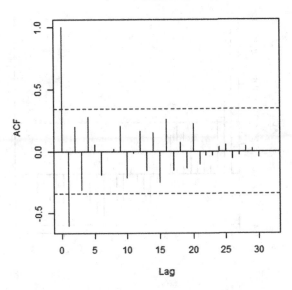

**Fig. 13.** Public order (i9) Autocorrelation graph

**Series  i9diff2**

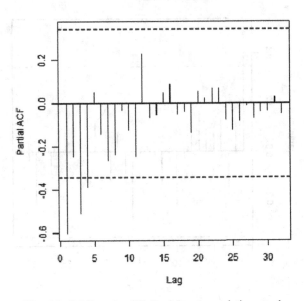

**Fig. 14.** Public order (i9) Partial autocorrelation graph

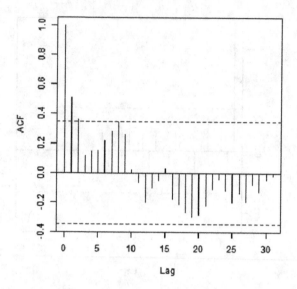

**Fig. 15.** Robbery (i10) Autocorrelation graph

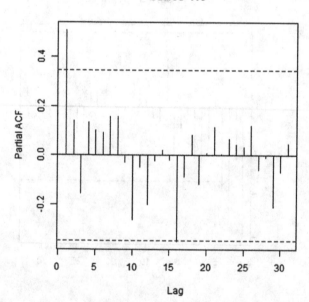

**Fig. 16.** Robbery (i10) Partial autocorrelation graph

**Shoplifting**

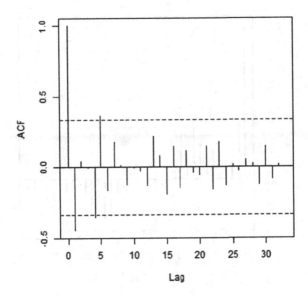

**Fig. 17.** Shoplifting (i11) Autocorrelation graph

**Series i11diff1**

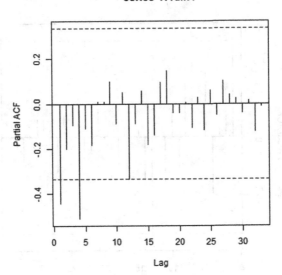

**Fig. 18.** Shoplifting (i11) Partial autocorrelation graph

**Theft.from.the.person**

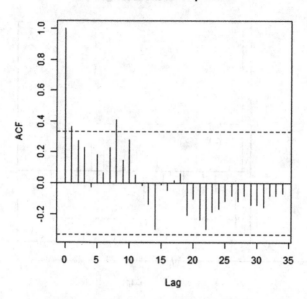

**Fig. 19.** Theft from the person (i12) Autocorrelation graph

**Series i12**

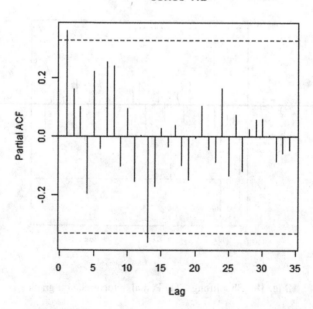

**Fig. 20.** Theft from the person (i12) Partial autocorrelation graph

**Violence.and.sexual.offences**

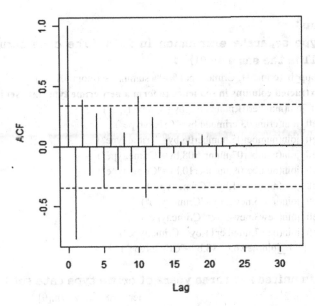

**Fig. 21.** Violence and sexual offences (i14) Autocorrelation graph

**Series i14diff2**

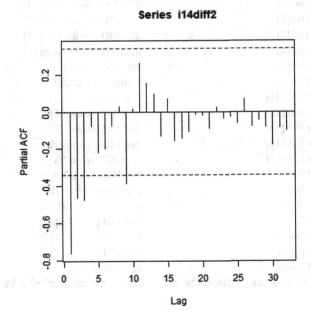

**Fig. 22.** Violence and sexual offences (i14) Partial autocorrelation graph

## A2. R Code

**Data cleanup:**
**Criminal type separate extraction in 2018 (The code structure of number02-11 is the same as 01):**
number01<- group_by(crime01, Crime.type) %>% summarise(month01=n())
**Combine the extracted columns in sequence to form a new crime type data set in 2018:**
new<-dplyr::left_join(number01,number02,by="Crime.type")
new1<-dplyr::left_join(crime03,crime04,by="Crime.type")
new2<-dplyr::left_join(number05,number06,by="Crime.type")
new3<-dplyr::left_join(number07,number08,by="Crime.type")
new4<-dplyr::left_join(number09,number10,by="Crime.type")
new5<-dplyr::left_join(new,new1,by="Crime.type")
new6<-dplyr::left_join(new5,new2,by="Crime.type")
new7<-dplyr::left_join(new3,new4,by="Crime.type")
new8<-dplyr::left_join(new7,number11,by="Crime.type")
crimedata<-dplyr::left_join(new6,new8,by="Crime.type")

**Crime type is united for three years of crime type data set:**

item16i1<-view16m[2]
item16i2<-view16m[3]
item16i3<-view16m[4]
item16i4<-view16m[5]
item16i5<-view16m[6]
item16i6<-view16m[7]
item16i7<-view16m[8]
item16i8<-view16m[9]
item16i9<-view16m[10]
item16i10<-view16m[11]
item16i11<-view16m[12]
item16i12<-view16m[13]
item16i13<-view16m[14]
item16i14<-view16m[15]
item17i1<-view17m[2]
item17i2<-view17m[3]
item17i3<-view17m[4]
item17i4<-view17m[5]
item17i5<-view17m[6]
item17i6<-view17m[7]
item17i7<-view17m[8]
item17i8<-view17m[9]
item17i9<-view17m[10]
item17i10<-view17m[11]
item17i11<-view17m[12]
item17i12<-view17m[13]
item17i13<-view17m[14]
item17i14<-view17m[15]
item18i1<-view2[2]
item18i2<-view2[3]
item18i3<-view2[4]
item18i4<-view2[5]
item18i5<-view2[6]
item18i6<-view2[7]
item18i7<-view2[8]
item18i8<-view2[9]
item18i9<-view2[10]
item18i10<-view2[11]
item18i11<-view2[12]
item18i12<-view2[13]
item18i13<-view2[14]
item18i14<-view2[15]

**Query missing values (The code structure of number2-14 is the same as 1):**
sum(is.na(itemi1))
**Delete missing values:**
itemi1=na.omit(itemi1)
itemi10=na.omit(itemi10)

**Data prediction :**

**Convert a crime type sequence to a time series (The code structure of numberi2–i14 is the same as i1) :**

```
i1<- ts(itemi1,start = c(2016),frequency=12)#
png(file="i1.png",width=398,height=417)
print(plot.ts(i1))
```

**Unit root test (The code structure of numberi2–i14 is the same as i1) :**

```
adf.test(i1)
```

**Pure randomness test (The code structure of numberi2–i14 is the same as i1) :**

```
Box.test(i1)
```

**Logarithm :**

```
i1log <- log(i1)
```

**First order difference :**

```
i1diff1 <- diff(i1log, differences=1)
i9log <- log(i9)
i9diff1 <- diff(i9log, differences=1)
i11log <- log(i11)
i11diff1 <- diff(i11log, differences=1)
i14log <- log(i14)
i14diff1 <- diff(i14log, differences=1)
```

**Unit root test :**

```
adf.test(i1diff1)
adf.test(i9diff1)
adf.test(i11diff1)
adf.test(i14diff1)
```

**Pure randomness test :**

```
Box.test(i1diff1)
Box.test(i9diff1)
Box.test(i11diff1)
Box.test(i14diff1)
```

**Second order difference**

```
i9diff2 <- diff(i9log, differences=2)
i14diff2<- diff(i14log, differences=2)
```

**Unit root test**

```
adf.test(i9diff2)
adf.test(i14diff2)
```

**Pure randomness test**

```
Box.test(i9diff2)
Box.test(i14diff2)
```

**Generate autocorrelation plots and partial autocorrelation plots and save (The code structure of rest item is the same as i2) :**

```
png(file="acfi2.png",width=398,height=417)
print(acf(i2, lag.max = 100))
dev.off()
```

## Automatic recognition of model order(The code structure of rest item is the same as i2) :

```
auto.arima(i2,trace=T)
```

**Forecast based on the best model(The code structure of rest item is the same as i2):**

```
i2arima <- arima(i2, order = c(0, 0, 0),seasonal=list(order=c(0,0,0),period=12),method="ML")
i2arima
i2arimaauto<-arima(i2,order= c(0,0, 0),seasonal=list(order=c(0,1,0),period=12),method="ML")
i2arimaauto
i2forecast <- forecast(i2arimaauto,h=5,level=c(99.5))
i2forecast
autoplot(i2forecast)
png(file="i2forecast.png",width=398,height=417)
print(autoplot(i2forecast))
dev.off()
```

## References

Adhikari, R., Agrawal, R.K.: An Introductory Study on Time Series Modeling and Forecasting. LAP Lambert Academic Publishing (2013). https://doi.org/10.13140/2.1.2771.8084

Balakrishnan, V.S.: Could Big Data Unlock the Potential of Predictive Policing? (2018). http://blogs.discovermagazine.com/crux/2018/10/02/big-data-predictive-policing/#.XETvilz7QRk. Accessed 30 Dec 2018

Bar, M.: Forecasting Times Series with R (2016). http://online.sfsu.edu/mbar/ECON312_files/Forecasting.html. Accessed 22 Dec 2018

Bartholomew, D.J.: Time series analysis forecasting and control. J. Oper. Res. Soc. **22**(2), 199–201 (1971). https://doi.org/10.1057/jors.1971.52

Bowers, K.J., Johnson, S.D., Pease, K.: Prospective hot-spotting: the future of crime mapping. Br. Criminol. **44**(2), 641–658 (2004)

Chatterjee, S.: Time Series Analysis Using ARIMA Model in R (2018). https://datascienceplus.com/time-series-analysis-using-arima-model-in-r/. Accessed 22 Dec 2018

Chu, L.D., Kraus, J.F.: Predictive fatal assault among the elderly using the national incident-based reporting system crime data. Homicide Stud. **8**(5), 71–95 (2008)

Clark, J.: Facing the threat: Big Data and crime prevention (2017). https://www.ibm.com/blogs/internet-of-things/big-data-crime-prevention/. Accessed 30 Dec 2018

Dalinina, R.: Introduction to Forecasting with ARIMA in R (2017). https://www.datascience.com/blog/introduction-to-forecasting-with-arima-in-r-learn-data-science-tutorials. Accessed 29 Dec 2018

Data.police.uk: About data.police.uk (n.d.). https://data.police.uk/about/. Accessed 9 Dec 2018

Dowle, M.: na.omit.data.table. R Documentation (n.d.). https://www.rdocumentation.org/packages/data.table/versions/1.11.8/topics/na.omit.data.table. Accessed 18 Dec 2018

Dussault, R.: Jack Maple: Betting on Intelligence. Government Technology (1999). http://www.govtech.com/magazines/gt/Jack-Maple-Betting-on-Intelligence.html. Accessed 20 Dec 2018

Kang, E.: Time Series: ARIMA Model (2017). https://medium.com/@kangeugine/time-series-arima-model-11140bc08c6. Accessed 28 Dec 2018

Methe, T.D., Hart, T.C., Regoeczi, W.C.: The conjunctive analysis of case configurations: an exploratory method for discrete multivariate analysis of crime data. J. Quant. Criminol. **24**(2), 227–241 (2008)

Prabhakaran, S.: Time Series Analysis (2016). https://r-statistics.co/. Accessed 22 Dec 2018

QuantStart Team: White Noise and Random Walks in Time Series Analysis (n.d.). QuantStart. https://www.quantstart.com/articles/White-Noise-and-Random-Walks-in-Time-Series-Analysis. Accessed 27 Dec 2018

Ryu, C.: Data Transformation (2018). https://cran.r-project.org/web/packages/dlookr/vignettes/transformation.html. Accessed 30 Dec 2018

Sangarshanan: Time series Forecasting—ARIMA models (2018). https://towardsdatascience.com/time-series-forecasting-arima-models-7f221e9eee06. Accessed 27 Dec 2018

Stephanie: Akaike's Information Criterion: Definition, Formulas Statistics How To (2015). https://www.statisticshowto.datasciencecentral.com/akaikes-information-criterion/. Accessed 28 Dec 2018

Trapletti, A., Hornik, K., LeBaron, B.: Time Series Analysis and Computational Finance (2018). CRAN https://cran.r-project.org/web/packages/tseries/tseries.pdf. Accessed 28 Dec 2018

University of Washington [UW]: 3 Time Series Concepts. 57–110 (2005). https://faculty.washington.edu/ezivot/econ584/notes/timeSeriesConcepts.pdf. Accessed 28 Dec 2018

# Research on Pedestrian Re-Identification Using CNN Feature and Pedestrian Combination Attribute

Mengke Jiang[1]([⊠]), Jinlong Chen[2], and Baohua Qiang[2]

[1] Guangxi Key Laboratory of Trusted Software,
Guilin University of Electronic Technology, Guilin, China
446084066@qq.com
[2] Guangxi Key Laboratory of Cryptography and Information Security,
Guilin University of Electronic Technology, Guilin, China

**Abstract.** Aiming at the problem that the existing pedestrian recognition technology re-identification effect is not good and the traditional method has low recognition effect. A feature fusion network is proposed in this paper, which combines the CNN features extracted by ResNet with the manual annotation attributes into a unified feature space. ResNet solved the problem of network degradation and multi-convergence in multi-layer CNN training, and extracted deeper features. The attribute combination method was adopted by the artificial annotation attributes. The CNN features were constrained by the hand-crafted features because of the back propagation. Then the loss measurement function was used to optimize network identification results. In the public datasets VIPeR, PRID, and CUHK for further testing, the experimental results show that the method achieves a high cumulative matching score.

**Keywords:** Pedestrian re-identification · ResNet · Pedestrian attribute

## 1 Introduction

Re-Id, also known as pedestrian re-identification, is a technology that uses computer vision technology to judge whether there is a same target pedestrian in a camera image that does not overlap. The research on pedestrian re-identification initially focused on the manual annotation of pedestrian feature extraction, the identification of different camera angles, and the learning methods based on distance measurement [1]. However, due to the cost of manual labeling and the lack of traditional methods, this research has not obtain great progress. With the development of machine learning and deep learning, in the ImageNet competition in 2012, the Hinton team [2] first tried to integrate the convolutional neural network into the pedestrian detection technology, and achieved good experimental results. Since 2014, researchers have attempted to incorporate deep learning into the issue of pedestrian re-identification. Deep learning provides a way to solve computer vision problems without the need for excessive manual annotation of image features. The back propagation algorithm dynamically

© Springer Nature Singapore Pte Ltd. 2019
R. Mao et al. (Eds.): ICPCSEE 2019, CCIS 1059, pp. 468–475, 2019.
https://doi.org/10.1007/978-981-15-0121-0_36

adjusts the parameters in the CNN so that the feature extraction and pairwise comparison processes are unified into a single network.

However in the real world, the appearance of a pedestrian will be largely affected by the difference of camera angle, illumination, height, etc. The manual annotation feature can solve the problem well, and the application can make the method more reliable in the pedestrian re-identification task. In order to effectively combine the artificial annotation features with the CNN features, a deep feature fusion network is proposed. The artificial annotation features are used to adjust the CNN process. The features extracted by CNN can also be used as a supplement to the manual annotation features.

Experiments on three challenging pedestrian re-identification attribute data sets (VIPeR, PRID, and CUHK) demonstrate the validity of the new features. Compared with the existing method, the recognition rate of rank-1 has been significantly improved. In summary, the manual annotation feature can improve the extraction process of CNN features and achieve a more robust image representation.

## 2  Related Work

Due to the outstanding performance of deep learning in pedestrian detection and recognition, the re-identification of pedestrians in recent years mainly focuses on deep learning. Many researchers at home and abroad have proposed different improved algorithms in this process. There are two main methods for pedestrian re-identification using deep learning. One is to use the end-to-end technology to extract pedestrian features using the convolutional neural network (CNN) to achieve pedestrian re-identification, such as DeepReID : deep filter pairing neural network for person re-identification; Another is to achieve pedestrian re-identification in combination with the high-level semantic features of pedestrians. For example, Li Jiali et al. [3] proposed to add multi-classification features based on human body structure detection based on deep learning features, and established a multi-feature fusion model with enhanced depth features; reference [4] not only identifies each type of pedestrian attribute separately, but also arranges and combines pedestrian attributes, and then jointly identifies multiple attributes and individual attributes; literature [5] jointly recognized pedestrian attributes and pedestrian IDs, making full use of the pedestrian's annotation information, and improving the accuracy of pedestrian recognition; paper [8] used color histogram features (RGB, HSV, YCbCr, Lab and YIQ) and texture features, combined with the features attracted by the traditional CNN, largely improved the recognition accuracy. These efforts have led to a leap-forward development in pedestrian recognition.

Combining feature extraction and image pair classification into a single CNN network, the comparison and symmetrical structure of paired pictures are widely used, but pairwise comparison demanded to form lots of pairs for each probe image and perform deep convolution on these pairs [6]. Inspired by [6], a feature fusion network is proposed that extracts depth features from a single image without using pairs of inputs. The artificially labeled pedestrian attribute features are combined and merged with the depth features to be unified into a single network, and then the loss measurement function proposed in this paper is used to optimize the network recognition results. The test results show that the proposed algorithm is superior to the current mainstream pedestrian recognition algorithm in public data sets, especially the first accuracy rate (rank-1).

## 3 Model

### 3.1 Network Structure

We use the fine-tuned feature fusion network to learn new features. The network structure consists of two parts, as shown in Fig. 1. The first part uses the ResNet network to extract features from pedestrian sample images. ResNet can solve the problem of network degradation and multi-convergence in multi-layer CNN training, and can extract deeper features. The second part deals with the hand-crafted feature of the same picture. Finally, the two sub-networks are combined to produce a complete image description. In the second part of the study, the results of the first part will be adjusted.

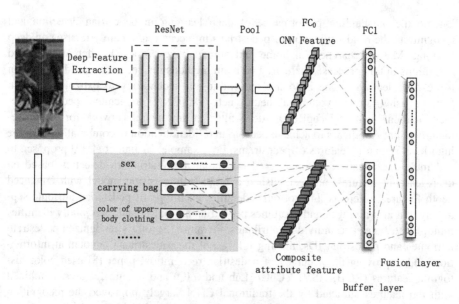

**Fig. 1.** Fusion feature network structure model

### 3.2 CNN Features

The extraction task of image features is using ResNet, the upper part of Fig. 1, which uses the ResNet-50 network, that is a residual network with a depth of 50 layers, and the network is mainly composed of convolutional layer, pooling layer and residuals. In the process of deep learning development, the researchers found that as the number of network layers deepens, the network will undergo gradient dispersion and gradient explosion during training, as a result, the network can't be able to converge during training. Differ from the general CNN, ResNet has a unique residual block structure, which avoid the network degradation and the convergence problem during training without introducing additional parameters and computational complexity by learning the residual function and realizes the identity mapping.

In our framework, by using back propagation, the parameters of the entire CNN network are affected by the hand-crafted features. Our goal is to combine features into a unified feature space. A feature fusion deep neural network is proposed to adjust CNN features by using hand-crafted features, so that CNN can extract complementary features and have a more complete feature representation. The buffer layer can be used as a bridge between the extracted CNN feature and the composite attribute feature to reduce the huge difference between different features and ensure the convergence of the fusion layer. If the input of the fusion layer is:

$$X = [Composite\_Attribute\_Features, \quad CNN\_Features], \tag{1}$$

Then the output of this layer is calculated by the following formula:

$$Z_{Fusion}(x) = h\big(W_{Fusion}^T x + b_{Fusion}\big), \tag{2}$$

Where $h(\cdot)$ indicates the activation function.

Existing deep re-identification networks for person re-identification adopt Deviance Loss or Maximum Mean Discrepancy as loss function [6]. Our goal is to effectively extract the depth features of each image, rather than comparing the image pairs through deep neural networks. So we use the softmax loss function, a more discriminative feature representation will get a lower softmax value.

## 3.3   Hand-Crafted Features

The recognition of pedestrian images by semantic attributes, such as gender, wearing, backpack color, etc., can be used as auxiliary information to improve pedestrian recognition accuracy. There are several advantages to using manual annotation of attribute features: First, because most people have similar appearances (such as clothing color, backpack, hair color, etc.) it is more difficult to manually mark the same pedestrian in a camera with low pixels. In contrast, the labeling of pedestrian attributes is simpler and more accurate. Second, the number of classification of pedestrian attributes is less than that of different pedestrians, since different pedestrians will have the same attributes.

The training data set contains N pedestrian images, and a pedestrian image has multiple attribute annotations (such as gender male, upperbody black), and we group these attributes. According to the paper [4], each image is marked by G attribute groups, e.g., Gender, Age, and every attribute group has a different number of attributes, denoted as K(g), e.g., group gender has male, female, group upperbody clothing has sweater, t-shirt, suit and so on. We assume that each attribute group can only have one attribute value. For example, a pedestrian's upperbody color is black and white. We only select one color, that is, one attribute value.

Since the categories of each attribute group are inconsistent, we define a weighted cross entropy loss function. The loss of output node j is calculated as follows:

$$P_{(y=j)} = \frac{e^{a_j}}{\sum_{j=1}^{T} e^{a_k}} \tag{3}$$

Where T represents the number of categories. The cross entropy loss function is as follows:

$$L = -\frac{1}{N^g} \sum_{j=1}^{N} \sum_{k=1}^{K^g} \frac{y_j log P_j}{N_{k(i)}^g} \tag{4}$$

Where N represents the number of pedestrian pictures, $N^g$ represents the number of pedestrian images in the g-th attribute group, $N_{k(i)}^g$ is the number of training samples of the k-th attribute in the g-th attribute group, and $P_j$ is calculated by the formula 3 inferred.

# 4   Experiment

## 4.1   Dataset

This paper uses the PETA (PEdesTrian Attribute) data set, which is the largest data set currently open for pedestrian attribute recognition tasks. The data set contains 8705 pedestrians for a total of 19,000 images (resolutions from 17*39 to 169*365). Each pedestrian is marked with 61 binary value and 4 multi-category attributes (binary values such as whether they are under 15 years old, and multiple categories such as upperbody colors can have multiple coexistences). In fact, the PETA data set is a collection of multiple smaller pedestrian re-identification data sets that are labeled by attributes. The partial data sets included are shown in the following Table 1:

**Table 1.**  Partial data set of PETA.

| Datasets | Images | Resolution |
| --- | --- | --- |
| 3DPeS | 1012 | From 31*100 to 236*178 |
| CAVIAR4REID | 1220 | From 17*39 to 72*141 |
| CUHK | 4563 | 80*160 |
| GRID | 1275 | From 29*67 to 169*365 |
| PRID | 1134 | 64*128 |
| VIPeR | 1264 | 48*128 |

For all manual annotation properties, we divided then into 8 groups, each group contains a different number of attribute characteristics, as shown in Table 2. Among them, some attribute will be removed if the number of them is less than 10. In addition, if an attribute group has two attribute values at the same time (for example, upperbody has black and white), and we randomly select one value as the attribute label.

**Table 2.** The group of the attribute

| Group | Attributes | Number |
|---|---|---|
| Gender | Male, Female | 2 |
| Age | Young, Teenager, Adult, Old | 4 |
| Hair | Long, Short | 2 |
| Upperbody Color | Black, White, Red, Purple, Yellow, Gray, Blue, Green | 8 |
| Upperbody Clothing | Long Sleeve, Short Sleeve, no Sleeve | 3 |
| Downbody Color | Gray, Black, White, Pink, Purple, Yellow, Blue, Green, Brown | 9 |
| Downbody Clothing | Dress, Pants | 2 |
| Other | Bag, Hat, Handbag, Backpack | 4 |

## 4.2 Setup

We use three pedestrian re-identification databases to evaluate the fine-tuned CNN features and implement our approach using the keras framework. We resize all training images into 256*128 pixels and add a pad with 10 pixels, then randomly crop 256*128 sub-windows. For test time, we resize all the input images to 256*128 pixels. The CNN parameters are all derived from the pre-trained model ResNet-50, we start the last fully connected layer from random weights. The batch size is set to 256, the initial learning rate is set to $\gamma = 0.0001$, and every 20,000 iterations is reduced as $\gamma_{new} = 0.1 * \gamma$.

The method proposed in this paper was tested on three data sets and compared with other methods, the experimental results are shown in Table 3. The results of rank-1 on the VIPeR, PRID, and CUHK datasets were 45.23%, 50.22%, and 49.20, separately. It can be seen that compared to the methods listed in the table, the accuracy tested in the three datasets used the method proposed in this paper has improved a lot, which proves the effectiveness of our method.

**Table 3.** Performance comparison with different methods

| Methods | VIPeR | | | | PRID | | | | CUHK | | | |
|---|---|---|---|---|---|---|---|---|---|---|---|---|
| | r = 1 | r = 5 | r = 10 | r = 20 | r = 1 | r = 5 | r = 10 | r = 20 | r = 1 | r = 5 | r = 10 | r = 20 |
| Ours | **45.23** | **73.88** | **85.23** | **92.34** | **50.22** | **68.89** | **83.23** | **90.20** | **49.20** | **73.34** | **86.34** | **95.31** |
| Deep Feature Learning [7] | 40.50 | 60.80 | 70.40 | 84.40 | – | – | – | – | – | – | – | – |
| Ahmed's Deep Re-id [8] | – | – | – | – | 34.81 | 63.72 | 76.24 | 81.90 | 47.53 | 72.10 | 80.53 | 88.49 |
| KISSME [9] | 24.75 | 53.48 | 67.44 | 80.92 | 36.31 | 65.11 | 75.42 | 83.69 | 14.02 | 32.20 | 44.44 | 56.61 |
| L2 - norm | 10.89 | 22.37 | 32.24 | 45.19 | 11.33 | 24.50 | 32.22 | 43.89 | 5.63 | 16.00 | 22.89 | 30.63 |
| L1 - norm | 12.15 | 26.01 | 32.09 | 34.72 | 25.50 | 25.33 | 51.73 | 53.07 | 10.80 | 15.51 | 37.57 | 35.57 |

## 5  Conclusion

The paper proposed a network structure that combines the processed manual annotation attribute features with CNN extraction features, and then uses the loss measurement function proposed in this paper to optimize the recognition results. The CNN extraction feature is based on the ResNet network, and combine the hand-crafted attribute features. The combine the two sub-network into a single network structure. According to the back propagation, the attribute feature can supplement the CNN features extracted by the network, and obtain a more complete feature representation to achieve more accurate pedestrian recognition results. Tested on three challenging public data sets (VIPeR, PRID, CUHK), the experimental results prove the effectiveness of the proposed method. Subsequent work will further study the pedestrian attributes and hope to further improve the accuracy of pedestrian re-identification.

## References

1. Bazzani, L., Cristani, M., Perina, A., Farenzena, M., Murino, V.: Multipleshot person re-identification by HPE signature. In: 20th International Conference on Pattren Recognition (ICPR) 2010, pp. 1413–1416. IEEE (2010)
2. Krizhevsky, A., Sutskerver, I., Hinton, G.E.: Imagenet classification with deep convolutional neural networks. In: Advances in Neural Information Processing System, pp. 1097–1105 (2012)
3. Li, J., Guo, J.: Research on pedestrian re-recognition algorithm based on enhanced depth feature fusion. Inf. Technol. **42**(320(07)), 23–27 (2018)
4. Matsukawa, T., Suzuki, E.: Person re-identification using CNN features learned from combination of attributes. In: Proceedings of IEEE Conference on Computer Vision and Pattern Recognition, pp. 2428–2433 (2017)

5. Roy, A., Sural, S., Majumdar, A.K.: Minimum user requirement in role based access control with separation of duty constraints. In: Proceedings of the 12th International Conference on Intelligent Systems Design and Applications, Washington D C, USA, pp. 386–391. IEEE Press (2013)

6. Wu, S., Chen, Y., Li, X.: An enhanced deep feature representation for person re-identification. In: WACV (2016)

7. Ding, S., Lin, L., Wang, G., et al.: Deep feature learning with relative distance comparison for person re-identification. Pattern Recognit. **48**(10), 2993–3003 (2015)

8. Ahmed, E., Jones, M., Marks, T.K.: An improved deep learning architecture for person re-identification. In: IEEE CVPR (2015)

9. Koestinger, M., Hirzer, M., Wohlhart, P., Roth, P.M., Bischof, H.: Large scale metric learning from equivalence constraints. In: IEEE CVPR (2012)

# Preliminary Study on Interpreting Stock Price Forecasting Based on Tree Regularization of GRU

Wenjun Wu, Yue Wang[✉], Jiaxiu Fu, Jiayi Yan, Zhiping Wang, Jiaqing Liu, Weifan Wang, and Xiuli Wang

School of Information, Central University of Finance and Economics, Beijing, China
yuelwang@163.com

**Abstract.** The gated recurrent unit (GRU) deep model is interpreted to predict price's falling or rising. By using a technique called Tree Regularization of Deep Models for Interpretability, a GRU network is converted to a decision tree (called GRU-Tree) to interpret its prediction rules. This approach was tested by experimenting on a few sample stocks (e.g., the Gree company) and a main stock market index (SSE Composite Index) in China. The discovered prediction rules actually reflect a general rule called Mean Reversion in stock market. Results show that the GRU-Tree is more effective (higher AUC) than the decision tree directly trained from the data for small and moderate average path length (APL) of trees. And the fidelity between GRU and its generated GRU-Tree is high (about 0.8).

**Keywords:** Tree regularization · Stock forecasting · Interpretability · Gated recurrent unit · GRU-Tree · Deep learning

## 1 Introduction

Forecasting stock prices has been one of the biggest challenge interested to both the finance community and the computer society.

Although the famous work by Fama in 1965 told the world "informationally efficient", that is, stock prices you noted have reflected all known information, and the price movement is in response to news or events. Since the news or events are unpredictable, stock price can not be forecasted. This is called Efficient Market Hypothesis (EMH) [1].

Actually, the market is not always that efficient [2,3], and the technical analysis is prevailing in the investment [4,5], which attempts to exploit recurring and predictable patterns in stock prices to generate superior investment performance. Nowadays, with the help of deep learning, it is not necessary to devise

W. Wu and Y. Wang contributed equally to the paper. This work is supported by National Natural Science Foundation of China (Project No. 61309030).
This work is also supported by Central University of Finance and Economics Year 2019 First-class Discipline Construction Project.

trading strategies manually based on the existing technical analysis rules, and people only need to throw the stock related parameters into a deep model and the model will learn the rules automatically. However, a deep model is a black box, and people do not know whether the model has learned the reasonable rules and thus can not trust it or rely on it confidently. Therefore, the interpretability of deep model is especially important for stock price forecasting.

Recently, Wu et al. proposed what they called a *tree regularization* method to interpret deep models [6], wherein regularized decision trees are trained to mimic deep time-series models (e.g., GRU) without sacrificing prediction accuracy. Because the corresponding decision trees are easy to interpret, the deep model under considered is interpreted this way.

In our work, we present using their method to interpret stock price prediction of deep models. To the best of our knowledge, we take the first step toward interpreting the stock forecasting results via deep models utilizing Wu's method. Especially, we make the following contributions.

1. First, we use gated recurrent unit (GRU) to predict tomorrow's price falling (0) or rising (1) based on the past days' prices for a few sample stocks and a main stock market index (SSE Composite Index) in China. The prediction accuracy on the test data can be slightly larger than 0.5 after parameters adjusting, indicating that the Chinese market is random but not fully random and a bit predicable (in other words, EMH is ideal compared with the fact).
2. Second, we use the regularized tree method to convert the GRU network into a decision tree. The performance of the tree (named *GRU-Tree* hereafter) is carefully investigated, such as accuracy, F1 score and AUC, and is compared with the GRU. The prediction power of GRU-Tree is also larger than 0.5, and the fidelity between the two models is high, about 0.8.
3. Third, we interpret the resulted GRU-Trees and find that some tree branches reflected a general rule called *mean reversion*, that is, the price will rise when it has dropped low, and the price will drop when it has risen high. This finding confirms the usability of the Wu's method in stock prices forecasting - to give the reasons behind the forecasting of the neural networks.

## 2   Background and Notation

We briefly introduce Wu's interpretability model [6]. We consider an input feature vector $x_n$ and an target output vector $y_n$. The $nth$ example sequence contains $T$ timesteps, and we can write $x_n = [x_{n1}, \ldots, x_{nT_n}]$ and $y_n = [y_{n1}, \ldots, y_{nT_n}]$. A neural network makes predictions $\hat{y}_n$ based on the current $x_n$ and parameters $W$ of the network. A typical time-series neural model is GRU-RNN (called GRU). Our goal is to learn $W$ via the following loss minimization objective:

$$\min_{W} \lambda \Psi(W) + \sum_{n=1}^{N} \sum_{n=1}^{T_n} loss(y_{nt}; \hat{y}_{nt}(x_n, W)), \tag{1}$$

where $\lambda$ is the weight of the regularization cost $\Psi(W)$ and $loss()$ is the loss function that describes the discrepancy (e.g., using cross entropy) of $\hat{y}_n$ and $y_n$ for all $N$ examples.

For each $\lambda$, we can train a GRU network, *GRU(L1)* or *GRU(L2)*, depending on using the L1 or L2 regularization. Then we use the GRU(L2) to train a decision tree, called *GRU-Tree*. Surprisingly, the GRU-Tree has a comparable and even better AUC (Area Under Curve) than the original GRU(L1) or GRU(L2), and it also has a much better AUC than a directly trained decision tree according to a lot of real-task testings. *This is the basic idea of Wu's tree-based interpretability model.*

Furthermore, a deep decision tree is not preferred, because it has complex human simulatability to allow people to audit predictions easily. In Wu's work, *average path length (APL)*, $\Omega(W)$, is used to measure the complexity of a GRU-Tree, which exactly counts the number of true-or-false boolean calculations needed to make an average prediction. And the authors proposed using $\Omega(W)$ to regularize a deep GRU-Tree. Because $\Omega(W)$ is not differentiable, they proposed a surrogate function $\hat{\Omega}(W)$ to approximate the real $\Omega(W)$, and $\hat{\Omega}(W)$ is trained by another neural network minimizing the loss function of $\hat{\Omega}(W)$ and $\Omega(W)$.

The theoretical reason why this way works was discussed in [7] by Hinton and his collaborators: "When we are distilling the knowledge from a large model into a small one, however, we can train the small model to generalize in the same way as the large model." In here, the first trained GRU network provides the class probabilities ("soft target") for each class prediction, which has "much more information per training case than hard targets and much less variance in the gradient between training cases". Thus, using a trained GRU network to train a decision tree will get better performance than training a decision tree directly from the training data.

## 3   Stock Forecasting Interpretability Using GRU

Currently, we use historical price-related daily data to predict ups and downs of tomorrow price. After the correlation analysis, we pick *seven features* which has a correlation coefficient larger than 0.5. They are listed in Table 1. Complying with the notation (2), $x_{nt}$ of the *nth* stock at day $t$ is $[x_{nt}(0), \ldots, x_{nt}(6)]$, and $y_{nt}$ is 0 (tmr down) or 1 (tmr up) where tmr (tomorrow) means day $t + 1$.

In order to unify dimension, we apply the $Z$ score to each variable (say $a$):

$$a_z = \frac{a - \mu}{\sigma}, \tag{2}$$

where $\mu$ is the mean and $\sigma$ is the variance for the sample of $a$. All the above variables are standardized by the $Z$ score. Thus, the 1 unit means one standard derivation of original variables.

The interpretability model follows Wu's method exactly. First, a GRU network is trained for the price time series of a given stock. Then, a GRU-Tree is generated from the GRU network, which provides certain interpretability and human simulatability, which can tell the decision rules of making prediction of stock prices.

**Table 1.** The seven features ($x$) of historical price-related daily data to predict ups and downs of tomorrow price ($y$).

| | |
|---|---|
| $x_{nt}(0)$ | Today opening price |
| $x_{nt}(1)$ | Today highest price |
| $x_{nt}(2)$ | Today lowest price |
| $x_{nt}(3)$ | Today closing price |
| $x_{nt}(4)$ | Yesterday closing price |
| $x_{nt}(5)$ | The day before yesterday closing price |
| $x_{nt}(6)$ | Past 5 day average closing price |
| $y_{nt}$ | 0 (tmr down), 1 (tmr up) |

In the operation, a time-series data is spitted into two parts: the training data set and the test data set. *For each lambda* (the GRU's regularization weight), an optimized GRU is obtained based on the training data. Then, using this GRU and the $x_n$'s in the training data, we train a GRU-Tree and calculate its average path length (APL). The AUC curves of the two models versus APL are calculated on the test data set and are then plotted. From the plotted figure, we can read off a suitable GRU-Tree of not large APL and still good AUC.

According to real and prediction results, the sample data is divided into four categories (Table 2). Here, we treat price rising (up) as positive examples and price falling (down) as negative examples.

**Table 2.** The four categories of real and prediction results.

| | |
|---|---|
| TP (true positive) | Prediction: up; reality: up |
| FP (false positive) | Prediction: up; reality: down |
| TN (true negative) | Prediction: down; reality: down |
| FN (false negative) | Prediction: down; reality: up |

Thus, we define the precision ($P$) and recall ($R$) of up and down:

$$P_{up} = \frac{TP}{TP + FP} \tag{3}$$

$$R_{up} = \frac{TP}{TP + FN} \tag{4}$$

$$P_{down} = \frac{TN}{TN + FN} \tag{5}$$

$$R_{down} = \frac{TN}{TN + FP} \tag{6}$$

And we calculate the F1 score for precision and recall, which is the harmonic mean of them and is sensitive to the lowest value.

$$F1_{up} = \frac{2}{\frac{1}{P_{up}} + \frac{1}{R_{up}}} \tag{7}$$

$$F1_{down} = \frac{2}{\frac{1}{P_{down}} + \frac{1}{R_{down}}} \tag{8}$$

$$F1 = \frac{F1_{up} + F1_{down}}{2} \tag{9}$$

Additionally, we can use the classification *accuracy* to measure the performance of stock price prediction, which is appropriate for the application scenario here. When we predict tomorrow's rising correctly, we can buy today to make money. When we predict tomorrow's falling correctly, we can sell today to make money too.

$$accuracy = \frac{TP + TN}{TP + FP + TN + FN} \tag{10}$$

Finally, the AUC (the area under the ROC curve) indicator is often used to measure the classification performance. In our experiments hereafter, we use AUC to measure and compare the classification performance, and we also show $F1$ and *accuracy* to show the prediction results.

We are also concerned with the fidelity. Fidelity is defined as the percentage of test examples on which the prediction made by a GRU-Tree agrees with its based deep model GRU(L2).

## 4    Experiments and Discussions

We obtain stock data from the CSMAR database, which is a famous database recording Chinese stock information.

### 4.1    The Gree Company Stock Price

The first illustrating example is the *Gree* company (code: 000651), the biggest air conditioning company in China. We obtain its price time-series from 31/12/2014 to 27/12/2018 (DD/MM/YYYY) of 1306 market days in total. We split the time series into two halves, the first for training and the second is for testing.

We plot the AUCs of GRU(L2), GRU-Tree and Decision Tree (Decision Tree here means decision trees trained directly from the data) in Fig. 1. We can see that the GRU-tree achieves the highest AUC of 0.556 at the APL of 7.47, but the APL is somewhat high and may harm interpretability. Therefore, we choose the GRU-Tree at the APL of 4.29 (*sweet point*) which can balance AUC (0.54) and tree complexity. Note that the AUC of the GRU-Tree is comparable to that of GRU(L2) and is higher than that of Decision Tree for moderate APLs (less than 6).

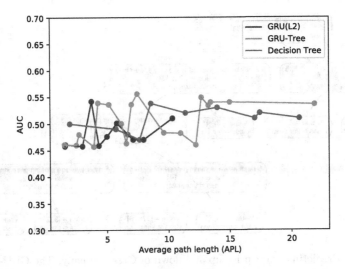

**Fig. 1.** Predicting stock price up and down of Gree company: The AUCs of GRU(L2), GRU-Tree and decision tree versus average path length (APL), respectively

We show the prediction performance in Table 3. We note that F1 and accuracy are only slightly larger than 0.5 because the market price is random and uncertain, which embodies the spirit of efficient market hypothesis [1]. And the fidelity between GRU-Tree and its based GRU shows a high value of 0.85, indicating the GRU-Tree approximates the GRU well. The strong learning ability of neural networks and high fidelity of the two models proves that converting a trained GRU to a decision tree can also learn meaningful decision rules.

**Table 3.** Prediction performance of the Gree stock.

| TP | FP | TN | FN | AUC | F1 | Accuracy | Fidelity |
|-----|-----|-----|-----|------|------|----------|----------|
| 120 | 113 | 226 | 194 | 0.54 | 0.52 | 0.53 | 0.85 |

We interpret the GRU-Tree shown in Fig. 2. The root node shows that there are 522 days of price down and 131 days of price up in the test data set (653 days in total), which is mainly a downturn market. Let us look at the leaf node larger than 10% of data, and list their branches:

1. Tomorrow price will be *up* with the prob. of 1 ($\frac{84}{0+84}$) when *5 days average closing price* $\leq -0.765$ and *today lowest price* $\leq -0.735$, indicating that stock price of Gree will rebound when two past price indicators drop bellow the mean (0 after normalization here) about $-0.7$ standard deviation.
2. Tomorrow price will be *down* with the prob. of 1 ($\frac{104}{104+0}$) when *the day before yesterday closing price* $> -0.8$ and *5 days average closing price* $> 0.635$,

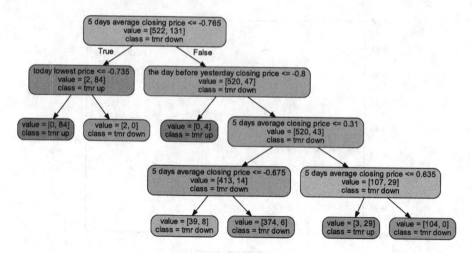

**Fig. 2.** Predicting stock price up and down of Gree company: The GRU-Tree

indicating that stock price of Gree will fall when two past price indicators are above their respective thresholds.

3. Tomorrow price will be *down* with the prob. of 98.4 ($\frac{374}{374+6}$) when 0.31 > *5 days average closing price* > −0.675 and *the day before yesterday closing price* > −0.8.

*The first two branches actually reflect the general mean reversion rule*, and the third is a specific pattern at that time period.

### 4.2    The SSE Composite Index

The second illustrating example is the SSE (Shanghai Stock Exchange) Composite Index from 31/12/2014 to 27/12/2018 (DD/MM/YYYY). Figure 3 shows similar patterns compared with Fig. 1, GRU-Tree still having better AUC than Decision Tree for moderate APL (less 6). Table 4 shows the prediction performance of GRU-Tree, and the fidelity of 0.79 is high.

**Table 4.** Prediction performance of SSE Composite Index.

| TP | FP | TN | FN | AUC | F1 | Accuracy | Fidelity |
|-----|-----|----|----|------|------|----------|----------|
| 227 | 179 | 87 | 77 | 0.53 | 0.52 | 0.55 | 0.79 |

However, the GRU-Tree of the SSE Composite Index in Fig. 4 shows somewhat a different pattern than that of the Gree stock. There are two dominant conditional branches:

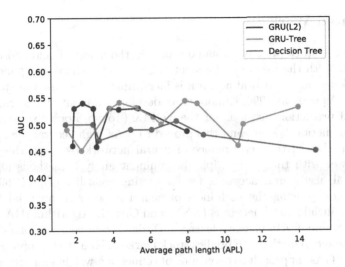

**Fig. 3.** Predicting stock price up and down of SSE Composite Index: The AUCs of GRU(L2), GRU-Tree and decision tree versus average path length (APL), respectively

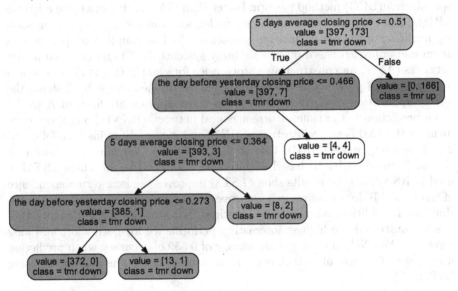

**Fig. 4.** Predicting stock price up and down of SSE Composite Index: The GRU-Tree

1. Tomorrow price will be *up* with the prob. of 1 ($\frac{166}{0+166}$) when *5 days average closing price* >0.51.
2. Tomorrow price will be *down* with the prob. of 1 ($\frac{372}{372+0}$) when *5 days average closing price* <0.364 and *the day before yesterday closing price* <0.232.

   *Again, they reflect the general rule of mean reversion.*

## 5    Related Work

Predicting the stock price is a common topic in the financial field and the computer field. With the rapid development of computer technology application in stock market analysis, neural network is increasingly widely used in stock price prediction. As early as 1990, Kimoto et al. developed a number of learning algorithms and prediction methods for the TOPIX (Tokyo Stock Exchange Prices Indexes) prediction system, and this proved to be of high accuracy [8]. In 1991, Yoon et al. showed the performance of neural network by comparing its predictive power with that of multiple discriminant analysis methods [9]. Later, Hamid et al. use neural networks for forecasting volatility of S&P 500 Index futures prices, proving the usefulness of neural network in financial forecasting [10]. Artificial neural networks (ANN) and Genetic Algorithm (GA) are two prominent techniques in machine learning. During recent years, scholars is trying to apply these two methods to predict stock markets. Kim et al. proposes genetic algorithms (GAs) approach to determine of connection weights for artificial neural networks (ANNs) to predict the stock price index, which can outperforms the other two conventional models [11]. Zhang et al. found bacterial chemotaxis optimization(BCO) method perform better than GA [12]. Pai et al. use a hybrid ARIMA and support vector machines model, and successfully apply it in solving nonlinear regression estimation problems [13]. Hadavandi et al. presents an integrated approach based on genetic fuzzy systems (GFS) and artificial neural networks (ANN) for constructing a stock price forecasting expert system, which outperforms all previous methods [14]. These mentioned research all shows the widely application of GA and ANNs. The most appropriate form of ANN for stock prediction is the time recurrent neural network (RNN). Long short-term memory (LSTM) is an architecture of RNN, and it has a simplified model called gated recurrent units (GRU). Gao et al. proposes a Convolutional Recurrent Neural Network (CRNN)-based architecture, ConvLSTM, in which LSTM is used in RNN. And the results shows LSTM improves the long-term dependence of traditional RNN and effectively improves the accuracy and stability of prediction [15]. In addition, Althelaya et al. noticed that a stacked LSTM architecture has demonstrated the highest forecasting performance for both short and long term [16]. And Nelson et al. got an average of 0.559 of accuracy when predicting if the price of a particular stock is going to go up or not in the near future using LSTM [17].

However, the specific process of deep learning resembles a "black box", and the architecture of end-to-end model and the working principle of optimizing model parameters are not familiar to most people. So it is vital to enhance the interpretability of neural work. At present, many visualization methods are used to improve the accuracy of neural network. Li et al. used a variety of visualization methods to study how the network structure affects the loss surface and how the training parameters affect the shape of the minimum point, greatly reducing the loss value of the neural network, making the model have better generalization performance and stronger interpretability [18]. Springenberg et al. found through the research of convolutional neural network that the convolutional layer could not always improve the expression ability of CNN, and replacing the maximum pooling with a convolutional layer did not lose the accuracy in image recognition and annotation [19]. In this paper, a new structure containing only convolution layer is proposed, and a new "deconvolution method" is introduced to analyze the structure of neural network, so as to visualize the features learned from CNN. Ribeiro et al. proposed that modular expandable interpretation technology LIME can learn interpretable models locally near the prediction [20]. By explaining different text and image recognition models, it has proved its flexibility and can explain the prediction of any classifier. Then Ribeiro again proposed model independent systems in 2018 [21]. Many Anchors are going to be used to help explain the behavior of complex models with high accuracy, even if the partial output of the specific algorithm is enough for the classifier to make predictions. Recently, Wu et al. proposed to add a tree regularization term into a general neural network and measure the complexity of the network by using the average path length of the decision tree, so that human users can quickly track and predict [6]. The training depth time series model enables the decision tree with a few nodes to depict the prediction process. It is concluded that the tree regularization model is easier to simulate and simpler than the L1 or L2 penalty term model without losing the prediction ability of the model.

The interpretability of using deep models for stock prediction is an almost unexplored area. The paper [22] provides a effective way to visualize and explain decisions made by deep stock market prediction models. And the paper [23] visually interprets text-based deep learning models in predicting stock price movements. So far as we know, this work is the first paper to generate decision rules of stock price prediction from GRU.

## 6   Conclusion

This work explores interpreting stock price movements using the GRU deep model based on Mike Wu's paper. The basic idea is to convert a trained GRU to a decision tree called GRU-Tree. The conclusion is positive that we can actually learn meaningful decision rules (mean reversion in the stock market) from the GRU model.

## References

1. Fama, E.F.: The behavior of stock-market prices. J. Bus. **38**(1), 34–105 (1965)
2. Malkiel, B.G.: Is the stock market efficient? Science **243**(4896), 1313–1318 (1989). New Series
3. Yao, S., Wang, Y.: An empirical study on the weak form market efficiency of Shanghai composite index in bull and bear markets. In: Proceeding of 12th International Conference on Management of e-Commerce and e-Government (ICMECG 2018), pp. 443–448 (2018)

4.  Bodie, Z., Kane, A., Marcus, A.J.: Investments, 10th edn. McGraw-Hill Education, New York (2014)
5.  Liu, Z., Wang, Y.: An empirical study on the forecasting effectiveness of price-based technical indicators in bull and bear cycles of China's Shanghai stock market. In: Proceeding of 12th International Conference on Management of e-Commerce and e-Government (ICMECG 2018), pp. 412–417 (2018)
6.  Wu, M., Hughes, M.C., Parbhoo, S., et al.: Beyond sparsity: tree regularization of deep models for interpretability. In: Proceeding of the Thirty-Second AAAI Conference on Artificial Intelligence (AAAI 2018), pp. 1670–1678 (2018)
7.  Hinton, G., Vinyals, O., Dean, J.: Distilling the knowledge in a neural network. Arxiv preprint. https://arxiv.org/abs/1503.02531 (2015)
8.  Kimoto, T., Asakawa, K., Yoda, M., Takeoka, M.: Stock market prediction system with modular neural networks. In: 1990 IJCNN International Joint Conference on Neural Networks, vol. 1, pp. 1–6 (1990)
9.  Yoon, Y., Swales, G.: Predicting stock price performance: a neural network approach. In: Proceedings of the Twenty-Fourth Annual Hawaii International Conference on System Sciences, Kauai, HI, USA, vol. 4, pp. 156–162 (1991)
10. Hamid, S.A., Iqbal, Z.: Using neural networks for forecasting volatility of S&P 500 index futures prices. J. Bus. Res. $57(10)$, 1116–1125 (2004)
11. Kim, K.-J., Han, I.: Genetic algorithms approach to feature discretization in artificial neural networks for the prediction of stock price index. Expert Syst. Appl. $19(2)$, 125–132 (2000). ISSN 0957–4174
12. Zhang, Y., Lenan, W.: Stock market prediction of S&P 500 via combination of improved BCO approach and BP neural network. Expert Syst. Appl. $36(5)$, 8849–8854 (2009)
13. Pai, P.-F., Lin, C.-S.: A hybrid ARIMA and support vector machines model in stock price forecasting. Omega $33(6)$, 497–505 (2005)
14. Hadavandi, E., Shavandi, H., Ghanbari, A.: Integration of genetic fuzzy systems and artificial neural networks for stock price forecasting. Knowl.-Based Syst. $23(8)$, 800–808 (2010)
15. Gao, S.E., Lin, B.S., Wang, C.: Share price trend prediction using CRNN with LSTM structure. In: 2018 International Symposium on Computer, Consumer and Control (IS3C), Taichung, Taiwan, pp. 10–13 (2018)
16. Althelaya, K.A., El-Alfy, E.M., Mohammed, S.: Stock market forecast using multivariate analysis with bidirectional and stacked (LSTM, GRU). In: 2018 21st Saudi Computer Society National Computer Conference (NCC), Riyadh, pp. 1–7 (2018)
17. Nelson, D.M.Q., Pereira, A.C.M., de Oliveira, R.A.: Stock market's price movement prediction with LSTM neural networks. In: 2017 International Joint Conference on Neural Networks (IJCNN), Anchorage, AK, pp. 1419–1426 (2017)
18. Li, H., Xu, Z., Taylor, G., Studer, C., Goldstein, T.: Visualizing the loss landscape of neural nets. In: International Conference on Learning Representations (2018)
19. Springenberg, J.T., Dosovitskiy, A., Brox, T., Riedmiller, M.: Striving for simplicity: the all convolutional net. In: International Conference on Learning Representations (2015)
20. Ribeiro, M.T., Singh, S., Guestrin, C.: Why should i trust you? Explaining the predictions of any classifier. In: Knowledge Discovery and Data Mining, pp. 1135–1144 (2016)
21. Ribeiro, M.T., Singh, S., Guestrin, C.: Anchor: high-precision model-agnostic explanations. In: AAAI Conference on Artificial Intelligence (2018)

22. Kumar, D., Taylor, G.W., Wong, A.: Opening the black box of financial AI with CLEAR-trade: a class-enhanced attentive response approach for explaining and visualizing deep learning-driven stock market prediction, arXiv preprint, arXiv: 1709.01574 (2017)
23. Shi, L., Teng, Z., Wang, L., Zhang, Y., Binder, A.: DeepClue: visual interpretation of text-based deep stock prediction. IEEE Trans. Knowl. Data Eng. **31**(6), 1094–1108 (2018)

# Simulation Analysis on the Rupture Trend of Intracranial Hemangioma

Dong Xiang[1], Yuehua Wang[2], Yingting Wang[1(✉)], Tianyi Bu[2],
Guangjun Chen[1], and Minhong Wu[1]

[1] School of Mechatronic Engineering,
Harbin Institute of Technology, Harbin 150080, Heilongjiang, China
YingTing_Wang@163.com
[2] Harbin Medical University, Harbin 150080, Heilongjiang, China

**Abstract.** The rupture of intracranial aneurysm would lead to serious conse-quences, and different shapes of aneurysm have different effects on the rupture. In this paper, a mechanical analysis method is proposed for calculating the rupture trend of intracranial aneurysm based on three-dimensional image reconstruction and hemodynamic numerical calculation and evaluation. Firstly, the modeling of actual intracranial aneurysms was conducted based on CT intracranial aneurysm angiography plane images. Secondly, the model was simplified from the perspective of fluid mechanics analysis to establish the mechanical model of different forms of intracranial aneurysms. Then the flow mechanics characteristics of each model in rigid boundary and fluid-structure coupling boundary were analyzed. The simulation research was conducted on the condition of normal pressure difference and pressure pulsation, and the most dangerous position of intracranial aneurysms rupture was analyzed. The results validate the correctness of theoretical analysis.

**Keywords:** Intracranial hemangioma · Non-newtonian blood · Aneurysm angiography · Rupture trend

## 1 Introduction

Intracranial aneurysms are abnormal local dilation of intracranial arteries caused by pathological defects in blood vessel walls. Rupture of aneurysms is one of the most serious consequences, which leads to high mortality and incidence of subarachnoid hemorrhage. According to statistics, the incidence of subarachnoid hemorrhage in intracranial cerebral aneurysms is about 15%–35% [1]. Although the mechanism of the occurrence and development of cerebral aneurysms has not been fully elucidated, the hemodynamic mechanism is considered to be the factor leading to the growth and rupture of aneurysms [2–4]. Some researchers has been used two-dimensional simu-lation, which can reflect the hemodynamic characteristics of intracranial aneurysms directly. Two-dimensional simulation studies can intuitively reflect the hemodynamic characteristics of intracranial aneurysms, the flow velocity of intracranial aneurysms and the velocity field morphology of aneurysm-related areas [5–7]. However, this

© Springer Nature Singapore Pte Ltd. 2019
R. Mao et al. (Eds.): ICPCSEE 2019, CCIS 1059, pp. 488–498, 2019.
https://doi.org/10.1007/978-981-15-0121-0_38

simulation study mode ignores the non-newtonian blood factors and the elasticity of blood vessel walls, which in pursuit of efficient simulation work smoothly progress.

In this paper, a three-dimensional model of cerebral aneurysm was reconstructed from 2D images of real cases, and the actual model was simplified and summarized according to the morphology. The mechanical model of intracranial aneurysm of different shapes is established based on the method of three-dimensional image reconstruction and hemodynamic numerical analysis, and the mechanical characteristics were analyzed by one-way fluid-solid coupling numerical simulation calculation, so as to predict the risk location of rupture of typical cerebral aneurysm. The validity of the mechanical analysis is verified by comparing the results of simulation analysis and simulation experiments.

## 2   Mechanical Model of Intracranial Aneurysm

### 2.1   The Model of Biological Intracranial Aneurysm

Relevant medical records were provided by the department of neurosurgery of the first affiliated hospital of Harbin medical university. The 2D image is stratified and refined by using the mimics medical imaging software to dispose the CT intracranial cerebral aneurysm angiography plane image of the patient as shown in Fig. 1.

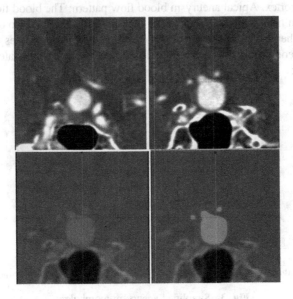

**Fig. 1.** The 2D intracranial images

The coverage area of hemangioma is extracted through continuous stratification to construct a 3D entity, as shown in Fig. 2.

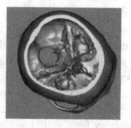

**Fig. 2.** The 3D model of cerebral aneurysm and whole brain

## 2.2    Simplified Model of Intracranial Aneurysm

The model shown in Fig. 2 provides boundary conditions for the numerical simulation model, which can be used for the numerical simulation of the characteristics of hemodynamic. However, to transform the whole model into a three-dimensional simulation model will lead to too much calculation, so the model should be simplified. Intracranial aneurysms can be divided into two types: lateral wall aneurysm blood flow pattern and apical aneurysm blood flow pattern. lateral wall aneurysm blood flow pattern: The blood flow which flows from the downstream side of the aneurysm neck orifice to the aneurysm, then back to the parent artery along the peripheral wall of the aneurysm, and then flows out from the upstream side of the aneurysm neck orifice, forms a major vortex. Apical aneurysm blood flow pattern: The blood flow directly act on the aneurysm from the inflow artery, and then flows back to the two outflow arteries along the peripheral wall of the aneurysm, forming two major eddies on both sides. Therefore, its geometric shape can be simplified into the following states according to its morphology:

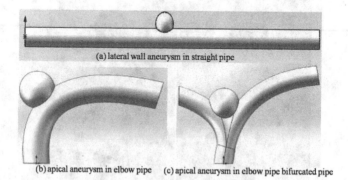

(a) lateral wall aneurysm in straight pipe

(b) apical aneurysm in elbow pipe    (c) apical aneurysm in elbow pipe bifurcated pipe

**Fig. 3.** Simplified aneurysm morphology

## 3    Mechanical Model of Intracranial Aneurysm

According to the morphology shown in Fig. 3, it can be further simplified into a flow model, as shown in Fig. 4.

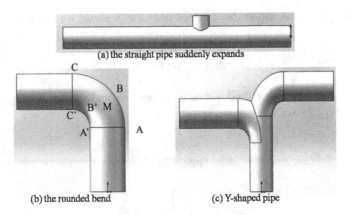

(a) the straight pipe suddenly expands

(b) the rounded bend                    (c) Y-shaped pipe

**Fig. 4.** Schematic diagram of blood vessel flow in each form

The characteristics of the vascular flow field were analyzed as follows:

### 3.1  The Straight Pipe Suddenly Expands

When the section of the pipe is suddenly enlarged, due to the effect of the inertia of the fluid, the motion of the fluid particles can not be changed in accordance with the shape of the pipe wall, so the streamline can only transit smoothly. There is a part of fluid between the outer surface of the expanded flow beam and the pipe wall does not flow forward with the main flow. The fluid particles near the surface of the main flow beam move in the direction of the main flow under the action of viscosity. Because of the expanding of the Section of the main flow area, the flow velocity decreases, the pressure increases along the flow direction, and a positive pressure gradient appears.

### 3.2  The Rounded Bend and Y Shaped Pipe

When the fluid flows through the elbow, the radius of lateral curvature is large, the fluid flows along ABC, and the velocity first decreases and then gradually increases. The inner fluid flows along A′B′C′, and the flow rate first increases and then decreases. Therefore, the pressure on the outside of the elbow increases gradually from A to B. Due to the action of viscous force at the side wall, the fluid slows down, which causes the fluid to break away from the side wall and form a vortex. When flowing from B to C, the flow rate gradually increases and then returns to the inlet flow rate. Similarly, in A′B′C′, the fluid also appears to break away from the side wall and form a vortex.

If the velocity of the fluid at point M is $v$, the centripetal acceleration will be $v^2/r$, according to Newton's second law:

$$pdA - (p+dp)dp - dmg\cos\theta = -dm\frac{v^2}{r} \tag{1}$$

According to the equation $dm = \rho dr dA$ and $cos\theta = dz/dr$, we can get:

$$\frac{d}{dr}\left(z + \frac{p}{\rho g}\right) = \frac{v^2}{gr} \tag{2}$$

It can be assumed that fluid viscosity is ignored and taking derivative of $r$ in Bernoulli's Eq. (3), we can get:

$$z + \frac{p}{\rho g} + \frac{v^2}{2g} = C \tag{3}$$

$$\frac{d}{dr}\left(z + \frac{p}{\rho g}\right) = -\frac{v}{g}\frac{dv}{dr} \tag{4}$$

According to the Eqs. (3) and (4), the Eq. (5) can be obtained.

$$\frac{dv}{dr} = -\frac{v}{r} \tag{5}$$

For round corner pipe and t-branch pipe, the velocity decreases with the increase of the curvature radius when the fluid flows along the curved channel. It can also be argued that: The differential of flow direction may lead to the change of hydrodynamic characteristics in the round corner pipe and t-branch pipe. The external appearance of hemangioma will be inconsistent. In this paper, only study from the perspective of simplified model.

From the analysis, it can be seen that the angle of the vascular branch will have an impact on the hemodynamic characteristics, which will lead to changes in the vascular wall shear stress, so the degree of impact on hemangioma rupture is also different.

It is found that the angle of the aortic neck significantly affected the velocity flow field in the model because the angle of the neck changed the surface curvature between the aortic inlet and aneurysm. The downstream flow field is affected. The results showed that the increased Angle of the aortic neck led to decreased flow recirculation in the aneurysm sac and the flow field to the center of the aorta.

# 4 Simulation Analysis

Due to the flow dynamics characteristics of straight pipe model are relatively simple, the finite element method is used to analyze rounded bend and Y shaped pipe. As shown in Fig. 5, the typical angle values of the branch of the blood vessel are taken, that is, the angle of the bend pipe is 30°, 45°, 60°, the t-branch pipe is 30° + 45°, 30° + 60°, 45° + 60°.

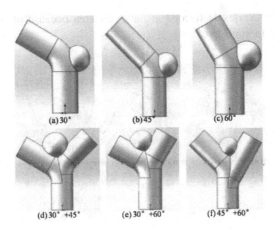

**Fig. 5.** Y shaped pipe with different angles

## 4.1 Rigid Boundary Simulation

It can be assumed that the vessel wall is rigid. The inlet pressure is 15007 Pa and the differential pressure is 1.5 Pa and the blood velocity is 0.5 m/s. The non-slip boundary condition is adopted for the pipe wall, and the iteration step is 500. The fluid parameters are set as follows: the blood viscosity coefficient is 0.004 Pa/s and the density is 1060 kgm$^{-3}$. The numerical model chooses laminar flow. The convergence criterion can be described that the residuals of each velocity component are less than $10^{-4}$. Friction occurs when blood flows through the vascular lumen. The shear stress nephogram of steady-state model of pipe shows in Fig. 6.

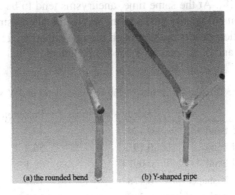

**Fig. 6.** Shear stress nephogram of steady-state model of pipe

Figure 6(a) is the shear stress distribution of the rounder bend. The arterial aneurysm area can be seen that the body and the top of the aneurysm have low shear stress. On the lateral side of the vascular bend, the shear stress is larger due to the impact of blood flow. On the contrary, the shear stress is lower in the medial part of the bend. A high-speed vortex is formed at the bend area because of blood flow flows outward under the inertia force.

(a) the rounded bend          (b) Y-shaped pipe

**Fig. 7.** The peak pressure shear stress nephogram of pulsating pipe model

Figure 6(b) shows the aneurysm area. The shear stress of the aneurysm top and body of the aneurysm are lower. Because the velocity of blood flow is faster, the shear stress is higher. The blood flow directly impacted the top of the aneurysm and formed two sections of reflux on both sides, then returned to the outflow artery. Therefore, the shear stress at the top of the aneurysm is higher than that at the abdominal muscular artery. Considering that the intravascular pressure is actually a fluctuating value, the dynamic simulation method is used to calculate the above model. The speed defines as $0.5 + 0.3 \times \sin(2\pi t)$ m/s. The peak pressure shear stress nephogram of pulsating pipe model as shown in Fig. 7.

Medically, it is generally believed that too low wall shear stress acting on endothelial cells in aneurysms will lead to degeneration and weakness of the wall tissues of aneurysms [8]. At the same time, aneurysms tend to be enlarged due to high pressure in the contents of aneurysms, and the top of the aneurysms is easy to rupture. Therefore the simulation results are approach the real situation. The calculation results of the rounded bend and Y shaped pipe show in Table 1.

**Table 1.** The calculation results of the rounded bend and Y shaped pipe

|  | Steady-state stress Pa | Pulsation stress Pa |
|---|---|---|
| 30° rounded bend | 0.124 | 47.3 |
| 45° rounded bend | 0.19 | 56.2 |
| 60° rounded bend | 0.0974 | 62.1 |
| 30° and 45° Y shaped pipe | 0.478 | 16.7 |
| 30° and 60° Y shaped pipe | 0.286 | 12.5 |
| 60° and 45° Y shaped pipe | 0.327 | 12.4 |

It is indicated that the impact of blood shock increase and rupture of the aneurysm is larger when the wall shear stress under pulsatile flow increase. Blood shock has a great influence on the growth and rupture of aneurysms from the calculation results in Table 1. Sudden elevation of blood pressure in patients with hypertension, the possibility aneurysm ruptures has a higher probability [9].

## 4.2    Fluid-Structure Interaction Numerical Simulation

Due to the elasticity of blood vessel, it will deform under the impact and pressure of blood flow. Therefore, in the fluid-solid coupling analysis, the biggest factor of blood vessel rupture should be the stress and deformation on the wall. Assuming that the vessel wall is an isotropic linear material with incompressible and it is uniform thickness, the density is 1150 kg · m², Poisson's ratio is 0.45, the thickness of the vessel wall is 0.2 mm, and the elastic model of the vessel wall is 2 MPa, the other simulation conditions are the same as in 4.1. The results are shown in Figs. 8 and 9.

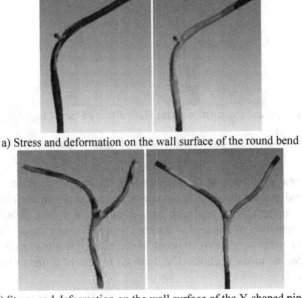

a) Stress and deformation on the wall surface of the round bend

b) Stress and deformation on the wall surface of the Y-shaped pipe

**Fig. 8.** Simulation results of fluid-solid coupling flow field in steady-state model

The simulation results of fluid-solid coupling flow field in pulsating mode show in Fig. 9. It can be seen from the Fig. 9 that the maximum deformation of the round bend model occurs in the neck of the aneurysm in the steady flow field. The Y-shaped pipe model occurs in the artery near the neck of the aneurysm. However, the deformation is mainly reflected in the bending of the artery caused by blood flow, while the part of the aneurysm is almost unaffected. In the pulsating flow field, the deformation of the model occurs when the pressure is a pulsating function.

a) Stress and deformation on the wall surface of the round bend

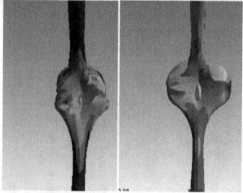

b) Stress and deformation on the wall surface of the Y-shaped pipe

**Fig. 9.** Simulation results of fluid-solid coupling flow field in pulsating model

The transverse expansion of the aneurysm is the largest deformation area of the whole model and the deformation is obviously increased. From the characteristics of biological tissue materials, the area with large deformation is easy to form shear injury in pulsating deformation. If this area is the area with maximum stress at the same time, the rupture of the aneurysm is easy to occur. It can be seen that the results of simulation analysis are close to medical clinic. The analysis results of different morphology are shown in Table 2.

**Table 2.** Fluid-solid coupling calculation results of rounded bend and Y shaped pipe

| | Steady-state stress Pa | Steady-state deformation m | Pulsation stress Pa | Pulsation deformation mm |
|---|---|---|---|---|
| 30° rounded bend | 1.863e6 | 0.32416 | 1.2526e6 | 16.559 |
| 45° rounded bend | 1.6708e6 | 0.09408 | 0.85644e6 | 13.839 |
| 60° rounded bend | 1.9233e6 | 0.01309 | 0.77301e6 | 9.214 |
| 30° and 45° Y shaped pipe | 87164 | 0.011489 | 2.8316e6 | 33.963 |
| 30° and 60° Y shaped pipe | 9533.9 | 0.00063 | 2.7681e6 | 16.557 |

From the calculation results in Table 2, the smaller the angle of rounded bend pipe, the greater the stress on the wall of the aneurysm under the same flows condition. Therefore it is more likely to rupture. The stress and deformation are larger under the pulsating flow. The impact of blood shock on the growth and rupture of the aneurysm has too much influence. Sudden elevation of blood pressure in patients with hypertension, the possibility aneurysm ruptures has a higher probability.

## 5 Conclusion

The model is simplified from the perspective of fluid mechanics analysis. The mechanical model of different forms of intracranial aneurysms has been established. Then the mechanics characteristics of each model in rigid boundary and fluid-structure coupling boundary can be simulated. The results show that the wall stress and deformation of the aneurysm are larger with the angle of rounded pipe decrease and it is more likely to rupture in the same condition. The deformation is larger when the wall stress increase under pulsating flow. It indicates that blood shock has a huge act on the growth and rupture of aneurysms. Thus it's easier to rupture.

In this paper, we can effectively analyze the main rupture location of intracranial hemangioma. It can effectively improve the efficiency of diagnosis. In practical application, the flow pattern inside the blood vessel can be analyzed. Accurately locating the rupture point is achieved. Analysis of wall thickness and blood flow velocity can also predict whether patients are at risk of vascular rupture. This method is still in the research process, and will be further studied in the future.

**Acknowledgements.** Dong Xiang and Yuehua Wang have the same contribution. They are all the first authors.

## References

1. Schroder, F., Regelsberger, J.: Asymptomatic cerebral aneurysms: surgical and endovascular therapy options. Wien. Med. Wochenschr. **147**, 159–162 (1997)
2. Fry, D.L.: Certain histological and chemical responses of the vascular interface to acutely induced mechanical stress in the aorta of the dog. Circ. Res. **24**, 93–108 (1969)
3. Caro, C.G., Fitz-Gerald, J.M., Schroter, R.C.: Arterial wall shear and distribution of early atheroma in man. Nature **223**(5211), 1159–1160 (1969)
4. Ku, D.N., Giddens, D.P., Zarins, C., Glagov, S.: Puslatile flow and atherosclerosis in the human carotid bifurcation: positive correlation between plaque location and low oscillating shear stress. Atherosclerosis **5**, 293–302 (1985)
5. Tateshima, S., Murayama, Y., Villablanca, J.P., Morino, T., Nomura, K., et al.: In vitro measurement of fluid-induced wall shear stress in unruptured cerebral aneurysms harboring blebs. Stroke **34**, 187–192 (2003)
6. Yiemeng, H., Scott, H.W., Minsuok, K., et al.: Validation of CFD simulations of cerebral aneurysms with implication of geometric variations. J. Biomech. Eng. **128**(6), 844–851 (2006)

7. Ford, M.D., Nikolov, H.N., et al.: PIV-measured versus CFD-predicted flow dynamics in anatomically realistic cerebral aneurysm models. J. Biomech. Eng. **130**, 021015-9 (2008)
8. Masaaki, S., Marie, O., Kiyoshi, T., et al.: Magnitude and role of wall shear stress on cerebral aneurysm. Stroke **35**, 2500–2505 (2004)
9. Jiang, D.F.: Analysis of the risk factors of induced rupture hemorrhage of intracranial aneurysm and its nursing countermeasures. Mod. Med. Health **23**(7), 955–956 (2007)

# Target Market Optimal Coverage Algorithm Based on Heat Diffusion Model

Jinghua Zhu[✉], Yuekai Zhang, and Bochong Li

School of Computer Science and Technology, Heilongjiang University,
Harbin 150080, China
zhujinghua@hlju.edu.cn

**Abstract.** The maximization of personalized influence is a branch of maximizing the influence of social networks, and the goal is to target specific social network users and mine the set of initial impact diffusion users that have made the most impact. However, most of the existing methods are based on the IC model and the LT model. The prediction of the impact of these two models on the nodes depends on the Monte Carlo simulation. In order to avoid Monte Carlo simulation time and simulate real life more, this paper introduces the heat diffusion model into the problem of maximizing the influence of personalization. The heat diffusion process was used to simulate the diffusion process of information influence. And the thermal energy was applied to measure the impact on the target users, and cluster candidate users. The cluster center as a seed node was proposed to spread information and maximizing the impact on specific users. The comparison experiments on real social networks show that the personalized maximization algorithm based on the thermal diffusion model has better time performance and diffusion effect than the traditional diffusion model.

**Keywords:** Social network · Influence maximization · Heat diffusion · Clustering

## 1 Introduction

In recent years, with the rapid development of information technology, the emergence of network information and social media had further changed the way people communicate with each other.

The problem of maximizing social network impact is to find the optimal set of users in a given social network, and the seed users initiate information diffusion and try to spread the information to as many users as possible. The problem originated from a viral marketing strategy in the marketing field. When a company promotes a new product or service, it usually provides a free product or service for a small number of trail users. It is hoped that these users will recommend the company's products to their friends. In this way by expanding the popularity of the product, is the so-called word-of-mouth effect [1–3]. Scholars have conducted in-depth research on the problem of maximizing influence and proposed CELF algorithm [4], New Greedy algorithm [5] and CELF++ algorithm [6], Degree Discount algorithm [6], LDGA algorithm [7] and simulated annealing algorithm [8].

© Springer Nature Singapore Pte Ltd. 2019
R. Mao et al. (Eds.): ICPCSEE 2019, CCIS 1059, pp. 499–510, 2019.
https://doi.org/10.1007/978-981-15-0121-0_39

However, in many cases, the traditional influence maximization problem cannot meet the needs of practical problems [9, 10]. For example, if a company wants to promote its own high-end sports car, if it uses traditional influence solutions, it will waste lots of resources to influence the people who cannot afford or are not interested. Before product information is passed on to potential customers, customers may be lost to competitors due to the consideration of irrelevant users. Nowadays, companies can predict which products the users are interested in from the information data they have mastered, and concentrate marketing resources on these target consumers through target marketing so that a small number of resources can be used to achieve the desired result. In reality, the target users of the company are often a user group or a plurality of different types of groups, so the problem of maximizing the impact of the research target coverage is more practical.

Since the traditional propagation model required a large number of Monte Carlo simulations to better simulate the propagation of information in real life, the time performance of the algorithm is reduced. To avoid this problem, this paper introduces a heat propagation model that uses heat flow to simulate information diffusion in social networks.

This paper proposes an algorithm for continuously updating cluster centers based on the heat propagation model. Firstly, K cluster centers are randomly selected, and the similarity between nodes is defined. Then the target nodes are clustered, and the nodes with the most influence on the cluster are found from each cluster, and the clusters are continuously updated. The center can maximize the total impact. Since the nodes that can have a greater influence on the target market will be in a certain range of subspace of the target node, the breadth-first search method can be used to obtain the subspace to improve the performance of the algorithm.

The main contributions of this paper are:

(1) We introduce the heat diffusion model into the study of the problem of maximizing the impact of personalization.
(2) We propose a clustering algorithm based on the thermal propagation model is proposed to solve the problem of maximizing the impact of the target group. The similarity and total cost of the clustering algorithm are defined and optimized.
(3) We conduct comparative experiments in real social networks, analyze and verify the effectiveness of the proposed algorithm.

Section 2 of this paper introduces related work; Sect. 3 introduces preparatory knowledge; Sect. 4 introduces the known influence maximization algorithm of the target population based on the thermal propagation model and its optimization algorithm; Sect. 5 introduces the experimental analysis of the algorithm; Sect. 6 is the conclusion.

## 2  Related Work

The problem of maximizing the impact is to select k initial users to maximize their range of propagation. Kempe et al. [11] modeled the process of influence propagation as a discrete stochastic process, introducing an independent cascade model (IC) and a linear threshold model (LT). They proved that the problem of maximizing impact is NP-hard, and proposed a classic greedy algorithm, but the implementation efficiency is low. Therefore, in order to improve the performance of the algorithm, Jung et al. [12] proposed the IRIE algorithm based on the traditional model, first estimated the influence ranking of each node, and then estimated the gain influence of the node by the linear method, and got good results.

With the rapid development of networks and technologies, a large amount of data information has become a "wealth". In order to achieve a better impact intensity and less time consumption, the problem of maximizing individualized influence was proposed in 2013. However, most of the algorithms are based on linear threshold models or independent cascade models. In order to make the simulation results closer to reality, it is necessary to increase the number of Monte Carlo simulations, improve the seed quality and reduce the efficiency of the algorithm.

In 2013, Ma et al. [13] proposed a social network marketing model called heat diffusion model, which applied the theory of physics—heat diffusion to describe the influence of communication in social networks. This method avoids the Monte Carlo simulation process. In 2014, Doo et al. [14, 15] proposed an excitation algorithm based on the heat diffusion model. In 2017, Yang et al. [16] proposed a target hot greedy algorithm for a specific node based on the heat diffusion model. The algorithm can quickly calculate the influence of a particular target, but its algorithm is still based on traditional greedy ideas.

Search for the best subsets by using clustering, because homogeneous users are more likely to form a whole. Zhang et al. [17] proposed a maximum coverage algorithm for information oriented to the target market. Based on the independent cascade model, the algorithm performs cluster analysis on users and finds out the representative users of each cluster. Finally, this algorithm can get high-quality seed nodes, and the better the effect is as the number of nodes increases.

We introduced the heat propagation model to remove the time cost of a Monte Carlo simulation in the traditional model and propose a set coverage algorithm for a specific user group. This method uses the clustering method to continuously update the cluster center. When the central node is optimally affected, the seed set can be obtained.

# 3 Preliminary Knowledge

## 3.1 Heat Propagation Model

In nature, the transfer of heat from a relatively high temperature position on the medium to a lower temperature position is a physical phenomenon. Also in social networks, information is also transmitted by users who are affected earlier to users who have not yet been affected. Users with higher levels of influence are disseminated to users with lower levels of influence [13]. Many research scholars have proposed a variety of thermal propagation models. In order to be practical, we use a heat propagation model based on directed network graphs.

The social network is represented by a directed graph G = (V, E), where V represents a set of nodes, and E represents a set of directed edges between nodes. The heat propagation model is used to simulate the information dissemination of social networks.

## 3.2   The Principle of Heat Propagation

**Definition (Thermal Conductivity).** The heat transferred by a unit temperature gradient per unit time through a unit heat transfer surface, also known as the thermal conductivity. It can reflect the ability of the physical medium to conduct heat, expressed as $\alpha$ [13].

For the node $v_i$ in the network, the propagation process of the heat propagation model is as follows:

1. At time $t_0$, the heat of node $v_i$ is denoted as $h_i(0)$; at time $t$, the heat possessed by $v_i$ is denoted as $h_i(t)$. The heat of all nodes in the network at time $t$ is represented by a vector as follows:

$$h(t) = [h_1(t), h_2(t), h_3(t), h_4(t). \ldots . h_n(t)]^T \tag{1}$$

2. If $v_j$ points to the edge $e_{ji}$ of $v_i$, at time $t$, node $i$ receives the heat from $v_j$ in $\Delta t$ time as: RH = (i, j, t, $\Delta t$). Where RH is proportional to the heat of $\Delta t$ and $v_j$; if there is no edge between $vi$ and $v_j$, then RH is equal to zero.

   The heat that $v_i$ can get from neighboring nodes is calculated as follows:

$$RH = \alpha \cdot \Delta t \cdot \sum_{j:v_j \in N^+(v_i)} \frac{h_j(t)}{d_j} \tag{2}$$

   ($N + (v_i)$ represents the set of edges where $v_i$ is connected to $v_j$).

3. The heat that $v_i$ flows out to its neighbor nodes is represented by DH = (i, t, $\Delta t$). If there is no edge starting from $v_i$, the heat of $v_i$ does not propagate outward. DH is proportional to $\Delta t$ and proportional to the heat of $v_i$; The formula for calculating DH is as follows:

$$DH = \tau_i \cdot \alpha \cdot \Delta t \cdot h_i(t) \tag{3}$$

   $\tau_i$ is a sign of the heat output of $v_i$, indicating whether node $vi$ outputs heat outward. When $d_i$ ($d_i$ is the degree of the node i) is greater than 0, $\tau_i$ is 1, $v_i$ heat can be transmitted to its successor through the interconnected edges, the total heat transferred is $\alpha \cdot \Delta t \cdot h_i(t)$; When $d_i$ is equal to 0, $\tau_i$ is 0, indicating that $v_i$ does not point to the network side of other nodes, and its heat cannot be transferred outward.

4. After $\Delta t$, the heat change of node $v_i$ is $h_i(t + \Delta t) - h_i(t)$:

$$h_i(t + \Delta t) - h(t) = RH - DH = \alpha \cdot \left[ \sum_{v_j \in N(v_i)} \frac{h_j(t)}{d_j} - \tau_i h_i(t) \right] \cdot \Delta t \tag{4}$$

Organize the various styles:

$$h(t) = e^{\alpha \cdot t \cdot H} \cdot h(0) = \left( 1 + \alpha \cdot t \cdot H + \frac{\alpha^2 \cdot t^2}{2!} H^2 + \frac{\alpha^3 \cdot t^3}{3!} H^3 + \cdots \right) \cdot h(0) \quad (5)$$

$$H_{ij} = \begin{cases} 1/d_j, & (v_j, v_i) \in E, \\ -1, & i = j, \ d_j > 0 \\ 0, & otherwise \end{cases} \quad (6)$$

Where H is an n-order matrix and e is a natural constant.

Figure 1 is an example. Observing the change process of the holding heat of each node from 0 to 10, the thermal conductivity $\alpha$ is 0.15, and the initial heat of the node v1 is 10 in the figure.

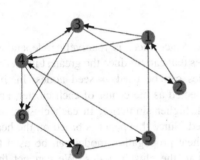

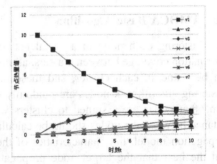

**Fig. 1.** Social network diagram    **Fig. 2.** Change in heat held by each node over time

The initial heat distribution of the network is expressed as:

$$h(0) = [10, 0, 0, 0, 0, 0, 0]^T$$

According to formula (5), each element of the main diagonal of matrix H takes a value of $-1$, two sides with $v_1$ as the starting point, d is 2, and the end point distribution is $v_3$, $v_4$, so H31 and H41 are both 1/2. In the same way, the matrix H is finally obtained as follows:

$$H = \begin{bmatrix} 1 & 1 & 0 & 0 & 1/2 & 0 & 0 \\ 0 & -1 & 1/2 & 0 & 0 & 0 & 0 \\ 1/2 & 0 & -1 & 1/3 & 0 & 0 & 0 \\ 1/2 & 0 & 0 & -1 & 1/2 & 0 & 0 \\ 0 & 0 & 0 & 0 & -1 & 0 & 1 \\ 0 & 0 & 1/2 & 1/3 & 0 & -1 & 0 \\ 0 & 0 & 0 & 1/3 & 0 & 1 & -1 \end{bmatrix} \quad (7)$$

According to formula (6), the change of heat held by each node with time can be calculated, as shown in Fig. 2:

As show in the figure, as time passes, heat flows from the node $v_1$ to the other nodes, and eventually tends to balance when the time t is 10.

# 4   TMOCA Algorithm

Studies have shown that similar users tend to be homogenous, these users may choose to establish a friendly relationship with people with the same interests, or may choose friends' preferences for some reasons. These users will form a more aggregated structure, so cluster analysis is used to explore influential nodes. This section proposes the Target Market Optimal Coverage Algorithm (TMOCA) based on the thermal propagation model and its optimization, algorithm.

## 4.1   TMOCA Basic Algorithm

We clustering each node in a social network. Since there is relatively independent information coverage between clusters, all nodes that can produce the greatest influence can be found in each cluster, and these nodes can be used as seed nodes. In this clustering algorithm, k nodes are randomly selected as the center of each cluster, and then other nodes are assigned to clusters with higher similarity. In each cluster, the distance between the nodes should be calculated. Since this paper is based on the heat propagation model, the characteristics of the heat propagation model can be used to find the heat that the target node propagates to the cluster nodes. We can get the influence of the target node to the cluster node, and the magnitude of the distance between the nodes is determined by the magnitude of the influence. If a target node has the largest amount of heat for a cluster center, then divide the node into this cluster. Use the formula to define the distance between nodes:

$$Heat(v, S_i) = h_{Si(t)} \tag{8}$$

In the above formula, $S_i$ is the i-th cluster center, and $h_{Si}(t)$ is the heat of the clustering $S_i$ of the target node v at time t. Firstly, the target node i is taken out from the target node set in turn, one at a time, the heat $Q_0$ is given to it, and the heat is propagated under the heat propagation model, thereby obtaining the heat of the node i to the k cluster centers. Add node i to the cluster that gets its most heat. If a node has the same impact on multiple clusters, the influence on the node is also added to other clusters.

Each node is assigned, then clusters are calculated, the total impact generated by each cluster center is obtained, and then the clustering scheme with the best total impact is found.

The total impact of formalizing the definition of clustering is:

$$Inf(S) = \sum_i \sum_{v \in S} Heat(S_i, v), \ i = 1, 2, \cdots, k. \tag{9}$$

Replace the cluster center with random nodes and calculate the total influence of the cluster until the best cluster center set is found. Then the k cluster centers are the seed nodes.

---

**Algorithm 1:** Target Market Optimal Coverage Algorithm(TMOCA)

> **Input:** Directed graph G(V, E), Target set T, Thermal conductivity α, Seed size k, The initial heat is $Q_0$, Propagation time t
> **Output:** Seed set s

---

1. Take $S \subseteq G[V]$ *and* $|S|$=k
2. randomly select k cluster centers
3.     FOR $v$ in T do
4.         $h_v(0)$=$Q_0$;
5.         FOR $S_i$ in S do
6.             compute Heat(v, $S_i$)             (use equation (7))
7.             Incorporate v into the cluster $S_i$ of the maximum frequency
8.         End
9.     End
10. Calculate the overall effect of species Inf(S)     (use equation (7))
11. FOR $S_i$ in S do
12.     $h_{Si}(0)$=$Q_0$;
13.     compute h(t)                  (use equation (4))
14.     compute Inf (S)           (use equation (8))
15. End
16. FOR $S_i$ in S do
17.     FOR $j \in V$
18.       IF $Inf(S - S_i + j) > Inf(S)$ then
19.           S. remove($S_i$);
20.           S. append(j);
21.       End;
22.     End
23. End
24. UNTIL Inf(S) no longer changes
25. RETUN S

---

**Complexity Analysis:** The clustering analysis in Algorithm 1 occupies the main time, the number of cycles is n, k is the size of the cluster, and |V| is the number of network nodes, and the complexity of the algorithm is O (n k |V|).

## 4.2 TMCOA Algorithm Improvement

Algorithm 1 finds k seed nodes from the entire social network as cluster centers, and the time overhead is large. The target node is known. In order to reduce the search of the node independent of the target node and improve the performance of the algorithm, this section proposes an improved algorithm, which uses the breadth-first traversal method to narrow the search space.

In Fig. 1, the node v1 is the initial node, the thermal conductivity α is 0.2, and the time t is fixed. The heat held by the different nodes at the time t can be obtained (Fig. 2). It is not difficult to find that the target node closer to the seed node is at the time t, the target node has the largest amount of heat; the farther away from the target node, the heat will be lost or flow elsewhere, and the influence will be weak. Therefore, the distance between the node that has a large influence on the target node and the target node is within a certain range.

We use the breadth-first traversal to narrow down the search space to improve the algorithm. The specific process is as follows: starting from the target node, from near to far, the access and the target node are in the same path and the path length is 1, 2, 3. Each node of ... is counted by the search step, and a limited set of steps is obtained to obtain a set of nodes having a certain influence on the target node.

Then we used the Algorithm 1 to computed this subs.

**Complexity Analysis:** The target seed set is much smaller than the reduced set of all users, so cluster analysis and selection of the optimal seed set still consume a lot of time, the algorithm complexity is $O(nk|U|)$, where n is the number of cycles, k is the set size, and $|U|$ is the number of nodes in the reduced subgraph.

# 5 Experimental Comparison and Analysis

In this section, the design experiments verify the proposed algorithm performance. We observe the influence range of the proposed algorithm and other algorithms on the target node set and the time-solving efficiency of different algorithms.

## 5.1 Experimental Data

This paper selects Wiki-Vote, com-Youtube and Ca-GrQc data sets to verify the performance of the proposed algorithm. Table 1 shows the nodes and sides information of the three data sets. The social networks represented by these three data sets all have the characteristics of a complex network.

**Table 1.** Experimental data set description.

| Datasets | WiKi-Vote | Com-Youtube | Ca-GrQc |
|----------|-----------|-------------|---------|
| Nodes    | 7115      | 1134890     | 5242    |
| Sides    | 103689    | 2987624     | 28980   |

Since the data set only contains relationships between nodes, the experiment randomly extracts the target nodes from each data set. The simulation experiments were carried out on different data sets, and the seed sets generated by different algorithms and the range of influence were recorded.

## 5.2   Parameter Settings

(1) The number of seed nodes k set the number of seed nodes from 1 to 15, and observe the change in the intensity of the affected nodes as the seed nodes increase.

(2) The higher the thermal conductivity α. According to the literature [12] and the multiple experiments selected α value of 0.15.

(3) For the experimental propagation unit time Δt, influence the final time t of propagation, and the unit time Δt takes a value of 0.5. The propagation result in the heat propagation model is related to the duration of the influence propagation. According to the literature [12] and the experimental measurement setting, the final time of the influence propagation is 50 Δt, so that the heat can be highly diffused.

(4) The breadth traverses the number of steps. Step takes values 3 and 6 respectively for comparison.

(5) The target seed set is randomly selected from 1000.

## 5.3   Analysis of Experimental Results

The target market optimal coverage algorithm (TMOCA) based on the heat propagation model proposed in this paper is compared with the following algorithm:

(1) THGA [16], a single-objective influence maximization algorithm based on heat propagation model.

(2) The KCC algorithm [17] is based on the target market information coverage maximization algorithm of the independent cascade model.

(3) The IRIE algorithm, based on the literature [12], the algorithm can calculate the influence ranking of each node, and uses a linear method to estimate the gain influence of the node, thereby selecting the seed node set. The time performance and seed quality of the algorithm are both good.

(4) TMOCA+ algorithm, the optimal algorithm for target market optimal coverage algorithm based on heat propagation model proposed in this paper.

In order to calculate the influence range of the seed node set by different algorithms, the Monte Carlo simulation information propagation process is used in this paper. Each node set is simulated for 2000 times, and then the average value is taken as the estimated value of the node set influence propagation value. Figure 3 compares Fig. 4 with the influence range of different algorithms in two different datasets. The x-axis is the number of seed nodes and the y-axis is the influence range.

Figures 3, 4 and 5 shows that the performance of the algorithms on different data sets is slightly different due to the different internal structure of the data set, but they tend to rise. For our given target group, when the number of seeds reaches 15, the growth tends to be flat, so we choose the number of seed nodes from 1 to 15 to compare the performance of different algorithms.

The KCC algorithm has little difference with the algorithm proposed in this paper when the number of seed nodes is increasing. This is because the KCC algorithm also adopts the clustering method to select the seed set. The performance of the THGA

508    J. Zhu et al.

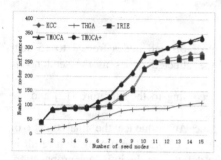

**Fig. 3.** Influence spread on WiKi-Vote

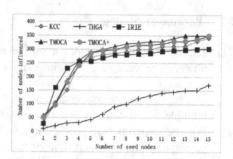

**Fig. 4.** Influence spread on com-Youtube

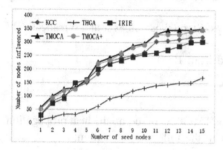

**Fig. 5.** Influence spread on Ca-GrQc

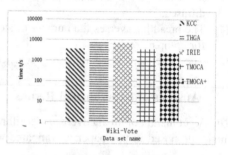

**Fig. 6.** Running time on Wiki-Vote

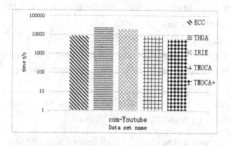

**Fig. 7.** Running time on com-Youtube

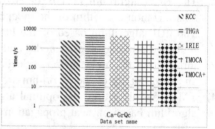

**Fig. 8.** Running time on Ca-GrQc

algorithm on the three data sets is not special. Well, because the THGA algorithm affects the single target, the performance of the algorithm can't be faster with the increasing seed set; the IRIE algorithm is relatively poor, probably because IRIE uses a linear method to estimate when a node joins Interference with the influence of other nodes after the seed node set. However, when it is affected by a given target node, it will be interfered by the unrelated nodes, which will reduce its performance. For any selected target node, the TMOCA algorithm proposed in this paper can get good results. This is because the algorithm is generated by clustering. Seed nodes are the most influential nodes in a certain area. As the number of selected seed nodes increases, the number of clusters increases, which leads to an increasing range of influence propagation. The optimization algorithm does not produce a larger range of influence, and the performance of the selected seed set is similar.

Figures 6, 7 and 8 shows the running time of different algorithms when 10 seed nodes are selected. Since the seed node is 10 in Figs. 3, 4 and 5, the effect of each algorithm tends to be stable. Therefore, select the seed node in this experiment. It is more representative of 10. It is obvious that the running time in Fig. 7 is long, and it may be because Fig. 7 is an experiment of each algorithm on the com-Youtube. The data set node is more related to the other two data sets, and the nodes are closely related to each other, resulting in an operation. The amount is too large. Comparing the five algorithms, it is obvious that the THGA algorithm takes the longest time because the target greedy algorithm is aimed at maximizing the single-objective influence. In this experiment, multiple loops are used to achieve the purpose of influencing the target group, resulting in its operation. The time is relatively high; the KCC algorithm runs slightly longer because the algorithm requires multiple Monte Carlo simulations to ensure the accuracy of the seed set; IRIE takes slightly longer than the KCC algorithm, possibly due to it's To seed selection. The TMOCA algorithm proposed in this paper has a slightly better influence on the propagation time than the KCC algorithm, but its optimization algorithm running time can reach a reasonable time requirement.

# 6 Conclusion

In this paper, when the target user group is known, we want to select a certain number of seed users in the social network to disseminate information, so that the propagation impact is optimal. This paper introduces the heat propagation model to simulate the propagation of information in social networks, so that the influence of nodes can be directly calculated, avoiding the high time consumption in the Monte Carlo simulation. In this paper, the TMOCA algorithm is proposed, and the clustering method is used to find k cluster centers with the greatest influence on the target users as seed nodes. In order to improve the performance of the algorithm, an optimization algorithm (TMOCA+) is proposed, and the breadth-first traversal method is used to narrow the social network map. Experiments on real data sets verify that the proposed algorithm can cover the target users to the greatest extent and has better time performance.

**Acknowledgment.** This work was supported in part by the Education Department of Heilongjiang Province (12531498).

# References

1. Wang, X., Yang, S.: Research on the maximization of the influence of social network personalization. Jiangxi University of Science and Technology, Jiangxi (2016)
2. Shen, C., Liang, W.: Research on the algorithm and propagation model of influence maximization. Dalian University of Technology, Dalian (2016)
3. Sheikhahmadi, A., Nematbakhsh, M.A.: Identification of multi-spreader users in social networks for viral marketing. J. Inf. Sci. **43**(3), 412–423 (2017)

4. Leskovec, J., Krause, A., Guestrin, C., et al.: Cost effective outbreak detection in network. In: ACM SIGKDD International Conference on Knowledge Discovery Data, KDD 2007, pp. 420–429 (2007)
5. Chen, W., Wang, Y., Yang, S.: Efficient influence maximization in social networks. In: Proceedings of ACM KDD, KDD, pp. 199–208 (2009)
6. Goyal, A., Lu, W., Lakshmanan, L.V.S.: CELF++: optimizing the greedy algorithm for influence maximization in social networks. In: International World Wide web Conference, pp. 47–48. ACM (2011)
7. Wang, C., Chen, W., Wang, Y.: Scalable influence maximization for independent cascade model in large scale social networks. Data Min. Knowl. Discov. 25(3), 545–576 (2012)
8. Jiang, Q., Song, G., Cong, G., et al.: Simulated annealing based influence maximization in social networks. In: AAAI Conference on Artificial Intelligence, pp. 127–132 (2011)
9. Alp, Z.Z., Oguducu, S.G.: Identifying topical influencers on twitter based on user behavior and network topology. Knowl.-Based Syst. 141, 211–221 (2018)
10. Chang, C.-W., Yeh, M.-Y., Chuang, K.-T.: Node reactivation model to intensify influence on network targets. Knowl. Inf. Syst. 54, 567–590 (2018)
11. Kempe, D., Kleinberg, J., Tardos, E.: Maximizing the spread of influence through a social network. In: Proceedings of the Ninth ACM SIGKDD International Conference on Knowledge Discovery and Data Mining, pp. 137–146. ACM (2003)
12. Jung, K., Heo, W., Cheng, W.: IRIE: scalable and robust influence maximization in social networks. In: Proceedings of the 12th International Conference on IEEE Data Mining (ICDM). Brussels, Belgium, pp. 918–923 (2012)
13. Ma, H., Yang, H., Lyu, M.R., et al.: Mining social networks using heat diffusion processes for marketing candidates selection. In: Proceedings of the 17th ACM Conference on Information and Knowledge Management, CIKM 2008, Napa Valley, California, USA, 26–30 October (2008)
14. Doo, M., Liu, L.: Probabilistic diffusion of social influence with incentives. IEEE Trans. Serv. Comput. 7(3), 387–400 (2014)
15. Doo, M., Liu, L.: Extracting top-k most influential nodes by activity analysis. In: Proceedings of the 2014 IEEE 15th International Conference on Information Reuse and Integration, pp. 227–236. IEEE, San Francisco (2015)
16. Yang, S., Wang, X., Liu, C.: A heat diffusion model based algorithm for personalized influence maximization in social network. J. High Speed Netw. 2–3(2), 137–147 (2017)
17. Zhang, B., Qian, Z., Wang, Q., et al.: Information maximum coverage algorithm for target market. J. Comput. Sci. 37(4), 894–904 (2014)

# Hybrid Model of Time Series Prediction Model for Railway Passenger Flow

Wei Sha, Shuai Qiu, Wenjun Yuan, and Zhangrong Qin$^{(\boxtimes)}$

Guangxi Key Lab of Multi-source Information Mining and Security,
Guangxi Normal University, Guilin 541004, Guangxi, China
qinzhangrong@gxnu.edu.cn

**Abstract.** Railway passenger flow forecasting is an important basis for scientific dispatching of railway transportation. In order to remedy the shortcomings of one single time series prediction method for passenger flow, a model of combining autoregressive integrated moving average (ARIMA) with extreme learning machine (ELM) based on wavelet transform, named WAADE is presented in this paper. Firstly, the complex railway passenger flow time series was decomposed into linear and non-linear components by wavelet transform. Then, the decomposed linear and non-linear components were predicted by using ARIMA and ELM respectively. Finally, the final prediction results were obtained through fusing the linear and nonlinear prediction results by wavelet transform once again. At the same time, considering the obvious seasonal and periodic regularity of the railway passenger flow data, a WAADES model was constructed combined the WAADE model with the seasonal model based on the entropy value method. The experimental results show that the prediction accuracy of proposed WAADE and WAADES model is higher than the one of the ARIMA or ELM or seasonal model when used alone. Because of the combination of seasonal characteristics, the prediction accuracy of WAADES model is higher than that of WAADE model. The effectiveness and superiority of the two combined models proposed are proved.

**Keywords:** ARIMA model · ELM model · Wavelet transform · Entropy value method · Seasonal model

## 1 Introduction

Passenger flow forecasting can provide an important reference for dispatching decision and travel planning. According to the historical data of railway passenger flow, the railway passenger flow is predicted to provides a scientific basis for passenger travel and train dispatch. In passenger flow forecasting, the time series method is the most common forecasting method. A time series is a sequence in which the values of the same statistical indicator are arranged in the order in which their time occurs (or spatial order), which reflects the law of variation of the data [1]. Time series forecasting has a

© Springer Nature Singapore Pte Ltd. 2019
R. Mao et al. (Eds.): ICPCSEE 2019, CCIS 1059, pp. 511–527, 2019.
https://doi.org/10.1007/978-981-15-0121-0_40

wide range of applications in various industries, including economic forecasting [2], market forecasting [3, 4], and traffic flow [5]. Railway passenger flow data is a typical complex time series data. By constructing models to learn historical statistics and predict future values, the predicted results can be used to guide scientific decision-making and rational planning in the transportation field [6], which can also produce good economic benefits. For a long time, time series prediction methods have been studied by many scholars and researchers [1, 7, 8]. So far, time series prediction methods such as passenger flow are mainly divided into linear prediction methods and nonlinear prediction methods. Among them, the ARIMA model shows good performance in many linear prediction models, while the ELM model excels in nonlinear data prediction. However, the linear prediction methods have weak predictive ability for time series data with nonlinear characteristics, lack of flexibility and generalization ability, and cannot accurately model and predict complex data [9, 10]. And the nonlinear prediction method still has shortcomings in network weight and network structure, which limits the accuracy of its prediction [11]. In general, time series data that is dominant for linear features is not as effective as linear prediction methods. In addition, a large number of studies have shown that a single model is difficult to achieve good results for all time series predictions [12], especially for complex time series is more difficult to obtain satisfactory prediction results. In order to further improve the accuracy of passenger flow data prediction and enhance the performance of prediction methods, Wang et al. [13–15] constructed an improved neural network railway passenger flow prediction model based on the neural network model and used the model to forecast the railway passenger flow. Han et al. [16] constructed a complex seasonal model to analyze the railway passenger flow by analyzing the seasonal characteristics of railway passenger flow and achieved good prediction results. Sun et al. [17] constructed a grey correlation analysis method to forecast and analyze passenger flow data by analyzing the influencing factors of intercity railway passenger flow. Theory and experiment prove that combining multiple prediction methods is a way to effectively improve the time series prediction performance of railway passenger flow [18].

By analyzing the research of related combination methods [9, 10, 18–20], and inspiring from the development trend of knowledge fusion and data fusion [21], this paper proposes a hybrid model of time series prediction model for railway passenger flow. A wavelet transform is used to decompose time series data into linear and nonlinear components and then use linear and nonlinear response models to learn and predict the corresponding components. Finally, the two sets of predicted values are synthesized by wavelet reconstruction and get the predicted value of the WAADE model. Further digging the seasonal and periodic characteristics of the railway passenger flow, using the seasonal model to predict the original time series, and then using the entropy value method to achieve the fusion of the seasonal model and the WAADE model prediction value, and obtain the predicted value of the WAADES model.

## 2  Methodology

In the exploration of many time series prediction methods for the combination of linear and nonlinear models [10], time series are usually regarded as linear and nonlinear structures. However, in practical applications, such as the prediction of railway passenger flow, the corresponding time series data is usually a combination of linear and non-linear, and even more complicated, so it is difficult to simply separate linear and nonlinear. Based on this, this paper decomposes the passenger flow data into high-frequency random components and low-frequency trend components according to the principle of wavelet transform to solve the problem of linear and nonlinear separation of time series data. The linear component is decomposed using a linear model, and the nonlinear component is modeled using a nonlinear model, so that the prediction values of linear and nonlinear are obtained. Since the wavelet transform also has the ability to reconstruct and restore the decomposed components, the wavelet prediction principle is used to synthesize the two sets of prediction values, and the final prediction result of the WAADE model is obtained. In order to further explore the data characteristics, the seasonal data is used to predict the original data, and the entropy method can be used to synthesize the seasonal model and WAADE model prediction results. Finally, the prediction results of the WAADES model are obtained.

### 2.1  Wavelet Transform

The concept of wavelet transform was introduced in 1984 by French physicist Morlet [22] in the process of processing and analyzing the massive data generated by earthquakes. As a new method for transform analysis of time series data, wavelet transform is based on the inheritance and development of the short-time Fourier transform localization idea, further improving the defect that the frequency change can not cause the window size to change. When the frequency changes, the "Time-Frequency" window will also change accordingly. So wavelet transform is an ideal tool for time-frequency analysis and processing of time series data. The wavelet transform can be represented as the following equation.

$$\mathrm{W}f(\omega, b) = \int_R f(t)\overline{\psi}_{ab}(t)dt \tag{1}$$

In this equation, $\psi_{ab}(t)$ can be represented as the following equation.

$$\psi_{ab}(t) = |a|^{\frac{1}{2}}\psi(a - b) \tag{2}$$

In Eq. (2), in order to enable the window to cover the entire time domain, the parameter b is used to adjust and control the moving window. In order to achieve the purpose of adaptive analysis of the high and low-frequency signals, the value of a needs to be adjusted and changed.

In the practical application of wavelet basis functions, various wavelet functions usually have different properties, and these properties between them cross each other and are limited. It is very difficult to find wavelet functions that satisfy all properties. After the wavelet basis function is selected, the time series can be decomposed by wavelet. Figure 1 is an example of wavelet multi-scale decomposition of time series data.

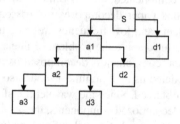

**Fig. 1.** Time series wavelet decomposition example

In Fig. 1, S is time series data, and A1, A2, A3, and D1, D2, and D3 respectively represent low-frequency trend components and high-frequency random components after time series data decomposition. The decomposed components are predicted by corresponding modeling and analysis methods, and two sets of prediction results are obtained. The wavelet transform can reconstruct the decomposed component and the predicted component, and finally restore the original time series data and obtain the predicted value. The following equation is the wavelet transform reconstruction equation.

$$L = a3 + d1 + d2 + d3 \tag{3}$$

In the Eq. (3), a3 is the decomposed low-frequency trend component, that is, the linear component, and d1, d2, and d3 are the decomposed high-frequency random components, and L is the result of the synthesis.

## 2.2    ARIMA Model

The ARIMA (Autoregressive Integrated Moving Average) model is further extended by ARMA. The ARIMA model appeared in the late 1960s, and this model was also called the Box Jenkins model, because Box and Jenkins systematically explained the ARIMA model in 1976. When predicting time series data, the model analyzes the data law according to the changes of past data, finds the moving average and periodic components of historical data, and realizes the prediction of time series data. In the face of complex and variable time series data, only the proper and reasonable use of the ARIMA model can achieve accurate, flexible and efficient prediction. The following is the parameter description.

p: p is the order (number of time lags) of the autoregressive model.
d: d is the degree of differencing (the number of times the data have had past values subtracted).

q: q is the order of the moving-average model.

Let y denote the difference of Y at time t.

If d = 0, then $y_t = Y_t$.

If d = 1, then $y_t = Y_t - Y_{t-1}$.

If d = 2, then $y_t = (Y_t - Y_{t-1}) - (Y_{t-1} - Y_{t-2}) = Y_t - 2Y_{t-1} + Y_{t-2}$.

The prediction model of ARIMA can be expressed as:

The predicted value of Y = constant u + weighted sum of Y of one or more recent times + predicted error of one or more recent times.

When p, q, and d are the best values, the ARIMA model can be expressed as the following equation.

$$\hat{y}_t = \mu + \phi_1 * y_{t-1} + \ldots + \phi_p * y_{t-p} + \theta_1 * e_{t-1} + \ldots \theta_q * \qquad (4)$$

Basic flow chart is shown in Fig. 2.

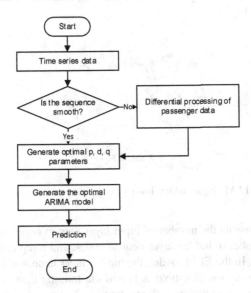

**Fig. 2.** ARIMA model flow chart

To correctly use the ARIMA model for time series data prediction, it is important to find the appropriate parameters and correlation coefficients. The ARIMA model uses autoregressive analysis to determine its correlation coefficients and models when making predictions. The actual data is used to verify the validity and feasibility of the model after obtaining the relevant model. This model considers the correlation of time series and reduces the interference of random volatility to some extent. Using this method to predict time series data can improve the accuracy of prediction results to some extent.

## 2.3  ELM Model

ELM (Extreme Learning Machine) is an easy-to-use and practical effective learning algorithm of single hidden layer feedforward neural network (SLFN), which is proposed by Associate Professor Huang Guangbin of Nanyang Technological University in 2004 [23]. Compared with the traditional neural network learning algorithm, the extreme learning machine only needs to set the number of hidden nodes in the network proposes a ARIMA-ELM and does not need to adjust the network weight and the hidden element offset. The algorithm automatically executes and generates a unique optimal solution, which has a fast learning speed, good generalization performance and other advantages.

During the execution of the ELM algorithm, the input layer weights and hidden network layer thresholds can be randomly selected, and the cost function is used to obtain the minimum weight of the output layer. This is why ELM has higher convergence speed and stronger generalization ability. The single hidden layer forward network training model used by the extreme learning machine in the training learning process is shown in Fig. 3:

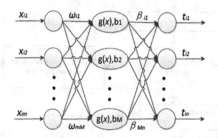

**Fig. 3.** ELM single hidden layer forward network training model

In Fig. 3, m represents the number of input layer neuron nodes in the ELM model, M represents the number of hidden layer neuron nodes, and n represents the number of output neuron nodes. In the ELM model, the hidden layer neuron excitation function is g(.), the hidden layer neuron threshold is b, and the training data is (xi, ti).

This paper presents a mathematical equation as shown in Eq. (5) to describe and express the network training model of the extreme learning machine.

$$\sum_{i=1}^{M} \beta_i g(\omega_i x_i + b_i) = o_i, \quad i = 1, 2, \ldots, N \tag{5}$$

In the Eq. (5), the input weight vector of the ELM model is $\omega_i = [\omega_{1i}, \omega_{2i}, \ldots, \omega_{mi}]$, the output weight vector is expressed as $\beta_i = [\beta_{i1}, \beta_{i2}, \ldots, \beta_{in}]^T$, and $o_i = [o_{i1}, o_{i2}, \ldots, o_{in}]^T$ represents the output value of the model.

The cost function E of the extreme learning machine is given a mathematical equation as shown in (6) for description and expression.

$$E(S, \beta) = \sum_{i=1}^{N} ||o_i - t_i|| \tag{6}$$

In the Eq. (6), $s = (\omega_i, b_i, i = 1, 2, \ldots, M)$ input weights and hidden layer thresholds are also included in the ELM network model. Based on the research of Huang et al., we can find that as an extreme learning machine, the cost function is minimized, and thus the ELM optimal output weight is found. This output weight is the error of the ELM minimizing the network output value and its corresponding actual value. In the specific experimental study, when the HH$^T$ is non-singular, we can get $\beta$ using Eq. (7).

$$\beta = H^T(\frac{1}{C} + HH^T)O \tag{7}$$

In Eq. (7), the matrix of the hidden layer of the ELM model is H; the vector of the predicted target data is O. According to the theory of ridge regression, increasing the normal number 1/C can make the solution result more stable and have better generalization ability. It can be expressed as Eq. (8).

$$f(x) = h(x)H^T(\frac{1}{C} + HH^T)O \tag{8}$$

Ridge regression is one of the complementary methods of the least squares method. By using this method, the risk of the matrix becoming a "morbid matrix" can be reduced. Although the losslessness is lost, the data with higher stability can be obtained.

## 2.4  RIMA-ELM Model

In order to eliminate the limitations of single model prediction and improve the accuracy and applicability of time series data prediction, this paper combines the advantages of linear ARIMA model and nonlinear ELM model, and uses the wavelet transform principle to achieve the organic combination of the two. Based on this, this paper proposes an ARIMA-ELM wavelet transform combination model. Specifically, the wavelet decomposition technique is used to decompose the time series data into low-frequency trend components reflecting linear features and high-frequency random components reflecting randomness. The low-frequency components are predicted by the ARIMA model with better linear characteristics. The high-frequency random components are predicted by the ELM model with strong random characteristics. Finally, the results predicted by the two are organically synthesized by wavelet reconstruction method, and the required ARIMA-ELM combined model prediction results are obtained. The ARIMA-ELM combination model is based on wavelet theory, after wavelet decomposition of the original data sequence, low-frequency (A3) using ARIMA model prediction, high-frequency (D1, D2, D3) using ELM model, and then using wavelet reconstruction to synthesize prediction results This model is referred to as "WAADE" for short in this paper.

In this way, the time series data is first decomposed according to the pattern of Fig. 1, and the obtained components can be parallelized [24] to improve efficiency, the low-frequency sequence is predicted by ARIMA, and the high-frequency sequence is predicted by the ELM model. According to Eq. (3) for organic synthesis, the WAADE model can be used to predict the railway passenger flow.

## 2.5  Entropy Method

In information processing, entropy is used to measure information uncertainty. The entropy weight represents the weight of an indicator or factor. The entropy method is based on the amount of information provided by the indicator or factor, and the weight of the indicator or factor is obtained. The entropy method can prevent the influence of subjective influence factors on the results. It is based on the information provided by the sample observations to judge the importance of the indicators. There are few human interference factors, and the weights of the indicators can be given objectively and fairly. The following are the steps for calculating the index weights and performing the weighted average calculation method by the entropy method:

Step 1: Measuring the contribution of sample indicators
Information entropy is used to determine weights. Suppose there is a multi-indicator decision matrix as shown in Eq. (9):

$$
M = \begin{matrix} A_1 \\ A_2 \\ A_3 \\ A_4 \end{matrix} \begin{bmatrix} x_{11} & x_{12} & \cdots & x_{1n} \\ x_{21} & x_{22} & \cdots & x_{2n} \\ \vdots & \vdots & \ddots & \vdots \\ x_{m1} & x_{m2} & \cdots & x_{mn} \end{bmatrix} \tag{9}
$$

Use $P_{ij} = \dfrac{x_{ij}}{\sum\limits_{i=1}^{m} x_{ij}}$ to indicate the contribution of the j to the Ai in the i scheme.

Step 2: Measure the total contribution of all programs to each indicator.

$$
E_j = -K \sum_{i=1}^{m} P_{ij} \ln(P_{ij}) \tag{10}
$$

In Eq. (10), $E_j$ is the total contribution of all schemes to index $X_j$, K is a constant, $K = 1/\ln(m)$, so $E_j$ has a value range of $0 \leq E_j \leq 1$, i.e. $E_j$ The maximum value is 1 and the minimum value is 0. By analyzing Eq. (10), it can be concluded that if each scheme has a consistent contribution to each scheme, then E tends to 1; in special cases, if they are all equal, there is no need to consider the role of indicators in decision-making. That is, the index has a weight of 0. In this way, it can be seen that the weight value corresponding to the indicator is determined by the size of all the scheme differences. Here $d_j$ is defined to indicate the degree of consistency of the $j^{th}$ indicator of all schemes: $d_j = 1 - E_j$.

Step 3: Calculate the weight corresponding to each indicator.

$$
W_j = \frac{d_j}{\sum\limits_{j=1}^{n} d_j} \tag{11}
$$

In Eq. (11), each indicator corresponds to a weight of $W_j$. When $d_j = 0$, the $j^{th}$ indicator can be culled, and its corresponding weight is 0.

Step 4: Calculate the weighted average.

The weighted average method is to give different weights and then average the data according to the importance of each attribute or indicator when averaging.

$$\bar{x} = \frac{x_1 W_1 + x_2 W_2 + x_3 W_3 + \cdots + x_m W_m}{W} \tag{12}$$

In Eq. (12), W is the sum of the weights. $\bar{x}$ is the weighted average. $x_1, x_2, \cdots, x_m$ are the corresponding indicator parameter values.

## 2.6 Seasonal Model

The seasons in the seasonal model do not only refer to the seasons that are usually considered, but the seasons here represent the development cycle of objective things. This cycle does not necessarily take a quarterly cycle. It may be in years, months, weeks, days, hours, minutes, etc., or even smaller time units. Through the statistics, we can get the seasonal index of things. The seasonal index is used to reflect changes in things during this time period.

Railway passenger flow is the research object of this paper. The seasonal model can be expressed as:

$$\hat{Y}_t = \bar{Y} \bullet f_i \tag{13}$$

$$f_i = \frac{\bar{a}}{\bar{b}} \tag{14}$$

In the Eqs. (13) and (14), the overall average of the time series is represented by $\bar{Y}$; the seasonal index is $f_i$; $\bar{a}$ is the average of the corresponding time of a statistical unit of railway passenger flow; and $\bar{b}$ is the average of each statistical data in a statistical cycle.

The seasonal model has a good application effect on data that is susceptible to seasonal and irregular changes.

## 2.7 Combined Modeling of Seasonal Features

The ARIMA-ELM hybrid model, the WAADE model, has been given in Sect. 2.4. On this basis, the seasonal and periodic characteristics of the railway passenger flow data series are further explored to construct a new combined model of the seasonal characteristics. This paper is called the WAADES model. Firstly, the WAADE model is used to predict the railway passenger flow. At the same time, the same railway passenger flow data sequence is predicted by using the seasonal model. After obtaining two sets of prediction results, the entropy method is used to measure the weights of the two prediction results, and the prediction results are weighted average. Finally, the final result of the WAADES combined forecasting model for railway passenger flow forecasting can be obtained. The principle of the WAADES hybrid model for railway passenger flow prediction is given below, as shown in Fig. 4.

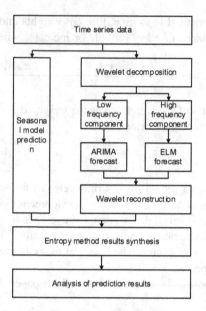

**Fig. 4.** WAADES hybrid model prediction schematic

The model realizes the linear, nonlinear and seasonal feature fusion of time series, and the theoretical prediction effect should be better.

## 3  Numerical Experiment

In order to verify the effectiveness of WAADE and WAADES, the monthly passenger traffic data is predicted and the results are analyzed and judged. The data source is the monthly data of the national railway passenger flow from January 2005 to May 2017 published by the Statistical Bureau of the People's Republic of China on the official website. The evaluation indicators seen have Root Mean Square Error (RMSE), standard deviation, Mean Absolute Deviation (MAD) and Mean Absolute Percentage Error (MAPE). Among them, the two uses of MAD and RMSE are more common. The MAD is the average of the absolute values of the difference between a single observation and an arithmetic mean. And the MAPE is the average of the absolute value of the error as a percentage of the true value of the data. The following Eqs. (15) and (16) are the expressions for the average absolute error and the average absolute percentage error.

$$MAD = \frac{1}{n}\left|X_{abs,i} - X_{modes,i}\right| \tag{15}$$

$$MAPE = \frac{1}{n}\sum_{i=1}^{n}\frac{\left|X_{abs,i} - X_{modes,i}\right|}{X_{modes,i}} \tag{16}$$

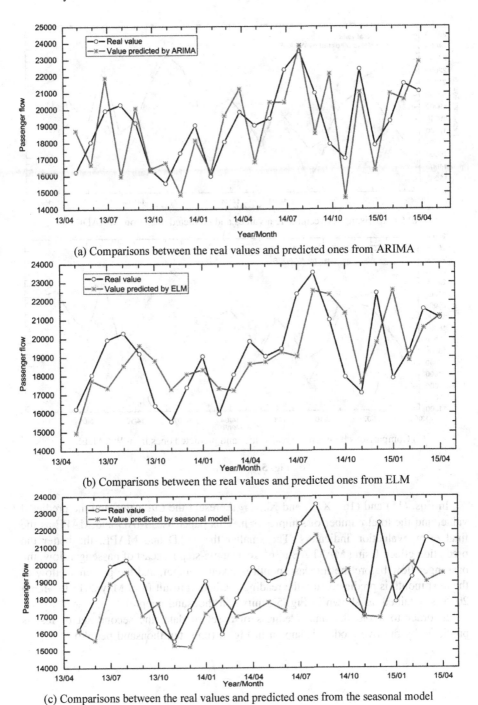

(a) Comparisons between the real values and predicted ones from ARIMA

(b) Comparisons between the real values and predicted ones from ELM

(c) Comparisons between the real values and predicted ones from the seasonal model

**Fig. 5.** Five model prediction results (the first group data)

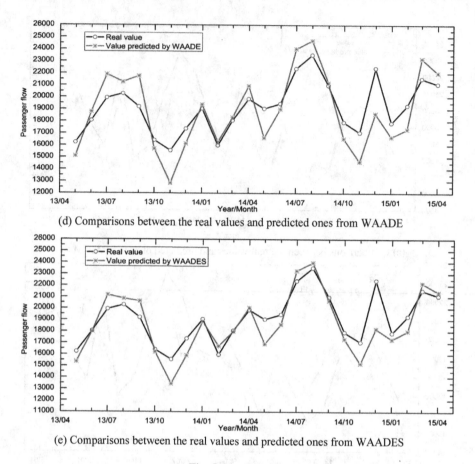

(d) Comparisons between the real values and predicted ones from WAADE

(e) Comparisons between the real values and predicted ones from WAADES

**Fig. 5.** (*continued*)

In Eqs. (15) and (16), $X_{abs,i}$ and $X_{modes,i}$ represent the true value and the predicted value, and the total number of samples is n. This paper uses MAD and MAPE as the final error evaluation indicators. The smaller the MAD and MAPE, the better the prediction effect. Using MATLAB to make a single-step forecast of passenger flow, the passenger flow before the forecast month is taken as input, and the passenger flow of the next month is predicted, and the reading prediction result from May 2013 to March 2015 is obtained, as shown in Fig. 5 (units: ten thousand people).

In order to avoid the one-sidedness of a set of data, the second set of results predicted by the five models is shown in Fig. 6 (unit: ten thousand people).

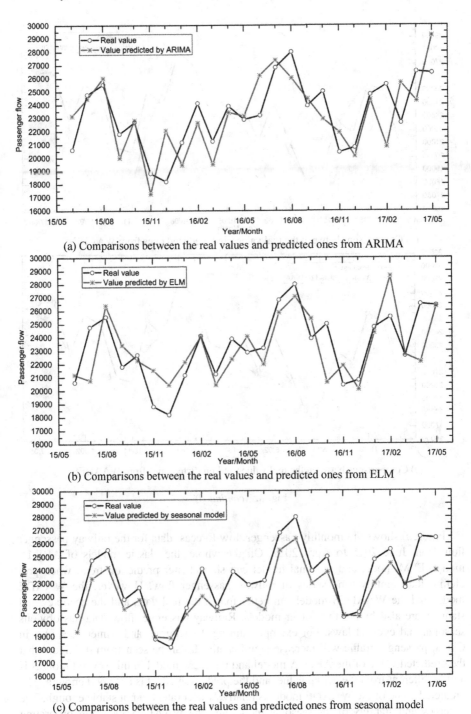

(a) Comparisons between the real values and predicted ones from ARIMA

(b) Comparisons between the real values and predicted ones from ELM

(c) Comparisons between the real values and predicted ones from seasonal model

**Fig. 6.** Five model prediction results (the second group data)

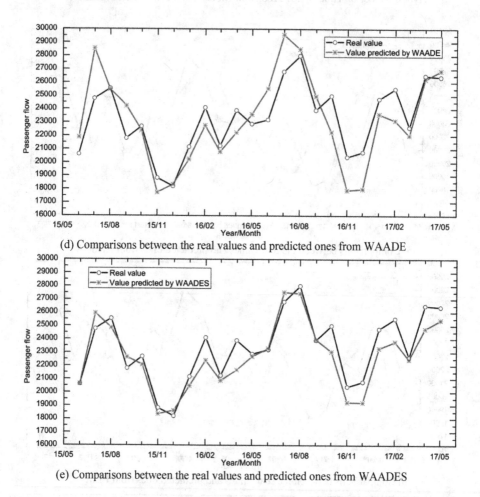

(d) Comparisons between the real values and predicted ones from WAADE

(e) Comparisons between the real values and predicted ones from WAADES

**Fig. 6.** (*continued*)

Figure 6 shows the monthly passenger flow forecast data for the railway passenger flow from June 2015 to April 2017. On the whole, the classic models of ARIMA model, ELM model and seasonal model can show better prediction trend when predicting the monthly time series of railway passenger flow. However, the WAADE model and the WAADES model can better match the real data, and the accuracy and stability are also higher than other models. Railway passenger flow data has obvious seasonal and cyclical laws. For example, during the summer and winter vacations in China, passenger traffic will increase significantly. It can be seen from the figure that the predicted value of the ARIMA model and the ELM model in this several months is a little different from the actual value, and the prediction effect is not good enough. The predicted value of the WAADE model is relatively accurate, but not stable enough. The accuracy and stability of the seasonal model and the WAADES model incorporating the seasonal model are significantly higher than those of other models. Therefore, the experiment proves that the WAADES model proposed in this paper has good validity and applicability in the prediction of railway passenger flow.

In order to facilitate the comparison between the WAADE and WAADES hybrid railway prediction models and the single model prediction results, the evaluation indicators table shown in this paper is shown in Table 1.

**Table 1.** Evaluation indicators for each model

| Model | First group MAD | First group MAPE | Second group MAD | Second group MAPE | MAD average | MAPE average |
|---|---|---|---|---|---|---|
| ARIMA model | 1697.62 | 0.089 | 1659.67 | 0.073 | 1678.64 | 0.081 |
| ELM model | 1433.75 | 0.076 | 1500.52 | 0.064 | 1467.13 | 0.7 |
| Seasonal model | 1671.57 | 0.085 | 1240.47 | 0.051 | 1456.02 | 0.068 |
| WAADE model | 1385.25 | 0.072 | 1401.04 | 0.06 | 1393.14 | 0.066 |
| WAADES model | 969.21 | 0.051 | 899.19 | 0.038 | 934.2 | 0.045 |

It can be seen from Table 1 that the average percentage errors of ARIMA, ELM and seasonal models are relatively small, indicating that they can reflect the overall change of time series data, and the prediction effect is good. At the same time thanks to the work of these people again. In addition, the WAADE and WAADES combined prediction model has a smaller MAD and MAPE than the other three models. It fully demonstrates that its prediction effect is better and its applicability is stronger.

In summary, the WAADE model constructed in this paper has higher accuracy for predicting time series data than ARIMA, ELM and seasonal models, which can improve the accuracy of railway passenger flow forecast to a certain extent. At the same time, it should be noted that in the railway passenger flow data experiment that is not shown, the prediction can also find that the combined model prediction accuracy is higher than the single model. The validity and superiority of the WAADES model for predicting railway passenger flow are confirmed by experiments.

# 4  Conclusion

This paper proposes a method for decomposing complex time series data into linear and nonlinear parts using wavelet transform. Based on this, the ARIMA-ELM combined model (the WAADE model) is proposed, which is suitable for modeling and predicting time series data. Specifically, the linear and nonlinear partial decomposition of time series data is realized by wavelet transform, the linear components of the decomposition are fitted and predicted by ARIMA, and the nonlinear components are fitted and predicted by ELM. The prediction results are realized by wavelet reconstruction, and the scientific synthesis of prediction results is realized. The time series of railway passenger flow is the research object. The prediction experiment results show that the prediction accuracy of ARIMA-ELM combined model WAADE is higher than that of single-use, which verifies the validity of the model. In order to further explore the seasonal and periodic laws of railway passenger flow, the entropy method is used to realize the fusion of the WAADE model prediction value and the seasonal model prediction value, and the WAADES combination model is obtained. Through experiments, it is verified that the accuracy of the model for railway passenger flow prediction is better than that of the previous four models.

**Acknowledgments.** This work was supported by the National Natural Science Foundation of China (Grant Nos. 11862003, 81860635, 11462003), the Key Project of Guangxi Natural Science Foundation (Grant No. 2017GXNSFDA198038), the Project of Guangxi Natural Science Foundation (Grant No. 2018JJA110023), the Project for Promotion of Young and Middle-aged Teachers' Basic Scientific Research Ability in Guangxi Universities (Grant No. 2019KY0084), Guangxi "Bagui Scholar" Teams for Innovation and Research Project.

# References

1. Huang, H.M.: Application time series analysis (2016)
2. Moiseev, N.A.: Forecasting time series of economic processes by model averaging across data frames of various lengths. J. Stat. Comput. Simul. **87**, 1–21 (2017)
3. Klæboe, G., Eriksrud, A.L., Fleten, S.E.: Benchmarking time series based forecasting models for electricity balancing market prices. Energy Syst. **6**, 43–61 (2015)
4. Wang, Y.: Stock market forecasting with financial micro-blog based on sentiment and time series analysis. J. Shanghai Jiaotong Univ. **22**, 173–179 (2017)
5. Dou, F., Xu, J., Wang, L., Jia, L.: A train dispatching model based on fuzzy passenger demand forecasting during holidays. J. Ind. Eng. Manag. **6**, 320–335 (2013)
6. Zhang, Y., Liu, J., Hao, J.: Application of regularized RBF network model in groundwater level prediction. J. Northwest A & F Univ. (Nat. Sci. Ed.) **39**, 204–208 (2011)
7. Fan, X., Lei, Y., Wang, Y.: Adaptive partition intuitionistic fuzzy time series forecasting model. Syst. Eng. Electron. (Engl. Ed.) **28**, 596 (2017)
8. Guo, Z., Zeng, Y., Yang, J., Ye, W.: Time series prediction of stock market index based on MPCA-RBF model. Appl. Res. Comput. **34**, 3299–3302 (2017)
9. Wedding II, D.K., Cios, K.J.: Time series forecasting by combining RBF networks, certainty factors, and the Box-Jenkins model. Neurocomputing **10**, 149–168 (1996)
10. Zhang, G.P.: Time series forecasting using a hybrid ARIMA and neural network model. Neurocomputing **50**, 159–175 (2003)
11. Zhang, Y., Lao, W., Jin, L., Chen, T., Liu, J.: Growing-type WASD for power-activation neuronet to model and forecast monthly time series. In: 2013 10th IEEE International Conference on Control and Automation (ICCA), pp. 1312–1317. IEEE (2013)
12. Chatfield, C.: What is the "best" method of forecasting? J. Appl. Stat. **15**, 19–38 (1988)
13. Zhao, S.Z., Ni, T.H., Wang, Y., Gao, X.T.: A new approach to the prediction of passenger flow in a transit system. Comput. Math Appl. **61**, 1968–1974 (2011)
14. Wei, Y., Chen, M.C.: Forecasting the short-term metro passenger flow with empirical mode decomposition and neural networks. Transp. Res. Part C Emerg. Technol. **21**, 148–162 (2012)

15. Wang, Y., Zheng, D., Luo, S.M., Zhan, D.M., Nie, P.: The research of railway passenger flow prediction model based on BP neural network. Adv. Mater. Res. **605–607**, 2366–2369 (2013)
16. Ma, J., Han, Z.: Research on complex season time series model. Stat. Decis. (6), 27–30 (2017)
17. Ma, J., Han, Z., Sun, W., Liu, G.: Research on railway intercity passenger flow forecasting model. J. Beijing Jiaotong Univ.: Nat. Sci. Ed. **29**, 84–87 (2005)
18. Makridakis, S.: Why combining works? Int. J. Forecast. **5**, 601–603 (1989)
19. Zhang, W., Huai, J.: Domain analysis and modeling research. J. Beijing Univ. Aeronaut. Astronaut. **30**, 1225–1230 (2004)
20. Jain, A., Kumar, A.M.: Hybrid neural network models for hydrologic time series forecasting. Appl. Soft Comput. **7**, 585–592 (2007)
21. Meng, X., Wang, J., Dong, X.: Data fusion and knowledge fusion topic foreword. J. Comput. Res. Dev. **53**, 229–230 (2016)
22. Goupillaud, P., Grossmann, A., Morlet, J.: Cycle-octave and related transforms in seismic signal analysis. Geoexploration **23**, 85–102 (1984)
23. Huang, G.B., Zhu, Q.Y., Siew, C.K.: Extreme learning machine: theory and applications. Neurocomputing **70**, 489–501 (2006)
24. Feng, Y., Wu, K., Xiong, Z., Wu, Z.: An efficient parallel high-dimensional clustering algorithm. Comput. Sci. **32**, 216–218 (2005)

# Classification Method of Encrypted Traffic Based on Deep Neural Network

Jing Wan, Libing Wu, Youhua Xia[✉], Jianzong Hu, Zhenchang Xia, Rui Zhang, and Min Wang

School of Computer Science, Wuhan University, Wuhan 430072, China
woshixiayouhua@126.com

**Abstract.** With the widespread use of network traffic encryption technology, the traditional traffic classification method has gradually become invalid, which increases the difficulty of network management and poses a serious threat to network security. This paper analyzes the traffic encrypted and transmitted by VPN and explores its classification method. By extracting the timing characteristics of the encrypted traffic, the classification model of the deep neural network was used to classify the traffic of seven different categories in the encrypted traffic, and compared with the commonly used naive Bayesian classification algorithm. At the same time, the batch size that affects the training of deep neural network models was studied. Experiments show that the classification ability of encrypted traffic classification model based on deep neural network is much better than the naive Bayesian method. During training, the batch size has different effects on the deep neural network model. When the batch size is 40, the deep neural network model has the best classification ability.

**Keywords:** Encrypted traffic classification · Deep neural networks · Deep learning · SSL/TLS

## 1 Introduction

With the development of network technology, the source of traffic in network traffic has become widespread, and the number of variety of traffic has increased dramatically, posing a huge challenge to network traffic management and network security. At the same time, as the user's network security awareness increases, a large number of applications use encrypted traffic to transmit data, and the types of traffic in the network become more diverse. The current network traffic contains not only traditional digital, image, file, and streaming media. Etc., but also contains a variety of encrypted traffic. Due to the lack of understanding of encrypted traffic and effective identification methods, encrypted traffic increases the difficulty of network management and poses a great threat to network security. The classification of encrypted traffic helps to understand the composition of network traffic, which is beneficial to network management and network security.

Traffic identification technology has important applications in network monitoring and management, traffic accounting, and user behaviour analysis. For example, intrusion prevention systems and firewalls use this technology to identify malicious traffic

© Springer Nature Singapore Pte Ltd. 2019
R. Mao et al. (Eds.): ICPCSEE 2019, CCIS 1059, pp. 528–544, 2019.
https://doi.org/10.1007/978-981-15-0121-0_41

and block malicious connections in a timely manner [1]; network service providers use it to analyze network traffic distribution to assist in QoS (Quality of Service) management; enterprises use traffic identification technology to control application access; The organization guesses user information and behaviour by identifying user mobile device traffic. In view of the above various requirements, achieving accurate and efficient traffic identification is extremely important.

Traditional methods for identifying traffic are based on packet content such as port number and load. In the early stages of the Internet, applications use the port number specified by IANA (The Internet Assigned Numbers Authority) to create a connection. For example, http uses port 80 to connect. With the development of port hopping and random port technology, port-based technologies are becoming ineffective [2]. The payload-based traffic classification method compares the traffic load content with the feature set constructed in advance. For example, the feature string of the http protocol includes GET and POST [3, 4]. With the widespread use of encrypted transport protocols, payload-based traffic classification methods also lose their effectiveness. The DIP (Deep Packet Inspection) technology proposed later identifies the traffic by detecting the load characteristics and has extremely high accuracy [5]. Therefore, the current traffic identification products are mainly based on DPI technology, and common traffic identification tools include PACE, Open DPI, NDPI, L7-filter, Libprotoident, and so on [6, 7]. However, as the increasing encryption traffic hides the load characteristics, making it impossible to identify using the DPI method, and packaging technologies such as tunnelling further limit its application. It is pointed out in the literature that the accuracy of using the DPI method to identify eMule encrypted traffic is only 30% to 70%. In addition, DPI technology has the problems of high computational complexity and infringement of user privacy.

Currently, machine learning methods based on network flow feature statistics are applied to traffic classification, usually using classical machine learning methods, such as naive Bayes and support vector machines. Then the classification ability of these algorithms depends on the selection of features by humans. It can't be widely used. Deep neural networks automatically select features during training without the need to select them in advance. Deep neural networks can still classify traffic well when new network traffic types occur or when existing traffic characteristics change. Deep neural networks are more capable of learning complex situations than traditional machine learning methods, and can better learn the nonlinear relationship between the original input and output. Therefore, the deep learning network is more suitable for encrypted traffic classification.

The contributions of this paper are as follows:

(1) Construct a classification model of encrypted traffic based on deep neural network.
(2) Exploring the impact of different batch sizes on the classification ability of encrypted traffic classification models based on deep neural networks.
(3) Comparing the encrypted traffic classification model based on deep neural network with the naive Bayesian classification algorithm, the experimental results show that the encrypted traffic classification model based on the deep neural network classification model can identify the encrypted traffic well.

The rest of this paper is organized as follows. In the second part, the main methods for encrypting traffic classification and the naive Bayesian classification model are introduced. In Sect. 3, the deep neural network classification model is constructed, and the principle is analyzed. In Sect. 4, we perform experimental results and analysis. Finally, we conclude the paper in Sect. 5.

## 2 Related Work

In recent years, with the rapid development of network technology, network security issues have received more and more attention. Some malware send confidential information to external networks such as botnets, Trojans, and Advanced Persistent Threats (APT) bypasses firewalls and intrusion detection systems through encrypted channel technology. APT attacks generally use unknown Trojans for remote control, avoiding network behaviours through hidden channels and encrypted channels, while attacks last for months or even years. Therefore, effectively identifying and detecting encrypted traffic is of great significance for maintaining network security operations. There are four types of current encrypted traffic identification methods: a method based on load randomness detection, a method based on payload identification, a method based on machine learning, and a hybrid method combining multiple strategies.

Literature [8] analyzes six common machine learning algorithms and demonstrates their respective performance in detecting malicious encrypted network sessions. It is found that the random forest integrated classifier is the most robust to this problem domain, which proves that the feature selection has a greater impact on performance. The enhanced feature set is created by augmenting the standard feature set used in the domain with the characteristics of the domain expert identity, which are specifically tailored for encrypted network sessions. All machine learning algorithms don't rely solely on features that are easy to collect, nor do they need to be discussed repeatedly with domain experts how to best represent the data, so their performance has improved significantly. Literature [9] proposed Blind-Box, which is the first system to provide both attributes. The Blind-Box method directly performs deep packet inspection on encrypted communication streams. Blind-Box implements this approach with a new protocol and a new encryption scheme. Demonstrates that Blind-Box supports applications such as IDS, ex-filtering detection, and parent filter, and supports true rule sets from open source and industrial DPI systems. Implemented Blind-Box and demonstrated that it is useful for long-term use of HTTPS connection settings. Its core encryption scheme is 3–6 orders of magnitude faster than existing related encryption schemes. In [10], random fingerprints are proposed for application flows transmitted in Secure Sockets Layer/Transport Layer Security (SSL/TLS) sessions. Due to the large difference in fingerprint parameters of the selected application, this method has better application recognition accuracy and provides the possibility to detect SSL/TLS session exceptions. Analysis of the experimental results shows that the acquisition of application discrimination mainly comes from incorrect implementation practices, misuse of the SSL/TLS protocol, various server configurations, and the nature of the application.

Literature [11] studied the validity of VPN-based time-related features to detect VPN traffic, and classified the encrypted traffic into different categories according to traffic types (such as browsing, streaming media, etc.). Two well-known machine learning techniques (C4.5 and KNN) were used to test the accuracy of the algorithm's characteristics. The experimental results show high accuracy and performance, confirming that the time-related features are good classifiers for cryptographic traffic characterization. Literature [12] proposed a second-order Markov chain-based attribute-aware encrypted traffic classification method. Some ways are first explored to further improve the performance of existing methods in terms of recognition accuracy, and observe that the application property bigram (composed of the certificate package length in the SSL/TLS session and the first application program data size) helps Application identification. In order to increase the diversity of applied fingerprints, a new method of introducing attribute two graphs into second-order homogeneous Markov chains is proposed. The simulation results show that compared with the most advanced Markov-based method, the method can improve the classification accuracy by an average of 29%. Literature [13] proposes a hybrid model that classifies computer network traffic based on computational intelligence-based classifiers, extreme learning machines (ELM), feature selection (FS), and multi-objective genetic algorithms (MOGA) instead of need to take advantage of load or port information. When using the four performance metrics (recall rate, accuracy, flow accuracy, and byte precision) to evaluate the UNIBS data set, the proposed model showed good results, with most of the ratios exceeding 90%. In addition, the best features and feature selection algorithms for a given problem are given along with the best ELM parameters. In [14], a new encrypted network traffic classification algorithm for N packets before sliding window is proposed. This method can significantly reduce the traffic feature dimension and reduce the number of packets in each traffic. The experimental results show that under the condition of reducing the feature dimension of the encrypted traffic flow and reducing the number of packets per stream, the average classification accuracy of the proposed N grouping algorithms before sliding window exceeds 95%. Compared with the existing methods, the traffic classification accuracy can be improved by about 3% overall.

Literature [15] proposed a software-intensive network (SDN)-HGW framework to better manage distributed smart home networks. The SDN controller implements efficient network quality of service management based on real-time traffic monitoring and core network resource allocation. An encrypted data classifier (represented as a data set) based on three deep learning schemes was developed. A data pre-processing scheme is proposed to process the original data packet and the test data set to create a data network. Literature [16] proposed an end-to-end encrypted traffic classification method based on one-dimensional convolution neural network. The method integrates feature extraction, feature selection, and classifier into a unified end-to-end framework that automatically learns the nonlinear relationship between the original input and the desired output. It is validated by the public ISCX VPN non-VPN traffic data set. In the four experiments, 11 of the 12 evaluation indicators of the experimental results were better than the current ones by the optimal representation of the flow and the fine-tuning of the model, indicating the effectiveness of the proposed method. Literature [17] uses Deep Learning (DL) as a feasible strategy to design traffic classifiers based on automatic feature extraction to reflect complex mobile traffic patterns. The DL technology from different technical states of TC is copied and analyzed, and it is set as a system comparison framework, including a performance evaluation workbench. Based on three real human user activity data sets, the performance of these DL classifiers is strictly studied, highlighting the defects, design criteria and openness of the DL classifier in mobile encryption TC. The literature [18] also shows various limitations for several recently proposed methods that focus on identifying fingerprints between various SSL/TLS applications. A Weighted Integration Classifier (WENC) was designed to address these limitations. WENC studies the characteristics of each substream during the HTTPS handshake and the subsequent data transfer period. In order to improve the fingerprint recognition rate, a second-order Markov chain model based on fingerprint variables is proposed. The model considers the packet length and message type in the HTTPS handshake process. The packet length sequence of the application data is modelled as a hidden Markov model with an optimal transmission probability. Finally, a weighted integration strategy is designed to accommodate the advantages of multiple methods as a unified approach. The experimental results show that the classification accuracy of this method reaches 90%.

Based on the previous studies, this paper can automatically select features in training based on deep neural networks, without the need to select features in advance. Deep neural networks can still classify traffic well when new network traffic types occur or when existing traffic characteristics change. Deep neural networks are more capable of learning complex situations than traditional machine learning methods, and can better learn the nonlinear relationship between the original input and output. The encryption traffic classification model based on deep neural network is proposed for the first time. The influence of different batch size on the classification ability of encrypted traffic classification model based on deep neural network is discussed. The encryption traffic classification model based on deep neural network and naive Bayesian classification algorithm are compared. The experimental results show that the encrypted traffic classification model based on the deep neural network classification model can identify the encrypted traffic well.

# 3 Encrypted Traffic Classification Based on Deep Neural Network

## 3.1 Deep Neural Network Classification Model

A deep neural network (DNN) is a multi-layer neural network composed of an input layer, a hidden layer, and an output layer. Deep neural networks can find the correct mathematical transformation between input and output, and calculate the probability of each output through the network layer. Figure 1 shows the classification model of deep neural networks.

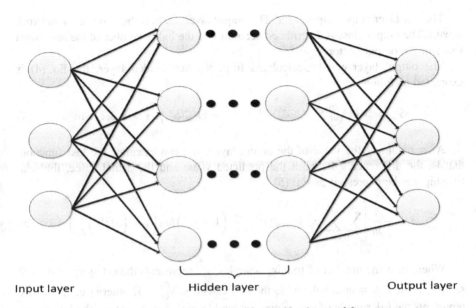

Input layer                    Hidden layer                    Output layer

**Fig. 1.** Deep neural network classification model

In the classification model based on deep neural network, the first layer is the input layer, and the input data is the timing-based feature of the pre-processed network stream. Layers 2 through 7 are six hidden layers that learn from the input vector to the output vector, each layer consisting of 256 neural network nodes.

There are m hidden layers, *L1, L2,..., Lm*, and there are *n* neurons in each hidden layer, and the neurons in the *L1* layer are *S1(1), S1(2),..., S1(n)*.

Then the output of the *i-th* neuron of the first hidden layer is expressed as Eq. (1)

$$S_1^i = V\left(\Theta_1^{(i,0)}x_0 + \Theta_1^{(i,1)}x_1 + \cdots + \Theta_1^{(i,n)}x_n\right), i = 1, 2, \ldots, m \qquad (1)$$

Among them, $\Theta\_1$ is a parameter matrix, x is the input eigenvector, V is the ReLU (Rectified Linear Unit) activation function, and ReLU is a commonly used activation function in neural networks. It is easy to calculate using the principle of pseudo-biological, and it is not saturated on a shallow gradient like sigmoid activation function [10], the relationship among them is expressed as Eq. (2)

$$V(t) = \max(0, t), \qquad (2)$$

Where t is the input to the activation function. After the output of the first hidden layer is obtained, it can be propagated forward to the next hidden layer until the last layer. The output of the *i-th* neuron of the *j + 1th* hidden layer is expressed as Eq. (3):

$$S_{j+1}^i = V\left(\Theta_{j+1}^{(i,0)}S_j^0 + \Theta_{j+1}^{(i,1)}S_j^1 + \cdots + \Theta_{j+1}^{(i,n)}S_j^n\right), i = 1, 2, \ldots, m \qquad (3)$$

The last layer is the output layer. The output layer predicts the class of the network stream. The output class is determined according to the index number of the maximum value in the output vector.

The output layer can be calculated from the last hidden layer, the Eq. (4) is expressed as follows:

$$S_{out}^i = V\left(\Theta_{out}^{(i,0)} S_m^0 + \Theta_{out}^{(i,1)} S_m^1 + \cdots + \Theta_{out}^{(i,n)} S_m^n\right), \; i = 1, 2, \ldots, m \tag{4}$$

After obtaining the value of the output layer, you can calculate the loss function, that is, the difference $J$ between the predicted value and the actual value, the relationship can be expressed as Eq. (5):

$$J = -\frac{1}{m}\left[\sum_{i=1}^{m}\sum_{k=1}^{7} y_k^{(i)} \log\left(S(x^i)\right)_k + \left(1 - y_k^{(i)}\right) log\left(1 - \left(S(x^{(i)})\right)_k\right)\right] \tag{5}$$

Where m is the number of training samples, $y_k^{(i)}$ represents the category of the *i-th* sample, if the *i-th* sample belongs to the *k-th* category, $y_k^{(i)} = 1$, otherwise, $y_k^{(i)} = 0$, $x_i$ represents the *i-th* sample of the training set, and $(S(x^i))_k$ is the value of the *k-th* output unit for the input $x^i$. $\Theta$ is the main parameter of the $(S(x^i))_k$ function. In the deep neural network algorithm, the loss function $J$ can be minimized by optimizing $\Theta$.

In this paper, Adam [11] optimization algorithm is used to optimize the parameter $\Theta$. Adam is a first-order optimization algorithm that can replace the traditional stochastic gradient descent process. It can update the neural network weight based on iterative training data. Adam's computational efficiency and low memory requirements make it ideal for solving optimization problems with large amounts of data and parameters.

## 3.2  Naive Bayesian Classification Model

This paper compares the deep neural network classification algorithm with the naive Bayesian classification algorithm [19]. The naive Bayesian algorithm describes the probability of an event based on the prior knowledge of the event, estimates the class prior probability $P(c)$ based on the training set $D$, and estimates the conditional probability $P_{(x_i|c)}$ for each attribute.

Let $D_c$ denote the set of the c-type samples in the training set $D$. If there are sufficient independent and identically distributed samples, the prior probability of the class can be easily estimated, the Eq. (6) is expressed as follows:

$$P_{(c)} = \frac{Dc}{D} \tag{6}$$

For discrete attributes, let $D_{c,x_{(i)}}$ denote a set of samples whose $D_c$ takes the value of $x_i$ on the *i-th* attribute, then the conditional probability $P_{(x_i|c)}$ can be estimated, the Eq. (7) is expressed as follows:

$$P_{(x_i|c)} = \frac{|D_{c,x_i}|}{|D_c|} \qquad (7)$$

## 3.3 Time Complexity Analysis

For the deep neural network classification algorithm, suppose $N$ is the number of training samples, $d$ is the dimension of each sample, $H$ is the number of hidden layers, and $i$ is the number of hidden layer neurons, then the time complexity is $O(Ndi + N(H-1)i^2 + Ni)$.

## 3.4 Mini-Batch Descent Method

Goyal [18] found that mini-batch SGD can effectively improve the efficiency of model learning. Mini-batch refers to training randomly selected small subsets from the training set. These subsets are called mini-batch. The number of samples in each subset is called mini-batch size, assuming each mini-batch size is m, the number of samples is x, then $1 < m < x$, then the optimized parameter $\Theta'$ is expressed as Eq. (8):

$$\Theta'_i = \Theta_i - \alpha \sum_{j=t}^{t+m-1} (S_\Theta(x_0^j, x_1^j, \dots x_n^j) - y_j)x_i^j \qquad (8)$$

This article explores the impact of the model's ability to classify when mini_batch size takes different values.

---

**Deep Neural Network Algorithm**

Input: Training Set $D = \{(x_i, y_i)\}_{i=1}^n$ , learning rate $\eta$, epoch s, mini-batch size s

Input: Initialize network weights $\Theta$
for $j$ in range(epochs):
   random.shuffle(D)
   mini-batches = [D[k:k+s]]
   for k in range(0,n,s):
      for mini-batch in mini-batches:
$$loss = J(mini\_batch\_y, \theta, mini\_batch\_x)$$
        Adam($\eta$, loss)
        Update($\Theta$)
      end for
   end for
end for
Output: Network weights $\Theta$

---

# 4  Experimental Results and Analysis

Deep neural network algorithms are written in Python. Third-party software and APIs include: Pycharm, tensorflow, numpy, sklearn. The naive Bayesian algorithm uses the Weka implementation. The configuration of the host is: Dell XPS13, the CPU is Intel Core i5-8250U 1.6 GHZ, the memory is 8G, the hard disk is 225 GB; the operating system is win10 family Chinese version. In this paper, the classification algorithm of deep neural network is compared with the naive Bayesian classification algorithm.

The network packet is pre-processed to generate a network stream, and the timing-based feature generation vector in the network stream is extracted. The user vector is divided into a training set and a test set, and the data of the training set is trained and generated based on the deep neural network model, and the generated classification model is used to predict the classification of the encrypted traffic in the test set. Figure 2 shows the overall architecture of an encrypted traffic classification model based on deep neural networks.

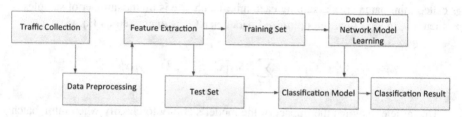

**Fig. 2.** The overall architecture of the encrypted traffic classification model based on deep neural network

## 4.1  Network Data Set

The experimental data set uses the ISCX VPN non-VPN [20] data set, which contains various types of traffic generated via Open VPN communication. The data set directory and number statistics are shown in Table 1. The data set contains the following 7 types of traffic.

Web Browsing (Browsing for short): Traffic when a user browses a web page using the https protocol.

Email: The traffic when the user sends and receives emails.

Chat: Traffic when users use instant messaging software.

Streaming: Traffic when a user uses a multimedia application.

File Transfer (FT for short): The traffic generated by the network to send and receive files.

VoIP: Traffic generated by voice applications.

P2P: A file sharing protocol that recognizes Bittorrent.

**Table 1.** Dataset catalogue and number statistics

| Traffic category | Number | Traffic source |
|---|---|---|
| Browsing | 1621 | Firefox and Chrome |
| Email | 569 | SMPTS, POP3 and IMAPS |
| Chat | 4546 | ICQ, AIM, Skype, Facebook and Hangouts |
| Streaming | 1144 | Vimeo and Youtube |
| FT | 1794 | Skype, FTPS and SFTP and an external service |
| VoIP | 11008 | Facebook, Skype and Hangouts voice calls (1 h duration) |
| P2P | 709 | uTorrent and Transmission (Bittorrent) |

## 4.2 Feature Extraction

The network stream consists of the same data packet as the sequence {source IP, destination IP, source port, destination port, protocol}. In this paper, ISCXFlowMeter [11] is used to generate the network stream and extract the timing-based features of the stream. ISCXFlowMeter can generate bidirectional streams. The first packet determines whether the direction of the stream is forward (source to destination) or backward (destination to Source) [21], respectively, statistics of the time-dependent characteristics of the flow forward and backward [22]. Some features [23] are shown in the list of time series related features in Table 2.

**Table 2.** List of time-related features

| Feature | Description |
|---|---|
| Duration | Duration of the stream |
| Fiat | Forward packet arrival interval |
| Biat | Backward packet arrival interval |
| Flowiat | P Active time of the stream |
| Active | Active time of the stream |
| Idle | Free time of the stream |
| fb_psec | Number of bits per second stream |
| Fp_psec | Number of packets per second stream |

## 4.3 Experimental Results and Analysis

This paper uses two evaluation indicators: Classification Accuracy (Precision) and Classification Recall Rate (Recall) to evaluate the performance of the algorithm.

$$Precision = \frac{TP}{TP + FP} \tag{9}$$

$$Recall = \frac{TP}{TP + FN} \tag{10}$$

TP is the number that is correctly classified as this class, FP is misclassified into the number of such classes, and FN is the number that is incorrectly divided into non-classes. The experimental results are showed as follows: Fig. 3 is the comparison of the VOIP classification accuracy rate and classification recall rate when the mini-batch size changes, respectively. Figure 4 is the comparison of the Chat class classification accuracy and classification recall rate when the mini-batch size changes, respectively. Figure 5 is the comparison of the FT class classification accuracy and classification recall rate when the mini-batch size increases, respectively. Figure 6 is the comparison of the Browsing class classification accuracy and classification recall rate when the mini-batch size changes, respectively. Figure 7 is the comparison of the Streaming class classification accuracy and classification recall rate when the mini-batch size changes, respectively. Figure 8 is the comparison of the P2P class classification accuracy and classification recall rate when the mini-batch size changes, respectively. Figure 9 is the comparison of the classification accuracy of the deep neural network classification model and the naive Bayes classification algorithm when the mini-batch size is 40, respectively.

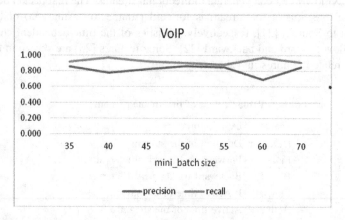

**Fig. 3.** VOIP class classification accuracy rate and classification recall rate

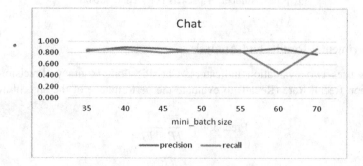

**Fig. 4.** Chat class classification accuracy rate and classification recall rate

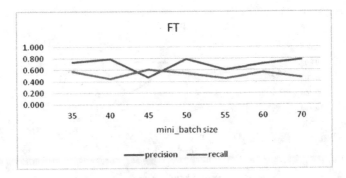

**Fig. 5.** FT class classification accuracy rate and classification recall rate

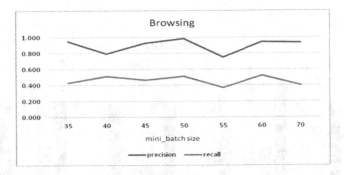

**Fig. 6.** Browsing class classification accuracy rate and classification recall rate

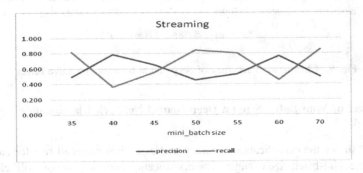

**Fig. 7.** Streaming class classification accuracy rate and classification recall rate

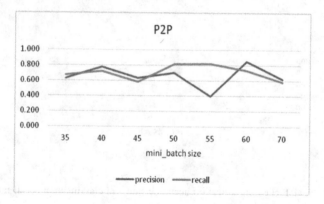

**Fig. 8.** P2P class classification accuracy rate and classification recall rate

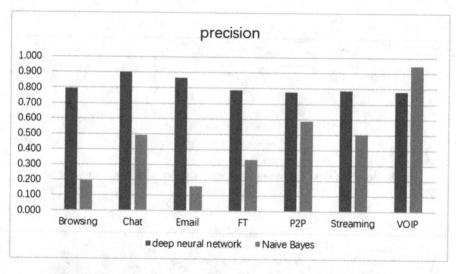

**Fig. 9.** Classification accuracy of deep neural network and naïve Bayes

## 4.4 Effect of Mini-Batch Size on Deep Neural Network Classification Model

Figure 10 shows the classification accuracy and classification recall rate for each category as the mini-batch size changes. The classification accuracy rate of VoIP class and Chat class with the largest number of samples is higher than other categories, and the impact of mini-batch size is also small. The maximum difference is 0.168 and 0.126 respectively. The classification accuracy of P2P classes with a small sample size is most affected by mini-batch size, the highest is 0.855, the lowest is 0.396, and the maximum difference is 0.459. It can be seen that the mini-batch size change will have different degrees of impact on the deep neural network classification model, and the degree of impact of each category is related to the proportion of samples in the category. When the mini-batch size is 40, the classification model has the highest classification accuracy and the classification model has the best classification ability. When the mini-batch size is 50, the overall classification of the classification model has a higher recall rate.

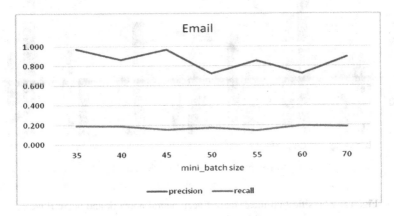

**Fig. 10.** Email class classification accuracy and classification recall rate

Although the classification accuracy of each category changes with the mini-batch size, when the mini-batch size is unchanged, the classification ability of the deep neural network classification model for each category remains almost unchanged. The deep network classification model has higher classification accuracy for VoIP, Chat, Browsing, and Email classes, while the classification of VOIP and Chat classes has higher recall rate. The results show that the proportion of sample classes still affects the classification ability of the model. For samples [24] with a high proportion of samples, the model has a stronger ability to predict the classification of the category.

### 4.5 Comparison Between Deep Neural Network Classification Model and Naïve Bayesian Classification Algorithm

Figure 11 and Table 3 are comparisons of the classification accuracy and classification recall ratios of the deep neural network classification model and the naive Bayes classification algorithm when the mini-batch size is 40, respectively. It can be seen from Fig. 11 that, except for the VOIP class, the classification accuracy of the other categories in the deep neural network classification model is much higher than that of Naïve Bayes. The naive Bayesian classification algorithm has the best VOIP classification ability than the depth. The neural network classification model, while the deep neural network classification model has much better classification ability for other samples than the naive Bayesian classification algorithm. In Table 3, the recall rate of the deep neural network classification model in Browsing, Chat, FT, P2P, VOIP is better than that of Naive Bayes, and the recall rate of the simple Bayesian classification algorithm for Email and Streaming. Better than the deep neural network classification model. In general, the classification ability of the deep neural network model is better than the naive Bayesian classification algorithm.

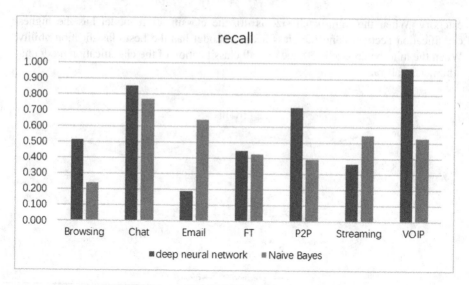

**Fig. 11.** Classification recall rate of deep neural network and naive Bayes

**Table 3.** Average classification accuracy and average classification recall rate of each category

| Traffic category | Ave-precision | Ave-recall |
|---|---|---|
| VoIP | 0.815 | 0.925 |
| Chat | 0.845 | 0.789 |
| FT | 0.696 | 0.524 |
| Browsing | 0.897 | 0.458 |
| Streaming | 0.597 | 0.669 |
| P2P | 0.659 | 0.705 |
| Email | 0.856 | 0.175 |

Table 3 shows the average classification accuracy and average classification recall rate for each category. It can be seen that the deep neural network classification model has the highest classification accuracy in the Browsing category and the lowest classification accuracy for the P2P category. At the same time, the deep neural network classification model has a much higher recall rate for VoIP than other categories. However, the classification accuracy of Email is high, but the classification recall rate is much smaller than other categories. The deep neural network classification model has the worst recognition ability for Email.

Since the data set is an unbalanced sample data set, the number of samples in each category varies greatly, which has a negligible impact on the experimental results. The experimental results show that the deep neural network classification model has stronger prediction ability for VOIP and Chat with more samples.

# 5 Conclusions

This paper uses the encrypted traffic classification model based on deep neural network to classify encrypted traffic such as Browsing, Chat, Email, FT, P2P, Streaming and VOIP transmitted via VPN. It is found that the encrypted traffic classification model based on deep neural network can effectively classify encrypted traffic. The mini-batch size also affects the classification ability of the deep neural network model. It is found that when the mini-batch size is 40, the deep neural network classification model has the best classification ability. At the same time, the deep neural network-based encrypted traffic classification model proposed in this paper is compared with the traditional naive Bayesian classification algorithm. The results show that the classification of the encrypted traffic classification model based on deep neural network is better. Due to the imbalance in the number of samples in each category of the data set, the classification results have a non-negligible impact.

**Acknowledgements.** This work was supported by the National Natural Science Foundation of China (No. 61772377, 91746206), the Natural Science Foundation of Hubei Province of China (No. 2017CFA007), Science and Technology planning project of ShenZhen (JCYJ2017081811 2550194), and Fund of Hubei Key Laboratory of Transportation Internet of Things (WHUTIOT-2017A0011).

# References

1. Korczynski, M., Duda, A.: Markov chain finger printing to classify encrypted traffic. In: 2014 Proceedings of IEEE INFOCOM, pp. 781–789 (2014)
2. Shen, M., Wei, M., Zhu, L., Wang, M.: Classification of encrypted traffic with second-order Markov chains and application attribute bigrams. IEEE Trans. Inf. Forensics Secur. 12(8), 1830–1843 (2017)
3. Zhang, Z., Li, J., Manikopoulos, C.N., Jorgenson, J., Ucles, J.: HIDE: a hierarchical network intrusion detection system using statistical pre-processing and neural network classification. In: Proceedings of IEEE Workshop on Information Assurance and Security, pp. 85–90 (2001)
4. Wang, W., Zhu, M., Wang, J., Zeng, X., Yang, Z.: End-to-end encrypted traffic classification with one-dimensional convolution neural networks. In: 2017 IEEE International Conference on Intelligence and Security Informatics (ISI), pp. 43–48 (2017)
5. Lotfollahi, M., Zade, R.S.H., Siavoshani, M.J., Saberian, M.: Deep packet: a novel approach for encrypted traffic classification using deep learning. arXiv preprint arXiv:1709.02656 (2017)
6. Li, W., Moore, A.W.: A machine learning approach for efficient traffic classification. In: 15th International Symposium on Modelling, Analysis, and Simulation of Computer and Telecommunication Systems, pp. 310–317. IEEE (2007)
7. Sun, R., Yang, B., Peng, L., Chen, Z., Zhang, L., Jing, S.: Traffic classification using probabilistic neural networks. In: 2010 Sixth International Conference on Natural Computation, vol. 4, pp. 1914–1919. IEEE (2010)
8. Anderson, B., Mcgrew, D.: Machine learning for encrypted malware traffic classification: accounting for noisy labels and non-stationarity. In: The ACM SIGKDD International Conference, pp. 1723–1732. ACM (2017)

9. Sherry, J., Lan, C., Popa, R.A., Ratnasamy, S.: Blind-box: deep packet inspection over encrypted traffic. ACM SIGCOMM Comput. Commun. Rev. **45**(4), 213–226 (2015)
10. Draper-Gil, G., Lashkari, A.H., Mamun, M.S.I., Ghorbani, A.A.: Characterization of encrypted and VPN traffic using time-related features. In: The International Conference on Information Systems Security and Privacy, pp. 94–98 (2016)
11. Schmidhuber, J.: Deep learning in neural networks: an overview. Neural Netw. **61**, 85–117 (2015)
12. Goyal, P., et al.: Accurate, large mini-batch SGD: training ImageNet in 1 hour. arXiv preprint arXiv:1706.02677 (2017)
13. Liu, Y., Chen, J., Chang, P., Yun, X.: A novel algorithm for encrypted traffic classification based on sliding window of flow's first N packets. In: 2017 2nd IEEE International Conference on Computational Intelligence and Applications (ICCIA), Beijing, China, 8–11 September 2017, pp. 463–470 (2017)
14. Wang, P., Ye, F., Chen, X., Qian, Y.: DataNet: deep learning based encrypted network traffic classification in SDN home gateway. IEEE Access **6**, 55380–55391 (2018)
15. Wang, W., Zhu, M., Wang, J., Zeng, X., Yang, Z.: End-to-end encrypted traffic classification with one-dimensional convolution neural networks. In: 2017 IEEE International Conference on Intelligence and Security Informatics (ISI), 22–24 July 2017, Beijing, China, pp. 43–48 (2017)
16. Aceto, G., Ciuonzo, D., Montieri, A., Pescapé, A.: Mobile encrypted traffic classification using deep learning. In: 2018 Network Traffic Measurement and Analysis Conference (TMA), Vienna, Australia, 26–29 June 2018, pp. 1–8 (2018)
17. Pan, W., Cheng, G., Tang, Y.: WENC: HTTPS encrypted traffic classification using weighted ensemble learning and Markov chain. In: 2017 IEEE Trustcom/BigDataSE/ICESS, Sydney, NSW, Australia, 1–4 August 2017, pp. 1723–1732 (2017)
18. Cireşan, D., Meier, U., Masci, J., Schmidhuber, J.: Multi-column deep neural network for traffic sign classification. Neural Netw. **32**, 333–338 (2012)
19. Nascimento, Z., Sadok, D., Fernandes, S., Kelner, J.: Multi-objective optimization of a hybrid model for network traffic classification by combining machine learning techniques. In: 2014 International Joint Conference on Neural Networks (IJCNN), Beijing, China, 6–11 July 2014, pp. 2116–2122 (2014)
20. ISCX. https://www.unb.ca/cic/datasets/vpn.html. Accessed 10 Mar 2019
21. Sun, G., Chen, T., Su, Y., Li, C.: Internet traffic classification based on incremental support vector machines. Mob. Netw. Appl. **23**(4), 789–796 (2018)
22. He, Z.B., Cai, Z.P., Han, Q.L., Tong, W.T., Sun, L.M., Li, Y.S.: An energy efficient privacy-preserving content sharing scheme in mobile social networks. Pers. Ubiquit. Comput. **20**(5), 833–846 (2016)
23. Sun, G., Dong, H., Li, A., Xiao, F.: NTCA: a high-performance network traffic classification architecture. Int. J. Future Gener. Commun. Netw. **6**(5), 11–20 (2013)
24. Ding, X.O., Wang, H.Z., Gao, Y.T., Li, J.Z., Gao, H.: Efficient currency determination algorithms for dynamic data. Tsinghua Sci. Technol. **22**(3), 227–242 (2017)

# Key Technologies of Traditional Chinese Medicine Traceability Based on Internet of Things

Lei Yu[1,2(✉)], Fangliang Huang[1], Yong Yang[3], Qunshan Tao[4],
Tongping Shen[1], and Luyao Zhang[1]

[1] School of Medical Information Technology,
Anhui University of Chinese Medicine, Hefei 230012, China
fishstonehfut1006@163.com
[2] Institute of Computer Application in Traditional Chinese Medicine,
Anhui Academy of Chinese Medicine, Hefei 230012, China
[3] School of Pharmacy, Anhui University of Chinese Medicine,
Hefei 230012, China
[4] School of Medical Economics and Management,
Anhui University of Chinese Medicine, Hefei 230012, China

**Abstract.** Compared with western medicine, there are many complicated factors affecting the intrinsic quality of traditional Chinese medicine, because its production needs to be planted, harvested, processed, transported, stored, and sold, etc. Therefore, the internet of things is integrated into the traceability of traditional Chinese medicine, and its key technologies are studied. An XML-based traceability information exchange model was constructed for traditional Chinese medicine and modeled the traceability process by the finite state machine (FSM). Furthermore, the specific electronic product code (EPC) coding scheme of traditional Chinese medicine was proposed based on the EPC coding structure model. Finally, the effectiveness of the above models and schemes is verified by an example of a traditional Chinese medicine traceability prototype system.

**Keywords:** Traditional Chinese medicine · Internet of Things · Traceability · Finite state machine · Electronic product code

## 1 Introduction

In recent years, with the continuous improvement of people's living standards and the gradual enhancement of their health awareness, traditional Chinese medicine, which is good at health care and prevention of diseases, has been favored by more and more people, so the demand for traditional Chinese medicine is steadily on the increase. However, due to the lack of industry traceability, the manufacture and sale of fake products in the circulation of traditional Chinese medicine often results in shoddy products. Compared with western medicine production, traditional Chinese medicine has many links that affect its internal quality due to the particularity of its source, from the environment of origin, cultivation and planting, harvesting and processing, to

© Springer Nature Singapore Pte Ltd. 2019
R. Mao et al. (Eds.): ICPCSEE 2019, CCIS 1059, pp. 545–555, 2019.
https://doi.org/10.1007/978-981-15-0121-0_42

packaging, transportation, storage and final marketing. Therefore, the establishment of Chinese medicine traceability system to realize the whole process control and management from the cultivation of raw materials to consumers has been an urgent problem to be solved in the traditional Chinese medicine industry at present.

The Internet of Things is a network for identification, location, tracking, monitoring and management [1, 2], which connects any article with the Internet according to the agreed protocol through information sensing equipment such as radio frequency identification (RFID), infrared sensors, global positioning system, laser scanner, etc., and carries out information exchange and communication to realize intelligence. The Internet of Things technology can carry out RFID identification on each link of traditional Chinese medicine planting, production and circulation, and use this as an index to query and update relevant information in real time, which can locate and track the traditional Chinese medicine in each link at the same time. In the meanwhile, it can also sense environmental information such as temperature, humidity, brightness, etc. in the process of traditional Chinese medicine planting, production, storage and circulation in real time, so we have achieved timely and accurate collection and sharing of traditional Chinese medicine information, providing a new technical means for ensuring stable and reliable quality of traditional Chinese medicine products.

Based on the above ideas, literature [3] gives the application prospect of Internet of Things technology in the Chinese medicine industry from the aspects of authentic medicine protection, production, circulation and quality control of Chinese herbal pieces and Chinese patent medicines. Literature [4] proposes a quality traceability framework for Chinese herbal pieces based on the Internet of Things, and gives the workflow and specific application scheme of the framework. Literature [5, 6] based on RFID technology and taking the circulation of traditional Chinese medicine as the research object, a visual tracing system is constructed to realize the whole process of traditional Chinese medicine from planting, processing, circulation and drinking. Literature [7] proposes a Chinese medicine traceability model and overall framework suitable for collectivized enterprises. Literature [8] through standardized information collection and two-dimensional code running through the chain, a set of circulation tracing system for the industrial chain of Chinese herbal medicine industry has been established from the aspects of Chinese herbal medicine planting, medicine distribution, Chinese herbal medicine market, Chinese herbal medicine decoction piece enterprises, decoction piece distribution and hospital pharmacy.

From the above documents, it can be seen that at present, most researches on the traceability of traditional Chinese medicine based on the Internet of Things focus on the system framework structure and function realization, and no relevant researches are carried out from the perspective of key technologies of application. Therefore, this paper focuses on the application of Internet of Things in Chinese medicine traceability from the following three aspects: The first is Chinese medicine traceability information exchange model, the second is Chinese medicine traceability process modeling, and the third is Chinese medicine EPC specific coding scheme, which is verified by specific application cases.

## 2  Exchange Model of Traditional Chinese Medicine Traceability Information

In order to realize the timely and convenient exchange of traceability information between the traceability center database and each node database, XML language which is independent of software and hardware and has self-description is adopted to describe the traceability information and a traceability information exchange model is established. On the basis of the business process of each circulation node in the Chinese herbal medicine supply chain, traceability information is divided into a tree structure according to the hierarchical relationship between circulation nodes and node information. Figure 1 shows some contents of the Chinese herbal medicine traceability information tree.

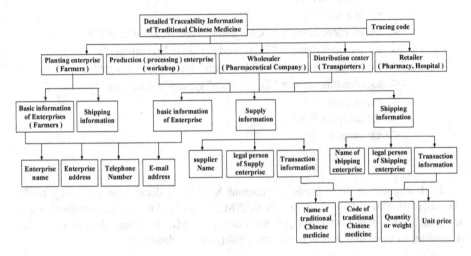

**Fig. 1.** Traceability information exchange model of traditional Chinese medicine

This system uses XML Schema to define the data format of traceability information exchange for traditional Chinese medicine products. According to the traceability information structure shown in Fig. 1, different levels of elements, attributes, sub-elements, sub-element order, number of sub-elements, default values and limit values of elements and attributes are defined. At the top level is the root element, which is detailed traceability information and has traceability code attribute. This attribute is unique and must be declared as follows:

$<$ xs: attribute name = "traceability _ id" type = " xs: integer" use = " required"/$>$

Below the root element are complex types of sub-elements such as planting enterprises (farmers), production enterprises (small workshops), wholesalers, distribution centers (transporters, third-party logistics, which can be ignored), retailers (pharmacies, hospitals) and other sub- elements, which are circulation node elements. Each circulation node contains sub-node elements composed of information to be uploaded by the node, which are recursively formed layer by layer in turn to form an XML document–traditional Chinese medicine traceability information exchange model. Then we take the information element of Chinese medicinal materials under the transaction information of Chinese medicinal materials as an example. The information element of Chinese medicinal materials includes 4 sub-elements such as the name of Chinese medicinal materials, the code of Chinese medicinal materials, the weight of Chinese medicinal materials, and the unit price of Chinese medicinal materials. The statement fragment is as follows:

```
< xs: element name = "TCM_info" >
< xs: complexType >
< xs: sequence >
< xs: element name = "TCM_name" type = "xs: string" />
< xs: element name = "TCM_code" type = "xs: integer" />
< xs: element name = "TCM_weight" type = "xs: integer" />
< xs: element name = "TCM_price" type = "xs: integer" />
</ xs: sequence >
</ xs: complexType >
</ xs: element >
...
```

Each circulation node needs to generate XML data documents according to the document structure defined by the above XML Schema. After XML is transmitted to the traceability data center through the network, XML document data is converted according to rules and stored in the traceability center database.

## 3    Traceability Process Modeling of Traditional Chinese Medicine

In order to better respond to consumers' query requests for traceability information of traditional Chinese medicine, and also to provide theoretical basis for further research on behavioral description and structural design of traditional Chinese medicine traceability, a finite state machine model [9–11] is used to model the traceability process of traditional Chinese medicine.

The operation state of the Chinese medicine traceability system is changed by the inquiry of the operation behavior of the consumer and the response state of the server. The system operation state and driving events defined in this paper are shown in Table 1.

**Table 1.**  System status and event description.

| State | State description | Events | Event description |
|-------|-------------------|--------|-------------------|
| $q_0$ | Not logged in/consumer | $I_1$ | Scanning QR code |
| $q_1$ | Getting basic information and traceability code | $I_2$ | Clicking "Traceability Code" to inquire |
| $q_2$ | After the user inputting the tracing code | $I_3$ | Inputting traceability code |
| $q_3$ | Tracking information | $I_4$ | User login |
| $q_4$ | Subsystem operation | $I_5$ | Enterprise login |
| $q_5$ | Information operation | $I_6$ | Operation |
| $q_6$ | Waiting for server response | $I_7$ | Submitting information |
| $q_7$ | Success | $I_8$ | Response |
| $q_8$ | Failure | $I_9$ | Response failing |

According to Table 1, the finite state machine model defining the traditional Chinese medicine traceability system is as below:

$$M = \left(K, \sum, f, q_0, F\right)$$

where $K = \{q_0, q_1, q_2, q_3, q_4, q_5, q_6, q_7, q_8\}$,
$\sum = \{I_1, I_2, I_3, I_4, I_5, I_6, I_7, I_8, I_9\}$,
$F = \{q_7, q_8\}$
and $f$ is defined by the following state transition equation:

$$f(q_0, I_1) = q_1 \quad f(q_0, I_4) = q_3 \quad f(q_0, I_5) = q_4 \quad f(q_1, I_2) = q_2$$
$$f(q_3, I_6) = q_5 \quad f(q_4, I_6) = q_5 \quad f(q_2, I_3) = q_6 \quad f(q_5, I_7) = q_6$$
$$f(q_6, I_8) = q_7 \quad f(q_6, I_9) = q_8$$

Figure 2 shows the state transition diagram of the finite state machine model for the Chinese medicine tracing process.

The tracing process can be analyzed by establishing the finite state machine model $E$. We take the symbol $E$ to represent the event set of finite automata $M$, and the language defined on the event set $E$ can be represented as a string. Languages generated by finite automata $M$ are represented by symbols $L(M)$. The language $L(M)$ expresses all possible directed paths in the state transition diagram. If the language $L(M)$ contains some illegal character strings, it will definitely violate the access rights stipulated by the Chinese herbal medicine traceability system. For example, inquiries made by ordinary consumers about the inventory and sales of traditional Chinese medicines should be rejected.

When an ordinary user is not logged in or is a consumer, basic information and tracing codes are obtained by scanning the two-dimensional code, and then relevant tracing information is inquired through the tracing codes. After logging in, the system will move to the next page to enter another state according to the user's permissions and nature. If the login person is a government administrator, he can query the planting information of traditional Chinese medicine (such as planting base, planting responsible

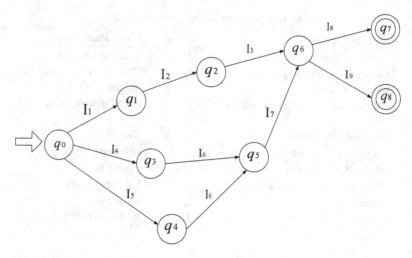

**Fig. 2.** Finite state machine model of traditional Chinese medicine tracing process

person, planting time, soil information, etc.), processing information (such as processing enterprise information, processing responsible person information, process information, processing time, etc.), and circulation information (such as supply chain information, distributor information, transaction information, etc.), in other words, he can trace and query all levels and types of information. If the login person is a worker of a processing enterprise, he can inquire the relevant enterprise information, product information, structure information and process information of the Chinese herbal medicines at the corresponding level. For consumers, information on enterprises, products and processes related to the purchase of batches of Chinese herbal medicines is provided.

## 4   Traceability Code Design Scheme of Traditional Chinese Medicine

There are many kinds of traditional Chinese medicine, and the design of traceability code is very important, which is the key technology to realize the traceability system of traditional Chinese medicine. The traceability code consists of three parts: the subject code, EPC code of medicinal materials and transaction serial number. The overall structure is shown in Fig. 3.

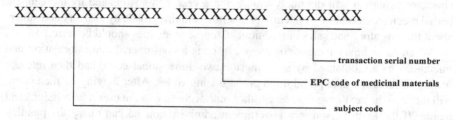

**Fig. 3.** General structure of traceability code

The subject code is the code used to identify the identity of the subject in the circulation of traditional Chinese medicine, which can be divided into the subject code of operator and the subject code of circulation node. The subject code consists of two parts: the administrative division code of the place of filing and the record number of the circulation node/operator (the party buying the traditional Chinese medicine). The administrative division code adopts the 6-digit code of GB/T2260, and the record number consists of 7 digits according to the filing sequence number of the circulation and management node in the competent department. The schematic diagram of the body code is shown in Fig. 4.

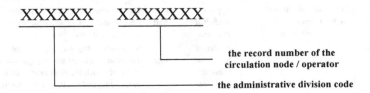

**Fig. 4.** Schematic diagram of subject code

EPC coding system is an ideal reference model for RFID radio frequency tag coding system in traditional Chinese medicine industry [12]. According to the "Printing Specification for Drug Electronic Regulatory Codes" issued in 2010, and in combination with the special requirements of the Internet of Things electronic coding system in the traditional Chinese medicine industry, EPC-96-bit universal identifier coding scheme is adopted to classify identification objects. The coding scheme is shown in Fig. 5, and the specific coding allocation scheme and its explanation are shown in Table 2.

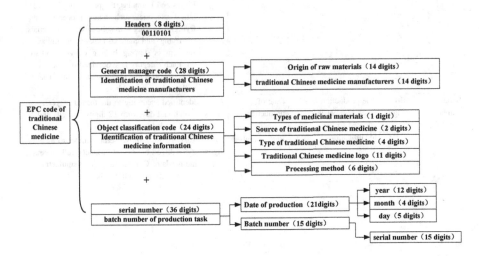

**Fig. 5.** EPC coding scheme of traditional Chinese medicine

**Table 2.** Description of EPC code allocation of traditional Chinese medicine

| Name | Digit | Effect | Specific composition | | |
|---|---|---|---|---|---|
| | | | Name | Digit | Specific coding instructions |
| Headers | 8 digits | | | | EPC 96 - bit universal identifier (GID-96) coding scheme is adopted, and the header field value is fixed as 00110101 |
| General manager code | 28 digits | Identification of the origin of raw materials and manufacturers of traditional Chinese medicine | Origin of raw materials | 14 digits | According to the "Administrative Division Code of the People's Republic of China", the codes of provincial and municipal level consist of 4 decimal digits |
| | | | Chinese medicine manufacturers | 14 digits | Enterprise registration number granted by the industrial and commercial department to the registration of Chinese medicine manufacturers |
| Object classification code | 24 digits | Identification of Chinese medicine information | Types of medicinal materials | 1 digit | The medicinal materials are classified into planting and wild, in which the planting code is 0 and the wild code is 1 |
| | | | Source of traditional Chinese medicine | 2 digits | According to drug sources, they can be divided into plant drugs, animal drugs and mineral drugs, of which the plant drug code is 01, the animal drug code is 10 and the mineral drug code is 11 |
| | | | Type of traditional Chinese medicine | 4 digits | The types of traditional Chinese medicine can be roughly divided into 4 categories: slice, segment, silk and block. Each category can be subdivided into some small categories. Specifically, the codes of "extremely thin slice and slice" are 0000, "thick slice and slice" are 0001, "segment" are 0010, "block" are 0011, "silk" are 0100, "ground powder and powder" are 0101, "fresh medicine" are 0110, and other codes are 0111 |
| | | | Traditional Chinese medicine logo | 11 digits | According to the Chinese Pharmacopoeia, each traditional Chinese medicine included in it is coded in sequence to establish a unique code |
| | | | Processing method | 6 digits | According to the classification method of processing in Code Rules for Chinese Herbal Pieces and Formulas, the processing is divided into 9 categories, such as non- processing, stir-frying, stir-frying with solid auxiliary materials, stir-frying, and charcoal making, calcimine, steaming or stewing, boiling, copying and so on. Each category is subdivided into several sub-categories and coded sequentially. The list will not be repeated here |
| Serial number | 36 digits | The production task batch number is adopted for identification | Date of production | 21 digits | Identified according to the format of year, month and day, with 12 digits, 4 digits and 5 digits coded in sequence |
| | | | Batch number | 15 digits | Take the form of the serial number to identify $2^{15} = 32768$ batches of products, fully meeting the needs of Chinese medicine manufacturers for one-day production tasks |

Transaction serial number is a 7-digits code automatically generated according to the transaction time sequence. The specific circulation process of traditional Chinese medicine is embodied by "transaction serial number". The subject code of the party to be bought may not be added, because the EPC code of traditional Chinese medicine is unique. According to the traceability of the EPC code of traditional Chinese medicine, all circulation transaction records can be queried by date.

## 5  Application Cases

On the basis of the above, a prototype system for tracing traditional Chinese medicine based on the Internet of Things is built, and a mixed architecture of C/S mode and B/S mode is comprehensively applied. And the medicine planting base, testing organization and supervision department use C/S structure to complete the production information management, data collection and transmission, while supervision department and consumers use B/S structure to realize the supervision and inquiry of medicine quality. The supervision department mainly analyzes the implementation of the enterprise's product standards, the source of raw materials and the flow direction of products according to the recorded traceability information, and judges the quality of medicinal materials. Consumers can learn the circulation information of the purchased traditional Chinese medicines at any time and place by means of mobile phone short messages, telephones, networks, mobile tracing terminals and other means, so that the traditional Chinese medicines within the scope of the platform can truly realize the whole tracing chain with traceable sources, traceable whereabouts and traceable responsibilities. The specific architecture and business process are shown in Fig. 6.

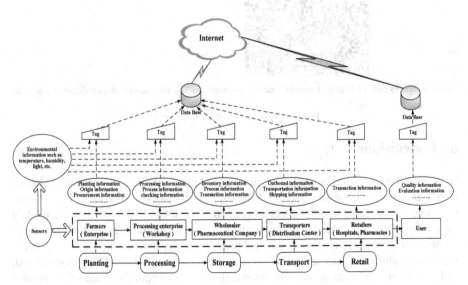

**Fig. 6.** A prototype system for tracing traditional Chinese medicine based on internet of things

The following is an example of pueraria lobata decoction pieces to show the actual tracing process. After retrieving the traceability data uploaded to the traceability center database from the links of pueraria lobata planting, processing, distribution, retail, the tracing system platform of traditional Chinese medicine carries out label coding on the traditional Chinese medicine, and consumers will obtain a two-dimensional code label as shown in Fig. 7(a) after purchasing the required traditional Chinese medicine.

Consumers can scan the two-dimensional code with devices with two- dimensional code scanning function such as smart phones to obtain the basic information and tracing code of traditional Chinese medicine as shown in Fig. 7(b). With this traceability code, consumers can send requests for detailed traceability information to the traceability system platform. After analyzing the traceability code, the system first queries whether the product data exists or not in the database. If not, the system will return an error message indicating that the drug is fake and inferior, and consumers can refuse to buy it. If so, the system will extract traceability information corresponding to the traceability code according to different requests and feed it back to the consumer.

(a) Two-dimensional code of traditional Chinese medicine    (b) Scanning results of Two-dimensional code

**Fig. 7.**   System application example

## 6   Conclusions

In view of the many and complicated production links of traditional Chinese medicine, the quality of traditional Chinese medicine needs to be strictly controlled. The Internet of Things is integrated into the traceability of traditional Chinese medicine, and the key technologies for its application are studied. The traceability information exchange model of traditional Chinese medicine based on XML and the traceability process model of traditional Chinese medicine based on finite automata FSM are constructed, and based on EPC coding structure model the specific coding scheme of traditional Chinese medicine EPC is proposed. On this basis, a prototype system for tracing traditional Chinese medicine is given. Based on the analysis of its functional framework and business process, the actual tracing process is demonstrated by taking pueraria lobata decoction pieces as an example. The results show that the prototype system can realize the whole process supervision from production to sales of traditional Chinese medicine, and fundamentally ensure the quality of traditional Chinese medicine products, which is feasible to adopt the above model and scheme.

**Acknowledgment.** This study was supported by the Natural Science Foundation of China (Grant No. 61701005), the Domestic Visiting Research Project of Excellent Young Backbone Talents from Anhui Higher Education Institutions (Grant No. gxgnfx2019009), the Key Project of Humanities and Social Sciences Research in Anhui Higher Education Institutions in 2019 (Grant No. SK2019A0242), the Quality Project Foundation of Anhui Province (Grant No. 2017mooc220, No. 2018zhkt079, No. 2015sxzx011, No. 2012sjjd025), the Key Project of Outstanding Young Talents Support Program of Anhui Higher Education Institutions (Grant No. gxyqZD2016128), the Key Project of Natural Science Research in Anhui Higher Education Institutions (Grant No. KJ2015A054, No. KJ2019A0437), the Key Teaching and Research Project of Anhui University of Chinese Medicine (Grant No. 2017xjjy_zd011), the Natural Science key Foundation of Anhui University of Chinese Medicine (Grant No. 2019zrzd11, No. 2018zryb06), the National Innovation and Entrepreneurship Training Program for College Student (Grant No. 20181036 9021, No. 201810369022, No. 201610369044, No. 201710369052).

# References

1. Wang, B.: Summary of research on internet of things technology. J. Electron. Meas. Instrum. **23**(12), 1–7 (2009)
2. Atzori, L., Iera, A., Morabito, G.: The Internet of Things: a survey. Comput. Netw. **54**, 2787–2805 (2010)
3. Li, L., Liu, X., Yin, F., et al.: Application prospect of Internet of Things technology in traditional Chinese medicine industry. J. Nanjing Univ. Tradit. Chin. Med. (Soc. Sci. Ed.) **11** (3), 170–173 (2010)
4. Zheng, X., Xie, J., Hu, C.: Architecture of quality tracing system for Chinese herbal pieces based on Internet of Things. Fujian Comput. (6), 14–16 (2013)
5. Zhang, C., Li, S.: Design of RFID-based traceability system for Chinese herbal medicines. J. Sanming Univ. **32**(4), 65–69 (2015)
6. Wang, X.: Traceability of Chinese medicinal materials based on supply chain perspective. Shizhen Guo Yi Guo Yao **26**(10), 2521–2522 (2015)
7. Xiao, Y., Shen, S.: Research on the establishment of traceability system of Chinese medicine supply based on collectivization. Chin. Herbal Med. **47**(21), 3925–3930 (2016)
8. Huang, Y.: Design and implementation of Chinese medicine circulation tracing system of Sichuan Zhengyuan Zhongsu Co., Ltd. University of Electronic Science and Technology, Chengdu (2015)
9. Jin, L.: Research on the traceability of seafood quality and safety under the Internet of Things system. Bohai University, Jinzhou (2015)
10. Gao, R.: Research on pork traceability and price early warning model based on Internet of Things. University of Electronic Science and Technology of China, Chengdu (2011)
11. Nan, C.: Research on traceability system of pork products based on internet of things. Comput. Appl. Softw. **33**(3), 96–99 (2016)
12. Wang, R., Hu, C., Zhang, W.: Research on electronic code coding scheme of products in Chinese herbal medicine industry based on RFID technology. World Sci. Technol.-Explor. Mod. Chin. Med. **17**(1), 296–300 (2015)

# Study on Measurement of Colloidal Liquid Concentration Based on Tyndall Phenomenon

Cunbo Jiang, Tiantian Zhu[✉], and Yaling Qin

College of Information Science and Engineering,
Guilin University of Technology, Guilin 541004, China
1554718594@qq.com

**Abstract.** In order to more easily measure the concentration of solution with the Tyndall phenomenon in the outdoor, a portable instrument using image-processing for collecting and processing images of Tyndall phenomenon is proposed. The software and hardware of light measurement module of the portable instrument are described in this paper, including the selection of the incident light and the image sensor. Then the optical path of the captured picture was extracted and the light intensity value was calculated. Through the standard sample, the linear relationship between the light intensity in the Tyndall phenomenon and the concentration was fitted to determine the concentration of the colloidal liquid to be tested. Theoretical analysis and simulation results show that the method of obtaining the concentration of colloidal liquid by using the relationship between the RGB mean of the image and the light intensity of Tyndall can control the error within 10%, which meets the preliminary test requirements.

**Keywords:** Light intensity · Image processing · Tyndall phenomenon · Colloidal liquid concentration

## 1 Introduction

Concentration is an important characteristic parameter used to characterize the properties of liquid in everyday life and industrial production. Quantitative measurement of solution concentration is required in many fields. At present, the general concentration detection methods mainly include chemical titration, physical spectroscopy, spectrometry, optical rotation method, resonance method, etc. [1]. Chemical titration is achieved by titrating a known solution to an unknown concentration until the change is realized [2]. This method is simple and easy to operate, but the accuracy is not high [3]. The physical spectroscopy method mainly uses light to irradiate the solution through different methods and angles, and uses a spectrometer to collect the pipe trench and analyze it [4]. This method is not easy to be popularized in the process of measuring different concentration solutions, and the precision is not high [5]. In addition, the real-time and operation of these methods have their limitations in real-time measurement of outdoor colloidal fluids. Studies have shown that for the colloidal liquid with a particle diameter of 1 to 100 nm, the Tyndall phenomenon occurs due to the scattering effect of the particles on the incident light [6]. The Tyndall phenomenon plays an important role

© Springer Nature Singapore Pte Ltd. 2019
R. Mao et al. (Eds.): ICPCSEE 2019, CCIS 1059, pp. 556–563, 2019.
https://doi.org/10.1007/978-981-15-0121-0_43

in the analysis of the optical properties, and it is also one of the judgments for discriminating the colloidal properties. And the intensity of the scattered light has a specific relationship with the concentration of the solution [7, 8].

The color digital image contains information related to the relative brightness, which can be used to measure the relative change in the relative light intensity of the image [9]. Some researchers conducted microscopic analysis by taking pictures under a microscope, and using different average absorbances of different particles, then the light intensity values were calculated from the average gray value of the image [10–12]. But for field water quality measurements, this method requires sample collection and preparation, which is uneconomical and time consuming. This study analyzes the images of Tyndall intensity of multiple solutions under the test environment of equal intensity incident light, and provide a method for achieving concentration measurement of a sample to be tested using a Tyndall color digital image.

## 2 Measurement Methods

### 2.1 Relationship Between Light Intensity of Tyndall and Concentration of Colloidal Liquid

According to the theoretical analysis of the existing literature, for the constant incident light, the intensity of the Tyndall phenomenon caused by the scattering of light is approximately linear with the concentration of the colloidal liquid [7]. This study assumes that it is fitted to a linear relationship. The concentration of the sample to be tested is determined by studying the relationship between the light intensity and the concentration of the standard sample. That is, for a colloidal liquid with a concentration of $c_k$, its expected values of Tyndall light intensity are:

$$I_{DEk} = a + b \cdot c_k \tag{1}$$

Assume that there are six samples with known concentrations of $[c_1, c_2, c_3, c_4, c_5, c_6]$ and two samples with concentrations to be tested. The intensity of Tyndall measured for six samples of known concentration is $[I_1, I_2, I_3, I_4, I_5, I_6]$. For a colloidal liquid with a concentration of $c_k$, the deviation between the actual measured value and the expected value is

$$\varepsilon_i = I_{Di} - I_i = a + b \cdot c_i - I_i \tag{2}$$

By constructing the objective function shown in the formula (3), the values of $a$ and $b$ when the error is minimized are obtained.

$$\xi = \sum_{i=1}^{6} \varepsilon_i^2 = \sum_{i=1}^{6} (a + b \cdot c_i - I_i)^2 \tag{3}$$

Using the least squares method, $a$ and $b$ values are:

$$a = \frac{\sum_{i=1}^{6} c_i^2 \sum_{i=1}^{6} I_i - \sum_{i=1}^{6} I_i c_i \cdot \sum_{i=1}^{6} c_i}{6 \sum_{i=1}^{6} c_i^2 - \left( \sum_{i=1}^{6} c_i \right)^2} \qquad (4)$$

$$b = \frac{6 \sum_{i=1}^{6} I_i c_i - \sum_{i=1}^{6} c_i \sum_{i=1}^{6} I_i}{6 \sum_{i=1}^{6} c_i^2 - \left( \sum_{i=1}^{6} c_i \right)^2} \qquad (5)$$

## 2.2  Basic Method of Measurement

Tyndall optical path image measurement principle is shown in Fig. 1. Among them, 1–6 are the standard samples of known concentration, and 7–8 are the samples to be tested. Keeping the incident light and the test environment unchanged, 8 samples are rotated and positioned to the shooting station under the control of the control device. At the shooting station, a constant light source is injected perpendicularly from the bottom of the sample, and a light path of the Tyndall phenomenon is generated in the vertical direction. After the sample is accurately positioned and stabilized, the camera is controlled to take a digital image containing the Tyndall light path. The station positioning device ensures that the optical path of the eight samples does not change at the center of the picture.

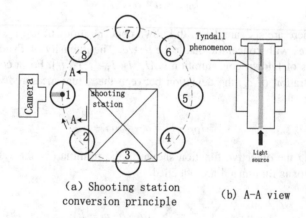

(a) Shooting station conversion principle

(b) A-A view

**Fig. 1.**  Measurement schematic diagram.

(1)  Selection of light source and image sensor

As one of the key parts of the image measurement module, the light source is mainly used to project light onto the object to be measured in a suitable way to generate the Tyndall phenomenon and highlight the optical path information required for image detection. Appropriate illumination intensity, stability of light source intensity, and uniformity of the light source have an important impact on the quality of the acquired image and the precise extraction and localization of the edges of the image. The generation of the Tyndall phenomenon of the colloidal liquid requires the wavelength of the incident light be larger than the diameter of the particles of the dispersion medium.

Through theoretical analysis, the light source selected in this paper is a red laser with a power of 50 mW and a wavelength of 650 nm.

The camera uses PTC08, its communication interface is RS232 (three-wire TX, RX, GND), and can also be changed to TTL level; image sensor type is CMOS 1/4 inch, image pixel is 300,000; pixel size is 5.6 μm * 5.6 μm, the output format is standard JPEG/M-JPEG; the image size is 640 * 480.

Because what wanted is the relationship between the concentration and the intensity of Tyndall phenomenon, and the images can be processed in a series of ways, the camera can meet the requirements.

(2)  Image acquisition

Connect image sensor to computer through converter and TTL interface level, then the computer sends out instructions to take pictures and save images to the specified folder. After all the required images are captured, these images are processed and analyzed. Because the device is fully enclosed, the influence of external light source is avoided.

## 2.3  Method for Realizing Tyndall Effect Light Intensity Measurement by Using Image-Processing

According to the foregoing method, the captured image of the Tyndall light path is as shown in Fig. 2. For each sample, the Tyndall optical path is collected under the same test environment and incident light source. The center position and picture size of each light path are consistent.

The Canny edge detection operator is used to extract the image path of the captured image, that is, the Tyndall phenomenon, and the RGB mean value of the optical path is detected to realize the conversion measurement of the light intensity. The detection principle of the Canny edge detection operator is:

(1)  Select the parameter $\sigma$ (the root mean square difference of the Gaussian function), and calculate the Gaussian filter template according to the 2D Gaussian smoothing filter impulse response function of formula (6).

$$h(x,y) = \frac{1}{2\pi\sigma^2} e^{-\frac{x^2+y^2}{2\sigma^2}} \tag{6}$$

(2) The image $f(x, y)$ is smoothed by a Gaussian filter template to obtain a filtered image.

$$S(x,y) = h(x,y) * f(x,y) \tag{7}$$

(3) Find the gradient of the filtered image.
(4) Control the non-maximum suppression of the gradient amplitude.
(5) Double threshold algorithm is used to detect and connect edges.

The RGB average value is calculated for the captured image, and the light intensity value is calculated by the formula (8), and the obtained light intensity value is considered to be the Tyndall light intensity value.

$$I = 0.299 \times R + 0.587 \times G + 0.114 \times B \tag{8}$$

## 3 Experimental Simulation

### 3.1 Experimental Data Measurement

A picture of the Tyndall phenomenon collected by the simple device shown in Fig. 1 is shown in Fig. 2. Subgraph a–h are 100%, 70%, 60%,50%, 40%, 30%,10%, and 8% nanogold content separately.

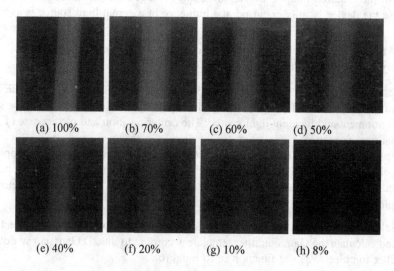

(a) 100%        (b) 70%        (c) 60%        (d) 50%

(e) 40%        (f) 20%        (g) 10%        (h) 8%

**Fig. 2.** Pictures of the Tyndall phenomenon.

## 3.2    Processing Method of Experimental Data

### Experimental Data of Reference Sample

The image of the Tyndall phenomenon is generated by photographing the colloidal liquid, and the optical path of the Tyndall phenomenon is intercepted and the RGB value of the optical path is detected to realize the conversion measurement of the light intensity. The image is simulated and analyzed by matlab, and the RGB mean of the image is obtained. The average light intensity value $I$ of the Tyndall phenomenon is obtained by the formula (8).

Taking the solutions with 100%, 70%, 50%, 40%, 10%, and 8% nanogold content as the standard samples, the images of the Tyndall phenomenon generated by them were analyzed and processed. Correspondence between average light intensity value and concentration as shown in Table 1 were obtained.

**Table 1.**  Correspondence between average light intensity value and concentration

| Concentration | Light intensity |
|---------------|-----------------|
| 100%          | 0.3103          |
| 70%           | 0.2456          |
| 50%           | 0.2354          |
| 40%           | 0.2047          |
| 10%           | 0.1005          |
| 8%            | 0.0387          |

### Analysis of Results

For the constant light source illumination, the theoretical analysis can be considered that the light intensity $I$ of the Tyndall phenomenon is linear with the colloidal liquid concentration $c$. The relationship shown in Fig. 3 can be obtained by the values of Table 1.

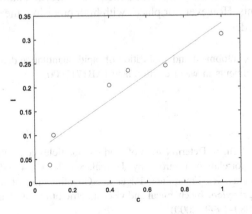

**Fig. 3.**  Linear relationship between concentration $c$ and light intensity value $I$

It can be calculated from Fig. 3 that the relationship between the concentration $c$ and the light intensity value $I$ is as shown in formula (9).

$$c = 3.7411 \times I - 0.2282 \tag{9}$$

**Experimental Verification of the Sample to be Tested**
The test samples with 60% and 30% concentration were validated, and the error was shown in Table 2.

**Table 2.**  Relative error of sample to be tested

| Actual concentration | Measured concentration | Relative error |
|---|---|---|
| 60% | 65.2% | 8.67% |
| 30% | 28.7% | 4.33% |

As shown in Table 2, 60% and 30% concentration samples were verified, and the errors were all within 10%, meeting the initial requirements.

## 4  Conclusion

Since the current methods of concentration measurement are mostly chemical methods, or image processing under a microscope, it is not convenient for outdoor measurements. In order to solve the inconvenience of outdoor liquid concentration measurement, this paper proposes to make a portable instrument through image processing. The canny edge detection operator is used to extract the optical path, and the light intensity value in the optical path of the Tyndall phenomenon is calculated by the RGB value of the optical path image. Then the relationship between the concentration and the light intensity value is calculated by the standard sample of the known concentration, thereby the concentration of the sample to be tested is calculated. Since the main purpose of this paper is to judge water quality by measuring the concentration of liquid sol, this method makes it easy for researchers to perform outdoor real-time water quality measurements. However, for places with high precision, the relative error needs to be further reduced.

**Acknowledgment.** Development and application of rapid quantitative detection technology for heavy metal-free instruments in water environment (AB17129003).

## References

1. Liu, F., Dong, C., Ren, J.: Determination of gold nanoparticles concentration in the solution with fluorescence correlation spectroscopy. J. Anal. Sci. **34**(04), 496–500 (2018)
2. Sclavons, M., Legras, R., Franquinet, P., et al.: Quantification of the maleic anhydride grafted onto polypropylene by chemical and viscosimetric titrations, and FTIR spectroscopy. Polymer **41**(6), 1989–1999 (2000)

3. Qiu, B., Zhahg, H., Xu, Z., et al.: Research progress on analysis of polysaccharides. Sci. Technol. Food Ind. **39**(06), 327–333 (2018)
4. Brun, N., Ponçot, M., Bourson, P.: Raman correlation spectroscopy: a method studying physical properties of polystyrene by the means of multivariate analysis. Chemom. Intell. Lab. Syst. **128**, 77–82 (2013)
5. Qin, X.-L., Zhang, Y., Li, Z.-Y.: Determination of dyeing wastewater by snapshot imaging spectrometer. Chin. J. Anal. Chem. **45**(11), 1635–1640 (2017)
6. Miyazawa, S., Iwasaki, H.: Light scattering from ferroelectric domains in LiTaO$_3$. Mater. Res. Bull. **13**(5), 511–518 (1978)
7. Aspanut, Z., Yamada, T., Lim, L.W., et al.: Light-scattering and turbidimetric detection of silica colloids in size-exclusion chromatography. Anal. Bioanal. Chem. **391**(1), 353–359 (2008)
8. Li, W.-J., Zhang, J., Gu, X.-Y.: Tyndall phenomenon of colloidal sol under different concentrations. Chin. J. Chem. Educ. **37**(07), 77–81 (2016)
9. Gang, Z., Zhi-Jian, X.: Effective measurement of optical density and analysis on its influencing factors. J. Third Mil. Med. Univ. **23**(7), 785 (2001)
10. Song, Y.K., et al.: A comparison of microscopic and spectroscopic identification methods for analysis of microplastics in environmental samples. Mar. Pollut. Bull. **93**(1–2), 202–209 (2015)
11. He, J., Tong, X., Zhao, Y.: Photoresponsive nanogels based on photocontrollable cross-links. Macromolecules **42**(13), 4845–4852 (2009)
12. Xiaohan, X., Guo, X., Shiqun, J., et al.: Design of imaging detection system for fluorescent immune-chromatographic test strip. Chin. J. Lasers **45**(04), 287–294 (2018)

# Neural Network Model for Classifying the Economic Recession and Construction of Financial Stress Index

Lujia Shen[1(✉)], Tianyu Du[2], and Shouling Ji[2]

[1] Northeastern University, Bsoton, MA 02120, USA
shen.luj@husky.neu.edu
[2] Zhejiang Univerisity, Hangzhou, Zhejiang, China

**Abstract.** In this paper, a C5.0 decision tree and neural network models are proposed to classify recessions in the US with 12 common financial indices and new financial stress indices inferred from the neural network models are created. A detailed experiment is presented and demonstrates that the neural network models with proper regularization and dropout achieve 98% accuracy in the training set, 97% accuracy in validation set and 100% accuracy in test accuracy. The financial stress indices outperform other existing financial stress indices in many scenes and can accurately locate crisis events even the most recent 2018 US Bear Market. With these models and new indices, contraction can be detected before NBER's announcement and action could be taken as early as the situation get worse.

**Keywords:** Financial crisis · Neural network · Financial stress index

## 1 Introduction

Economic market is imperfect, the fluctuation will be reflected on the change of employment, GDP, income and lots of other factors. The 2007–2008 financial crisis, also known as the global financial crisis, is the most severe one since the Great Depression of the 1930s. After this crisis, investors, financial analysts, and even stock traders, security dealers are concerning more and more about the latent stress in the market which may contribute to the downturn in the economy. Therefore, to construct a stress index that can accurately reflect the status of the current market timely and predict whether it is in recession or boom is important. To define the economic recession and boom, "Business Cycle" created by National Bureau of Economic Research (NBER) is used in this paper. The NBER's Business Cycle Dating Committee maintains a chronology of the U.S. business cycle which contains the dates of peaks and troughs in economic activity [9].

To monitor the uncertainty of the market, lots of financial stress indices are constructed, like KCFSI [12], STLFSI [17], FMSI [3], etc. The higher the index,

© Springer Nature Singapore Pte Ltd. 2019
R. Mao et al. (Eds.): ICPCSEE 2019, CCIS 1059, pp. 564–578, 2019.
https://doi.org/10.1007/978-981-15-0121-0_44

the high the stress. But neither of them can indicate an economic recession or downturn clearly. In Figs. 1 and 2, both indices show high market stress near 2008, but neither of them has strong evidence to show the early 2000s recession. Even in 1994, the KCFSI shows negative stress, however, STLFSI shows positive stress. In addition, most recession forecasting use a probit model which is a logistic regression with past data to predict future recession. One problem for forecasting recession is some crisis events occur without any sign, like the economic effects arising from the 911. Furthermore, the NBER committee's goal is to determine the dates of peaks and troughs as definitively as possible which commonly determined by real GDP and GDI measured quarterly or annually and the committee is careful to avoid premature judgments [9]. Therefore, the committee took a long time to determine a peak or a trough. However, time is precious in investing, and investor will not wait a long time to know a peak or a trough.

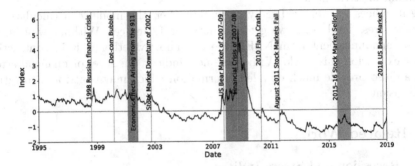

**Fig. 1.** St. Louis Fed Financial Stress Index (STLFSI).

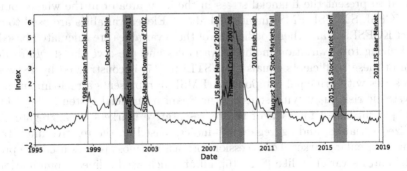

**Fig. 2.** Kansas City Financial Stress Index (KCFSI).

To settle the problems stated above, this paper proposes new models to directly detect the expansion and contraction in "real-time", and 2 new indices that reflects the market stress accurately and clearly is constructed. A total of

1154 instances from 12 indices representing different aspects of the economic market which can be easily obtained from the internet are used to build models. Models is validated and tested, and achieves 98% accuracy in validation set and predicts test set with 100% accuracy and 2 new indices are constructed from these models. With these models and indices, investors can make the next decision as soon as they get the result of new data from the model.

The main contribution of this paper are as follows:

1. We adopt neural network to build the models which has't been used before and turn out to be an excellent classifier for financial contractions. Neural networks can learn to capture such uncaptured relationship between features as most of the current financial stress indicators and recession forecasting models are constructed by a linear combination of various indices. The indices we built by neural network models show the powerful ability to precisely determine the crisis event date in the past which outperforms other existing financial stress indices.

2. All indices are released daily or weekly, then the model can calculate the new stress level when we get new data for 12 indices. Without waiting for the announcement from NBER, we can know whether it is in contraction or expansion this week. With these new indices, the level of financial stress can also provide much detailed information which may contribute to future decisions.

## 2   Related Work

### 2.1   Recession and Stress Indices

A lot of works have been done to visualize the stress in the financial market or predict the next recession. Many Financial Stress Indicators (FSI) have been created to present the financial stress in the local area or in the whole country like KCFSI, STLFSI, FMSI mentioned above. Eleven variables are used to construct KCFSI [12] and then coefficients of these variables are calculated based on maximizing total variation in the 11 variables and preserve the standard deviation of these coefficients equals to 1. STLFSI [17] is constructed by 18 weekly data series with principal components. FMSI [3] represents a real-time measure of systemic risk quantifying stress in the Spanish financial system. They obtain a unique sub-index for each of the representative markets by aggregate standardized variables, and aggregate sub-indices based on the relative importance of this particular market. For recession forecasting, most papers used the probit model and its variants, like [8,13,16], all of which used a linear combination of indices to predict recession.

### 2.2   Decision Tree

Many machine learning techniques have been used in plenty of areas in economy for different purpose like classification, regression, clustering, recommendation system, etc. Decision trees are one of the most important models in finance, which aim to help people make decisions in financial market. In [7], four popular decision tree algorithms (CHAID, C5.0, QUEST and CART) were implemented to analyze the impact of financial ratios on firm performance. In [6], Chen et al. used decision trees (C5.0, CART, and CHAID) and logistic regression techniques to implement the financial distress prediction model. In [11], an non-parametric CART decision tree has been applied to financial distress prediction and compared with each other. In [10], Elsayyad et al. presented both multilayer perceptron neural network and C5.0 decision tree model to classify marketing campaigns from customers' data. Three statistical measures-classification accuracy, sensitivity, and specificity are used to measure model performance.

## 2.3  Neural Network

The neural network has made great progress in many fields of machine learning and its application in financial domain shows a prominent success. In [14], Lam investigates the ability of neural networks to integrate fundamental and technical analysis for financial performance prediction. In [18], Yasin et al. used Probabilistic Neural Network (PNN) to classify company performance. In this paper, we mainly focus on using neural networks to classify whether it is a contraction or expansion at a particular day and construct new stress indices.

# 3  Methodology

## 3.1  Features Selection

**Dependent Feature.** The term "Business Cycle" is usually referred to as the ascending and descending movements of GDP [15]. The peak and trough in a "Business Cycle" are determined by The Business Cycle Dating Committee of NBER [9]. A peak marks the end of an expansion and is the highest point in one business cycle. A trough, on the contrary, is the beginning of expansion as well as the end of a contraction and is the lowest point of a business cycle. Therefore, in one business cycle, it contains one expansion starting from a trough and ending at a peak and one contraction starting from a peak and ending at a trough. Therefore, the dependent feature used in this paper is a Boolean function where 1 and 0 are used to indicate whether the financial market is in a contraction or in an expansion respectively. We want the result of the final model close to 1 when it is in a contraction and close to 0 when it is in an expansion. We take two recent contractions, from March 2001 to November 2001, and from July 2007 to June 2009 into consideration. Earlier contractions can also be considered if the span of the feature indices contains this period of time.

**Independent Indices.** Ideally, indices that can represent a variety aspect of the entire financial system are considered as our independent indices. However, the financial system can be represented as a large complex network where financial institutions like banks are interconnected to each other [2]. For a better selection of indices, the important element should be examined, for example, stock futures market, bond and mortgage interest rates, the yield curve, foreign exchange rates stock, and commodity prices.

To be more accurate, it is better to have more contractions used in building the model. There are two recent contractions, from Mar. 2001 to Nov. 2001 and from Dec. 2007 to Jul. 2009 respectively. The independent features should be chosen that cover these two time periods. There are many ways to choose the independent features, and to simplify the problem, this paper refers to the indices used in STLFSI [17]:

(1)  Effective federal funds rate (FF)
(2)  2-year Treasury (WGS2)
(3)  10-year Treasury (WGS10)
(4)  Baa-rated corporate (WBAA)
(5)  Merrill Lynch High-Yield Corporate Master II Index (BAIV)
(6)  Merrill Lynch Asset-Backed Master BBB-rated (BAEY)
(7)  Yield curve: 10-year Treasury minus 3-month Treasury (T10Y3)
(8)  Corporate Baa-rated bond minus 10-year Treasury (DBS10)
(9)  3-month London Interbank Offering Rate (US6N)
(10)  3-month Treasury-Eurodollar spread (TED)
(11)  3-Month AA Financial Commercial Paper Rate (DC3M)
(12)  CBOE Volatility Index (VIX)

There are 12 indices in total and these indices capture some aspect of financial stress but may not capture all the factors that affect the financial market. Therefore, when the financial stress in the market changes, the above indices are likely to have some fluctuations or trends. The aim is to find the relationship between these indices with financial expansion and contraction.

## 3.2  Model Design

**Decision Tree Model.** First, the C5.0 algorithm improved from C4.5 is one of the most efficient and accurate decision tree models. In this section, C5.0 pruned with $max\_leaf\_nodes = 10$, $min\_samples\_leaf = 10$, $max\_depth = 5$ is applied to the data and achieved an accuracy of 99.12% on validation data and it predicts 110 test data all correctly into expansions. The model of the C5.0 algorithm is presented in Fig. 3. From this Figure, Merrill Lynch Asset-Backed Master BBB-rated (BAEY) plays a critical role in this model, and when $BAEY \leq 0.403$ all the dates are in expansion. Additionally, it is also obvious to see that 735 out of a total of 738 contraction data located at the range in $BAEY \geq 0.403$, $WGS2 \leq 0.635$, $TED \geq 0.022$ and $US6N \leq 0.746$, where all of these data are normalized to $[0, 1]$ at first. The validation data obtain an accuracy of 96.20

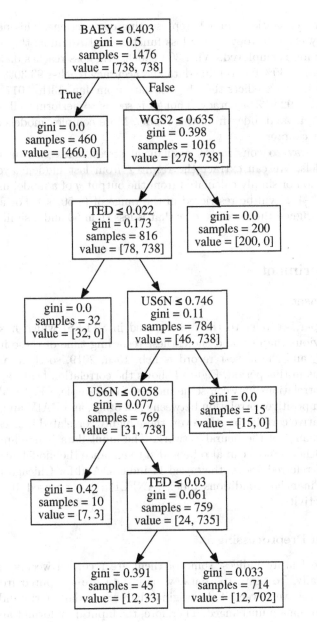

**Fig. 3.** C5.0 decision tree model.

**Neural Network Model.** A simple neural network with only two hidden layers which has 3 nodes and 6 nodes respectively is used first for demonstration. Instead of using the deep neural network, this shallow network is intended to prevent overfitting. The output layer consists of 1 node with a response of 0, representing expansions, and 1, representing contractions. This neural network uses binary cross-entropy as the loss function and Adam as the optimizer, and early stopping is employed. With 300 epochs, the loss reaches 0.0338 and accuracy reaches 96.40% for training data and accuracy reaches 93.30% for validation data. This model predicts the 110 test data from Jan. 13th, 2017 to latest Feb. 15, 2019 with 92.91% accuracy. Though it seems to perform well, it is not good for building a valid index from this model. More complex models are evaluated in the next chapter.

A new way to construct the financial stress index can be inferred from these models. We can extract the result $x$ from last hidden layer before sigmoid function or simply calculated from the output $y$ of a model using function: $x = \log\left(\frac{y}{1-y}\right)$. $x$ will be restricted in the range of $(-50, 50)$. For any $x \geq 50$, $x$ will be reassigned the maximal of $x$ that is less than 50 and a similar way to any $x \leq -50$.

## 4 Experiment

### 4.1 Dataset

In this paper, we refer to the indices used in STLFSI which has been stated in the previous chapter, and the earliest recording time of these indices is Jan. 10th, 1997, and the newest record is Feb. 15th, 2019, so there are two major contractions in this period. Figure 4 shows the correlation heatmap where there are high correlations between some indices, for example, FF, DC3M and US6N have a high positive correlation between each other, and BAIV and WBAA have a high negative correlation. However, these highly correlated indices won't affect the performance of the neural network. The input data is not limited to these features. Other indices can also be added to enhance the model accuracy. As for the KCFSI referred above, they used 11 indices. And for Chicago Fed Adjusted National Financial Conditions Index (ANFCI) [1], they used 105 measures of financial activity.

### 4.2 Data Preprocessing

In all these feature indices, some of them are recorded weekly and some are recorded daily. To unify the data, weekly data are computed from averaging the daily data every week if the original data are in daily form and the original weekly data remain unchanged. Therefore, the input data format are in a weekly format.

There are a total of 1154 instances, and first 1044 instances are used as our training and validation set, of which 922 expansion instances, 122 contraction instances. The latest 110 instances are considered as the test data. To help the neural network to find the minimal faster, independent features are all normalized to range [0, 1] using all the instances. Then 1044 instances are separated into 70% training set which contains 638 expansion instances and 92 contraction instances and 30% validation set. The data is skewed due to the imbalance

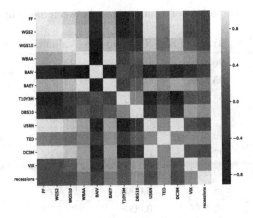

**Fig. 4.** Correlation heatmap of indices

of the number of contraction instances and expansion instances. Thus, the over-sampling method is applied to our data. Synthetic Minority Over-sampling Technique (SMOTE) [5] helps to over-sample our training data into 738 expansion instances and 738 contraction instances, and validation data remain unchanged.

### 4.3 Evaluation Setting

To illustrate the robustness of the neural network model, we used 5 folds cross-validation to demonstrate that the model is accurate and stable. Cross-validation results are combined to make the model more general and accurate. Sometimes the model may fall into local minima, and retrain the model will get ride of this result.

We compared the training accuracy, training loss, validation accuracy and test accuracy between models with different amounts of nodes in hidden layers and different amount of hidden layers. The accuracy and loss are calculated by the average of 5 folds cross-validation results. We also add the regularization and dropout to improve the performance of models and indices. Other hyper-parameter are remain unchanged (Optimizer: Adam, epochs: 300, Activation function: ReLU).

### 4.4 Model Evaluation and Index Analysis

Table 1 demonstrate that the models with fewer parameters and no regularization or dropout have slightly lower training accuracy and validation accuracy but a high test accuracy. The more complex the model, the lower the test accuracy. Indices return from each fold varies, but with the figure of 5 indices from 5 folds of cross-validation data in Figs. 5, 6, 7, 8 and 9, we can see that the more complex the model, the more similar the 5 indices. This means that the complex models have more robustness. From Figs. 5 and 6, we can see that some part of

**Table 1.** Comparison of accuracy, loss between different models.

| Hidden nodes | Training acc | Training loss | Validation acc | Test acc |
|---|---|---|---|---|
| (3, 6) | 96.40% | 0.0338 | 93.30% | 92.91% |
| (5, 10) | 95.59% | 0.0376 | 93.48% | 91.09% |
| (10, 15) | 98.40% | 0.0167 | 97.02% | 71.63% |
| (20, 30) | 98.74% | 0.0104 | 97.99% | 64.73% |
| (50, 100) | 9922% | 0.0069 | 98.76% | 66.91% |
| (3, 6, 10) | 97.06% | 0.0259 | 94.92% | 79.64% |
| (10, 15, 30) | 99.06% | 0.0082 | 98.18% | 71.64% |
| (20, 30, 50) | 98.86% | 0.0093 | 97.70% | 081.45% |
| (100, 200) | 99.19% | 0.0068 | 98.57% | 73.27% |
| (100, 200, 300) | 98.86% | 0.0093 | 97.70% | 81.45% |

**Table 2.** Comparison of accuracy, loss between different models.

| Hidden nodes | Training acc | Training loss | Validation acc | Test acc |
|---|---|---|---|---|
| Model 1 (Table 3) | 95.40% | 0.0892 | 95.21% | 100% |
| Model 2 (Table 4) | 98.21% | 0.0360 | 97.41% | 95.82% |
| Model 3 (Table 5) | 98.12% | 0.0351 | 97.51% | 96.91% |
| Model 4 (Table 6) | 98.29% | 0.0350 | 97.51% | 96.73% |
| Model 5 (Table 7) | 96.45% | 0.0802 | 95.88% | 100% |
| Model 6 (Table 8) | 96.81% | 0.0795 | 95.59% | 100% |

the index has flattened with no information of the movement of stress level and the difference between each index is large. While in Figs. 7, 8 and 9, 5 indices are much similar to each other where model (100, 200, 300) has the highest similarity. As the model becomes more complex the 5 indices will converge to one uniform index which is the ultimate index we want to get. One problem with the complex models are the low test accuracy and high variance of the index.

Cross-validation is also employed to C5.0 decision tree model and it achieved an average accuracy of 99.03% on validation set and predicted all 110 test data correctly for 5 folds. Decision tree model in this problem is extremely capable of classifying future data.

When using more complex models or more epochs for training, the indices will have a greater range of stress values, because trained values are much closer to 1 when it is contraction and much closer to 0 when it is expansion. Thus, the values before the sigmoid function will be very large. To build more applicable indices and prevent the indices goes too large (both positive and negative), we apply regularization, dropout, and early stopping to reduce the potential overfitting problem and help to control the weights being not too big.

**Table 3.** Model 1

| Layer | Nodes | Regularizer | Dropout |
|---|---|---|---|
| Input | 12 | / | / |
| Hidden 1 | 100 | / | 0.5 |
| Hidden 2 | 200 | 0.01 | 0.5 |
| Output | 1 | 0.01 | / |

**Table 4.** Model 2

| Layer | Nodes | Regularizer | Dropout |
|---|---|---|---|
| Input | 12 | / | / |
| Hidden 1 | 100 | / | / |
| Hidden 2 | 200 | / | 0.5 |
| Hidden 3 | 200 | 0.01 | 0.5 |
| Output | 1 | 0.01 | / |

**Table 5.** Model 3

| Layer | Nodes | Regularizer | Dropout |
|---|---|---|---|
| Input | 12 | / | / |
| Hidden 1 | 50 | / | / |
| Hidden 2 | 100 | / | 0.5 |
| Hidden 3 | 50 | 0.01 | 0.5 |
| Output | 1 | 0.01 | / |

**Table 6.** Model 4

| Layer | Nodes | Regularizer | Dropout |
|---|---|---|---|
| Input | 12 | / | / |
| Hidden 1 | 100 | / | / |
| Hidden 2 | 200 | / | 0.5 |
| Hidden 3 | 300 | 0.01 | 0.5 |
| Output | 1 | 0.01 | / |

In experiments, $L_2$ regularizer are implemented. With one regularizer on the output layer, one on the hidden layer, the indices are more converged and the values are restricted as in Fig. 10. In Table 2, all models show a high test accuracy, as well as training and validation accuracy, remain high. Two generated indices are examined in this section, one from model 1 where details provided in Table 3, another from model 2 where details provided in Table 4. Figures 11 and 12 show two indices with various economic crisis event where greyed areas are two contractions used as dependent variable, red areas denote two another crisis lasting long period but in expansion periods and red vertical lines denote the peaks of one crisis events. Though trained using different hyperparameters, two indices are almost identical. Spikes in two indices have the same dates. The only difference is the scale of two indices, the index from model 2 has a larger range of stress level comparing to the index from model 1.

There is no visible difference between the retraining model and changing the hyperparameters of the model unless using fewer nodes in the model, which shows the robustness of this model. From Fig. 13, 4 indices created from 4 different models in Tables 5, 6, 7 and 8 are presented. They all have the exact same pattern whose only difference lies in the range of levels.

One of the most important things is that the index can show other crisis events. Taking the index from model 1 as an example, it shows a high stress level in both two contractions. It also detects the high stress level of the stock market downturn of 2002 [4]. In the period of August 2011 stock market fall and 2015–2016 stock market sell-off, it is obvious that the index has protrusions which are significantly higher than before. For any other events, we can clearly

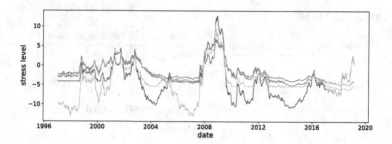

**Fig. 5.** Index from model (3, 6) for 5 folds cross validation.

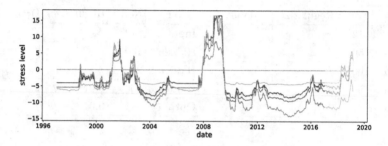

**Fig. 6.** Index from model (3, 6, 10) for 5 folds cross validation.

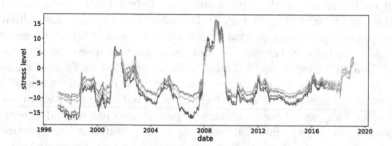

**Fig. 7.** Index from model (50, 100) for 5 folds cross validation.

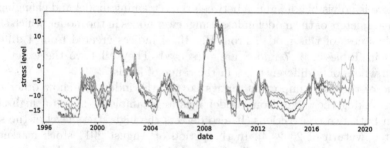

**Fig. 8.** Index from model (100, 200) for 5 folds cross validation.

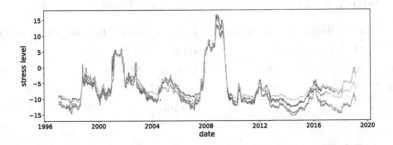

**Fig. 9.** Index from model (100, 200, 300) for 5 folds cross validation.

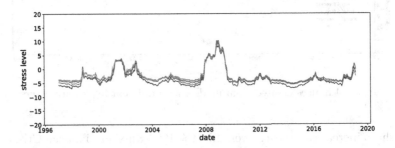

**Fig. 10.** Index from model 4 for 5 folds cross validation.

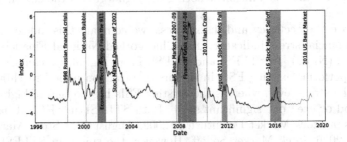

**Fig. 11.** Index from model 1.

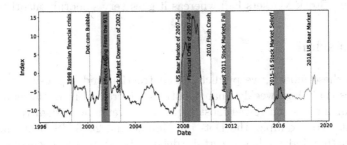

**Fig. 12.** Index from model 2.

**Table 7.** Model 5

| Layer | Nodes | Regularizer | Dropout |
|---|---|---|---|
| Input | 12 | / | / |
| Hidden 1 | 100 | / | 0.5 |
| Hidden 2 | 500 | 0.01 | 0.5 |
| Output | 1 | 0.01 | / |

**Table 8.** Model 6

| Layer | Nodes | Regularizer | Dropout |
|---|---|---|---|
| Input | 12 | / | / |
| Hidden 1 | 200 | / | 0.5 |
| Hidden 2 | 600 | 0.01 | 0.5 |
| Output | 1 | 0.01 | / |

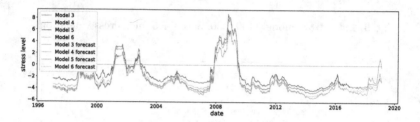

**Fig. 13.** 4 indices created from models with different hyperparameters.

see a sharp increase after each event like the 1998 Russian Financial Crisis, Dot-com Bubble, 2010 Flash Crash and 2018 US Bear Market. Both indices from model 1 and model 2 have a spike after the 2018 US Bear Market, but index from model 2 is more significant denoting the high stress of the current financial market.

To prove the accuracy and distinctness, we compare this index with other existing financial stress indicators, like Chicago Fed National Financial Conditions Index (NFCI) [1], STLFSI and KCFSI. From Fig. 14, we can see that our index outperforms existing FSIs in the first contraction period reaching a high point in the stress level, whereas the existing FSIs has no sign of high stress in that period or even one significant spike. Both STLFSI and KCFSI have obvious spikes in Stock Market Downturn of 2002, August 2011 Stock Markets Fall and 2015–2016 Stock Market Selloff. However, the spikes in STLFSI are not significant as those in KCFSI as well as the STLFSI remain a high variance in non-crisis periods. NFCI only has noticeable spikes at second contraction and August 2011 Stock Markets Fall whereas it goes very smoothly at other events.

## 5   Discussion

Firstly, in these models, we predict the contraction and expansion every week and indices that show weekly financial stress are created from the same model. Though our indices can reveal most high-stress events, it is slightly lagging for some events. For example, the Dot-com Bubble and 2018 US Bear Market are two events that indices has a certain delay relative to the occurrence of the events. The delay of the index is critical for those investors who investigate

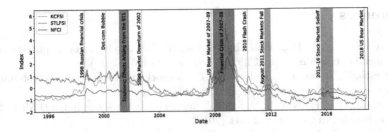

Fig. 14. KCFSI, STLFSI, NFCI.

depends on the stress index, thus, a precise forecasting model should be added to compensate for the lag of the index.

Moreover, models are built only depend on two recent contractions, as there are more contractions in the past. Feature indices that cover more contractions can replace indices with a short time span. The more contractions are used to build the model, the better it can predict future data. The reason for the events not predicted correctly may lies on the independent features selection, as their impact does not reflect on our independent features. More important crisis events and more relevant indices may be considered in constructing a better stress indicator.

Finally, as all FSIs show a spike near the 1998 Russian Financial Crisis, however, the financial crisis in Russia did not affect the US market, and no major crisis happened in the US at that time. This spike is unexpected and detailed research needed to find out what happened.

## 6    Conclusion

In this paper, we present two methods to predict contractions and expansions in the financial markets, one is C5.0 decision tree and another is neural network. Decision tree model has already achieved a great accuracy for our data, and two neural network models are applied to the data and are carefully examined. New indices that show the stress level in the market are constructed. From the experiment results, simple models achieve high accuracy in test set but do not produce valid indices. Complex models cannot achieve high accuracy in test set due to overfitting, but produce robust indices. Then regularization and dropout are applied to prevent overfitting and control the range of our indices to avoid them being too large. Hence, our new indices can accurately and distinctly predict past crisis events and outperforms existing FSIs in many scenes.

Finally, these models and the new indices can help investors decide whether they should invest or sell assets, to act before the market getting worse. Thus, they may minimize future losses or increase return on investment.

# References

1. Brave, S.A., Kelly, D.L., et al.: Introducing the Chicago fed's new adjusted national financial conditions index. Chicago Fed Letter (386) (2017)
2. Caccioli, F., Barucca, P., Kobayashi, T.: Network models of financial systemic risk: a review. J. Comput. Soc. Sci. **1**(1), 81–114 (2018)
3. Cambón, M.I., Estévez, L.: A Spanish financial market stress index (FMSI). Span. Rev. Financ. Econ. **14**(1), 23–41 (2016)
4. Carlson, J.B., Pelz, E.A., et al.: A retrospective on the stock market in 2000. Economic Commentary, 15 January 2001 (2001)
5. Chawla, N.V., Bowyer, K.W., Hall, L.O., Kegelmeyer, W.P.: SMOTE: synthetic minority over-sampling technique. J. Artif. Intell. Res. **16**, 321–357 (2002)
6. Chen, M.Y.: Predicting corporate financial distress based on integration of decision tree classification and logistic regression. Expert Syst. Appl. **38**(9), 11261–11272 (2011)
7. Delen, D., Kuzey, C., Uyar, A.: Measuring firm performance using financial ratios: a decision tree approach. Expert Syst. Appl. **40**(10), 3970–3983 (2013)
8. Dueker, M.J., et al.: Regime-dependent recession forecasts and the 2001 recession. Rev.-Fed. Reserv. Bank St. Louis **84**(6), 29–36 (2002)
9. National Bureau of Economic Research: Us business cycle expansions and contractions (2010). https://users.nber.org/cycles/cyclesmain.html
10. Elsalamony, H.A., Elsayad, A.M.: Bank direct marketing based on neural network and C5.0 models. Int. J. Eng. Adv. Technol. (IJEAT) **2**(6), 392–400 (2013)
11. Gepp, A., Kumar, K.: Predicting financial distress: a comparison of survival analysis and decision tree techniques. Procedia Comput. Sci. **54**, 396–404 (2015)
12. Hakkio, C.S., Keeton, W.R., et al.: Financial stress: what is it, how can it be measured, and why does it matter? Econ. Rev. **94**(2), 5–50 (2009)
13. Khomo, M.M., Aziakpono, M.J.: Forecasting recession in South Africa: a comparison of the yield curve and other economic indicators. S. Afr. J. Econ. **75**(2), 194–212 (2007)
14. Lam, M.: Neural network techniques for financial performance prediction: integrating fundamental and technical analysis. Decis. Support Syst. **37**(4), 567–581 (2004)
15. Madhani, P.M.: Rebalancing fixed and variable pay in a sales organization: a business cycle perspective. Compens. Benefits Rev. **42**(3), 179–189 (2010)
16. Nyberg, H.: Dynamic probit models and financial variables in recession forecasting. J. Forecast. **29**(1–2), 215–230 (2010)
17. Federal Reserve Bank of St. Louis: St. Louis Fed Financial Stress Index [STLFSI]. https://fred.stlouisfed.org/series/STLFSI. Accessed 13 Dec 2018
18. Yasin, H., Arifin, A.W.B., Warsito, B.: Classification of company performance using weighted probabilistic neural network. J. Phys: Conf. Ser. **1025**, 012095 (2018)

# Using Augmented Reality Techniques to Simulate Training and Assistance in Electric Power

Zifeng Tang[1], Jiyao Deng[1], Hao Gao[1], Jiageng Song[1], Qinlun Li[1], Peixin Zhang[1], and Minghui Sun[2(⊠)]

[1] College of Software, Jilin University, Changchun 130022, China
[2] College of Computer Science and Technology, Jilin University, Changchun 130022, China
42600225@qq.com

**Abstract.** In combination with the augmented reality (AR) technology and Hololens device, an electric power work assist system is developed to provide a more realistic and convenient training experience for electric power training. The training effect was improved, and help guidance information was provided in the actual operation of electric power work. Experimental results show that difficulty of electric power work is reduced and the efficiency of actual work is improved.

**Keywords:** Electric power simulation training ·
Electric power work assistance · Augmented reality · Image recognition

## 1 Introduction

In recent years, science and technology have developed rapidly, and people's living standards have been continuously improved. All of this is inseparable from the rapid development of the power industry. Nowadays, electricity, as the basis of almost all technologies, has long been an indispensable thing. Therefore, the importance of power companies that provide electricity and maintain electricity supply is self-evident. As a basic industry, power companies are prerequisite for the normal development of many other industries. Therefore, power companies must shoulder the heavy responsibility and develop steadily forward. And a sustainably developing company cannot do without are skilled technicians. Thus it's important for the companies to do well in training novice technicians and quickly acquire skilled technicians [1].

For the traditional training methods, there are two choices, the first is based on the actual operation and another one is textbook-based. Because the electric power project has the characteristics of not being able to easily power off and has high risk, the actual-operation-based training, once the operation is wrong, has the unpredictable consequence of personnel safety and property, so traditional electric power training is mainly

---

J. Deng and H. Gao—Authors contributed equally to this work.

based on theoretical training, using textbook-based method. Such training methods are difficult to provide practical training opportunity, and many technicians are not satisfied with these methods and they think it is dull and not real enough [2]. Thus the technical personnel usually have less opportunity to practice, and can only slowly accumulate experience in a long period of practice, it is difficult to master the operation skills in a short period of time, which leads to long period of technicians training, resulting in the situation of technicians shortage.

In order to improve the efficiency of the training process, in addition to improve the effectiveness of the traditional training, we can also find ways to reduce the difficulty of entry level for beginner by the newest technology. Some research results showed that by integrating AR technology into the instruction, the students took on a more positive autonomous learning attitude [3].

The development of AR (Augmented Reality) technology and emergence of Hololens device, combined with existing machine-learning-based image recognition technology, can solve both problems well at the same time. Hololens is a Windows 10 based smart eyewear product. It features advanced sensors, a high-definition 3D optical head-mounted full-angle lens monitor and surround sound effect. It makes the user interface in augmented reality to communicate with the user through eyes, voice and gestures. By using Hololens, user can see the real world like normal glasses, and have a scanning function that can scan and recognize real-world surroundings, with space recognition and positioning. Most importantly, it can load virtual information, use the high-transparency display on the both sides of the glasses, use retinal imaging technology to form a visual 3D effect in real space, and use image recognition technology to analyze the relative space of the corresponding components. The location, and the simulation data is combined with the data scanned by Hololens, users can see that the virtual model exists in the real world coordinate system [4].

During the training process, technicians can use the Hololens to perform more realistic operation training. At the same time, technicians can also see the virtual guidance information attached to the objects through the glasses, which provide great convenience for training. To mare training more efficient and achieve better training results. In the actual operation process, using image recognition technology, Hololens identify the corresponding positions of different components (such as main circuit breakers) on the actual objects(such as distribution box), combined with augmented reality technology, can provide some help information for technicians. Greatly reduced the difficulty of getting started in the installation and maintenance of the power equipment. This article will explain how to build an electric power work assist system based on image recognition technology, 3D modeling technology and augmented reality technology of Hololens.

# 2 Methods

See Figs. 1 and 2.

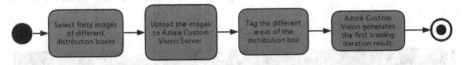

**Fig. 1.** The activity diagram of classification and training of images.

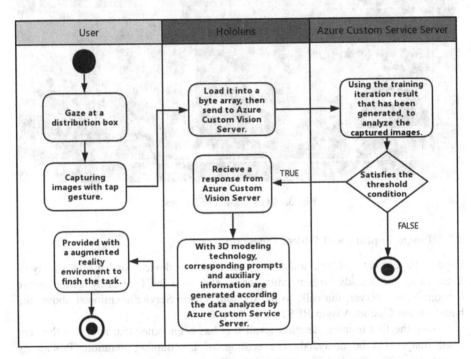

**Fig. 2.** Swim lane flowchart of image capture and object detection.

## 2.1 Classification and Training of Images

Build a custom image classifier using Microsoft's service, Azure Custom Vision. First create a new project, select forty images of the distribution box (no less than 15 images), after uploading successfully, then tag the different areas of the distribution box. To tag the images, use the mouse. As user hover over an image, a selection highlight will aid user to drawing a selection around the specific object automatically. If it is not so accurate, user can draw by himself. It is finished by holding left-click on the mouse, and then dragging the selection region to encompass the specific object. After that, a prompt will inform the user to add region tag. Then user can select the created tag (such as 'main breaker' and 'other breakers'). Repeat the same procedure mentioned above until all the images is tagged.

As shown, the main breaker and the other breakers in the distribution box are tagged. When all the images have been tagged in turn, based on the tagged images as training set, Azure Custom Vision will generate the first training iteration result (Fig. 3).

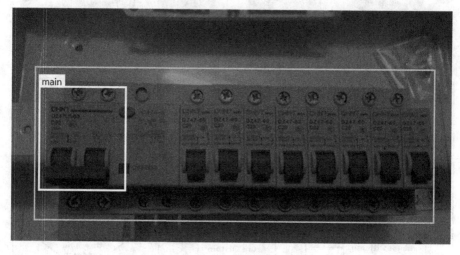

**Fig. 3.** Classification of images.

## 2.2  Image Capture and Object Detection

Capture the current real-time image with the Hololens device, load it into a array of bytes using the already written unity script, and send the byte array to the Azure Custom Vision server, the calls to the Custom Vision Service mentioned above, are based on the Custom Vision REST API.

Using the first training iteration result that has been generated, and then the captured image will be analyzed. The settings of the training iteration: Probability Threshold: 50%. Overlap Threshold: 30%. When the analysis result of the image satisfies the threshold condition, it will respond in the form of a JSON string, and data would be deserialized and passing the resulting prediction to a SceneOrganiser class, to determine the specific location of the main breaker and all other circuit breakers relative to the Hololens device (Fig. 4).

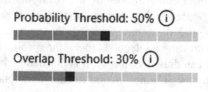

**Fig. 4.** Setting the threshold.

## 2.3   Image Capture and Object Detection

After determining the relative position of the main breaker and all other circuit breakers and Hololens device, the corresponding 3D model can be constructed based on the position information mentioned above and using Unity3D according to the specific application situation and added in the training exercise and actual operation. Corresponding guidance and auxiliary information to help trainee and electrical power industry practitioners to complete their learning and work tasks more efficiently.

First, the unity script would instantiate a corresponding virtual object (such as guidance and auxiliary information mentioned above) in the scene, (which at that moment is still invisible to the user). It places a quad at the same time (also invisible) where the image is placed and overlaps with the real surroundings. It is very important since the box coordinate retrieved from the Azure Custom Service server after analysis are traced back into this quad to confirm the object's location approximately in the real surroundings.

Finally, after receiving the analysis data from Azure Custom Service, the unity script in Hololens would set the virtual object with the tag of the prediction which has the highest confidence. Then it would call the specific method to calculate the bounding box on the quad object, positioned previously, and the virtual objects would be placed in the surroundings. To have a better effect to mix the virtual reality and real world, the unity script would even adjust the label depth, using a Raycast toward the virtual object's bounding box, which would collide the object in the real surroundings.

## 3   Methods

Using the aforementioned related methods, we finally built a mixed reality application that leverage Azure Custom Vision and the corresponding object detection API, which recognizes objects from the image and the provides an approximate location for the objects in real world.

The user performs the following operations: the user gaze at a distribution box (as shown in Fig. 5); then the user captures the image of the scene in front of the eye through a Tap gesture (Fig. 6); the application will send the image to Azure Custom Vision. The server analyzes the image. If the analysis meets the set threshold condition, that is the main breaker or other breakers is detected, Hololens will then receive a response from Azure Custom Vision, which will through the space tracking function of Hololens, corresponding prompts and auxiliary information are generated at the detected position of the circuit breaker (Fig. 7). Novice technicians can follow the virtual guidance information which is mixed with the real world to finish the task more efficiently (then talk something about error.)

The test mentioned above was performed on a actual distribution box to evaluate the effect of mixing true world with AR and confirm the accuracy of the whole system. The results show a mean Euclidean distance of 2.5 cm with a maximum error of 4 cm to detect the specific location of the distribution. The presented system created a more intuitive guidance system. With the development of AR technology, there will be more research to optimize the effect of visualization and a greater reduction in the error. Lots of electric power industry operation would benefit from the convenience of AR guidance.

**Fig. 5.** Gaze at a distribution box.

**Fig. 6.** Capturing images with Tap gesture.

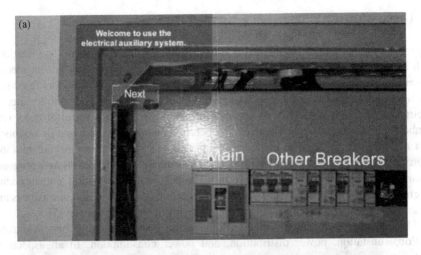

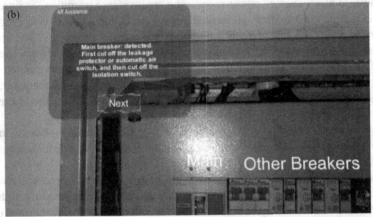

**Fig. 7.** (a–c) AR guidance.

# 4 Discussion

After the introduction of augmented reality technology, there are many applications, such as medical, manufacturing, maintenance, military training [7–9]. In the electric power industry, with the expansion of the power business and commissioning of new intelligent equipment, the manufacturing, installation, operation and maintenance of electrical equipment is faced with shortage of skilled technicians to fulfill tasks with cumbersome operation steps. The existing operation instructions are based on textbook. It is not intuitive to provide guidance to operators, nor can it provide the required information in real time in combination with the operating environment and progress. Electricity worker are inefficient in performing tasks, finding information, understanding information and communicating with each other due to different scope of business and capabilities. Even serious injuries or deaths may result from negligence [5].

The electric power industry involves many links such as power generation, transmission, substation, power distribution, and power consumption. In all aspects of electric power industry such as power monitoring, maintenance operation, engineering design, and simulation training, augmented reality technology has a wide range of applications. In the process of overhauling operations, due to many risks, in the maintenance of power systems, prevention of personal accidents is always the top priority of all work. How to familiarize with the complex maintenance environment before the start of the inspection work, to drill the specific maintenance operation, and identify the specific risks in the maintenance, is an urgent problem for the electricity workers.

Augmented reality technology combined with machine-learning-based image recognition technology is a training method. Using the Hololens, novice technician can have a better training effect and can get assistance in practical work.

The auxiliary images and prompt information generated by real-time scanning in Hololens can be well integrated with the actual object images transmitted through the glasses. The images obtained by Hololens enable the technician to obtain various parts of the power system during operation. Kinds of information, and intelligently identify the information in the environment, thus greatly assisting the technicians, can achieve the role of prompts, thereby helping inexperience technicians to distinguish each part that needs to operate and guide the operation.

The specific operation method is that the technician uses the Hololens device to activate the recognition function, and call the corresponding API to perform the recognition operation in the background, then Hololens will give a prompt information and fall on the recognized position. Through this basic method, the technicians can obtain sufficient prompt information without too much actual working experience, thereby facilitating the maintenance process.

Based on the research above, we can add more equipment information including the appearance information of the equipment from the outside, whether there is deformation caused by external force, cylinder oil leakage, and internal information, including temperature, pressure, etc., all of the information mentioned above is expressed in a visualized way.

**Fig. 8.** Equipment maintenance simulation [6].

In the equipment maintenance work, the maintenance technician face a large number of types of equipment, and need to carry out a large number of data query preparation work in advance. Using the registration tracking technology, the maintenance operation instructions of the 3D model are attached to the device, and the maintenance technicians can directly view and follow the instructions on the actual equipment, and quickly complete the maintenance work according to the steps. In the process of overhaul, if the basic guidance content cannot solve the actual problems on site, you need to consult the corresponding experts. The most flexible way for presenting remote imagery is via virtual reality [11]. An example of this is recent work by Lindlbauer and Wilson, who explored the concept of remixed reality, whereas user's viewpoint can be moved to an arbitrary location, while the scene's objects can be copied, moved, and removed on demand [12]. You can use the remote assistance method of augmented reality to share the work site perspective with the remote experts. The virtual guidance information such as "Error spotting" can be attached to actual equipment and presented to on-site maintenance technician to avoid error-prone language descriptions and improve communication efficiency. Equipment maintenance simulation is shown in Fig. 8 [6].

The appliance mentioned above could be also used in another specific condition, when the technician's direct view of an object is the object may be observed from some other perspective, using the augmented reality technology [11]. For example, endoscopic inspection cameras are commonly used in construction to provide a view inside of walls or confined spaces. Similarly, finger [13–15] and body-mounted [13, 16] cameras have been used to provide remote perspectives. We envision camera and sensor technology to shrink further, allowing systems to collect real-time visual data from anywhere the user wants to interact. With such data, occluded objects can be rendered into the user's visual field through augmented reality (AR) headsets.

Milgram and his fellows propose the concept of VR continuum to further define the augmented reality technology, that is, there are two kinds of technologies: augmented reality and enhanced virtual between the real environment and the virtual environment to form a whole area [10]. Integrating Hololens augmented reality technology and machine-learning-based image recognition technology into electric power simulation training is a trend of technology development, which will become a milestone in the intelligent development of the electric power industry.

**Acknowledgement.** This study has been partially supported by Program of Science and Technology Development Plan of Jilin Province of China (20190302032GX) and the Fundamental Research Funds for the Central Universities, JLU.

# References

1. Lin, D.F.: Some suggestions on talents development and training in electric power industry. Enterp. Reform Manag. (23), 84–85 (2018)
2. Hao, T., Liu, X., Li, J., Xiong, S., Feng, W., Li, X.: Design of three dimensional training system for electric power enterprises based on virtual reality technology. Autom. Instrum. (10), 91–93 (2018)
3. Chang, R.-C., Yu, Z.-S.: Using augmented reality technologies to enhance students' engagement and achievement in science laboratories. IJDET **16**(4), 54–72 (2018). https://doi.org/10.4018/ijdet.2018100104. Web. 9 Mar. 2019
4. Shi, L., et al.: Preliminary use of HoloLens glasses in surgery of liver cancer. J. Cent. South Univ. (Med. Sci.) https://doi.org/10.11817/j.issn.1672-7347.2018.05.007
5. Chen, H., Yan, L., Chen, X., et al.: A survey of virtual reality technology and its applications in electric power industry. Electr. Power Inf. Commun. Technol. **15**(5), 16–21 (2017)
6. Hadar, E., Shtok, J., Cohen, B., et al.: Hybrid remote expert-an emerging pattern of industrial remote support (2017)
7. Caudell, T.P., Mizell, D.W.: Augmented reality: an application of heads-up display technology to manual manufacturing processes
8. Azuma, R.T.: A survey of augmented reality. Presence: Teleoperators Virtual Environ. **6**(4), 355–385 (1997)
9. Davis, M.C., Can, D.D., Pindrik, J., et al.: Virtual interactive presence in global surgical education: international collaboration through augmented reality. World Neurosurg. **86**, 103–111 (2016)
10. Milgram, P., Takemura, H., Utsumi, A., Kishino, F.: Augmented reality: a class of displays on the reality-virtuality continuum. In: Proceedings of Telemainpulator and Telepresence Technologies, vol. 2351, no. 34, pp. 282–292 (1994)
11. Lilija, K., Pohl, H., Boring, S., Hornbæk, K.: Augmented reality views for occluded interaction. In: Proceedings of the 2019 CHI Conference on Human Factors in Computing Systems - CHI 2019, Paper no. 446, pp. 1–12. Association for Computing Machinery (2019, Accepted/in press). https://doi.org/10.1145/3290605.3300676
12. Lindlbauer, D., Wilson, A.D.: Remixed reality: manipulating space and time in augmented reality. In Proceedings of the 2018 CHI Conference on Human Factors in Computing Systems-CHI 2018, pp. 129:1–129:13. ACM Press, New York (2018). https://doi.org/10.1145/3173574.3173703

13. Kim, D., et al.: Digits: freehand 3D interactions anywhere using a wrist-worn gloveless sensor. In: Proceedings of the 25th Annual ACM Symposium on User Interface Software and Technology (UIST 2012), pp. 167–176. ACM, New York (2012). https://doi.org/10. 1145/2380116.2380139

14. Stearns, L., DeSouza, V., Yin, J., Findlater, L., Froehlich, J.E.: Augmented reality magnification for low vision users with the Microsoft Hololens and a Finger-worn camera. In: Proceedings of the 19th International ACM SIGACCESS Conference on Computers and Accessibility - ASSETS 2017, pp. 361–362. ACM Press, New York (2017). https://doi.org/ 10.1145/3132525.3134812

15. Yang, X.-D., Grossman, T., Wigdor, D., Fitzmaurice, G.: Magic finger: always-available input through finger instrumentation. In: Proceedings of the 25th Annual ACM Symposium on User Interface Software and Technology (UIST 2012), pp. 147–156. ACM, New York (2012). https://doi.org/10.1145/2380116.2380137

16. Kurata, T., Sakata, N., Kourogi, M., Kuzuoka, H., Billinghurst, M.: Remote collaboration using a shoulder-worn active camera/laser. In: 2004 Eighth International Symposium on Wearable Computers, ISWC 2004, vol. 1, pp. 62–69. IEEE (2004)

# Research on the Motion Track of Ocean Towing System

Jie Yuan[1], Feng Sun[1], Guofeng Tang[1], Zubin Chen[1],
Lingling Zheng[2(✉)], and Xinran Yang[1]

[1] College of Instrumentation and Electrical Engineering, Jilin University,
Changchun 130062, China
[2] The State Key Laboratory of Automotive Simulation and Control, Jilin
University, Changchun 130061, China
zhenglingling@jlu.edu.cn

**Abstract.** In this paper, a method for predicting the position of towline is presented. The location of the finite node is determined by installing fixed spaced attitude sensors in towline arrays, then the appropriate objective functions are selected for water depth profile and course profile respectively, and the interpolation fitting method is combined with the determined predicted positions. Through the hydrodynamic analysis of the existing towing cable's underwater motion, the position of the towing cable under the steady state motion is obtained as the reference basis, and two methods are put forward, which are improved spline interpolation method, polynomial fitting method and multivariate nonlinear regression analysis. In the case of steady state motion, the two methods are compared and compared with the hydrodynamic simulation results. Finally, a more suitable method is selected as the basis of cable location inversion and applied to deep-sea towing operations.

**Keywords:** Towing position · Stable state of motion · Attitude data ·
Curve-fitting

## 1 Introduction

As the largest resource in the world, the ocean is rich in mineral resources and oil and gas resources. In the exploration activities of various marine resources, the ocean towing system, as an effective underwater exploration platform, has been widely used in many fields such as marine research, monitoring and acoustic confrontation, and is playing an increasingly important role. In order to improve the detection effect and accurately determine the spatial position information of the towing cable, the modern towing system has widely adopted the active control system to adjust the position [1] and attitude of the towing cable body in real time, so as to keep its working state stable. Otherwise, the positioning accuracy and detection effect of the system will be greatly affected.

© Springer Nature Singapore Pte Ltd. 2019
R. Mao et al. (Eds.): ICPCSEE 2019, CCIS 1059, pp. 590–609, 2019.
https://doi.org/10.1007/978-981-15-0121-0_46

In the ocean, the marine environment always interferes with the normal operation of the underwater towing system. The underwater part of the towing cable, the towing body or the towing equipment are also inevitably affected by the current, the disturbance of seawater and the internal wave, etc. Moreover, there are complex interactions among surface towing vessels, towing cables, towing bodies and towing equipment. At this time, it is not only difficult for surface tug to ensure stable and accurate speed and attitude, but also undesired motions such as heave, pitch and yaw, which cause underwater towing cable or towing body to deviate from the expected position, which will seriously affect the detection effect of the system. In order to improve the detection effect, it is necessary to accurately determine the spatial position information of the towing cable, obtain the position and attitude of the towing cable and the towing body in real time, and keep its working state stable. Otherwise, the positioning accuracy and detection effect of the system will be greatly affected.

For the system composed of underwater towing cable and towing body, the steady and dynamic response characteristics of the towing system should be understood to better predict the position and track of towing cable and towing body, so as to ensure the reliability and safety of towing system equipment, which not only has theoretical rationality and innovation, but also has important engineering application value [2, 3].

Based on existing methods and actual conditions, this paper, on the one hand, through dynamics and kinematics research on underwater towing cable, constructs the steady state model of towing cable under no external force, and obtains the track positions of towing body and towing cable under ideal circumstances. On the other hand, in practical deep-sea testing, many parameters and variables cannot be estimated and confirmed, so it is difficult to analyze them from the perspective of dynamics and kinematics through forward modeling, and the accuracy is difficult to guarantee. We install attitude sensors in the towing body and cable, obtain the sensor Angle of fixed node position, and obtain the position of a limited number of nodes with a certain mathematical method. Through mathematical interpolation and fitting, invert the position and trajectory changes of the towing cable under steady and dynamic conditions.

## 2  Related Work

Underwater towing cable motion is characterized by time-varying hydrodynamics and strong nonlinear motion. Although the continuous method has high modeling accuracy and accurate description of towing cable motion in water, its analytical solution is difficult. The discrete model is simple, intuitive and easy to calculate. In the process of modeling the towing cable motion, the discrete model can be obtained from the continuous model [4]. It can also be implemented by discrete first and then modeling. The main idea of constructing and calculating the motion model of the towing cable is to discretization of the towing cable in space and time and to transform the abstract [5] and complex partial differential equations into ordinary differential equations which are easy to calculate based on the dynamic condition and continuity condition [6–13].

Zoysa established the three-dimensional steady-state equilibrium equation of underwater towing cable, gave various fluid resistance calculation models according to different conditions, and numerically solved them by using shooting method, studied the motion of tug ship-submarine trailer system under the action of uniform ocean current [14]. Mr Friswell studied the steady-state motion characteristics of circular and streamlined towing cables under water by means of target shooting [15]. Leech and Tabarrok have studied the steady-state space shape and tension distribution of two-dimensional towing cables with an analytical method. However, in order to obtain the analytical form of steady-state governing equation, the tangential resistance of towing cables has been neglected. Based on the results of previous studies, Chai etc. considered the effects of bending moment, bottom contact on towing cable and torque. The motion characteristics of anchor chain, riser and towing cable are studied by numerical simulation.

Through numerical calculation, luyi zhang obtained the influences of various towing elements (speed, cable length, cable weight and drag coefficient, etc.) on the towing motion, providing an important basis for selecting appropriate towing cable in the design of towing system. Based on the finite difference method, the motion of underwater aperture sonar towing system is studied and analyzed by li yinghui. Cheng nan used three-dimensional concentrated mass method to establish a deep-sea mooring model and calculate the configuration and tension of the mooring cable. The mooring cable is considered to be composed of many concentrated mass points and massless elastic elements, and it is assumed that all forces acting on the towing cable are acting on the concentrated mass points.

All the above methods are based on the underwater kinematics and dynamics of the towing cable, and the underwater position of the towing cable is obtained through the physical and environmental parameters of the towing cable. In addition to the above methods, Rowe and Gettrust 1993 used readings from several depth sensors set along the hydrophone array to estimate the approximate inclination of the tow cable. Ferguson BG proposed a constrained optimal (adaptive) beamforming device based on the observation (signal plus noise) cross-spectrum matrix inversion of the hydrophone output, which can infer the spatial distribution of the outlet hydrophone along the array, but requires at least one sound source in the far field [16]. E.C. van Ballegooijen proposed an inversion of the arrival time of sound waves based on the sound source deployed by two frigates and towing ships, without making any assumptions about the shape of the array. Another technique USES interpolation scheme to obtain the remaining array positions, and USES quartic spline to approximate the spatial curve based on interpolation technique [17–19].

In the practical application and test of the project, based on existing methods and actual conditions, we analyzed and constructed the steady state motion model of the towing cable, with this as a reference and comparison, inversion is carried out according to the angular attitude data of fixed node position, Finally we can get the track curve of the whole cable algorithm design.

# 3  Algorithm Framework

By analyzing and studying existing methods and combining with the actual situation and environment of this project, we designed the following scheme for the attitude and position research of the towing system (Fig. 1):

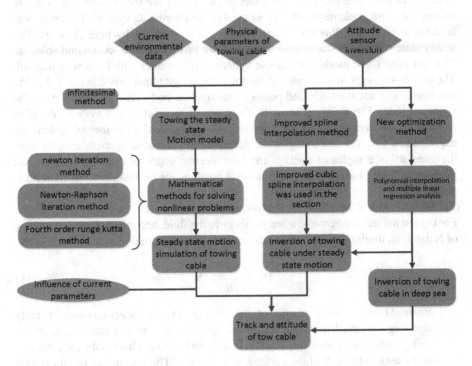

**Fig. 1.** Research scheme of track and position of towing system

## 3.1  Analysis of Towing Cable Trajectory Based on Kinematics and Dynamics

(1)  Selection and transformation of coordinate system
In order to facilitate motion analysis, this chapter establishes three Cartesian coordinate systems, namely: inertial (system) coordinate system, towing cable local coordinate system and trailer/trailer motion coordinate system. The inertial coordinate system $(x\,y\,z)$ is a spatially fixed Cartesian coordinate whose origin can be set on the water surface, $z = 0$. The mathematical model is based on this coordinate system, and

all calculations are converted to the current coordinate system for calculation. In order to distinguish other coordinate systems, the coordinate system is hereinafter referred to as a system coordinate system. (i, j, k) is the unit coordinate vector of the coordinate system in three directions, where the vector k is vertically upward. The local coordinates (b, t, n) of the towing cable are attached to the towing cable, where the axis t represents the tangential direction of the towing cable and the direction is the growth direction of the cable length s. The Euler angles $(\varphi, \theta)$ are the attitude angles of the towing cable microelements relative to the system coordinate system. The three unit coordinate vectors of this coordinate system are represented by bold (b, t, n). The steady-state solution of the towing cable will use this coordinate for integral solution. The underwater tow motion coordinate system $(\xi \, \eta \, \varsigma)$ moves with the tow or tugboat. The origin is usually at the center of the tow. The longitudinal axis $\xi$ is parallel to the baseline of the traction body and points to the traction body. The axis $\eta$ points to the starboard side. The axis conforms to the rules of the right hand system. Pointing downward, the coordinates of the origin of its coordinates in the inertial system are expressed as $[-\xi \, \eta \, \varsigma]$. Referring to the definition of submarine maneuvering motion, the three attitude angles of traction are: bow steering angle $\psi$, right turn to positive; pitch angle $\vartheta$ lift bow (or trim by the stern) is positive; the heel angle $\varphi$, the right tilt is positive.

(2) Motion control equations of the towing cable micro-element

For the towing cable micro-ds, when it moves in the fluid, according to the second law of Newtonian motion, we can get the dynamic governing equation of the towing cable:

$$M\ddot{x} = \frac{\partial T}{\partial s} + \sum F \qquad (1)$$

Where M is the mass matrix of the towing cable micro-component, usually including the actual mass and additional mass; $\ddot{x}$ is the motion acceleration of the towing cable micro-element; T is the drag chain tension, which is the only internal force along the length of the cable; F is a function. The vector sum of all external forces on the towing cable micro-element typically includes the gravity, buoyancy, drag, and external pull of the towing cable micro-element.

(3) Research on steady-state motion of towing cable

Under the influence of external disturbance, the underwater towing cable will appear as a stable spatial curve. The steady state solution is generally the basis for studying the motion of the towed system as an initial condition for dynamic motion simulation. When the traction system runs in a straight line, the effects of bending moment and towed cable torque are neglected, and the towing cable is in equilibrium under the action of gravity, buoyancy and fluid resistance. A mathematical model is established by using the micro-element method to control the steady motion of the towing cable (Fig. 2).

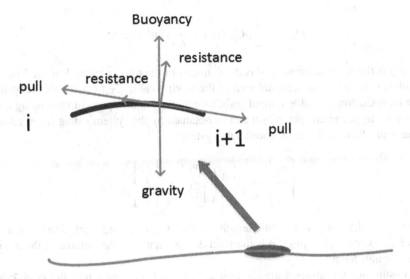

**Fig. 2.** The micro-element method constructs a mathematical model for the steady-state motion of the towing cable

Starting from the motion control equation of the towing cable microelement, we can all get the steady-state motion control equation of the towing cable:

$$\frac{\partial T}{\partial s} + w + D = 0 \tag{2}$$

where T is the towing cable tension, always pointing to the tangential direction of the towing cable, w is the wet weight (buoyancy and gravity) per unit length of the towing cable, and D is the fluid resistance it receives.

For the above control equation, the cable length T is differentiated from the cable length and deployed in the local coordinate system of the towing cable, thereby obtaining:

$$\frac{\partial T}{\partial s} = \frac{\partial (Tt)}{\partial s} = \frac{\partial T}{\partial s} t + T \cdot \frac{d\phi}{ds} n + T\cos\phi \cdot \frac{d\theta}{ds} b \tag{3}$$

The seawater resistance affected by the towing cable can be derived from the Morison formula: but for the actual test case, the towing cable is a smooth cylindrical towing cable, so we can simply confirm that the towing cable resistance to seawater is:

$$D_T = -\frac{1}{2}\rho\pi C_T d\sqrt{1 + \frac{T}{E\sigma}}|u_t|u_t \tag{4}$$

$$D_N = -\frac{1}{2}\rho C_N d\sqrt{1 + \frac{T}{E\sigma}}\sqrt{u_n^2 + u_b^2}u_n \tag{5}$$

$$D_B = -\frac{1}{2}\rho C_N d\sqrt{1+\frac{T}{E\sigma}}\sqrt{u_n^2+u_b^2}u_b \tag{6}$$

where $\rho$ is the seawater density, d is the diameter of the towing cable, E is the Young's modulus, $\sigma$ is the cross-sectional area of the towing cable $C_T$ towing cable tangential resistance, $C_N$ towing cable normal resistance $u_t$ $u_n$ $u_b$ is the speed of the towing cable relative to the ocean current, which can be obtained by the system's drag speed relative to the water flow to the local coordinate system:

$$\begin{bmatrix} u_b \\ u_t \\ u_n \end{bmatrix} = [A]^{-1}\left(\begin{bmatrix} v_x \\ v_y \\ v_z \end{bmatrix} - \begin{bmatrix} J_x \\ J_y \\ J_z \end{bmatrix}\right) \tag{7}$$

where w is the weight per unit length of the water towing cable, which can be expressed as w = $(m - \rho v)g$, $\rho$ is the fluid density, and v is the volume of the towing cable per unit length.

Substituting the above formula with gravity and buoyancy into the equilibrium equation and expanding it along the direction of each coordinate axis in the local coordinate system of the towing cable, the equilibrium equation can be written as a scalar form:

$$dT/ds - w\sin\phi - \frac{1}{2}\rho\pi C_T d\sqrt{1+\frac{T}{E\sigma}}ds \tag{8}$$

$$T \cdot d\phi/ds - w\cos\phi - \frac{1}{2}\rho C_N d\sqrt{1+\frac{T}{E\sigma}}\sqrt{u_n^2+u_b^2}u_n ds = 0 \tag{9}$$

$$T \cdot \cos\phi \cdot d\theta/ds - \frac{1}{2}\rho C_N d\sqrt{1+\frac{T}{E\sigma}}\sqrt{u_n^2+u_b^2}u_b ds = 0 \tag{10}$$

The above differential equation constitutes the steady-state motion control equation of the towing cable. As long as we give the state value of the towing cable at the end point or any position, the fourth-order Runge-Kutta method can be used as the initial value problem for integral solution.

## 3.2    Towing Cable Track Inversion Based on Sensor Attitude Data

In the deep sea test, although the dynamic trajectory of the streamer can also be obtained from the hydrodynamic model of its motion, the process requires more detailed physical property parameters, and many of the parameters and variables are not predictable and validated [20]; Moreover, the towline is affected by the natural environment such as the speed of the ship, the operation of the towed rope, and the current flow. Therefore, it is difficult to ensure the accuracy from the perspective of dynamics and kinematics by the forward method.

In some oil exploration industries, its advanced array system uses GPS and high frequency acoustic systems to accurately position the streamer array. However, this technology has gone beyond the application of many marine seismic studies [21]. A more general case is to increase the depth and heading sensor distributed along the array spacing in the tow array [22]. However, the general depth and heading sensors are costly and used in sonar detection or marine military applications. In the field of marine earthquakes, it is relatively simple. Marine seismic exploration is usually designed for linear surveying of ships in the area of interest, and most of them are reduced to two-dimensional space, that is only study the position of the water depth profile.

Therefore, based on the above situations, we adopted different methods to add attitude sensors distributed along the length of the cable. Data collected from attitude sensor nodes on towing cables are used for inversion. On the basis of ensuring the accuracy of attitude Angle, the attitude position of the towing cable can be accurately estimated, and the three-dimensional space position of the underwater towing cable can be obtained. Meanwhile, its trajectory can be quickly obtained based on the real-time uploaded attitude data.

For collecting and uploading attitude data of sensor, We use EKF algorithm for processing. On the basis of ensuring the accuracy of attitude Angle, an appropriate mathematical model is built by analyzing and combining the motion characteristics of the towing cable under water.

(1)  Sensor selection and installation

In the method test, we selected MTi series sensors designed and manufactured by Xsens company, which are based on micro MEMS inertial sensing technology and widely used for motion, direction and position measurement. Its low cost, high precision, these characteristics make it can be used for small vehicles, aircraft control, stability and navigation, navigation and other fields, and in a leading position, showing good performance.

In addition, we chose to use the MPU9250 attitude sensor, which is the first digital sensor in the world to integrate the 9-axis motion attitude detection, eliminating the problem of inter-axis difference of multi-sensor combination, reducing the volume of the sensor and reducing the power consumption of the system. Its internal integration of three axis accelerometer, three axis magnetometer, three axis gyroscope, transmission rate up to 400 kHz/s, angular velocity measurement range up to $\pm 2000(°)$/s, with good dynamic response characteristics.

In this test project, attitude sensors, including MTI sensors and MPU sensors, are installed in the fixed position of the towing cable. The towing cable is a smooth cylindrical towing cable about 210 m long, whose starting end is a towing body. We installed 4 MTI attitude sensors and 12 MPU attitude sensors respectively on the towing cable, and also installed 1 attitude sensor in the towing body. Four MTI attitude sensors on the towing cable are spaced 50 m apart, and 12 MPU attitude sensors are spaced 12.5 m apart (Figs. 3 and 4).

**Fig. 3.** Tow cable physical map

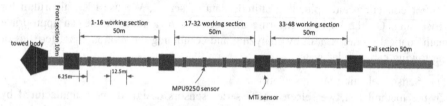

**Fig. 4.** Schematic diagram of the installation of the towed body and streamer attitude sensor

(2)  Sensor data type and operation

Common methods to describe attitude include Euler Angle method, quaternion method, direction cosine method and equivalent rotation vector method. Four MTI attitude sensors and 12 MPU attitude sensors are respectively installed on the towing cable. The data uploaded by each sensor is 9-axis attitude data, including three-axis accelerometer, three-axis magnetometer and three-axis gyroscope data. Therefore, the attitude Angle information calculated by gyroscope with better dynamic performance, accelerometer with better static performance and magnetometer is fused, and the fusion algorithm adopts Kalman filter algorithm. The attitude Angle information solved by gyroscope is used as the system prediction information of Kalman filter algorithm and the attitude Angle information solved by accelerometer and magnetometer is used as the measurement information of Kalman filter algorithm.

Through EKF operation of attitude data, the influence of gyroscope random drift on attitude estimation is reduced, and the attitude Angle accuracy of each towing cable node position is effectively improved, which serves as the basis for further improvement of towing cable track inversion (Fig. 5).

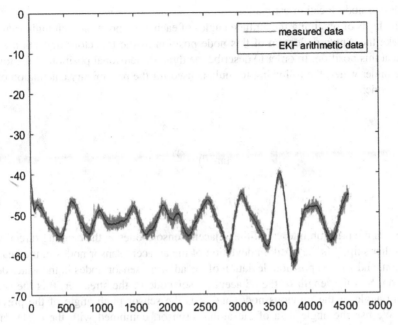

**Fig. 5.** Operation results of attitude sensor EKF

The horizontal axis is the data collection sequence and the vertical axis is the single-axis Angle data of the attitude sensor. It can be found that the noise of attitude sensor is eliminated and the accuracy of Angle data is improved.

(3) Attitude data inversion method
(1) Coordinate system determination
A three-dimensional space coordinate system is established for underwater towing bodies and cables. The towing body is taken as the coordinate origin, the cable length direction is the positive direction of x axis, the water depth profile direction pointing to sea surface is the positive direction of z axis, and the Y-axis direction is determined by the right hand rule of space coordinate system. Based on this coordinate system, the position changes of each cable node are studied, and the track changes of the cable are described, so that the attitude tracks of the cable in the water depth profile and the course profile can be observed intuitively and obviously. We could pre-estimate the limited node positions.

(2) Finite node location estimation

On the basis of obtaining the attitude angles of each node position with high accuracy, we take the Angle deflection of this node position as the trajectory deflection of the cable at this position. In order to describe the three-dimensional position of the towing cable under water, the following formula is used for the preliminary calculation of the finite node:

$$x^2 + y^2 + z^2 = s^2 \tag{11}$$

$$\tan(\theta) = \frac{y}{x} \tag{12}$$

$$\tan(\varphi) = \frac{z}{x} \tag{13}$$

where x is the position interval of the adjacent sensor nodes in the heading direction of the mother ship, y is the position deviation of the adjacent sensor nodes in the heading direction, and z is the position deviation of the adjacent sensor nodes in the water depth direction. S is the length of the adjacent sensor node of the streamer, $\theta$ is the sensor heading angle of the previous node, and $\varphi$ is the sensor pitch angle of the previous node. We use the angle data of the attitude sensor, combined with the cable length interval of each attitude sensor node position, and calculate the position coordinates of the fixed stream node (px, py, pz) at the default initial tow position coordinate (0, 0, 0) the simplest shape of the streamer track is formed.

(3) Position interpolation of towing cable

For the finite nodes obtained, there are many methods for interpolation fitting. Here we propose two methods: improved spline interpolation, polynomial fitting and multivariate nonlinear regression analysis.

Here, two methods are used for analysis and comparison. One is an improved fitting method based on cubic spline interpolation, in order to better describe the three-dimensional spatial coordinates of towing cable nodes. In this method, because the Angle accuracy of MTi sensor is relatively high, we only use its Angle data to estimate the position of finite nodes. The towing body is the origin of coordinates, and the coordinates of the four MTi nodes on the towing cable can be estimated in advance. Based on the basic principle of cubic spline interpolation, the goal is to derive a third order polynomial on the interval between every two adjacent data points.

$$f_i(x) = a_i x^3 + b_i x^2 + c_i x + d \tag{14}$$

Therefore, for n + 1 data points (i = 0, 1, 2..., n), there are n intervals, so 4n unknown constants need to be calculated, so 4n conditions are needed to calculate these unknowns, and the required conditions are not explained here; In order to better use of the Angle of the node position to describe the trajectory of towing is located in the node, the target function structure between adjacent data points to much higher levels of four polynomial, which introduce a new parameter variables, the whole curve to calculate 4 (n + 1), the Angle of the paragraphs added data point data, also known conditions into 4 (n + 1), as a result, the input of the towline limited node location coordinates at the same time, also the attitude of the corresponding node Angle input data, thus realized the path of towing two-dimensional curve fitting.

The improved spline interpolation method is applied to the depth profile and the course profile respectively to obtain the three-dimensional spatial coordinate position of the towing cable (Fig. 6).

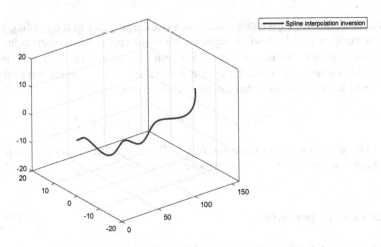

**Fig. 6.** Result graph based on improved spline interpolation

By the method of the above, the steady state motion of towing path analysis, and compared with towing steady-state kinetics analysis, the problems are as follows: 1, towing trajectory are greatly influenced by the known position node attitude Angle, to the point of view of real-time acquisition data, small Angle changes result in a marked change trajectory, poor robustness. 2, for underwater towing under the condition of the steady state motion, through the steady state dynamics, kinematics analysis, for towing length of 210 m, the whole of the towing path curve is relatively flat, there will not be too extreme value point, and adopt the method of spline interpolation, due to its essence is to the whole curve segmentation, between the adjacent nodes is higher number of polynomial, so its entire towing path more twists and turns, and this does not conform to the actual situation.

In view of the problems arising from the above algorithms, we propose a new solution.

1. First, make full use of all the attitude sensors installed in the streamer. The more the angle and attitude data of the nodes we know under the finite length, the more between the adjacent nodes, the smaller the distance difference, the smaller the error and the higher the accuracy of fitting when we perform curve fitting or interpolation.

2. Construct the model in the space coordinate system described above, and calculate the node positions of the four MTi and twelve MPU attitude sensors on the tow by the preprocessing of the finite nodes.

3. In order to realize the three-dimensional space fitting of the tow, we first fit in one direction. Through the analysis of the above tow kinematics and dynamics, we know that the tow is in the steady-state motion is lower, and there is generally no case where an angular extreme point occurs. Therefore, we use a cubic polynomial as the objective function in a section and perform interpolation fitting by the least squares curve fitting principle.

$$y = f(x) = ax^3 + bx^2 + cx + d \tag{15}$$

Thus, we obtain the trajectory position (x, y) of the streamer in the heading profile.
4. For the position z of the towline depth profile in the space state, construct the space quadratic equation as the objective function. Based on the above steps, use MATLAB's inline function to perform multivariate nonlinear regression analysis to determine the position of z that towline depth profile.

$$z = F(x, y) = ax^2 + by^2 + cxy + dx + ey + f \tag{16}$$

5. Through the analysis and calculation of the above steps, we finally get the position trajectory coordinates (x, y, z) of the streamer underwater.


# 4   Practical Experiment

In view of the above different methods and their characteristics, we use the simulation results of the towing cable under steady-state motion as a reference. By using the improved spline interpolation method and polynomial fitting and multivariate nonlinear regression analysis, the dragging of this state Cable trajectory and position are inverted.

## 4.1   Test Environment Construction

As shown in the figure below, in order to study the trajectory and position of the towing cable under steady-state motion, we drag the towing cable on the surface to maintain the stable motion state of the towing cable, and analyze the dynamics of the towing cable motion to obtain its trajectory. The locations are reference for subsequent inversion methods (Fig. 7).

## 4.2   Experiment Result

In view of the above several methods, we carried out simulation tests based on the test environment of the towing cable in steady state and the acquired attitude sensor. The first is the comparison between the steady-state simulation of the towing cable at a single moment and the improved spline interpolation method and the polynomial fitting method of the towing cable; Under steady-state conditions, the simulation results of the improved spline interpolation method and polynomial fitting and multivariate nonlinear regression analysis of the towing cable over a period of time (Figs. 8 and 9).

**Fig. 7.** Towing cable steady state test

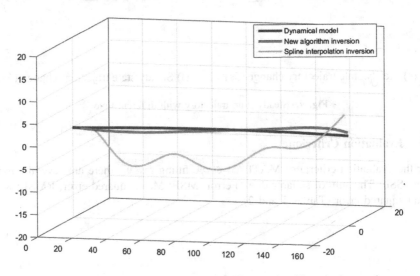

**Fig. 8.** Single-time Comparison of effects of different algorithms under steady state

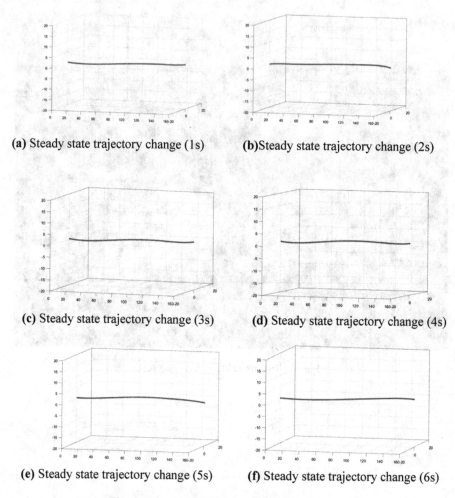

(a) Steady state trajectory change (1s)    (b)Steady state trajectory change (2s)

(c) Steady state trajectory change (3s)    (d) Steady state trajectory change (4s)

(e) Steady state trajectory change (5s)    (f) Steady state trajectory change (6s)

**Fig. 9.** Steady state trajectory with different time

## 4.3   Evaluation Criterion

For the evaluation criteria of MATLAB curve fitting results, there are several parameters: SSE: The sum of squares due to error; MSE: Mean squared error; RMSE: Root mean squared error (Tables 1 and 2).

**Table 1.** Evaluation of each algorithm under steady state

| Arithmetic | SSE | | MSE | | RMSE | |
|---|---|---|---|---|---|---|
| | xy profile | xz profile | xy profile | xz profile | xy profile | xz profile |
| Steady state dynamics | 0 | 0 | 0 | 0 | 0 | 0 |
| Improved spline interpolation | 5915.6 | 46726.5 | 3.7 | 29.4 | 2.0 | 5.4 |
| Polynomial fitting and multivariate nonlinear regression analysis | 201.0 | 890.0 | 0.1 | 0.6 | 0.2 | 0.7 |

**Table 2.** Evaluation of the inversion effect of the new algorithm at different time in steady state

| Time | SSE | | MSE | | RMSE | |
|---|---|---|---|---|---|---|
| | xy profile | xz profile | xy profile | xz profile | xy profile | xz profile |
| 1S | 516.67 | 378.63 | 0.3207 | 0.2350 | 0.5663 | 0.4848 |
| 2S | 486.82 | 293.13 | 0.3022 | 0.1820 | 0.5497 | 0.4266 |
| 3S | 689.21 | 590.21 | 0.4278 | 0.3664 | 0.6541 | 0.6053 |
| 4S | 1459.71 | 981.44 | 0.9061 | 0.6092 | 0.9519 | 0.7805 |
| 5S | 3147.32 | 1011.21 | 1.9537 | 0.6277 | 1.3977 | 0.7923 |
| 6S | 565.39 | 847.49 | 0.3510 | 0.5261 | 0.5924 | 0.7253 |

① SSE

The statistical parameter calculates the sum of the squares of the differences between the fitting data and the corresponding points of the original data, and the calculation formula is as follows:

$$SSE = \sum_{I=1}^{N} w_i(y_i - \hat{y}_i)^2 \tag{17}$$

The closer the SSE is to 0, the better the model selection and fit, and the more successful the data prediction. The next MSE and RMSE are the same as SSE, so the effect is the same.

② MSE

The statistical parameter is the mean of the sum of the squares of the differences between the predicted data and the corresponding points of the original data, that is, SSE/n, and SSE is not much different, and the calculation formula is as follows:

$$MSE = \frac{SSE}{N} = \frac{1}{n} \sum_{I=1}^{N} w_i (y_i - \hat{y}_i)^2 \tag{18}$$

③ RMSE

The statistical parameter, also called the standard deviation of the regression system, is the square root of the MSE, even if the formula is as follows

$$RMSE = \sqrt{MSE} = \sqrt{SSE/n} = \sqrt{\frac{1}{n} \sum_{I=1}^{N} w_i (y_i - \hat{y}_i)^2} \tag{19}$$

## 4.4    Interpretation of Result

After the analysis of the above several methods, it is not difficult to find that the streamer should be a nearly straight position trajectory. Although the results based on kinematics and dynamics are more accurate, due to the large number of variables such as the physical parameters of the streamers and the surrounding environmental parameters, the roots are complex in actual ocean testing and work, and we can't get some parameters. Therefore, it has certain limitations and applicability; for the improved spline interpolation method, the angle value is small, and the attitude data changes weakly. It also has certain significance. However, when the angle attitude data has obvious changes, the inversion result shows obvious deviation or even error. It is difficult to achieve the desired result; for the new method selected last, based on the selection of as many angle pose data as possible, polynomial fitting and multivariate nonlinear regression analysis are now used to improve the accuracy of the inversion. The robustness of the simulation algorithm is more practical than the first two methods.

Based on the above analysis and research, we conducted a test study on the operation streamer with a water depth of 2000 m. The results are as follows (Fig. 10):

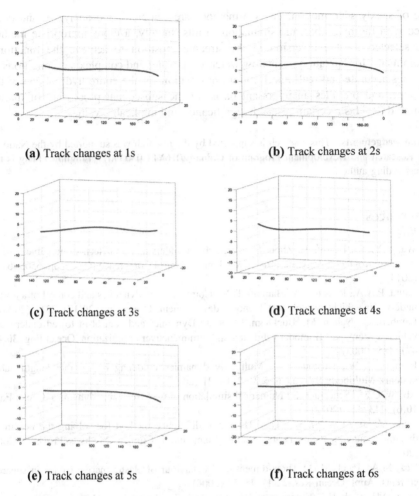

(a) Track changes at 1s                    (b) Track changes at 2s

(c) Track changes at 3s                    (d) Track changes at 4s

(e) Track changes at 5s                    (f) Track changes at 6s

**Fig. 10.** Track changes at depth of 2000 m with different time

## 5  Conclusion

This paper is based on the analysis of the existing towline hydrodynamics methods for the actual project testing needs. Abandon the existing kinetic analysis methods for the actual conditions of the project and the existing conditions. By installing a fixed-interval attitude sensor in the streamer array, the position of the finite node is pre-estimated, and the appropriate objective function is selected in the water depth profile and the heading profile, and the interpolation fitting method is combined with the determined estimated position. Taking the hydrodynamic analysis of the existing streamer underwater movement, the position under the steady motion of the streamer is obtained as a reference. Two methods are proposed, namely improved spline inter-polation method, polynomial fitting and multiple nonlinear regression analysis. In the case of steady state motion, the two methods are compared and analyzed, and compared with the hydrodynamic simulation results. Finally, a more appropriate method was selected for the inversion of the streamer position to achieve the functional requirements for operation in shallow water, deep water and complex sea conditions. It provides accurate and real-time location information for the entire hydrophone of the streamer, and provides an important basis for the subsequent processing of marine seismic data. Has a certain research significance and applicable value.

**Acknowledgements.** This research is supported by the research was supported by the National Key Research and Development Program of China (2016YFC0303901). Lingling Zheng is the corresponding author.

# References

1. Wang, Y., Li, L., Fu, T., Zhou, S.: Research on underwater vehicle floating cable towing speed control technology. In: 2014 33rd Chinese Control Conference (CCC), pp. 6505–6508 (2014)
2. Laura, P.A.A., Rossit, C.A., Bambill, D.V.: Comments on: dynamics and control of a towed underwater vehicle system, part 1: model development. Ocean Eng. **30**(17), 2311 (2003)
3. Lambert, C., Nahon, M., Buckham, B., et al.: Dynamics and control of towed underwater vehicle system, part II: model validation and turn maneuver optimization. Ocean Eng. **30**(4), 471–485 (2003)
4. Kamman, J.W., Huston, R.L.: Multibody dynamics modeling of variable length cable systems. Multibody Syst. Dyn. **5**(3), 211–221 (2001)
5. Ablow, C.M., Schechter, S.: Numerical simulation of undersea cable dynamics. Ocean Eng. **10**(6), 443–457 (2012)
6. Yuan, Z., Jin, L., Chi, W., Tian, H.: Finite difference method for solving the nonlinear dynamic equation of underwater towed system. Int. J. Comput. Methods **11**(4), 1350060 (2014)
7. Vaz, M.A., Patel, M.H.: Three-dimensional behaviour of elastic marine cables in sheared currents. Appl. Ocean Res. **22**(1), 45–53 (2000)
8. Jung, D.H., Park, H.I., Koterayama, W.: A numerical and experimental study on dynamics of a towed low-tension cable. Appl. Ocean Res. **25**(5), 289–299 (2003)
9. Bi, G., Zhu, S., Liu, J., et al.: Dynamic simulation and tension compensation research on subsea umbilical cable laying system. J. Mar. Sci. Appl. **12**(4), 452–458 (2013)
10. Wang, Y.B., Li, L., Fu, T.H., et al.: Research on underwater vehicle floating cable towing speed control technolog. In: Control Conference. IEEE (2014)
11. Sanders, J.V.: A three-dimensional dynamic analysis of a towed system. Ocean Eng. **9**(5), 483–499 (1982)
12. Pan, G., Yang, Z.D., Du, X.X.: Research on dynamic simulation of AUV launching a towed navigation buoyage. Appl. Mech. Mater. **433–435**, 5 (2013)
13. Kamman, J.W., Huston, R.L.: Multibody dynamics modeling of variable length cable systems. Multibody Syst. Dyn. **5**(3), 211–221 (2001)
14. Zoysa, A.P.K.D.: Steady-state analysis of undersea cables. Ocean Eng. **5**(3), 209–223 (1978)
15. Friswell, M.I.: Steady-state analysis of underwater cables. J. Waterw. Port Coast. Ocean Eng. **121**, 98–104 (1995)
16. Bucker, H.P.: Beamforming a towed line array of unknown shape. J. Acoust. Soc. Am. **63**(5), 1451 (1978)

17. Jing, W., Tong, D., Wang, Y., Wang, J., Liu, Y., Zhao, P.: MaMR: high performance MapReduce programming model for material cloud application. Comput. Phys. Commun. **211**, 79–87 (2017)
18. Miao, Q., Jing, W., Song, H.: Differential privacy–based location privacy enhancing in edge computing. Concurr. Comput.: Pract. Exp. **31**(8) (2019)
19. Jing, W., Likun, H., Shu, L., Mukherjee, M., Hara, T.: RPR: recommendation for passengers by roads based on cloud computing and taxis traces data. Pers. Ubiquitous Comput. **20**(3), 337–347 (2016)
20. Dosso, S.E., Riedel, M.: Array element localization for towed marine seismic arrays. J. Acoust. Soc. Am. **110**(2), 955 (2001)
21. Milinazzo, F., Wilkie, M., Latchman, S.A.: An efficient algorithm for simulating the dynamics of towed cable systems. Ocean Eng. **14**(6), 513–526 (1987)
22. Kuo, E.Y.T., Casati, M.J.: Physical splines, application to towed array dynamics, Naval Underwater System Center, New London, CT (1990)
23. Riley, J.L., Gray, D.A., Ferguson, B.G.: Estimating the shape of a towed array of hydrophones using both acoustic and non-acoustic sensor techniques. In: Moura, J.M.F., Lourtie, I.M.G. (eds.) Acoustic Signal Processing for Ocean Exploration. ASIC, vol. 388, pp. 225–230. Springer, Dordrecht (1993). https://doi.org/10.1007/978-94-011-1604-6_22

# Using Case Facts to Predict Penalty with Deep Learning

Yu Li, Tieke He[✉], Ge Yan, Shu Zhang, and Hui Wang

State Key Laboratory for Novel Software Technology, Nanjing University,
Nanjing 210093, China
hetieke@gmail.com

**Abstract.** With the promotion of Wisdom Court construction and the increasing completeness of judicial big data, the combination of judicial and artificial intelligence attracted more and more attention. The Judicial document is the most common textual information in cases. Due to the development of text analysis and processing techniques, we can mine more information from the judicial text and apply it in judgment. In this paper, we use the popular deep learning text classification algorithms to predict the imprisonment based on the fact of cases, which is expected to assist judges and staffs of procuratorate on sentencing. The result of our experiments shows the feasibility and utility of our method.

**Keywords:** Smart court · Judicial data analysis · Text classification · fastText · TextCNN

## 1 Introduction

In recent years, various industries and fields have paid more and more attention to big data and artificial intelligence algorithms. On the basis of the construction of the judicial information system, the government has also begun to vigorously promote the construction of intelligent court, intelligent procuratorial and intelligent judicial administration. At first, most of our tasks focused on using artificial intelligence to assist in the processing of simple, mechanical and repetitive tasks. Such as applying speech recognition techniques in converting the oral language to text automatically [10], which can assist staff in generating the court record. Alternatively, use text recognition and other techniques to convert the litigant evidence materials of the parties into electronic documents to facilitate later work such as filing tables, retrieval, and statistics.

With the emergence of more analysis and data mining methods, our focus has gradually shifted to using AI to assist judicial trials, such as social hazard assessment, analysis of case situation, the characteristics of parties analysis and statistical analysis of judgments. However, most of them stay at the stage of analysis instead of further using the information.

© Springer Nature Singapore Pte Ltd. 2019
R. Mao et al. (Eds.): ICPCSEE 2019, CCIS 1059, pp. 610–617, 2019.
https://doi.org/10.1007/978-981-15-0121-0_47

We choose the fact as the only source of our method, using deep learning methods to predict the imprisonment result of the case. Facts are the fundamental and indispensable part of the case, which is recorded as text. Imprisonment is part of the final judgment from the court. It is divided into three categories: death penalty, life imprisonment and fixed-term imprisonment. For the fixed-term imprisonment, we divided it into different categories according to the distribution of the data and set it as the label of the fact. So far, we have converted the prediction of imprisonment based on facts into the text classification problem.

There are many methods to handle text classification problem. The traditional machine learning algorithm generally takes two steps. Firstly, convert the preprocessed text into a fixed-dimensional vector, which is also called feature engineering. Commonly we use bag-of-words, TF-IDF [2], word2vec [8], LDA, etc. Secondly, input the feature vector into the classifier. Frequently used machine learning classification algorithms include LR, SVM, GBDT, etc.

With the development of deep learning technology, the step of text representation has been integrated into the training period. It can extract features of input text automatically while retaining more information to make classifier more accurate. Commonly used deep learning algorithms include TextCNN, TextRNN, BiLSTM and so on.

However, most deep learning algorithms inevitably face the problem of time complexity, which also limits the use of deep learning methods on large-scale data. In order to ensure the classification accuracy while minimizing the time complexity, Facebook AI research proposed fastText in 2016 [3], which uses the idea of hierarchical softmax to construct the classifier. It dramatically improves the overall efficiency of the model, and also introduces the idea of n-grams to capture the order relationship between characters.

The deep learning classification method has been widely applied to various label prediction and sentiment analysis problems, but it has few been combined with judicial aspects. And most of the existing judicial research is based on many manually labeled data while the label can not cover all the information of the case. So we choose to take the simplest and most common facts in the case as the data set and use fastText and textCNN to generate a predictive model of the sentence. As for a new-come case fact, our model can provide a prediction of penalty as an assistant to the procurator.

## 2   Related Work

### 2.1   Machine Learning Text-Classification

The traditional text classification algorithm is mainly divided into two steps. The first step is text representation, which is to convert the text into the fixed-dimensional vector. The earliest model is bag-of-words, which only uses a collection of words to represent the text. The simplest one-hot model or using the number of tfidf to instead 1 [6] all belong to the bag of word model. However, it ignores the order and relevance between words which leads to a large loss of information during the representation process. The improved bag of n-grams [3] takes into account some of the contextual information but faces the problem of large vector dimensions.

The emergence of word embedding solves the problem. It maps each word into a high-dimensional vector through a large amount of corpus training, so that each word itself has a vector representation which can almost express the features. The commonly used word2vec is generally implemented based on CBOW and Skip-Gram algorithms. The input of CBOW is the word embedding of the context of the target word. The output is the embedding of the target word [8]. While the input and output of Skip-Gram are opposite. However, both of them are base on neural network, whose input layer is different according to the input vector and the output layer has vocabulary sized neurons [8]. When training, we can use the DNN's backpropagation algorithm to obtain parameters of the DNN model, meanwhile with the embedding of all words. When testing, we use softmax to choose the top-ranked words calculated by DNN forward propagation. The Fig. 1 shows the models of two algorithms. The second step is to use machine learning algorithms to train the classification model. Most machine learning algorithms can be applied in text-classification while the results are different in different scenarios. The Naive Bayesian algorithm assumes that the attributes are independent and its training time is short. For a large number of training data in different categories, Naive Bayes have better performance. However, it does not perform well for categories defined with only a few features [1,7]. Different from Naive Bayes, decision tree based algorithm is more suitable for classification problems with few features instead of a large number of features. Moreover, SVM has good results in two-category text classification problem, while the performance on the multi-classification problem is not satisfactory. It should be noted that when using machine learning algorithms, it is often necessary to choose parameters through a large number of experiments to get the best classification results, and different feature extraction and reduction methods can also have a great impact on the accuracy of the classifier.

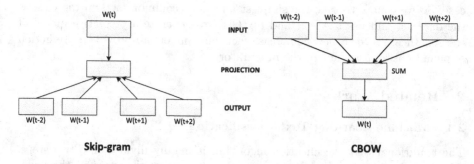

**Fig. 1.** The three-layer models of two word2vec algorithms. The left one is Skip-gram and the right is the CBOW.

## 2.2  Deep Learning Text-Classification

As the vector representation of text is transformed from the original high-latitude and high-sparse format to continuous dense data like images and speech, combined with the strong data migration of the deep learning algorithm itself, many deep learning algorithms previously used in the image field are now also used in the field of text.

The first is CNN, which has achieved great success in the field of image. Besides, the convolutional neural network has the function of extraction of local features which can also be used to extract context features in the sentence. TextCNN is proposed by Kim in [4] to classify text using CNN. Since CNN requires fixed-length input and output, we need to preprocess the input vector through truncating the exceeding length or completing the insufficient byte with 0. The supplemental 0 will not affect the final result, because max-pooling will only output the maximum value, and the 0-filled item will be filtered out. Actually, we can choose static or non-static word embedding when using. The static one means using the pre-trained word embedding like word2vec directly and no longer update during training, which is suitable for the small volume of data, while the non-static volume allows changing the original word embedding to accelerate convergence.

TextCNN has excellent performance in many tasks, but due to fixed filter-size, it is hard to model a longer sequence. Besides, the adjustment of parameters is also complicated. Comparatively, the recurrent neural network is more suitable for capturing the context of natural language [5]. Because of the temporal characteristics of text, BiLSTM is more common in text modeling tasks. Moreover, in order to improve the interpretability of text categorization tasks, the researchers also introduced attention mechanisms to preserve the long-distance dependence information in the text, and intuitively give the contribution of each word to the results [9].

## 3  Experiment

### 3.1  Dataset

We use the small scale of CAIl2018 dataset which contains 15.5w pieces of training data and 3.3w pieces of test data [11]. Each row of the dataset corresponding to a case and all of them are stored in the type of dictionary. The information used in our article contains the $'fact'$ and the $'term_of_imprisonment'$ field, in which the imprisonment is divided into the death penalty, life imprisonment and term of imprisonment. We extract and recombine these two fields as the dataset of our experiment.

### 3.2  Preprocess

First of all, according to the classification of the original dataset, we divide the imprisonment into three categories: death penalty, life imprisonment, and fixed-time imprisonment. And we further analysis the fixed-time imprisonment in months. Based on the frequency of each month, we conducted mathematical statics on the data. We can find that distribution is particular uneven and the under-48 months accounts for about 85%. In combination with the felony level in US law, we divide the imprisonment into 12–48 months, 48–84 months, 84–180 months and larger than 180 months, corresponding to the level E, D, C, B. In addition, 0–12 months belongs to the misdemeanor and the death penalty add life imprisonment are divided into the A level. So far, we have divided the data into five levels. For each case, we classify it into a specific level and use the level as the only label of the case.

For the facts of the case, we first use the precise mode of Jieba tokenizer for word segmentation, and then use part-of-speech filtering to screen out nouns, verbs, gerunds and other words that are helpful to the case, and also delete words less than 1 in length. Besides, we also find that the difference between the facts of the case and the conventional text is that in order to protect the privacy of the parties, there are a large number of words such as someone and so on, which are not helpful to the prediction of the sentence. So we delete all these words in the text processing, and the result is the word sequence of the fact.

### 3.3    Experiment

**FastText:** The model of fastText is very similar to CBOW of word2vec. They are both based on Hierarchical Softmax and three-layer architecture, which are input layer, hidden layer and output layer as Fig. 2. For the input word sequence, it first generates the word embedding of the text, adding the character-level n-gram vector as an additional feature to consider the context of the word. Then the composed feature vectors are mapped to the hidden layer through the linear transformation. Based on the weights and parameters of each category, the Huffman tree is constructed and used to classify unlabeled text.

**TextCNN:** TextCNN is to migrate the traditional CNN to the text classification problem, so its basic architecture is the same as the original CNN, which is divided into embedding layer, convolution layer, max-pooling layer and full-connected layer as the Fig. 2. The embedding layer encodes the natural language input into a distributed representation. During this period, we can choose to use a pre-trained word embedding or train a personal one. And for the pre-trained one, we can choose whether to adjust the parameters of the word embedding or not while training. The convolution layer is to extract different n-gram features of the text. We set the number of kernels as 256 and each kernel size as 5. The maximum pooling layer takes the maximum value of several one-dimensional vectors obtained after convolution and then splices them together as the output of this layer. Last but not least, the full-connection combine another layer with the max-pooling layer as the output to improve the learning ability of the network. The model we use contains two full-connection layers, and each layer consists of 128 neurons. In the iterative process, we used cross-entropy as loss function and optimized with AdamOptimizer. Figure 2 shows the structure of textCNN.

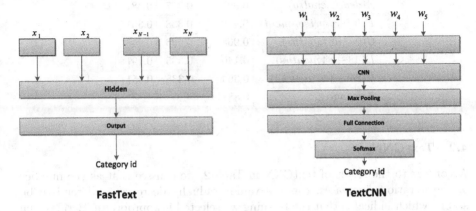

**Fig. 2.** Structure of the model used in our experiment. The left one is the three-layer fastText model and the right is the simplified model of textCNN. We use two full-connection layers before softmax in our experiment.

**Experiment Environments:** The configuration of our experimental machine is Windows10 system, 8 GB of memory. Because fastText is released on MacOS and Linux, we use the Ubuntu environment in Windows Subsystem for Linux(WSL), and the version of Ubuntu is 18.04, c++ compiler version is gcc 7.3.0. TextCNN runs under the Windows system.

## 4  Result

### 4.1  FastText

According to the experiment result in Table 1, we can see that the average accuracy of fastText is 88.75%, which is quite higher in prediction. But different performance in different categories is also the problem of fastText. For the category with more data, the classifier has better performance. For example, the category of 1–48 months gets the best performance among all categories which is 0.968 in accuracy, 0.926 in recall and 0.946 in F1Score. For the death penalty and life imprisonment with less data, the accuracy is 0.397 and 0.452 respectively. But the E and F categories with very little data get poor performance in prediction, and even no text is classified into the category F in our testing period. As a result, using fastText to predict imprisonment according to facts is feasible, and the larger the amount of data of each category is input, the more accurate of the prediction will be.

**Table 1.** Result of fastText in predicting the label of imprisonment

| Label of imprisonment | Precision | Recall | F1Score |
|---|---|---|---|
| $A(death_penalty)$ | 0.397 | 0.377 | 0.387 |
| $B(life_imprisonment)$ | 0.452 | 0.338 | 0.351 |
| $C(0-48\,months)$ | 0.968 | 0.926 | 0.946 |
| $D(48-84\,months)$ | 0.140 | 0.335 | 0.197 |
| $E(84-180\,months)$ | 0.364 | 0.338 | 0.351 |
| **Average** | 0.887 | – | – |

## 4.2    TextCNN

According to the result of textCNN in Table 2, we can see that as the number of the iterations increases, the loss value gradually decreases and tend to be stable, which indicates that the learning we selected is appropriate. Besides, the accuracy of our model on the test dataset is also gradually increased and to be stable after about 200 iterations, and the maximum value is 88.76%, indicating the feasibility of using textCNN to train imprisonment prediction model.

**Table 2.** Result of textCNN in predicting the label of imprisonment

| Iter | Train loss | Train Acc | Val loss | Val Acc |
|---|---|---|---|---|
| 0 | 1.8 | 42.19% | 1.8 | 36.34% |
| 100 | 0.44 | 87.50% | 0.43 | 87.75% |
| 200 | 0.42 | 87.5% | 0.43 | 87.75% |
| 300 | 0.42 | 85.16% | 0.4 | 88.03% |
| ... | ... | ... | ... | ... |
| 1500 | 0.36 | 87.50% | 0.34 | 88.76% |
| ... | ... | ... | ... | ... |
| **Maximum** | 0.36 | 87.50% | 0.34 | 88.76% |

# 5    Conclusion and Future Work

In this paper, we use facts of the case in judgment documents as the sole data source to predict the imprisonment and use fastText and textCNN to train the deep learning models. In our experiments, the accuracy of fastText is 88.75 and textCNN is about 88.76, which show that the textCNN is slightly better than fastText in the result. However, considering the efficiency of the algorithm, fastText is much faster than textCNN, which is more suitable in the large-scale dataset. It is also needed to note the data problem caused by the unevenness of each kind of data.

Our future study is divided into three directions. First, we will try to use more neural network algorithms to get better performance in prediction. Second, we will add more information from judgment documents in addition to the facts, such as combining the label of accusation and information of law articles, which concerning the term of imprisonment in law. Third, we will add more feature engineering techniques in the period of text preprocessing, such as LDA or LSI, or try more pre-trained word embedding such as BERT and Glove.

**Acknowledgement.** The work is supported in part by the National Key Research and Development Program of China (2016YFC0800805) and the National Natural Science Foundation of China (61772014).

# References

1. Qiang, G.: An effective algorithm for improving the performance of Naive Bayes for text classification. In: Second International Conference on Computer Research and Development (2010)
2. Joachims, T.: A probabilistic analysis of the Rocchio algorithm with TFIDF for text categorization. In: International Conference on Machine Learning (1996)
3. Joulin, A., Grave, E., Bojanowski, P., Mikolov, T.: Bag of tricks for efficient text classification. arXiv preprint arXiv:1607.01759 (2016)
4. Kim, Y.: Convolutional neural networks for sentence classification. Eprint Arxiv (2014)
5. Liu, P., Qiu, X., Huang, X.: Recurrent neural network for text classification with multi-task learning (2016)
6. Martineau, J.C., Finin, T.: Delta TFIDF: an improved feature space for sentiment analysis. In: Third international AAAI Conference on Weblogs and Social Media (2009)
7. Meena, M.J., Chandran, K.R.: Naïve bayes text classification with positive features selected by statistical method. In: International Conference on Advanced Computing (2009)
8. Mikolov, T., Chen, K., Corrado, G., Dean, J.: Efficient estimation of word representations in vector space. CoRR abs/1301.3781 (2013)
9. Pappas, N., Popescu-Belis, A.: Multilingual hierarchical attention networks for document classification (2017)
10. Rusko, M., et al.: Slovak automatic dictation system for judicial domain (2011)
11. Xiao, C., et al.: CAIL2018: a large-scale legal dataset for judgment prediction. arXiv preprint arXiv:1807.02478 (2018)

# Integrated Modeling Framework to Guide Novel Insole Designs for Stress Redistribution at the Human Knee Joint

Yi-Heng Cai and Wen-Ming Chen[✉]

Department of Biomedical Engineering,
University of Shanghai for Science and Technology, Shanghai, China
chenwm@usst.edu.cn

**Abstract.** An insole design with variable stiffness properties could be an effective conservative treatment option for knee osteoarthritis (OA). However, limited to the experimental conditions, stress distribution inside the cartilage of a knee joint is often difficult to be obtained. In this study, a finite element (FE)-based computational model of the human knee-foot-ankle complex is presented to investigate the redistribution of the internal stress of the knee joint using a variable stiffness insole. Based on relationship between the insole parameters and the resulting tissue stress data, additive manufacture technology (AM) could be a fast deliver of an optimal insole that meets patient characteristics. Insoles of four different material combinations were designed. Von-Mises stress declined significantly at the meniscus when the ratio of the lateral and medial stiffness of the insole changes. It suggests that the intervention of variable stiffness insole contributes to the rehabilitation of medial knee osteoarthritis (OA) and provide the guidance for the design of therapeutic insole.

**Keywords:** Osteoarthritis · Variable stiffness insole · FE analysis · Additive manufacturing · Data-driven healthcare

## 1 Introduction

Knee osteoarthritis (OA) has been deemed to a dominating cause of disability over age 60 [1, 2], the reason is the medial compartment of a diseased knee bears about ten times more loads than the lateral compartment [3]. Abnormal loads at the medial compartment of the knee can accelerate the degradation of joint cartilage [4–7]. Non-surgical mechanical interventions such as laterally wedged insoles and variable stiffness insoles can reduce the adduction moment in patient's knee joint [8]. In previous studies, variable stiffness insole had been considered by many researchers to be better interventions shoes due to smaller side effects [9]. However, different parameters of the intervention insole have inconsistent therapeutic effects on Knee osteoarthritis (OA) patients. The use of finite element (FE) technology can predict the stress distribution in patients' knee joint under different intervention insoles [10]. Saunders et al. established the assembly model of foot and heel insole and performed finite element simulation, found individually designed insole can reduce the peak von-Mises stress at the heel [11]. Zhang Ming et al. used finite element technology to analyze the effects of

© Springer Nature Singapore Pte Ltd. 2019
R. Mao et al. (Eds.): ICPCSEE 2019, CCIS 1059, pp. 618–625, 2019.
https://doi.org/10.1007/978-981-15-0121-0_48

full-contact insoles and arch insoles on the pressure of the soles of feet, their results showed that the softer material insole could reduce the plantar pressure more effectively [12]. FE analysis has proven to be a powerful and fast evaluation tool to guide the insole designs for knee osteoarthritis [13].

Several previous researches have demonstrated that the medial compartment loading of the knee joint depends on the peak knee adduction moment [14, 15]. The magnitude of the knee joint adduction moment reflects the severity, rate of disease progression and treatment outcome of the medial compartment knee osteoarthritis (OA) [6, 7, 16, 17]. Therefore, the aim of the intervention insole is to decrease the magnitude of peak knee adduction moment. However, there is no gold standard exists for the design of the intervention insole at the moment, in particular for the variable stiffness insole. In the past, it is generally considered that the lateral stiffness is 1.1 to 1.8 times the medial stiffness of the insole is effective [9], but the individualized parameters that can meet different patient characteristics are not realized.

In addition, additive manufacture (AM) technology has gradually received increasing attention, apparently, AM overcomes traditional manufacturing constraints and has been used to build lightweight structures with controlled mechanical properties [18]. Thus, based on the stress value obtained from FE analysis, additive manufacturing (AM) could be a fast generator for personalized insoles that meet the patient's pathological characteristics [19].

In this study, we attempted to analyze the stress distribution at meniscus under different stiffness designed insoles. The specific aims were: (1) to establish a finite element model of the lower limb, involving bones, soft tissues, meniscus, ligaments and fascia which with large deformations; (2) to find optimal stiffness designed insole that minimum stress distribution on the meniscus.

## 2　Materials and Methods

### 2.1　Construction of a Knee-Ankle-Foot Computational Model

A clinical CT machine (iCT256, Philips, Netherlands) was used to scan a healthy man' left lower limb at a slice thickness of 0.67 mm and a slice spacing of 0.43 mm (in-plane resolution at 768 × 768 pixels). The CT images comprise bones, associated soft tissue, ligament, fascia and meniscus in the knee joint. A total of 1161 CT images were reconstructed by segmentation software Mimics Research 19.0 (Materialise, Belgium) and were threshold-divided by setting the threshold range to 226–1821. A preliminary geometric model was established by calculating 3D afterwards (Fig. 1).

Further repaired in Geomagic 12.0 is necessary cause of the burrs and holes on the surface geometrical model needed to be removed. Mesh doctor was used firstly and then repaired the small burrs and holes manually, in the end to generate the NURBS surface of the model (Fig. 1).

## 2.2    FE Mesh Generation

The mesh generation process was completed in Hypermesh 13.0. Linear tetrahedron elements (C3D4) [20] were used to mesh the bone component. The soft tissue was meshed using linear hybrid tetrahedron elements (C3D4H) and the meniscus was meshed with quadratic modified tetrahedron elements (C3D10 M) to get the optimal stress distribution value. The insole assembled to the lower limb was designed in the Mechanical Design software SolidWorks2016 (Dassault Systems) and was divided into two sides of equal size medial and lateral (Fig. 1). Both the medial and lateral were meshed with linear reduced integration hexahedral elements (C3D8R). The joint ligament and plantar fascia were simulated with one-dimensional truss element (T3D2), which not only avoids computational convergence caused by geometric sharp angles, but also decrease computation costs. A mesh convergence analysis was performed to get the appropriate element size. The element size was deemed to be acceptable when a further decrease in element size did not change the output variables by more than 5%.

## 2.3    Assigning Material Parameters

The entire FE model consists of six major components (Fig. 1): bones, soft tissue, meniscus, ligaments, planta fascia and insole. The material property of bone, ligament and planta fascia was assumed as linear elastic. The soft tissue attached to the bone is modeled as the hyperelastic Ogden 1st formulation, the strain energy function is [21]:

$$U = \sum_{i=1}^{N} \frac{2\mu_i}{\alpha_i^2} \left( \bar{\lambda}_1^{\alpha i} + \bar{\lambda}_2^{\alpha i} + \bar{\lambda}_3^{\alpha i} \right) + \sum_{i=1}^{N} \frac{1}{D_i} (J^{el} - 1)^{2i}$$

$\bar{\lambda}_i$ represent the deviatoric principal stretches, N is the order and was set to 1 (1st-order), $J^{el}$ is the elastic volume ratio. The hyperelastic material constants are represented by $\mu_i$, $\alpha_i$, and $D_i$.

The constitutive model of meniscus was defined as isotropic Mooney-Rivlin [22]. Model parameters were simulate nearly incompressible (Poisson's ratio of 0.48) behavior.

Four different material combinations were created for the variable stiffness insole by controlling the elastic modulus of the lateral and medial (Fig. 1) at 2 MPa with 2 MPa (lateral is 2 MPa, medial is 2 MPa), 2 MPa with 1.54 MPa (lateral is 2 MPa, medial is 1.54 MPa), 2 MPa with 1.33 MPa (lateral is 2 MPa, medial is 1.33 MPa), 2 MPa with 1.177 MPa (lateral is 2 MPa, medial is 1.177 MPa). The ratio of the lateral and medial elastic modulus is 1, 1.3, 1.5 and 1.7 respectively.

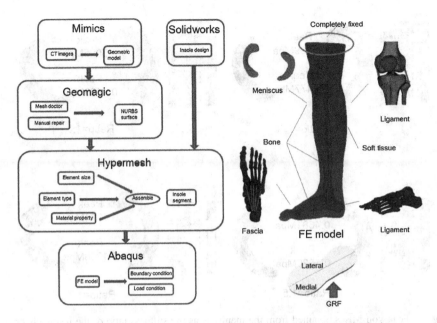

**Fig. 1.** The integrated modeling framework for therapeutic insole design.

## 2.4  Loading and Boundary Conditions

The stance phase of walking was simulated in order to obtain the distribution of stress at the meniscus of the knee joint with four different variable stiffness insoles. The bottom of the insole was coupling with reference point respectively and was defined concentrate force to simulate the ground reaction force. The bones and soft tissue were tied and the distal of the FE model was completely fixed (Fig. 1). Surface-to-surface contact condition was defined at the foot-insole interface, and the coefficient of friction was set to be 0.6 [23].

## 3  Results

Figure 2 depicts the von-Mises stress distribution pattern at the meniscus, with a ratio of lateral and medial stiffness of variable stiffness insole is 1, 1.3, 1.5, and 1.7, respectively. The peak con-Mises stress of the meniscus under the four ratio conditions occurred on the medial. As the stiffness of the medial insole gradually decreased, the peak von-Mises stress on the medial meniscus decreased. Compared with the insole with a ratio of lateral and medial stiffness of 1, when the ratio is 1.3, the peak von-Mises stress on the medial meniscus reduced by about 5%, the peak von-Mises stress on the lateral meniscus reduced by about 1%; when the ratio is 1.5, the peak von-Mises stress on the medial meniscus reduced by 10%, and the peak von-Mises stress on the lateral meniscus reduced by 2%. When the ratio is 1.7, the peak von-Mises stress on the medial meniscus reduced by 15%, and the peak von-Mises stress on the lateral meniscus reduced by 3%.

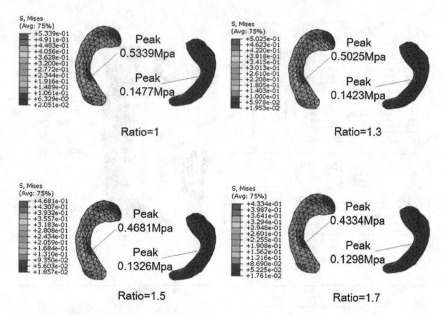

**Fig. 2.** Peak von-Mises obtained from the meniscus as the stiffness ratio of the insole changes.

## 4  Discussion

Due to the anatomical complexity of the human lower limb, only the distal femur part was considered in the model. In addition, the 3D structure of the ligaments and the plantar fascia was simplified; they were modeled based on the anatomical location and simulated with 2D truss element. In many cases, the complexity of the ligaments and fascia in geometric features may cause the analysis hard to converge. In the current model, a total of 80 ligaments and 5 bundles of plantar fascia are established as truss element. Apparently, setting appropriate material properties and cross-sectional area can effectively simulate the role of ligaments and fascia on the lower limbs. The material properties of the bone were simplified to isotropic linear elastic materials, but in reality the bones were anisotropic [24]. Furthermore, the stress values on the FE model will vary depending on the element size and element type, and the convergence behavior may also depend on the software solver (implicit or explicit) [25]. Nevertheless, establishing a finite element model is indeed a necessary condition for analyzing musculoskeletal diseases of the lower limbs. Combined with the development of additive manufacturing technology, the corresponding relationship between the mechanical properties of the insole with the stress values in the joint can be established for a specific patient and then provide guidance for the design of the insole.

A major goal of this study is to provide an integrated modeling framework that reconstructs human limbs based on the Digital Imaging and modifies the geometric model, finally obtain a FE model by assigning element parameters and material properties. The FE model can effectively simulate the human limb biomechanical properties, and reproduce various load conditions (not limited to the stance phase that was simulated in the study). For the past decade, FE analysis has gained popularity in the musculoskeletal biomechanics, clinical orthopedics and even sport injury community. Khaja Moiduddin et al. established a patient's mandibular finite element model, and designed the mandibular-sized implants that fit the patient by computer-assisted design, obtained the best material and shape that can withstand chewing load condition by finite element analysis, finally printed the implant by additive manufacturing [26]. Young-Eun Lim et al. designed a heel with 3D conformal lightweight structure based on the tetrahedral mesh by FE analysis [19]. Obviously, additive manufacturing in medical applications makes it possible to personalize 3D printed insole due to their customizability and fast delivery [26].

Internal loads imposed onto the meniscus of the knee joint remain highly relevant in the severity of Knee osteoarthritis (OA) [27]. Abnormal internal tissues load has been suggested as a key factor for the disease progression of knee osteoarthritis but could be hard to measure experimentally. The prediction in this FE simulation demonstrated that the percentage change of von-Mises stress on the lateral meniscus was much larger than that on the medial meniscus (Fig. 2), which is consistent with the point that the knee load transmitted by the lateral compartment is much smaller than that of the medial compartment [28]. Comparing with other similar computational model predictions, Liu et al. simulated the stress of the meniscus at the medial compartment of the knee joint with three different lateral wedge insoles and found that the peak stress on the meniscus with 5° and 10° wedge angles decreased by about 12.6% and 15.4%, respectively [29]. This suggests that the effect of a variable stiffness shoe with a lateral stiffness of 1.7 times the medial stiffness is approximately equal to 10° lateral wedge insole. However, the location of the peak von-Mises stress on the meniscus was different. Such difference might be reasonable because the inconsistent knee geometry between our subject and the sample from Liu's study. Therefore, the development of FE analysis is a necessary step for personalized design of the insole.

While the optimal stiffness ratio of 1.7 is within the range of 1.1 to 1.8, this is limited to one specific patient, and the optimal stiffness ratio may vary between patients. By adjusting the parameters of the insole, the biomechanical response under different mechanical properties could be obtained by means of FE simulation. In summary, finite element predictions indicate that a variable stiffness insole can significantly reduce the load on the medial compartment of the knee joint. With the further improvement of the model, the use of variable stiffness shoes can prevent the onset and progression of knee osteoarthritis (OA), and together with additive manufacturing provides a useful platform for the biomechanical effects of orthopedic insoles.

**Acknowledgements.** The current work is supported by Shanghai PuJiang Program (17PJ1 407100) and National Key Research and Development Project, China (2018YFC2001500).

# References

1. Felson, D.T., Lawrence, R.C., Dieppe, P.A., et al.: Osteoarthritis: new insights. part 1: the disease and its risk factors. Ann. Intern. Med. **133**(8), 635–646 (2000)
2. Lawrence, R.C., Hochberg, M.C., Kelsey, J.L., et al.: Estimates of the prevalence of selected arthritic and musculoskeletal diseases in the United States. J. Rheumatol. **16**(4), 427–441 (1989)
3. Repicci, J.A., Hartman, J.F.: Unicondylar minimally invasive approach to knee arthritis. In: Hozack, W.J., et al. (eds.) Minimally Invasive Total Joint Arthroplasty. Springer, Berlin (2004). https://doi.org/10.1007/978-3-642-59298-0_27
4. Andriacchi, T.P., Mündermann, A., Smith, R.L., et al.: A framework for the in vivo pathomechanics of osteoarthritis at the knee. Ann. Biomed. Eng. **32**(3), 447–457 (2004)
5. Astephen, J., Deluzio, K., Caldwell, G., et al.: Biomechanical changes at the hip, knee, and ankle joints during gait are associated with knee osteoarthritis severity. J. Orthop. Res. **26**(3), 332–341 (2008)
6. Baliunas, A.J., Hurwitz, D.E., Ryals, A.B., et al.: Increased knee joint loads during walking are present in subjects with knee osteoarthritis. Osteoarthr. Cartil. **10**(7), 573–579 (2002)
7. Sharma, L., Hurwitz, D.E., Thonar, E.J., et al.: Knee adduction moment, serum hyaluronan level, and disease severity in medial tibiofemoral osteoarthritis. Arthritis Rheum. **41**(7), 1233–1240 (1998)
8. Cheung, J.T., An, K.N., Zhang, M.J.F., et al.: Consequences of partial and total plantar fascia release: a finite element study. Foot Ankle Int. **27**(2), 125 (2006)
9. Fisher, D.S., Dyrby, C.O., Mündermann, A., et al.: In healthy subjects without knee osteoarthritis, the peak knee adduction moment influences the acute effect of shoe interventions designed to reduce medial compartment knee load. J. Orthop. Res. **25**(4), 540–546 (2010)
10. Sun, B.: Finite element analysis on foot and application. J. Jiangsu Polytech. Univ. (2009)
11. Saunders, M.M., Schwentker, E.P., Kay, D.B., et al.: Finite element analysis as a tool for parametric prosthetic foot design and evaluation. Technique development in the solid ankle cushioned heel (SACH) foot. Biomech. Biomed. Eng. **6**(1), 75–87 (2003)
12. Cheung, J.T., Zhang, M., An, K.N.: Effect of achilles tendon loading on plantar fascia tension in the standing foot. Clin. Biomech. **21**(2), 194–203 (2006)
13. Cheung, J.M., Zhang, M.: A 3-dimensional finite element model of the human foot and ankle for insole design. Arch. Phys. Med. Rehabil. **86**(2), 353–358 (2005)
14. Crenshaw, S.J., Pollo, F.E., Calton, E.F., et al.: Effects of lateral-wedged insoles on kinetics at the knee. Clin. Orthop. Related Res. **375**(375), 185 (2000)
15. Schipplein, O.D., Andriacchi, T.P.: Interaction between active and passive knee stabilizers during level walking. J. Orthop. Res. **9**(1), 113–119 (1991)
16. Miyazaki, T., Wada, M., Kawahara, H., et al.: Dynamic load at baseline can predict radiographic disease progression in medial compartment knee osteoarthritis. Ann. Rheum. Dis. **61**(7), 617–622 (2002)
17. Prodromos, C.C., Andriacchi, T.P., Galante, J.O.: A relationship between gait and clinical changes following high tibial osteotomy. J. Bone Joint Surg. Am. Vol. **67**(8), 1188 (1985)
18. Campbell, I., Bourell, D., Gibson, I.: Additive manufacturing: rapid prototyping comes of age. Rapid Prototyping J. **18**(4), 255–258 (2012)
19. Lim, Y.E., Park, J.H., Park, K., et al.: Automatic design of 3D conformal lightweight structures based on a tetrahedral mesh. Int. J. Precis. Eng. Manuf. - Green Tech. **5**(4), 499–506 (2018). https://doi.org/10.1007/s40684-018-0053-2

20. Pegg, E.C., Gill, H.S.: An open source software tool to assign the material properties of bone for ABAQUS finite element simulations. J. Biomech. **49**(13), 3116–3121 (2016)
21. Mahnken, R.: An inverse finite element algorithm for parameter identification of thermoelastic damage models. Int. J. Numer. Methods Eng. **48**(7), 1015–1036 (2015)
22. Sibole, S.C., Erdemir, A.: Chondrocyte deformations as a function of tibiofemoral joint loading predicted by a generalized high-throughput pipeline of multi-scale simulations. PLoS One **7**(5), e37538 (2012)
23. Tadepalli, S.C.: Comparison of hexahedral and tetrahedral elements in finite element analysis of the foot and footwear. J. Biomech. **44**(12), 2337–2343 (2011)
24. Petre, M., Erdemir, A., Panoskaltsis, V.P., et al.: Optimization of nonlinear hyperelastic coefficients for foot tissues using a magnetic resonance imaging deformation experiment. J. Biomech. Eng. **135**(6), 61001 (2013)
25. Cheung, T.M., Zhang, M., An, K.N.: Effects of plantar fascia stiffness on the biomechanical responses of the ankle–foot complex. Clin. Biomech. **19**(8), 839–846 (2004)
26. Moiduddin, K.: Implementation of computer-assisted design, analysis, and additive manufactured customized mandibular implants. J. Med. Biol. Eng. **38**(5), 744–756 (2018). https://doi.org/10.1007/s40846-018-0370-5
27. Scott, T., Ahmet, E., James, W., et al.: What has finite element analysis taught us about diabetic foot disease and its management? a systematic review. PLoS One **9**(10), e109994 (2014)
28. Boyer, K.A., Federolf, P., Lin, C., et al.: Kinematic adaptations to a variable stiffness shoe: Mechanisms for reducing joint loading. J. Biomech. **45**(9), 1619–1624 (2012)
29. Liu, X., Zhang, M.: Redistribution of knee stress using laterally wedged insole intervention: finite element analysis of knee – ankle – foot complex. Clin. Biomech. **28**(1), 61–67 (2013)

# Movie Score Prediction Model
# Based on Movie Plots

Hui Xie[1,2], Haomeng Wang[1,2], Chen Zhao[1,2], and Zhe Wang[1,2(✉)]

[1] College of Computer Science and Technology, Jilin University,
Changchun 130012, China
wz2000@jlu.edu.cn
[2] Key Laboratory of Symbolic Computation and Knowledge Engineering,
Ministry of Education (Jilin University), Changchun 130012, China

**Abstract.** With the rapid development of the movie industry, it is vital to evaluate and predict a movie's quality. In this paper, a movie score prediction model is proposed based on the movie plots. Movie data was processed with the word2vec method, and the linear regression model and back propagation neural network algorithm were employed to establish the movie score prediction model. The high-quality classic movie plots of high-scoring movies summed up by big data contributed to a high synthesis of the wonderful content of the film. Experimental results show that it is effective in terms of movie evaluation and prediction, and helpful in understanding people's preferences for movie plots.

**Keywords:** Movie bridge plot · Movie score prediction ·
Linear Regression Model · Back propagation neural network

## 1 Introduction

As an important part of cultural life, the movie not only enriches people's spare time and the spiritual world but also is an important medium for cultural exchanges between countries and regions with different cultural backgrounds. With the formation of a new subject in movie science, the movie has gradually evolved into a merchandise form from the pure art form [1, 2]. The movie industry is booming, and people's requirements for movie content and quality are getting higher and higher. As an evaluation part of the movie ecosystem, movie scores are directly presented by quantified scores, potentially affecting people's judgments on the movie, and even the scoring of the website is once magnified as the only measure to determine whether the quality of a movie is recognized by the audience. The movie score will largely determine the fate of the movie. Therefore, what characteristics should a movie with a high score had become a concern of more and more people, and also become the focus attention of movie investment institutions. As early as the 1930s, movie reviews were an important part of the movie industry. Until the end of the 20th century, audiences' contact with

---

H. Xie and H. Wang—These authors contributed equally to this work and should be considered co-first authors.

movie reviews was basically through the carrier of newspapers. The comments and scores of this period are given by the famous movie critics in the true sense. They are relatively more professional, but at the same time, they can only reflect the preferences of a few elites. With the rapid development of online movie commenting systems, movie online reviews have become an important source of information about movie consumers. By mining the score data, you can analyze the user's hobbies, purchase intentions and factors affecting the user's purchase behavior.

There have been many studies on movie scoring prediction both domestic and international. At present, most of the prediction models are combined with factors such as director's influence, scriptwriter's influence, actor's influence, release date, technical effect, movie type (such as comedy, thriller) to predict movie scoring, and these factors are mostly external factors [3–8]. The existing movie scoring prediction models did not introduce the attributes of the movie itself, such as the development of the movie plots as an influencing factor, the plot structure and setting of the movie itself will certainly have an impact on the audience's viewing experience.

After research, it is found that the existing models seldom use one of the key attributes of the movie itself, that is, the relevant data of the movie bridge plots as an impact factor to predict. The score of a movie is closely related to the number of bridge plots used in the movie and whether the sequence of use is reasonable. Because bridge plots are an essential attribute of the movie itself, classic bridge plots can easily make the audience resonate when watching, thus making a deep impression on the movie. The project is mainly to construct a new movie scoring prediction model with the introduction of movie bridge plots factors. The high-quality classic movie bridge plots in high-scoring movies summed up by big data can effectively help understand people's thoughts about the preferences and requirements of the movie bridge plot, and make the movie quality prediction model more accurate, thus alleviates the bottleneck problem of movie development. At the same time, can also provide some suggestions for movie script creation. Bridge plots are the segment construction and setting of the movie itself. If the scriptwriter properly introduces the classical bridge segment to design the movie plot and arranges the sequence of the bridge plots, it will effectively improve the quality of the movie. It has strong practicability to introduce the movie bridge plot as the influencing factor.

## 2 Proposed Work

### 2.1 Overall Work

During the research, different from the existing models for predicting movie scores by factors such as directors, actors, and genres, how to represent a large number of movie bridge plots and how to apply the movie bridge plots to a predictive model was the important problems.

In order to predict the movie scores with the movie bridge plots, firstly, the movie bridge plots and the movie scores on the relevant website were got. Then by using Word2Vec, the movie bridge plots were sorted and digitized. Finally, the movie scores were predicted by using the Linear Regression Model and Back Propagation Neural Network based on the collated data (Fig. 1).

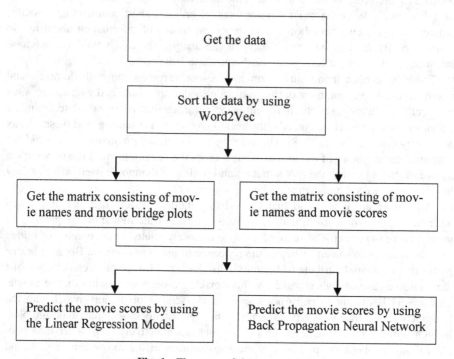

**Fig. 1.** The steps of data processing.

## 2.2    Word2Vec

Word2Vec [9] is a tool of natural language processing (NLP). It can turn natural language into a dense vector which can be understood by the computer. By using Word2Vec, the relationship between words and words can be quantitatively measured and mined, so as to express the similarity between different words more accurately.

Word2Vec mainly includes two models. The one is a continuous bag of words (CBOW) and the other is skip-gram. CBOW is a three-layer neural network. After inputting a known context, the model outputs a prediction of the current word. The learning goal is:

$$\alpha = \sum_{w \in C} logp(w|Context(w)) \tag{1}$$

With w represents any word in corpus C.

Skip-gram reverses the causal relationship of CBOW. The model can predict the context with the current word.

Word2Vec also includes two methods for efficient training. The one is negative sampling and the other is hierarchical softmax.

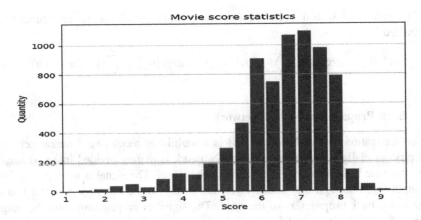

**Fig. 2.** The statistics of movie scores.

## 2.3  Linear Regression Model

The Linear Regression Model [10] is a predictive function which uses the linear combination of attributes.

$$y = w_0 x_0 + w_1 x_1 + \ldots + w_m x_m = \sum_{i=0}^{n} w_i x_i = w^T x \tag{2}$$

With $w_0$ indicates the intercept on the y-axis when $x_0 = 1$ and $w$ indicates the importance of the attributes in the forecast.

Here is training data $D = \{(X_1, Y_1), (X_2, Y_2), \ldots, (X_n, Y_n)\}$, and the data should be represented as a matrix.

$$X = \begin{bmatrix} 1, x_1^1, x_2^1, \ldots, x_n^1 \\ 1, x_1^2, x_2^2, \ldots, x_n^2 \\ \cdots \cdots \\ 1, x_1^n, x_2^n, \ldots, x_n^n \end{bmatrix} \tag{3}$$

$$XW = h_w(x^i) \tag{4}$$

In order to solve W according to X, the cost function is designed. The least square method is used to solve the parameters of the regression curve based on the mean square error minimization so that the sum of the squares of the vertical distance from the regression curve to the sample point is the smallest.

Here is the loss cost function:

$$J(W) = \frac{1}{2M} \sum_{i=0}^{M} (h_w(x^i) - y^i)^2 = \frac{1}{2M}(XW - y)^T(XW - Y) \tag{5}$$

The values of w and b can be solved when the result of the cost function is minimized.

$$(w^*, b^*) = argmin_{(w,b)} \sum_{i=1}^{m} (f(x_i) - y_i)^2 = argmin_{(w,b)} \sum_{i=1}^{m} (y_i - wx_i - b)^2 \quad (6)$$

## 2.4  Back Propagation Neural Network

Back Propagation Neural Network [11] is a multilayer feedforward neural network. The process of Back Propagation Neural Network is mainly divided into two stages. The first stage is the forward propagation of the signal. The signal is propagating from the input layer through the hidden layer and finally to the output layer. The second stage is the back propagation of the error. The signal is propagating from the output layer to the hidden layer and finally to the input layer. The weights and offsets of the hidden layer to the output layer and the weights and offsets of the input layer to the hidden layer are adjusted in turn.

The network connection weight between every two neurons is initialized with a small random number and each neuron has an offset value which is also initialized to a random number. The input layer of the network is provided according to the training sample x and the output of each neuron is obtained by calculation.

$$O_j = \frac{1}{1 + e^{-S_j}} = \frac{1}{1 + e^{-(\sum_i \omega_{ij} O_i + \theta_j)}} \quad (7)$$

With $\omega_{ij}$ is the network weight which from the unit i of the upper layer to the unit j and $O_i$ is the output of unit i of the previous layer.

Finally, the actual output is obtained at the output layer and the error of each output unit can be obtained by comparison with the expected output.

$$E_j = O_j(1 - O_j)(T_j - O_j) \quad (8)$$

With $T_j$ is the expected output of the output unit.

The error obtained needs to be propagated from the back to the front. The error of the previous layer unit can be calculated by the error of all the units in the subsequent layer connected to it:

$$E_j = O_j(1 - O_j) \sum k\omega_{jk} E_k \quad (9)$$

The error of each neuron from the last hidden layer to the first hidden layer can be obtained in turn.

The weights are adjusted from the connection weights of the input layer and the first hidden layer and are sequentially performed backward. Each connection weight is adjusted by the following formula:

$$\omega_{ij} = \omega_{ij} + \Delta\omega_{ij} = \omega_{ij} + (l)O_i E_j \quad (10)$$

With l is the learning rate which is usually taking a constant between 0 and 1.

The method of adjusting the neuron bias is to update each neuron with the following formula:

$$\theta_j = \theta_j + \Delta\theta_j = \theta_j + (l)E_j \tag{11}$$

For each sample, if the final output error is less than the acceptable range or the number of iterations reaches a certain threshold, the next sample will be taken and the output of each neuron will be restarted. Otherwise, the number of iterations will be incremented by one and then the output of each neuron using the current sample will be continued to calculate.

## 3  Experiment

Original movie data is stored separately according to the release year. Each movie in the test data has a corresponding actual score. The scores of all the movie data were calculated. As shown in Fig. 3, the test data scores were distributed over multiple intervals, with most scores ranging from 5–8. At the same time, it can be seen that some movie scores are 0, and these movie data will be rejected in subsequent operations.

Remove the space and lowercase the bridge plots and name of each movie. Keep only the letters and numbers in the movie names which are in the movie scores, and change to lowercase. The movie names in the movie scores are compared with the movie names in the movie bridge plots and if the movie names are the same and the movie scores are not 0, the movies' information will be saved (Table 1).

**Table 1.** The data of movie scores.

| No. | Movie name | Movie score |
|-----|------------|-------------|
| 1   | The garage | 5.0         |

**Table 2.** The data of movie bridge plots.

| No. | Movie name | Movie bridge plot 1 | Movie bridge plot 2 | Movie bridge plot 3 | ... |
|-----|------------|---------------------|---------------------|---------------------|-----|
| 1   | The garage | Acrobatic           | The alcoholic       | The alleged car     | ... |

The sequence of bridge plots for each movie could be regarded as a sentence and each bridge plot could be regarded as a word. Each bridge plot was converted into a vector by using Word2Vec so that a movie could be represented as a combination of multiple vectors.

Two matrices could be got. One matrix was consisting of movie names and movie bridge plots. The other matrix was consisting of movie names and movie scores.

The Linear Regression Model and The Back Propagation Neural Network were used to predict separately.

Firstly, The Linear Regression Model was used. The vectors which represent movies were used as the training set and test set of the Linear Regression Model. In the Linear Regression Model, each component of the vector was treated as a term in the polynomial and the movie score was treated as the result of the polynomial. The Linear Regression Model was trained with the data from the training set. After that, the movie scores of the test set were got and were compared with the original movie scores of the test set. Finally, the accuracy of the forecast was calculated.

Next, Back Propagation Neural Network was used. The matrix consisting of movie names and movie bridge plots were taken as input and the matrix consisting of movie names and movie scores were taken as output so that the model could be trained. Then the model was used to predict (Table 2).

## 4   Results

Prediction Model Based on the Linear Regression Model:

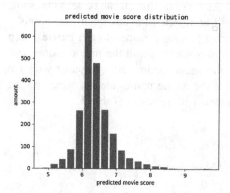
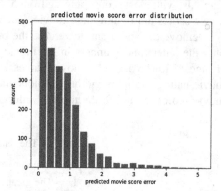

**Fig. 3.** Predictive score of linear regression mode

**Fig. 4.** Predictive score and real score error.

Figure 3 shows the movie scores predicted by the linear regression model. The predicted scores are also distributed over multiple intervals, but the scores between 5-8 are more. This is in good agreement with the true score data (as shown in Fig. 2), which is mainly distributed among 5–8 points. Figure 4 shows the difference between the predicted movie score and the actual score. It can be seen that the difference between the predicted value and the true value is mostly around 1 point, indicating that the prediction result is relatively accurate (Table 3).

Mean Error Statistics of Prediction Model Based on Linear Regression Model:

**Table 3.** Mean Error Statistics1.

| MSE | RMSE | MAE | Prediction accuracy |
|---|---|---|---|
| 1.17477 | 1.08387 | 0.81058 | 0.70520 |

Prediction model based on the Back Propagation Neural Network Algorithm:

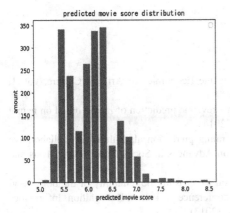
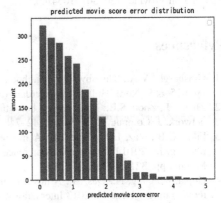

**Fig. 5.** Predictive score of BP neural network    **Fig. 6.** Predictive score and real score error

Figure 5 shows the results of a movie prediction score using back propagation neural network to construct a predictive model. The predicted score is mainly between 5–8 points. More consistent with the real score data. The error between the predicted value and the true value shown in Fig. 6 shows that the difference between the predicted value and the true value is mostly about 1 point, indicating that the prediction result is relatively accurate (Table 4).

Mean Error Statistics of Prediction Model Based on Back Propagation Neural Network:

**Table 4.** Mean Error Statistics1.

| MSE | RMSE | MAE | Prediction accuracy |
|---------|---------|---------|---------------------|
| 1.79843 | 1.34105 | 1.07163 | 0.53389 |

## 5   Conclusion

Through comparison, it can be seen that the accuracy of the prediction model constructed by the linear regression model is higher than that constructed by back propagation neural network algorithm, but on the premise of only referring to the factor of movie bridge plot, the accuracy of the two prediction models has reached the expected goal, which shows that the movie bridge plot has a greater influence on movie score.

Of course, under the premise of taking only the bridge segment, the prediction accuracy is not close to 100%, because, in reality, the film score is also affected by other factors (such as starring, director, etc.). Therefore, it is planned to introduce other factors affecting the film score as influence factors in the subsequent work, develop the current forecasting model, and improve the forecasting accuracy.

**Acknowledgments.** This work was supported by Natural Science Foundation of Jilin Provincial Science and Technology Department (20180101016JC); Science and Technology Development Plan of Jilin Province (20180101054JC).

# References

1. Ginsburgh, V.A., Throsby, D.: Handbook on the Economics of Art and Culture, vol. 1, pp. 615–659. North-Holland, Amsterdam (2006)
2. Zheng, J., Zhou, S.B.: Modeling on box-office revenue prediction of movie based on neural network. J. Comput. Appl. **34**(3), 742–748 (2014)
3. Fikir, O.B., Yaz, I.O., Özyer, T.: A movie rating prediction algorithm with collaborative filtering. In: 2010 International Conference on Advances in Social Networks Analysis and Mining, pp. 321–325. IEEE (2010)
4. Gupta, R., Garg, N., Das, A.: A novel method to measure the reliability of the bollywood movie rating system. In: 2013 International Conference on Pattern Recognition, Informatics and Mobile Engineering, pp. 340–345. IEEE (2013)
5. Changkaew, P., Kongkachandra, R.: Automatic movie rating using visual and linguistic information. In: 2010 First International Conference on Integrated Intelligent Computing, pp. 12–16. IEEE (2010)
6. Liu, C.L., Hsiao, W.H., Lee, C.H., et al.: Movie rating and review summarization in mobile environment. J. IEEE Trans. Syst. Man Cybern. Part C **42**(3), 397–407 (2012)
7. Li, J., Xu, W., Wan, W., et al.: Movie recommendation based on bridging movie feature and user interest. J. Comput Sci. (2018)
8. Çizmeci, B., Ögüdücü, Ş.G.: Predicting IMDb ratings of pre-release movies with factorization machines using social media. In: 2018 3rd International Conference on Computer Science and Engineering, pp. 173–178. IEEE (2018)
9. Tang, M., Zhu, L., Zhou, X.C.: Document vector representation based on Word2-Vec. Comput. Sci. **43**(6), 214–217 (2016)
10. Liu, Z.Z.: Analysis of domestic tourism consumption based on multiple linear regression model. J. Chongqing Univ. Technol. (Nat. Sci.) **30**(6), 167–170 (2016)
11. Zhao, L., Wang, S.: Research of public opinion heat trend simulation model based on BP neural network. J. China Soc. Sci. Tech. Inf. 989–999 (2016)

# Combining Multiple Factors of LightGBM and XGBoost Algorithms to Predict the Morbidity of Double-High Disease

Yingying Song[1], Xueli Jiao[2], Sen Yang[1], Shuangquan Zhang[1], Yuheng Qiao[1], Zhiyong Liu[3], and Lin Zhang[2(✉)]

[1] College of Computer Science and Technology, Jilin University, Changchun, China
[2] College of Software, Jilin University, Changchun, China
Zhang_lin@jlu.edu.cn
[3] Information Science and Technology College, Northeast Normal University, Changchun, China

**Abstract.** In recent years, cardiovascular and cerebrovascular diseases have seriously jeopardized people's health. Double-high (hyperlipidemia and hypertension) is one of the main causes of cardiovascular and cerebrovascular diseases in clinical practice. To diagnose cardiovascular and cerebrovascular diseases early, a reliable prediction system should be developed to assist doctors. In this paper, the different elements and evaluate the weight of these elements on double-high diseases are analyzed by machine learning method. The LightGBM algorithm and XGBoost algorithm were employed to construct the prediction models, respectively. Significantly, the proposed model was trained by real physical examination data and five meaningful and useful biochemical indicators were selected to encoding the raw physical examination data to numerical vector. The selected features are systolic blood pressure, diastolic blood pressure, serum triglyceride, serum high-density lipoprotein and serum low-density lipoprotein. The mean square error (MSE) after calculating the logarithm of the predicted value and the true value was introduced to assess the prediction model. Results show that this model can effectively predict cardiovascular and cerebrovascular diseases in advance.

**Keywords:** LightGBM algorithm · XGBoost algorithm · Double-high biochemical index prediction

## 1 Introduction

According to statistics, chronic diseases have became the primary reason of leading to people's death in China. These chronic diseases are increasing the risk of humans' death around the world, such as cardiovascular and cerebrovascular diseases [1–3]. The Hyperlipidemia and Hypertension are collectively referred to as double-high [4–6]. Clinically, hyperlipidemia refers to high blood lipid levels, which can cause diseases such as atherosclerosis and coronary heart disease. Hypertension refers to high blood pressure and it is prone to stroke, heart failure and other diseases.

© Springer Nature Singapore Pte Ltd. 2019
R. Mao et al. (Eds.): ICPCSEE 2019, CCIS 1059, pp. 635–644, 2019.
https://doi.org/10.1007/978-981-15-0121-0_50

In this paper, we first preprocess the real medical examination data. The size of the data to be processed is very big, which involves more than 8 million medical examination data of more than 50,000 people. There are five predicted target values: systolic blood pressure, diastolic blood pressure, triglyceride, high-density lipoprotein cholesterol, low-density lipoprotein cholesterol [7]. Since the LightGBM algorithm [8] and the XGBoost algorithm [9] have achieved good results in a variety of scenarios, in this paper, we use the LightGBM algorithm and the XGBoost algorithm to construct the predictive models, which effectively predict cardiovascular and cerebrovascular diseases. Meanwhile, we compare the effects of some models' parameters on the prediction results, which is intended to obtain a model with good generalization.

## 2    Background

Many people in the world are suffering from double-high disease and the causes of the disease are complicated. Therefore, it is important to establish an effective model to achieve double-high early screening and timely intervention, which will delay the dominance of the disease and prevent chronic diseases from becoming more severe. In this section, we introduce two common machine learning methods: the LightGBM algorithm and the XGBoost algorithm.

### 2.1    LightGBM Algorithm

LightGBM [10] is a gradient lifting framework based on decision tree algorithm, which is open source code by Microsoft on GitHub. Compared with commonly used machine learning algorithms, the LightGBM algorithm [8] has the following advantages: faster training efficiency, lower memory consumption, parallel learning, and large-scale data processing. LightGBM approximates the residual by the Taylor expansion of the loss function, and uses the regularization term to control the complexity of the model. It has the following two main features:

Firstly, LightGBM is a decision tree algorithm based on Histogram. The pre-sorting algorithm saves the feature values of the data and also saves the results of the feature sorting, while the histogram algorithm only saves the discretized values of the features, which takes up less memory. In addition, the pre-sorting algorithm needs to calculate the gain of a split every time it traverses a feature value, whereas the histogram algorithm only needs to traverse k times (k is a constant), which greatly accelerates the training speed of the model.

Secondly, LightGBM replaces XGBoost's level-wise splitting strategy by using a leaf-wise splitting strategy. It only selects the node with the largest split gain for splitting, which avoids the overhead caused by some nodes with smaller gains.

In order to get good results using a leaf-wise tree, these are some important parameters [11]: max_bin, num_leaves, min_data_in_leaf, min_sum_hessian_in_leaf, feature_fraction, max_depth, etc.

## 2.2  XGBoost Algorithm

The XGBoost algorithm [9] is an iterative decision tree algorithm, which is improved by Dr. Chen Tianqi based on the GBDT algorithm [12]. The XGBoost algorithm provides parallel tree enhancement, which can quickly and accurately solve many data science problems. The XGBoost algorithm has many advantages over the traditional GDBT algorithm. The GBDT algorithm only uses CART as the base classifier, while XGBoost also supports linear classifier and logistic regression classifier. The GBDT algorithm uses only the first derivative to optimize, while the XGBoost algorithm uses the first and second derivatives, which specifies the default gradient direction for the loss function, which indirectly improves the model accuracy and speed. The XGBoost algorithm adds a regular term to the loss function, which reduces the variance of the model and prevents overfitting. The XGBoost algorithm has many tunable parameters that have great customization flexibility, such as: num_early_stop, num_rounds, silent, max_leaf_nodes, min_child_weight, max_depth, colsample_bytree, etc.

## 3  Modeling Process and Evaluation Indicators

In this section, we first describe the physical examination data and introduce the data preprocessing process. Then, we give the defined evaluation indicators. Finally, we introduce the modeling process.

**Table 1.** The description of the dataset.

| vid | Table_id | Field_results |
|---|---|---|
| 002d1e4859fafd9ded2a2e1e7c839b62 | 2403 | 72.9 |
| 92dd479df5e30ab6a0a1cf85ac53efc3 | 102 | Liver: Fatty liver (moderate) gallbladder, pancreas, spleen, left kidney, right kidney did not find obvious abnormalities |
| c643a744e2e94f3ff354d920958bd37b | 409 | Medical history: History of hypertension (in treatment), history of diabetes (in treatment), no abnormalities in medical history of cerebrovascular accident history |
| 6bb59d517c4c70f8f50844d24fbd0355 | 201 | No abnormalities |
| 0ebb42adae512906f7e1135da734ea63 | 731 | Not checked |
| 6b5d8275d74c20d56f21bb8a51129af4 | 425 | Null |
| ebe7811e919109c42c092abbd98b4ca6 | 539 | Vagina: wall mucosa congestion cervix: congestion |
| 589b1feb7f1a5921749afc284ba91a99 | 102 | Liver: fatty liver (mild) gallbladder unexplored (reported has been removed) pancreas, spleen, right kidney, left kidney no significant abnormalities |
| fac42d33b77bca41183c7d692c2550e5 | 1402 | Bilateral cerebral artery blood flow velocity is slightly slowed down |
| f0b9cd99f3168049655d72f924319648 | 2501 | No intraepithelial lesions or malignant cells |

## 3.1 Data Discription

The dataset used in this paper are cellected from the Aliyun Tianchi Competition's C-Health AI Competition, a double-high disease risk forecast. The download URL is https://tianchi.aliyun.com/competition/entrance/231654/information. The dataset is from real medical examination data, involving more than 8 million medical examination data of more than 50,000 people. The raw dataset contains 8104368 samples, and the dataset information is shown in Table 1. The vid indicates the identity of the patient, the Table_id represents the identity of the patient's medical examination item, and the Field_results represents the physical examination result of the patient corresponding to Table_id.

## 3.2 Data Preprocessing

Since the raw data contains large amount of text data, the data needs to be preprocessed. Firstly, according to the different vid of the patient, multiple feature attributes of the same patient are spliced, de-duplicated, etc. For the pure numeric data, we need to remove the outliers, for example, change "0.49." to "0.49". For the simple enumeration-like text data, we convert it to a value by one-hot encoding, such as "negative", "no exception", "sinus". If the data contains both numeric values and text, we extract the numeric values from it. For some long text data, we first need to extract the keywords from the text feature column. For example, in the historical case, the keyword "sugar, blood lipid, blood pressure" needs to be extracted. Then we segment the text features using the jieba R package [13], and then use TF-IDF to count the word frequency of each segment phrase to convert the text data into word vectors. After the data preprocessing, the sparse matrix of the training set and the test set is finally obtained, the former data size is (30520, 23462) and the latter is (7630, 23462).

The data preprocessing process is as shown in Fig. 1.

## 3.3 Modeling Process

In this paper, we predict the five biochemical indicators, they are, systolic blood pressure, diastolic blood pressure, triglyceride, high-density lipoprotein cholesterol, low-density lipoprotein cholesterol, on the processed dataset mentioned above and we use the predicted results as the target variable to determine the possibility of suffering from double-high diseases. Based on the XGBoost algorithm and the LightGBM algorithm, we build and train the XGBoost and LightGBM models, and then analyze the stability of the two models on the influence of each parameter.

After data preprocessing, we use the XGBoost algorithm and the LightGBM algorithm to train the model, respectively. Then we analyze the stability of the model, which is influenced by the tree depth and feature fraction. The parameter tree depth is selected from the set $\{2, 3, 4, 5, 6, 7\}$ and the parameter feature fraction is selected from the set $\{0.6, 0.7, 0.8, 0.9, 1.0\}$. The remaining parameters of models are obtained by means of a grid-search method [14].

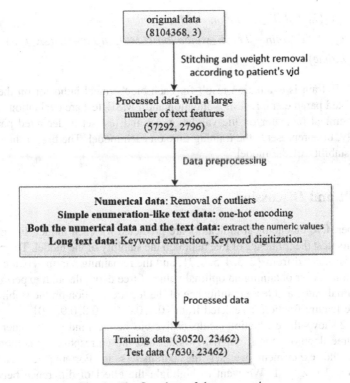

**Fig. 1.** The flowchart of data processing.

## 3.4 Evaluation Indicators

Take LMSE as the scoring criteria. LMSE is the mean square error after the logarithm of the predicted value and the true value. The formula is as follows:

$$lmse_j = \frac{1}{m}\sum_{i=1}^{m}\left(\log\left(y_i' + 1\right) - \log(y_i + 1)\right)^2, \tag{1}$$

$$LMSE = \frac{1}{N}\sum_{n=1}^{N} lmse_n, \tag{2}$$

m is the total number of people, $y_i'$ is the predicted value of the j-th indicator of the i-th person, yi is the real value of the j-th indicator of the i-th person, N is the value of the biochemical indicator.

In order to select the optimal values of tree depth and feature fraction, we design a scoring system, which comprehensively considers the influence of the difference between the LMSE values on the training set, the validation set and the test set, the size of the LMSE values on the validation set and the test set, and the training time on model training. The formula is as follows:

$$Score = 1/(\lambda_1(LMSE\_validation * LMSE\_test) +$$
$$\lambda_2(LMSE\_train - LMSE\_validation)(LMSE\_train - LMSE\_test) + , \quad (3)$$
$$\lambda_3 time)$$

where LMSE_train is evaluation result for each biochemical indicator on the training set under fixed parameters, LMSE_validation and LMSE_test are evaluation results for each biochemical indicator on the validation set and test set under fixed parameters, respectively, time represents the training time on each model. The bigger the score, the better the stability of the model.

## 4  Result and Discussion

In this paper, five biochemical indicators are predicted for each model. For each indicator, we first test the influence of tree depth on the stability of the model. The parameter tree depth is selected from $\{2, 3, 4, 5, 6, 7\}$, and the remaining parameters are obtained by grid-search. After obtaining an optimal value of tree depth through experiments, we fix the optimal value and test the influence of the feature fraction on the stability of the model. The feature fraction is selected from $\{0.5, 0.6, 0.7, 0.8, 0.9, 1.0\}$.

Figure 2 shows the evaluation results of five biochemical indicators under different values of tree depth on the LightGBM model and the corresponding comprehensive scores. In the experiment we set the parameters in Formula (3) as follows: $\lambda_1 = 5$, $\lambda_2 = 10$, $\lambda_3 = 1$. We want to highlight the effect of difference between the LMSE values on the training set, the validation set, and the test set on the stability of the models, because the larger the difference, the more likely the model is over-fitting.

It can be seen in Fig. 2 that for the systolic pressure index, as the value of tree depth changes, the difference between the LMSE values on the training set, the validation set and the test set is gradually reduced. For example, when the value of tree depth is 7, the LMSE on the training set is smaller while the LMSE on the validation set and the test set is relatively large, which indicates that the model has been over-fitting. Meanwhile, the training time is so long. As the value of tree depth decreases, when the tree depth is 3, the LMSE values on the training set, the validation set and the test set are not much different and the LMSE values on the validation set and the test set are relatively small. And the training time is also greatly shortened. However when the tree depth is 2, the LMSE values on the validation set and the test set become larger. Therefore, we conclude that when the value of tree depth is 3, the LightGBM model does not over-fitting, the training effect and the training time is relatively good. Correspondingly, when the value of tree depth is 3, the score is the highest, which proves the correctness of our scoring formula. Again, we get the optimal values of tree depth for the remaining four indicators. After obtaining the optimal value of tree depth, we fix the tree depth value to the optimal value, and compare the stability of the model under different values of feature fraction on the LightGBM model. The result shown as Fig. 3. From the scores shown in Figs. 2 and 3, we get the optimal values of tree depth and feature fraction of the five biochemical indicators on the LightGBM model, which list in Table 2.

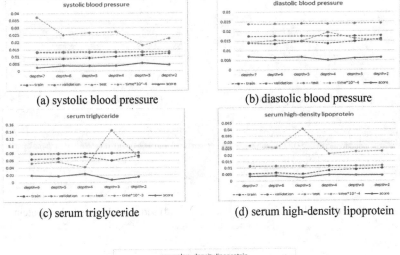

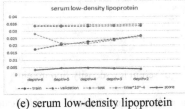

(e) serum low-density lipoprotein

**Fig. 2.** The evaluation results and scores of five biochemical indicators under different values of tree depth on LightGBM model.

We used the same method to obtain the evaluation results under different tree depths and feature scores of the XGBoost model. Since the time is too long, we multiply the time by $10^{-6}$ when calculating the scores. The results are shown in Figs. 4 and 5. The optimal values of tree depth and feature fraction of the five biochemical indicators on the XGBoost model are listed in Table 2.

Table 2 also lists the evaluation results and the CPU time of LightGBM and XGBoost on the verification set. We can see the training effect of the LightGBM algorithm is generally better than the XGBoost algorithm.

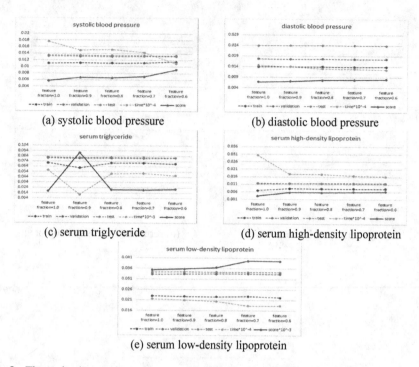

**Fig. 3.** The evaluation results and scores of five biochemical indicators under different values of feature fraction on LightGBM model.

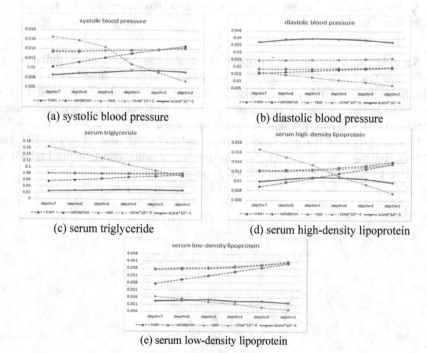

**Fig. 4.** The evaluation results of five biochemical indicators under different values of tree depth on the XGBoost model.

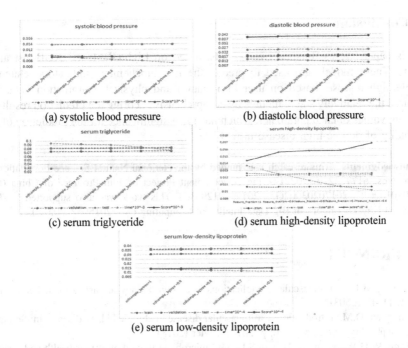

(a) systolic blood pressure

(b) diastolic blood pressure

(c) serum triglyceride

(d) serum high-density lipoprotein

(e) serum low-density lipoprotein

**Fig. 5.** The evaluation results of five biochemical indicators under different values of colsample_bytree on the XGBoost model.

**Table 2.** The optimal values of tree depth and feature fraction on the LightGBM model and the XGBoost model.

| | LightGBM model | | | | XGBoost model | | | |
|---|---|---|---|---|---|---|---|---|
| | Tree depth | Feature fraction | LMSE | Time | Tree depth | Colsaple_bytree | LMSE | Time |
| Systolic blood pressure | 3 | 0.6 | 0.013156 | 110 | 3 | 0.6 | 0.013888 | 65 |
| Diastolic blood pressure | 7 | 0.6 | 0.01753 | 125 | 5 | 0.6 | 0.018515 | 111 |
| Serum triglyceride | 4 | 0.9 | 0.079792 | 11 | 3 | 0.8 | 0.080156 | 114 |
| Serum high-density lipoprotein | 4 | 0.6 | 0.011658 | 162 | 5 | 0.7 | 0.012486 | 106 |
| Serum low-density lipoprotein | 4 | 0.7 | 0.03353 | 183 | 5 | 0.6 | 0.033545 | 189 |

# 5    Conclusion

In this paper, we established two models, based on LightGBM and XGBoost algorithms. These models are used to predict the five biochemical indicators related to double-high disease, based on user information and physical examination data. We acquired the relatively stable models by comprehensive scoring the prediction results of different values of each parameter, which has great significance for the prediction of the morbidity of double-high disease in real life.

**Acknowledgments.** This research was funded by the National Natural Science Foundation of China (Nos. 61772227,61702214), the Development Project of Jilin Province of China (Nos 20170101006JC, 20170203002GX, 20190201293JC). This work was also supported by Jilin Provincial Key Laboratory of Big Date Intelligent Computing (No. 20180622002JC).

# REFERENCES

1. Sarnak, M.J.: Cardiovascular complications in chronic kidney disease. Am. J. Kidney Dis. **41**, 11–17 (2003)
2. Charytan, D.M.: Introduction: cardiovascular disease in chronic kidney disease. In: Seminars in Nephrology, vol. 38, no. 6, p. 541. Elsevier (2018)
3. Tang, W.H.W., Kitai, T., Hazen, S.L.: Gut microbiota in cardiovascular health and disease. Circ. Res. **120**(7), 1183–1196 (2017)
4. Maqbool, M., Cooper, M.E., Jandeleit-Dahm, K.A.M.: Cardiovascular disease and diabetic kidney disease. In: Seminars in Nephrology. vol. 38, no. 3, pp. 217–232. WB Saunders, Philadelphia (2018)
5. Gupta, P., Gan, A.T.L., Man, R.E.K., et al.: Risk of incident cardiovascular disease and cardiovascular risk factors in first and second-generation Indians: the Singapore Indian eye study. Sci. Rep. **8**(1), 14805 (2018)
6. Njie, G.J., Finnie, R.K.C, Acharya, SD, et al.: Peer reviewed: reducing medication costs to prevent cardiovascular disease: a community guide systematic review. Prev. Chronic Dis. **12** (2015)
7. Onuegbu, A.J., Olisekodiaka, J.M., Udo, J.U., et al.: Evaluation of high-sensitivity C-reactive protein and serum lipid profile in southeastern Nigerian women with pre-eclampsia. Med. Princ. Pract. **24**(3), 276–279 (2015)
8. Ke, G., Meng, Q., Finley, T., et al.: LightGBM: a highly efficient gradient boosting decision tree. In: Advances in Neural Information Processing Systems, pp. 3146–3154 (2017)
9. Chen, T., Guestrin, C.: XGBoost: a scalable tree boosting system. In: Proceedings of the 22nd ACM SIGKDD International Conference on Knowledge Discovery and Data Mining, pp. 785–794 (2016)
10. https://github.com/Microsoft/LightGBM
11. https://github.com/Microsoft/LightGBM/blob/master/docs/Parameters-Tuning.rst
12. Friedman, J.H.: Greedy function approximation: a gradient boosting machine. Ann. Stat. **29** (5), 1189–1232 (2001)
13. https://github.com/fxsjy/jieba
14. Hsu, C.W., Lin, C.J.: A comparsion of methods for multiclass support vector machines. IEEE Trans. Neural Netw. **13**(2), 415–425 (2002)

# Control Emotion Intensity
# for LSTM-Based Expressive Speech
# Synthesis

Xiaolian Zhu[1,2](✉) and Liumeng Xue[1]

[1] School of Computer Science, Northwestern Polytechnical University,
Xi'an, Shaanxi, China
xiaolianzhu@mail.nwpu.edu.cn, lmxue@nwpu-aslp.org
[2] Public Computer Education Center, Hebei University of Economics and Business,
Shijiazhuang, Hebei, China

**Abstract.** To improve the performance of human-computer interaction interfaces, emotion is considered to be one of the most important factors. The major objective of expressive speech synthesis is to inject various expressions reflecting different emotions to the synthesized speech. To effectively model and control the emotion, emotion intensity is introduced for expressive speech synthesis model to generate speech conveyed the delicate and complicate emotional states. The system was composed of an emotion analysis module with the goal of extracting control emotion intensity vector and a speech synthesis module responsible for mapping text characters to speech waveform. The proposed continuous variable "perception vector" is a data-driven approach of controlling the model to synthesize speech with different emotion intensities. Compared with the system using a one-hot vector to control emotion intensity, this model using perception vector is able to learn the high-level emotion information from low-level acoustic features. In terms of the model controllability and flexibility, both the objective and subjective evaluations demonstrate perception vector outperforms one-hot vector.

**Keywords:** Emotion intensity · Expressive speech synthesis ·
Controllable text-to-speech · Neural networks

## 1 Introduction

Speech synthesis, also known as text-to-speech (TTS), is to convert the text sequence into speech waveform. It is one of the most important human-computer interfaces that has been used extensively in navigation devices, smart home, audio books and education. Thanks to the development of deep learning

This document is the results of the research project funded by Natural Science Foundation of Hebei University of Economics and Business (No. 2016KYQ05).

© Springer Nature Singapore Pte Ltd. 2019
R. Mao et al. (Eds.): ICPCSEE 2019, CCIS 1059, pp. 645–656, 2019.
https://doi.org/10.1007/978-981-15-0121-0_51

techniques, the quality of generated speech from TTS systems has been improved significantly. Intelligibility, naturalness and expressivity are the commonly used criteria for evaluating high-quality synthesized speech. Intelligibillity describes the clarity of the synthesized audio, specifically how well a listener is able to extract the original message. Naturalness describes the overall ease of listening, global stylistic consistency, regional or language level nuances [1]. Now, state of art speech synthesis systems is able to generate intelligible speech with good naturalness [1–5]. However, expressivity is still the main challenge to be overcome that TTS system achieves the variability of speech. So expressive speech synthesis with the goal of generating pleasant and emotional speech attracted increasing attention on the field of TTS.

The expression is defined as the vocal indicator of various emotional states that reflect on the speech waveform. The different emotions and speaking styles are also considered as expressions. Various techniques have been used for expressive speech synthesis of previous decades.

The earlier unit selection synthesis approach [6–8] is able to synthesize natural speech with high quality that sounds like an original speaker on condition that there is a very large corpus database. The disadvantage is that a limited-size speech database results in low quality and inconsistent voice (e.g. minor word become unclear). Moreover, it is short of controllability and flexibility and so we can not control the unit select synthesis system to change prosody or emotion.

Another general TTS is hidden markov model (HMM) based synthesis approach. Thanks to their adaptation capabilities, HMM has been significantly successful in the TTS task. It is still in use even nowadays for both high number of emotional classes (e.g. the basic categories: anger, happiness, disgust, fear, sadness and surprise) [9] and a low number of emotional classes (e.g. emphasis and neutral). It is able to produce consistent and intelligible speech with relatively little input data and to be adapted with a small amount of data even only 5 sentences [10]. Moreover, it can generate speaker-independent emotional speech models that learn linguistic information from different speakers and different emotions and provide high controllability and reliability. Some studies have been proposed from the style control technique of emotional classes (joyful and sad) [11] to emotion transplantation of 4 emotional classes [12,13].

Although HMM-based TTS has many advantages, it cannot synthesize natural speech because there are some problems of acoustic model low accuracy, vocoder quality, and over smoothing [14]. In recent years, long short-term memory recurrent neural networks (LSTM-RNNs) significantly outperform the HMM-based approach owing to modeling the deep representation of long time span contextual features [15]. To estimate the mapping function of output acoustic features given input linguistic features, LSTM-based TTS uses a multi-layer neural networks [16–19]. Furthermore, in order to improve expressiveness, some methods using input code have been made on emotion [20], speaker adaptation [21,22] and global prosodic [23].

Recent developments of the end-to-end speech synthesis have led to a renewed interest in encoder-decoder TTS architecture. Tacotron is a modern end-to-end framework that integrates a text analysis front-end, an audio synthesis module and an acoustic module into a comprehensive model [2]. It inputs a text sequence and outputs acoustic parameters sequence then reconstructs the speech waveform by Griffin-Lim algorithm. To control and transfer the speaking style, global style tokens (GST) based on Tacotron was proposed by Wang et al. [24]. Moreover, to improve sound quality, an updated version Tacotron2 was proposed by Skerry-Ryan et al. [3]. However, these expressive speech synthesis models are more likely to suffer from two problems in synthesizing speech with delicate-emotion-variation in emotion expression.

- Only generating discrete emotion. Current expressive speech synthesis systems cannot produce speech reflecting the delicate and complicated of the emotion state owing to the coarse-grained emotion modeling for six basic emotions.
- Lack of controllability and flexibility. Emotion-dependent models were usually trained using adaptation technology, which is difficult for users to control and generate speech with rich emotion-expressive information.

We investigate a novel expressive speech synthesis based on LSTM, which is proven to generate speech with different emotion intensity. Our contributions include: (1) Through audio analysis, we present emotion intensity as a fine-grained modeling unit to solve the problem of coarse-grained modeling on discrete emotion. It could improve model expressivity for synthesizing rich emotional information. (2) We propose a perception vector, a low-dimensional continuous variable vector which is extracted from high-dimensional emotion space, to control emotion intensity in TTS system. It can improve model controllability and flexibility to manipulate emotion intensity.

The rest of the paper is organized as follows: Sect. 2 describes the framework of our expressive speech synthesis model. Section 3 analysis emotion-related acoustic features, clusters emotion intensity level and introduces two control vectors: one-hot vector and perception vector. Section 4 explains the LSTM-based speech synthesis module. Section 5 presents the details of the experimental results and analysis. In Sect. 6, we draw some conclusions.

## 2   The Main Framework of Control Emotion Intensity TTS

In this section, we give an overall description of our expressive speech synthesis framework which could control emotional intensity. As shown in Fig. 1, the expressive TTS system is composed of a speech synthesis module and an emotion analysis module. The emotion analysis module demonstrates the procedure of audio processing to get emotion intensity control vector, which includes emotion feature extraction, emotion intensity level and emotion intensity control vector.

The speech synthesis module is to convert a text sequence to an audio sequence. It has a text frontend extracting various linguistic features, an acoustic

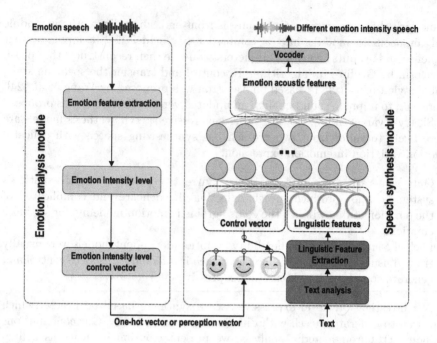

**Fig. 1.** The model architecture. The model composes of a speech synthesis module and an emotion analysis module.

feature prediction model and a complex signal-processing-based vocoder. We augmented a control vector with linguistic features as network input to represent emotion intensity.

## 3 Emotion Analysis Module

In order to design a powerful expressive TTS system can synthesis emotional speech, it is helpful to find the answers to three questions:

- Feature issue: what features can represent the emotion intensity?
- Domain issue: how to distinguish the emotion intensity levels?
- Control issue: what technique can control the emotion intensity levels?

### 3.1 Emotion Feature Extraction

To answer the first question, we investigated many acoustic features extracted from speech to find suitable features presenting emotion intensity. Pitch, energy, duration and speech rate are commonly used to represent an emotion. For instance, sad speech is spoken slowly and softly as well as is indicated with a lower pitch, narrower range of pitch, low speaking rate and low-frequency energy. On the contrary, happy speech is spoken quickly and loudly as well as is indicated with a higher pitch, wider range of pitch, high speaking rate and high-frequency energy.

In our study, we computed the statistics of pitch, energy, speech rate and use these features to represent emotion intensity. The statistics of pitch and energy include mean, range (maximum–minimum), variance. The speech rate is calculated in the number of syllables spoken per second on condition that the first silence and the last silence of the sentence were deleted, while the silence between the voiced segments was retained. The purpose of handling silence is to reduce incorrect judgment of the emotion intensity owing to the long silence. The total dimension of emotion features is 7.

## 3.2   Emotion Intensity Level

To answer the second question, we investigated how many intensity levels are reasonable for listeners to perceive. We used an unsupervised method, $k$-means clustering algorithm, to partition one emotion into $k$ emotion intensities because of the speech database without emotion intensity labels.

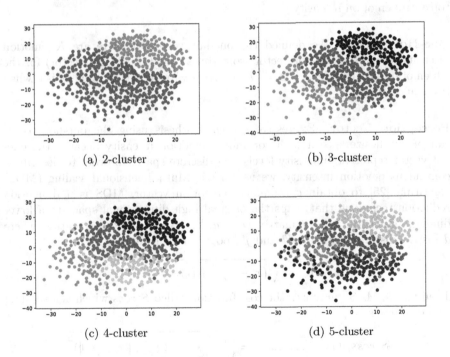

**Fig. 2.** Different cluster number of emotion strength for happiness.

As shown in Fig. 2, we compared 4 different partition methods for happy speech corpus. We prefer the 3-cluster approach to other three partition methods because the 3-cluster is not only clear distinguishable but also not overlap that denotes it can represent emotion intensity levels with fine-grained emotion Fig. 2(b). Although the 2-cluster is able to distinguish emotion intensity clearly (Fig. 2(a)), we believe the more clusters, the better fine-grained emotion intensity levels. But the 4-cluster Fig. 2(c) and the 5-cluster Fig. 2(d) are not good partition methods because they are overlapped. Hence, we divide one emotion into three intensity levels: weak, moderate and strong.

Although there are 3 emotion intensity levels, we are still unable to identify the weak, moderate and strong intensity levels of speech. To solve this problem, 100 utterances were randomly selected from each emotion intensity level and were assessed by 3 experts according to their perceptual evaluation. The experts made consistent judgments that emotion intensity is corresponding to the cluster. In Fig. 2(b), the emotion intensity enhances as the values increase. Therefore, the weak emotion intensity locates on the bottom left of Fig. 2(b) and the strong emotion intensity locates on the top right.

### 3.3   Emotion Intensity Level Control Vector

To answer the third question, we use the one-hot vector and perception vector to control emotion intensity.

**One-Hot Vector.** We defined the one-hot vector: if there are $K$ emotion intensity levels, the one-hot vector (sometimes called the 1-or-k vector) for the $i$th emotion intensity level is $e_i = (e_1, e_2, \ldots, e_K)$ where each value $e_k$ is 0 when $k \neq i$ and 1 when $k = i$.

**Perception Vector.** Expressive speech synthesis using the one-hot vector cannot synthesize speech with continuous emotion intensity because the one-hot vector represents intensity levels in a discrete space. In order to describe a continuous emotion intensity, we use metric Multidimensional scaling (MDS) algorithm [25] to obtain continuous perception vector. MDS is a dimension reduction technique that maps the original high dimensional space to a lower dimensional space. A $t \times t$ matrix $D = [d_{ij}]$ is called a distance matrix, where $d_{ij}$ is the distance between $i^{th}$ and $j^{th}$ point and is computed by

$$d_{ij} = \sqrt{(x_i - x_j)^2 + (y_i - y_j)^2} \tag{1}$$

Then metric MDS minimizes the cost function called *Stress* which is a residual sum of squares:

$$Stress_D(x_1, x_2, \ldots, x_N) = \sqrt{\sum_{i \neq j = 1, \ldots, N} (d_{ij} - \|x_i - x_j\|)^2} \tag{2}$$

or

$$Stress_D(x_1, x_2, \ldots, x_N) = \sqrt{\frac{\sum_{i,j} (d_{ij} - \|x_i - x_j\|)^2}{\sum_{i,j} d_{ij}^2}} \tag{3}$$

Finally, emotion features in 7-dimension space extracted beforehand are reduced to low 2-dimension space as continuous variable vector for controlling emotion strength.

## 4  Speech Synthesis Module

The schematic diagram of LSTM-based speech synthesis module is shown in the right part of Fig. 1. Speech production can be seen as a process to select spoken words, formulate their phonetics and then finally articulate output speech with the vocoder. Hence, it is a continuous physical dynamic process. LSTM can simulate human speech production by a layered hierarchical and wide in time scale structure to transform linguistic text information into its final speech output. In a TTS module, where usually a whole sentence is given as input, there is no reason not to access long-range context in both forward and backward directions.

In our experiments, rich contexts were used as input features, which contain the binary features for categorical contexts, e.g. phone labels, POS labels of the current word, and TOBI labels, and numerical features for the numerical contexts, e.g., the number of words in a phrase or the position of the current frame of the current phone. The output features are acoustic features like linear spectral pairs and fundamental frequency. Input features and output features are time-aligned, frame-by-frame by a well-trained LSTM model.

## 5  Experimental Results and Analysis

We carried out a series of experiments to evaluate the performance of the proposed controllable expressive speech synthesis, specifically focusing on emotion controllability by subjective evaluation (mean opinion score test and AB preference test) and model flexibility by objective evaluation.

### 5.1  Experimental Setup

**Dataset.** We used a speech corpus that contains twelve hours of high-quality speech which is recorded by a Chinese female speaker. The database contains 10,000 neutral recordings, 586 happy recordings and 628 angry recordings. All the recordings were down-sampled from 44K to 16K.

**Experimental Settings.** The model structure was three feed-forward layers, two bidirectional LSTM layers and a linear output layer. Each layer has 512 units. We used ReLU as the activation function. All models were optimized using Adam optimizer. The networks were trained with an initial learning rate of 0.0001. All the experiments were carried out using TensorFlow. We used STRAIGHT vocoder [26] to extract $F_0$ in log-scale at 5-ms frame shift. 41-dimensional line spectral pairs (LSP). The final acoustic features included 41-dimensional LSP, one extra binary voiced/unvoiced flag and 9-dimensional $F_0$ (previous 4 frames, current frame and proceeding 4 frames). The total dimension was 51.

## 5.2    Emotion Control Evaluation: Mean Opinion Score Test

Manipulating the control vector can enhance or decrease the emotion intensities. For the one-hot vector, (1, 0, 0), (0, 1, 0) and (0, 0, 1) referred weak, moderate and strong emotion intensities, respectively. We can assign perception vector real values in the 2-dimensional continuous space. In our experiments, the perception vectors were assigned $(-30, -30)$, $(0, 0)$ and $(30, 30)$ to represent weak, moderate and strong emotion intensities, respectively. The values were bigger, the emotion intensity was higher. The goal of the control evaluation was to estimate how much we can manipulate the emotional expression for TTS system.

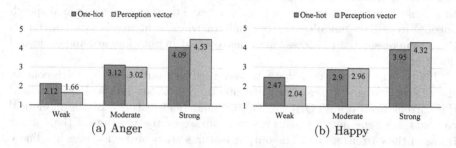

**Fig. 3.** Mean opinion score for emotion control evaluation.

Mean opinion score (MOS) is a measure used in the domain of quality of experience, which is usually gathered in a subjective quality evaluation test. We randomly chose 30 synthesized utterances from the testing set. Twenty Mandarin listeners including 5 females and 15 males were asked to rate the emotion intensities of generated speech from TTS system. In our experiments, the MOS is expressed as a single rational number, in the range 1–5, where 1 is no emotion, and 5 is the strongest emotion.

We selected two emotions, happy and angry, to conduct evaluation tests on. Figure 3 shows the MOS results that demonstrate both the perception vector and one-hot vector can control the emotion intensity. For angry emotion, the perception vector obtained 1.66 on weak intensity level and 4.53 on strong intensity level, respectively. And for happy emotion, the perception vector obtained 2.04 at weak intensity level and 4.32 at strong intensity level, respectively. This means that the perception vector has better controllability than the one-hot vector because the perception vector can be assigned arbitrary real values other than the one-hot vector just only can be 0 or 1. The goal of this study is to investigate controlling the emotion intensity so the moderate emotion intensity level is not considered.

## 5.3    Emotion Control Evaluation: AB Preference Test

An AB preference test was conducted to evaluate the model controllability of emotion intensities. We randomly selected 30 pairs of testing speech and asked

20 listener to rank them. For each pair, listeners were presented with two speech samples and asked to indicate which one has stronger emotion intensity. The question was "which one sounds stronger". The choices were: "A", "B" and "no preference".

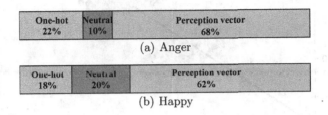

(a) Anger

(b) Happy

**Fig. 4.** The results of AB preference test: weak emotion intensity level.

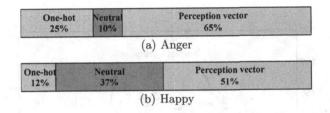

(a) Anger

(b) Happy

**Fig. 5.** The results of AB preference test: strong emotion intensity level.

From Figs. 4 and 5, we can see that the perception vector is superior to one-hot vector in terms of the performance of emotion controllability. Listeners significantly preferred the perception vector over one-hot vector with $p$-value $< 0.01$ on angry and happy emotion.

### 5.4  Model Flexibility Evaluation

From Fig. 6, we can see the 2-dimensional perception vectors (indicated by green, red and blue dots) that were used to train the expressive speech synthesis model. Six different perception vectors (indicated by black dots) were used to synthesize six repetitions of a single speech during the testing period. These perception vectors for testing were $(-50, -50)$, $(-30, -30)$, $(-10, -10)$, $(10, 10)$, $(30, 30)$ and $(50, 50)$, which evenly distributed along the diagonal line. The emotion intensity was enhanced as the values of perception vector increased.

The $F_0$ trajectories of a single sentence synthesized using six perception vectors are shown in Fig. 7. Although expressive speech synthesis model generated a single sentence, the $F_0$ trajectories of the speech are very different while manipulating perception vectors. When the values of perception vectors

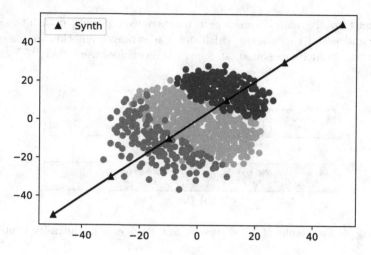

**Fig. 6.** Perception vector learned in training and used at synthesis time. (Color figure online)

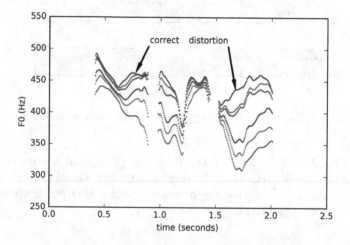

**Fig. 7.** $F_0$ of a single sentence, synthesized using six perception vectors.

increase, the values of $F_0$ also increase. However, the variation of $F_0$ trajectories is a complicated process other than a simple global shift along the vertical axis. It is important to note that even for the same $F_0$ trajectory the values change dramatically in some places, while in other places they change gently. For example, when perception vector is (50, 50), it plays a positive role in correcting $F_0$ contours and also cause negative effects to result in unwanted distortion of speech. In our experiments, when the perception vector was (30, 30), the expressive speech synthesis model generated strong emotional and natural speech. Moreover, in terms of the model flexibility, perception vectors are superior to one-hot vectors because perception vectors can be set to any values in continuous space while one-hot vectors just only are set 0 or 1 in discrete space.

## 6   Conclusions

In this paper, we investigated a novel expressive speech synthesis system that can effectively control the emotion intensity and synthesize emotional speech. The system is composed of an emotion analysis module with the goal of extracting control emotion intensity vector and a speech synthesis module responsible for mapping text characters to acoustic parameters. The perception vector is a continuous variable to control emotion intensities, which is extracted from the emotion analysis module. We compared the performance of two models using different control vector: the perception vector and one-hot vector. In terms of the model controllability and flexibility, both the objective and subjective evaluations demonstrate perception vector outperforms one-hot vector.

## References

1. Sotelo, J., et al.: Char2wav: end-to-end speech synthesis (2017)
2. Wang, Y., et al.: Tacotron: a fully end-to-end text-to-speech synthesis model. arXiv preprint (2017)
3. Skerry-Ryan, R., et al.: Towards end-to-end prosody transfer for expressive speech synthesis with tacotron. arXiv preprint arXiv:1803.09047 (2018)
4. Arik, S.O., et al.: Deep voice: real-time neural text-to-speech. arXiv preprint arXiv:1702.07825 (2017)
5. Li, N., Liu, S., Liu, Y., Zhao, S., Liu, M., Zhou, M.: Close to human quality TTS with transformer. arXiv preprint arXiv:1809.08895 (2018)
6. Hunt, A.J., Black, A.W.: Unit selection in a concatenative speech synthesis system using a large speech database. In: 1996 IEEE International Conference on Acoustics, Speech, and Signal Processing Conference Proceedings, ICASSP 1996, vol. 1, pp. 373–376. IEEE (1996)
7. Bulut, M., Narayanan, S.S., Syrdal, A.K.: Expressive speech synthesis using a concatenative synthesizer. In: Seventh International Conference on Spoken Language Processing (2002)
8. Pitrelli, J.F., Bakis, R., Eide, E.M., Fernandez, R., Hamza, W., Picheny, M.A.: The IBM expressive text-to-speech synthesis system for American English. IEEE Trans. Audio Speech Lang. Process. **14**(4), 1099–1108 (2006)
9. Cabral, J.P., Saam, C., Vanmassenhove, E., Bradley, S., Haider, F.: The ADAPT entry to the Blizzard challenge 2016. In: 2016 Proceedings of the Blizzard Challenge (2016)
10. Yamagishi, J., Onishi, K., Masuko, T., Kobayashi, T.: Acoustic modeling of speaking styles and emotional expressions in HMM-based speech synthesis. IEICE Trans. Inf. Syst. **88**(3), 502–509 (2005)
11. Nose, T., Kobayashi, T.: An intuitive style control technique in HMM-based expressive speech synthesis using subjective style intensity and multiple-regression global variance model. Speech Commun. **55**(2), 347–357 (2013)

12. Lorenzo-Trueba, J., Barra-Chicote, R., Watts, O., Montero, J.M.: Towards speaking style transplantation in speech synthesis. In: Eighth ISCA Workshop on Speech Synthesis (2013)
13. Lorenzo-Trueba, J., Barra-Chicote, R., San-Segundo, R., Ferreiros, J., Yamagishi, J., Montero, J.M.: Emotion transplantation through adaptation in HMM-based speech synthesis. Comput. Speech Lang. **34**(1), 292–307 (2015)
14. Zen, H., Tokuda, K., Black, A.W.: Statistical parametric speech synthesis. Speech Commun. **51**(11), 1039–1064 (2009)
15. Zen, H.: Acoustic modeling in statistical parametric speech synthesis-from HMM to LSTM-RNN. In: Proceedings of the MLSLP (2015)
16. Ze, H., Senior, A., Schuster, M.: Statistical parametric speech synthesis using deep neural networks. In: 2013 IEEE International Conference on Acoustics, Speech and Signal Processing, pp. 7962–7966. IEEE (2013)
17. Fan, Y., Qian, Y., Xie, F.-L., Soong, F.K.: TTS synthesis with bidirectional LSTM based recurrent neural networks. In: Fifteenth Annual Conference of the International Speech Communication Association (2014)
18. Uria, B., Murray, I., Renals, S., Valentini-Botinhao, C., Bridle, J.: Modelling acoustic feature dependencies with artificial neural networks: Trajectory-RNADE. In: 2015 IEEE International Conference on Acoustics, Speech and Signal Processing (ICASSP), pp. 4465–4469. IEEE (2015)
19. Zhu, X., Xie, L., Chen, X., Lou, X., Zhu, X., Tan, X.: Controlling expressivity using input codes in neural network based TTS. In: 2018 First Asian Conference on Affective Computing and Intelligent Interaction (ACII Asia), pp. 1–6. IEEE (2018)
20. Li, B., Zen, H.: Multi-language multi-speaker acoustic modeling for LSTM-RNN based statistical parametric speech synthesis. In: INTERSPEECH 2016, pp. 2468–2472 (2016)
21. Huang, Z., Tang, J., Xue, S., Dai, L.: Speaker adaptation of RNN-BLSTM for speech recognition based on speaker code. In: 2016 IEEE International Conference on Acoustics, Speech and Signal Processing (ICASSP), pp. 5305–5309. IEEE (2016)
22. Luong, H.-T., Takaki, S., Henter, G.E., Yamagishi, J.: Adapting and controlling DNN-based speech synthesis using input codes. In: 2017 IEEE International Conference on Acoustics, Speech and Signal Processing (ICASSP), pp. 4905–4909. IEEE (2017)
23. Watts, O., Wu, Z., King, S.: Sentence-level control vectors for deep neural network speech synthesis. In: Sixteenth Annual Conference of the International Speech Communication Association (2015)
24. Wang, Y., et al.: Style tokens: unsupervised style modeling, control and transfer in end-to-end speech synthesis. arXiv preprint arXiv:1803.09017 (2018)
25. Cox, T.F., Cox, M.A.: Multidimensional Scaling. Chapman and Hall/CRC, New York (2001)
26. Kawahara, H., Masuda-Katsuse, I., De Cheveigne, A.: Restructuring speech representations using a pitch-adaptive time-frequency smoothing and an instantaneous-frequency-based F0 extraction: possible role of a repetitive structure in sounds. Speech Commun. **27**(3–4), 187–207 (1999)

# License Plate Recognition Model Based on CNN+LSTM+CTC

Hangming Zhang[1], Feng Sun[1], Xiaopu Zhang[1], and Lingling Zheng[2(✉)]

[1] College of Instrumentation and Electrical Engineering, Jilin University,
Changchun 130062, China
[2] The State Key Laboratory of Automotive Simulation and Control,
Jilin University, Changchun 130061, China
zhenglingling@jlu.edu.cn

**Abstract.** With the continuous improvement of the social and economic level, the number of vehicles has exploded in the city, and traditional manual identification license plates have been unable to meet the demand. In this paper, a Convolutional neural network (CNN)-based license plate recognition system is designed. The recognition module uses the CNN+LSTM+CTC model to simplify the convolutional layer structure to adapt to the lightweight training mode. The two-way LSTM structure is used to learn from both sides of the license plate to enhance the end-to-end recognition effect. Compared with the traditional scheme, the CTC loss calculation method eliminates the need for character alignment, streamlines the steps, and improves the recognition accuracy. The experiment shows that the license plate recognition software system designed in this paper has a high recognition accuracy rate of 98.59%.

**Keywords:** License plates · Neural network · Model · LSTM · CTC · Recognition

## 1 Introduction

Since the car was invented, the control of the relevant traffic flow relied on manpower. Because the productivity at that time is not enough, the car's possession is not large, so manpower can solve the corresponding problem well. Nowadays, with the entry of the national economy into a new normal, a socialist well-off society is about to be fully completed, and the number of households owned by the nation has risen sharply. In the first decade of the 21st century, major first-line and second-tier cities suffered from serious traffic diseases, traffic congestion became increasingly serious, and traffic conditions continued to deteriorate. In order to alleviate traffic pressure, some cities have adopted measures such as restrictions and restrictions. However, traditional measures are not able to solve the corresponding problems well. Manual identification of license plates still has inefficiencies. Urban traffic management is in urgent need of new technology [1].

The smart transportation project that has recently emerged in major cities across the country is committed to improving the current state of the traffic environment with new technologies. The license plate recognition system is an indispensable part of the whole

© Springer Nature Singapore Pte Ltd. 2019
R. Mao et al. (Eds.): ICPCSEE 2019, CCIS 1059, pp. 657–678, 2019.
https://doi.org/10.1007/978-981-15-0121-0_52

process [2]. It is like the eyes of city managers. It can identify license plates under various road conditions; timely punish urban speeding and red light vehicles; combine with the Public Security Bureau database to analyze suspicious vehicles and arrest illegal elements; count the number of license plates for traffic warning and reduce the occurrence of congestion; charging fees, parking management, etc. in highway toll stations, community parking lots.

In the last century, the license plate recognition process first studied by foreign scholars was divided into four modules. First, the license plate is pre-processed, the image is converted into a suitable gray scale image, and then the relevant license plate area is found, then each character is segmented, and finally the identification license plate number is performed [3]. Although the foreign license plate recognition system has been developed very early in the country, the situation of the license plate itself varies widely [4], and the policies and regulations are different. The Chinese motor vehicle license plate contains the Chinese characters, which are more difficult characters. Etc., it is not possible to directly use foreign identification software, and it is necessary to independently develop a system suitable for the national conditions. The wave of deep learning that has erupted in recent years has given new technical support to the field of license plate recognition. The model of human key point detection, speech recognition, etc. will be transferred to the license plate recognition, which will achieve good results.

This paper combines the latest convolutional neural network model to design a license plate recognition software system, and carried out no less than 200 pictures test, the recognition accuracy needs to be higher than 95%, trying to help solve the current natural scene license plate Identify the problem.

## 2 Related Work

Traditional license plate recognition technology stems from OCR technology. The traditional OCR is based on digital image processing and statistical machine learning. Based on such technologies, developed countries have achieved good accumulation in the 1990s.

Since the new century, with the outbreak of deep learning technology, various effective, convenient and practical networks have emerged, such as SSD, RCNN, FAST RCNN, etc., which have been used in tasks such as target detection and semantic segmentation. The recognition accuracy is constantly refreshed in the license plate recognition task. At the same time, computer hardware, computer software technology continues to improve, camera clarity continues to improve, and the field of license plate recognition has been steadily improving.

Up to now, most developed countries have license plate recognition systems suitable for their own national conditions, such as the automatic identification system for the license of the IPI company called ANPR, the See/Car system of Hi-Tech, the VLPRS of Optasia of Singapore, Germany. Siemens' ARTEM7S system, etc. [6]. However, unlike the letters used in some developed countries, Chinese license plates contain hieroglyphic Chinese characters, and the license plates of different countries and their usage specifications vary widely. Therefore, the identification systems of other countries cannot be used directly. China needs independent research and development. Since the 1990s, China has been paying close attention to the development and application of license plate recognition systems, and has achieved initial results. Since the new century, computer vision-related companies have made some progress driven by the huge domestic market. At present, the relatively mature products are the license plate recognition system developed by Hangzhou Hikvision, the Hanwang eye of Hanwang Company of the Institute of Automation of Chinese Academy of Sciences, the Huiguang license plate number recognition system of Hong Kong Asia Vision Technology Co., Ltd., etc. Good performance [7], can achieve more than 95% recognition success rate.

In the traditional positioning scheme, the positioning module is in the first step position in the license plate recognition, and its success has a decisive influence on the subsequent identification. If the error is fully identified, such as when troubleshooting a non-license card area, positioning other areas similar to the license plate is catastrophic for the entire system. The traditional positioning scheme has many shortcomings in its implementation. First of all, the above situation often occurs. If the probability of occurrence of this time is to be minimized, complex algorithm preparation is required, which brings a huge workload. Using machine learning methods, as long as the database is large enough, it can theoretically solve such problems to a large extent. The following are related cases of positioning failure (Figs. 1, 2 and 3).

**Fig. 1.** The case where the tilt angle is large

**Fig. 2.** Misplacement to the logo of car

**Fig. 3.** Misplacement to a similar area of car license plate

The traditional scheme requires character segmentation. In the case where characters are stuck and the license plate is deformed, the segmentation is likely to cause great errors, which will bring great difficulties to subsequent character recognition.

Even if the pre-processing improves the image quality, the combination of individual modules will produce errors that cannot be ignored. In the same way, the optimization of each independent process is placed in the whole system, and the generalization ability will be greatly reduced. Therefore, the traditional recognition method has a low recognition success rate in complex natural scenes, and the recognition effect is limited.

# 3 Algorithm Design

## 3.1 Design of Positioning Scheme Based on MASK RCNN

### 3.1.1 Introduction to MASK RCNN Model

MASK RCNN is a model published by the HeKaiming team in 2018. Born out of the Faster RCNN and enhanced performance based on the original model. Faster RCNN was originally a framework for target detection, and it was not possible to perform instance segmentation, while MASK RCNN was able to achieve pixel-level recognition and segmentation. It was used at the beginning of its birth to identify various areas of the human body and achieved excellent results [21].

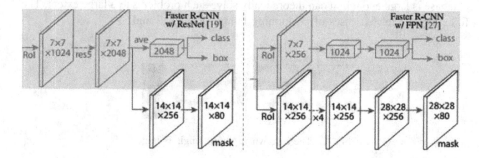

**Fig. 4.** MASK RCNN network structure

Figure 4 shows the MASK RCNN model invented by the HeKaiming team, evolved from the Faster RCNN. The Faster RCNN has two output branches, the category label and the bounding box offset. The innovation of MASK RCNN is that it adds the mask part, which can split the target area in the whole picture. At the same time, this new model has a problem that seems to be trivial and in fact significant to solve the problem of pixel misalignment. Based on ROI Pooling, a refined and practical RolAlign layer [21] was designed. In addition, RolAlign has an unexpected effect, it can link the program to extract features without intermediate batch processing [21].

The loss function of MASK RCNN is L = Lbox + Lcls + LMASK. The loss function on the category label is Lcls, LMASK is the loss function of the MASK part, the specific value is k * m * m, and the K categories can map a m * m mask [21]. Compared to the Faster RCNN, it adds a loss function evaluation by a mask.

MASK RCNN can achieve pixel level segmentation with fast speed and high efficiency. The training process is simple. It takes at least two days for each frame to run for 200 ms on a computer with 8 GPUs. This time-consuming does not seem to be suitable for this topic. However, considering that the COCO data set contains more than 90 types of objects, 328,000 pictures, and 2,500,000 labels, the complexity is far greater than the data set recognized by the license plate. Therefore, the model can be used for license plate detection in combination with the means of migration learning.

### 3.1.2    Data Set Preprocessing

The experimental data has a total of 1065 license plate photos. The weather conditions cover sunny and cloudy days. The time includes noon, dusk, night, etc. It also collects a certain number of angles of boring, approximate regional interference, and severely defaced photos. The image is first manipulated to a width of 800. After the collection is complete, the data set is divided into a training set and a test set. Figure 5 is part of the data used in this article.

**Fig. 5.**  Part of the data set used in this experiment

For the collected data set, the first task is to label set, this article uses VGG Image Annotator. Enter a picture in the tool and mark the specific location of the license plate in the image, as shown in Fig. 6. After the labeling is completed, the file with the format .json is output. The four-dimensional matrix information of each item shows the position of the marked license plate, which is the license plate coordinates and the length and width. The last output marker position information is shown in Fig. 7. When you mark the image, you need to use a picture type. Otherwise, when you input .json, the suffix will be repeated repeatedly, which will waste a lot of time. After the debug run, you need to set the train and val files. In the annotation file, you need to associate the two annotation files with the files in each folder. Otherwise, the running program will report an error.

**Fig. 6.** Using the VIA labeling process

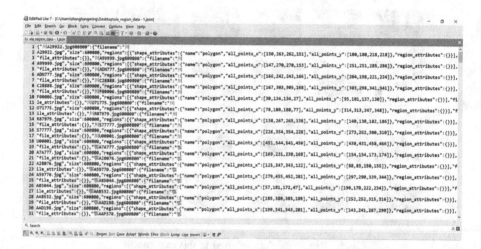

**Fig. 7.** Marked part of the license plate location

### 3.1.3    Pre-setting of Related Parameters

In the parameter category of ShapeConfig, because the original model uses the main component of Resnet's network structure, the requirements for the graphics card are very high, and the images are all set to be long by 800, which can be divisible by 2 to the 5th power. When IMAGE_PRE_GPU is set to 2, the training cannot be performed normally, and out of memory occurs, so the value is reduced to 1, and GPU_COUNT is also adjusted to 1. NUM_CLASSES is the category setting. Because the license plate area framed in this design can be regarded as foreground, and the parameter is set to mask 1, the area other than the license plate frame can be regarded as the background, and the parameter is set to mask 0, so the category value of the first category is Set to 2; in the second major category, the data of the four corners of the license plate should be considered, so it can be divided into four categories, which are the coordinates of the

XY axis of the center point of the rectangle and the length and width of the rectangle. In the MAX_DIM and MIN_DIM of the image, you can set the size of the image to meet the needs of the model.

### 3.1.4    Software and Hardware Environment and Data Summary

Hardware environment: graphics chip: NVIDIA GeForce GTX 1060 Max-Q; memory capacity: 6 GB; memory type: GDDR5; hard disk: 128 GB PCIe solid state drive +1 TB mechanical hard disk; memory: 8 GB DDR4 2666 MHz; CPU: Intel Core i5-8300H quad core Processor 4.0 GHz. Software environment: IDE development environment: PyCharm 5.0.3; operating system: Windows 10; cuda 8.0.61; python 3.6.3; tensorflow; OpenCV-3.4.3

### 3.1.5    MASK RCNN Based Positioning Model Training and Detection

Train the neural network on the gpu using tensorflow. If you are fully autonomous, it takes at least a few days, so you can use migration learning. First download the model weights trained using the COCO data set. Although there are no related tags for license plate location in this data set, many common features in everyday scenes have been covered. Then you can continue to train after making adjustments that suit your situation based on the weight of MASK_rcnn_coco.h5. Using this program can greatly reduce training time and improve efficiency. Because the amount of COCO dataset is large enough, the content is complex enough, and most of the datasets have similar underlying features, the weights trained by different datasets can be used together to greatly reduce workload, speed up training, and improve accuracy.

When the training is finished, open the log and find the .h5 file at the end of the log folder, which is the model obtained by the training. Find the bottom .h5 file, which is the latest round of training results. In general, this file is the best performing and highest accuracy model if no overfitting occurs. Model training effects can be evaluated in inspect_model.ipynd. If the weights in the COCO dataset are directly used, the correct positioning rate is only 34%. When the number of steps in each iteration is set to 30, the success rate is increased to 54%, and the number is increased to 100 steps. When the number of iterations is increased to 30, the loss is reduced to 0.0974, and the correct rate is 90.3%. Test 213 test sets, the coarse positioning accuracy rate reached 89.2%, and the coarse positioning success rate was calculated according to the following formula.

Success rate of rough positioning

$$= \frac{\text{MASK Basic number of license plates covering all characters}}{\text{the number of car license plates}} \times 100\%$$

$$(1)$$

### 3.1.6    Minimum Enveloping Rectangle Correction

However, MASK sometimes does not have a good correction effect on the corner points. Sometimes it is too tight, and the related characters are accidentally injured, resulting in too much cropping. This will bring great difficulties to subsequent character recognition and correct rate compliance. For the final recognition success rate to be above 95%, it must be fairly accurate during the positioning phase. Therefore, it is necessary to take measures to solve the problem that the edge characters are incorrectly tailored.

After many tests and experiments, it was found that the position of the character that was mistakenly cut was often in the depression of the surface. Therefore, a minimum enveloping rectangle correction algorithm is designed. The MASK mask is connected to the depression of the region to fill, and the approximate rectangle closest to the MASK region is obtained, thereby determining four vertices to continue the subsequent operations. The function minAreaRect is used. It belongs to the RotatedRect class function and is used to calculate the minimum rotation rectangle of the bounding point set. To ensure corrective effect, the extension of each side of the rectangle allows intersection with each border. Use InputArray points to get the corresponding mask coordinates of MASK. After the function is processed, adjust the process parameters to get the position of the handle plate vertices.

Divide the rectangle enclosed by the four points and import it into the folder to get the following results in Fig. 8. It can be seen that although the edges of the first and last characters of a very small license plate are still slightly divided, they retain their character characteristics and do not adversely affect subsequent character recognition.

**Fig. 8.** Output after correction

### 3.1.7 Perspective Transformation

When encountering a license plate with a large inclination angle, if the character recognition processing is directly performed, the success rate will be greatly reduced, and since the characters are distorted, the characteristics learned from the model will be deviated to some extent. Therefore, before the identification step, the relevant transformation processing is generally performed.

Perspective transformation is the mapping of existing image projections to new observation planes, which is often used in pattern recognition. This paper uses the cv2. warpPerspective function to estimate the best matrix. Then use random sampling to calculate to improve accuracy. Then input the image and transform it through the perspective transformation matrix to output the opposite positive license plate image [26]. The perspective effect is shown in Fig. 9.

**Fig. 9.** Perspective transformation correction effect

For some heavily inclined license plates, the algorithm can also play a good correction effect. The correction effects of the some samples are listed below (Fig. 10).

**Fig. 10.** Tilting license plate correction effect

For the case where the license plate is inclined and slightly curved, it also has excellent correction ability. The partial recognition effect is shown in the figure below (Fig. 11).

**Fig. 11.** Tilt and slightly curved license plate correction effect comparison

## 3.2  Design of Identification Scheme Without Segmentation

### 3.2.1  Design Purpose

At present, most commercial license plate recognition software still uses the technical route of first segmentation and recognition, so there are certain requirements for the conditions and environment for collecting license plate images, otherwise it cannot be identified. However, with the continuous maturity of end-to-end technology in the past two years, the idea of speech recognition can also be borrowed from the license plate recognition. The traditional binarization and vertical projection, and then the method of character recognition have gradually become obsolete. In the foreseeable future, applying the sequence recognition result of speech recognition to license plate recognition will greatly improve the accuracy of the algorithm while pursuing the improvement of recognition accuracy, which will be a very hot research direction.

Single character recognition does not require learning semantics, because it has no context to understand. When it comes to the recognition of sentences, the scheme of splitting and then performing single character recognition is very extravagant. Therefore, in the case of recognizing the orderly meaningful character string of the license plate, the LSTM which can extract the sequence features can be added after the convolutional neural network according to the good generalization performance of the RNN model, and then the corresponding characters are correspondingly corresponding. The loss calculation method for learning is replaced with the corresponding CTC calculation method, and the construction of a new model suitable for license plate recognition is completed.

### 3.2.2  License Plate Recognition Model Based on CNN+LSTM+CTC

Learn from the idea of speech recognition, and use the CNN+ structure that is most popular for text recognition indefinite length. This paper adopts the CNN+LSTM+CTC framework proposed by domestic scholars recently. Compared with the CNN+Seq2Seq +Attention framework, its model is simpler and more controllable. It is more suitable for student computers after further simplification. Two model structures are shown in Fig. 12.

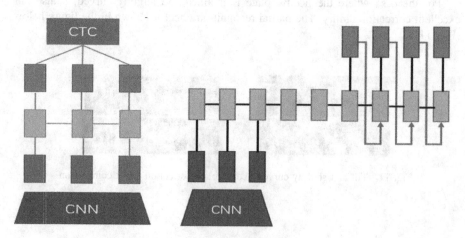

**Fig. 12.** CNN+LSTM+CTC frame and CNN+Seq2Seq+Attention frame

The CNN+LSTM+CTC framework can be applied to license plate recognition. Although some structural fine-tuning and parameter modification are needed, the overall idea is basically the same. The specific procedure is as follows:

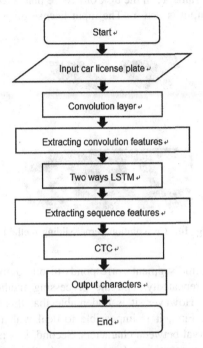

**Fig. 13.** License plate end-to-end identification process

As shown in Fig. 13, the CNN+LSTM+CTC network model is divided into three parts. The three-dimensional image displayed by the input size is height, width, and number of channels, and the number of channels is generally three channels of RGB. In the case of text recognition, the convolution layer extracts the features of the input image, changes the three-channel text size of 100 * 32 into a feature map of 512 * 25, and then inputs the LSTM to extract the sequence features. The license plate recognition needs to be based on the aspect ratio. Adjust the input size. In the speech recognition model, the commonly used LSTM is a one-way mode that extracts vectors from left to right or from right to left. However, when applied to license plate recognition, a two-way LSTM is used. Because both sides of the license plate are useful, they cannot be undirectionally transmitted in a specific time series like speech. In the transcription layer, the path with the highest probability is output. The result is then classified and decoded, and converted to the license plate number [30].

In the models widely used in traditional, the size of the input image needs to be fixed, but in the license plate recognition task, the parallel moving convolution kernel is used, so that it can be read for any length, and the license plate skew and distortion can be well solved. Figure 14 shows the convolution kernel sliding mode of the feature sequence to the corresponding license plate area.

The CTC algorithm is applied to license plate recognition. The specific strategy used is as follows: For input sequences and output sequences, CTC does not require minimum unit alignment, but the final alignment between X and Y is required. Multiple Xs can correspond to the same Y. In the task of license plate recognition, X is a license plate content, and the output is text Y. The input license plate is divided into several

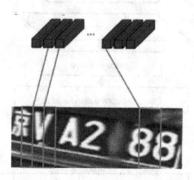

**Fig. 14.** Convolution kernel sliding method

time segments, and one time segment corresponds to one output. In the process, the same character appears repeatedly, and the processing method is to swallow the adjacent repeated words. However, it is undeniable that this technical solution has certain technical defects: First, it is impossible to deal with the situation of pause, corresponding to the interval between characters. Second, it is more difficult to handle the processing of character repetition. For example, in the case of "* GZZ100", the output is "* GZ10", and there is a problem in the actual application.

The way to solve the above problem is to introduce meaningless blanks. Referring to the pause processing operation in speech recognition, in the license plate recognition, the interval between characters can emulate the pause in the speech, which is represented here as #. Similarly, the CTC alignment operation in character recognition, in addition to merging adjacent repeating characters, also includes removing the introduced blank segments, as shown in Fig. 15.

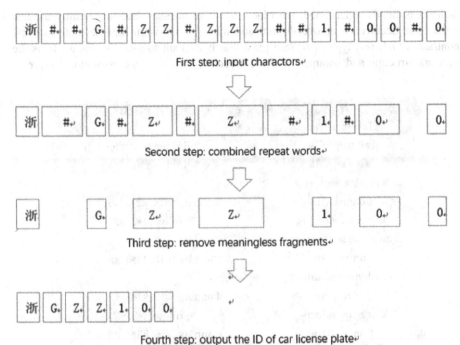

First step: input charactors

Second step: combined repeat words

Third step: remove meaningless fragments

Fourth step: output the ID of car license plate

**Fig. 15.** CTC alignment strategy used in this paper

There is a distinct characteristic between input and output: X and Y are different in length, and in general, the length of Y is less than the length of X; if X moves several time segments, Y also moves forward several time segments or remains in place. Do not move; multiple Xs can correspond to the same Y. This strategy can be applied to license plate recognition tasks well and achieve good learning results.

### 3.2.3 Improving Model Configuration
This section uses parametric configuration and is improved on the basis of the original model to suit the existing license plate recognition scheme. The first is to choose the appropriate convolution network. If the convolution module adopts an overly simple CNN model, it can achieve convergence within one day. But it also caused the corresponding problem. The reason is that the license plate number is equivalent to a person's ID number, belongs to the citizen's personal privacy, and cannot collect a large amount of data on the network, so the data set collected in this experiment is very limited. In this case, if the collected license plate picture features are too different, it is extremely easy to cause over-fitting. Subsequently, an attempt was made to perform the DenseNet model, and the effect was found to be very unstable. At some point, training 450 steps can achieve a correct recognition rate of 50%, but in some cases, training 3000 steps does not converge. Because the stability is too poor, the structure is eventually eliminated. After repeated attempts, it was decided to make improvements based on VGG-Very Deep CNN [30] to solve the current problem. When the training

data set is finally determined, the blue and white license plates are mainly used to reduce the occurrence of over-fitting, and the blue-and-white license plate has low contrast, which is a type of license plate that is difficult to identify. Figure 16 is the network structure and parameters of the model containing this convolutional layer.

| Types | Configuration |
|---|---|
| Transcription | |
| Two way LSTM | # Hidden unit: 256 |
| Two way LSTM | # Hidden unit: 256 |
| Mapping sequence | |
| Convolution | Mapping: 512, k2*2, s:1, p:0 |
| Maximum pooling | Window: 1*2, s2 |
| Batch normalization | |
| Convolution | Mapping: 512, k3*3, s:1, p:1 |
| Batch normalization | |
| Convolution | Mapping: 512, k3*3, s:1, p:1 |
| Maximum pooling | Window: 1*2, s2 |
| Convolution | Mapping: 256, k3*3, s:1, p:1 |
| Convolution | Mapping: 256, k3*3, s:1, p:1 |
| Maximum pooling | Window: 2*2, s2 |
| Convolution | Mapping: 128, k3*3, s:1, p:1 |
| Maximum pooling | Window: 1*2, s2 |
| Convolution | Mapping: 64, k3*3, s:1, p:1 |
| Input | W *32 image |

**Fig. 16.** Model hierarchy and parameters

By adopting a batch normalization operation and adding a batch operation between the convolution layer and the activation function, an effect similar to adding a Dropout can be obtained. The process of training has been greatly accelerated, and the over-fitting condition has been alleviated to some extent. At the same time, the fourth convolutional layer is deleted as appropriate, leaving only six convolutional layers; the pooling operation is added after the original fifth convolutional layer to further reduce the amount of data.

In terms of training time, because text recognition tasks are almost all Chinese characters, the convergence is slow. However, in the license plate recognition, the convergence speed of the numbers is much faster, which is also an unexpected surprise. In training, considering that the picture of the license plate is a long and narrow shape, the conventional convolution mode may not extract the features in the figure well. Therefore, the rectangular pooled sliding window is used instead of the traditional square, and the pooled window is set according to a ratio of 2:1, and the extraction feature is more efficient.

The difference between LSTM and other networks is that its output can be indefinitely long. If there are too many crops and Chinese is cropped, only six characters are output, one less. In theory, it is also possible to input a label of indefinite length [31]. In the example of license plate recognition, since the license plates are all seven characters, the training label labels are all seven. The label length is set in the label_len in the TextImageGenerator. All the data in this chapter is generated in the Class TextImageGenerator.

### 3.2.4  Data Preprocessing and Training Process

Use the cv2.resize function to resize the positioning image. The uniform aspect ratio needs to be considered when dimensioning. The license plate provided by the National Road Traffic Authority is 9 cm long and 4.5 cm wide. In most cases, there are 7 characters in the license plate, and the gap between the character and the character is counted as approximately 8 characters wide, which is about four times the length of a single character. Therefore, the normalized ratio is set. For an aspect ratio of 4:1. Setting the scale can eliminate the deformation of the characters caused by the elongation and widening of the license plate, and improve the accuracy of the training. On the other hand, it is possible to reduce the size as much as possible while reducing the parameter amount and training time while ensuring the aspect ratio.

Use the training set to train the model, use the test set to verify the accuracy of character recognition, and calculate the error. If the error persists for a certain period of time, the training process is terminated to prevent overfitting. The default learning rate set is 1, and convergence can be achieved in the experiment, but an unstable state occurs in the later stage, so the learning rate is gradually reduced, and the effect is good after reaching 0.1. As shown in Fig. 17, the loss map is drawn on the tensorboard. It can be seen that after 6000 steps, the loss rate changes very little and the training can be terminated.

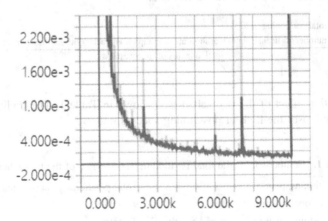

**Fig. 17.** Screenshot of the loss rate on Tensorboard

# 4  Experiment

## 4.1  Effect Analysis of Two Positioning Schemes

The license plate location algorithm designed in this paper is better than the traditional algorithm, and the steps are simpler. Directly using the trained MASK RCNN model, the success rate of coarse positioning is quite good, close to 90%. After positioning the license plate, use the minAreaRect function to calculate the minimum enclosing rectangle to get the four vertices of the rectangle. A perspective transformation is then performed to obtain a final locating license plate image.

In the experiment, the over-expanding scheme is also adopted, that is, the binarization is performed after the expansion, and after the binarization, the character contour is searched, the upper and lower sides are fitted, and then the horizontal projection is horizontally, and the Sobel operator in the conventional scheme is not used. Find the character outline using the findcontours function, and the fit is the fitline function. The effect is slightly worse than using the minimum bounding rectangle.

Under the traditional morphological positioning method, there is a case where the tilt angle is large and the positioning cannot be accurately performed. Even if several target frames are located, the interference cannot be well overcome in the case where some of the interference items and the license plate are extremely similar. For example, if there is a board at the upper end of the windshield of the vehicle, and the circumference of the board is very similar to the surrounding of the license plate, it is difficult to accurately eliminate such a situation. In this case, the program has excellent performance.

According to the principle of 4:1 allocation of test set and training set, 1065 pictures collected were divided to obtain 213 pictures of test set. After testing and evaluating 213 images in various scenes, the number of successfully positioned images reached 212, with a success rate of 99.53%. The total positioning success rate is calculated according to the following formula:

$$
\begin{aligned}
&\text{success rate of total positioning} \\
&= \frac{\text{The number of license plates that show all the complete characters and are basically correct}}{\text{total quantity of license plate}} \times 100\%
\end{aligned}
$$

(2)

The solution effect of the two schemes is shown in Table 1. The traditional positioning scheme uses the highest positioning success rate.

**Table 1.** Comparison of the total positioning success rate of the two schemes

| Positioning plan | Total positioning success rate |
|---|---|
| Traditional positioning plan | 98.95% |
| Positioning plan based on MASK RCNN | 99.53% |

## 4.2    Analysis of the Effect of the Identification Scheme

Compared with the newly proposed solution, the scheme of first segmentation and recognition needs to solve the correctness and stability of the two steps. If the segmentation state is a little unstable, it will have a very large impact on character recognition. Unfortunately, from this experiment, the license plate segmentation error is quite large when dealing with license plates with very large tilt angles, and the adjusted threshold-related parameters are not well adapted to most situations. Once a mismatch occurs, the division will result in a large deviation, and the identification is almost impossible.

In the end-to-end identification system designed in this paper, because there is no dragging of the split module, the recognition success rate of the system is greatly improved. In the detection of a total of 212 license plates successfully located in the previous chapter, only two cases of error were obtained. The specific recognition results are shown in Fig. 18. The first behavior recognition result of each small segment, the second behavior license plate actual number, and the third line are the total number of currently recognized license plates, the total number of currently correctly recognized license plates, the recognition accuracy rate, and finally the individual. The row shows the recognition accuracy of the total test set.

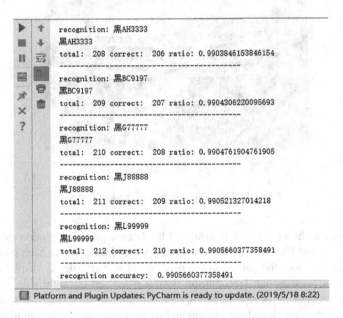

**Fig. 18.**    Identification accuracy display

As shown in the figure above, the recognition accuracy rate reached 99.06%. The recognition accuracy detection in this chapter is only for the end-to-end character recognition phase, and is tested on the basis of all successfully positioned images. In the next section, the overall recognition accuracy test for the entire system will be performed. The calculation formula for the recognition accuracy used in this paper is as follows:

$$\text{Recognition accuracy} = \frac{\text{Identify the correct number of license plates for all characters}}{\text{Total number of license plates}} \times 100\%$$

(3)

## 4.3 Overall Accuracy Test

The first part of the MASK RCNN-based positioning module is connected with the second part of the identification module that does not need to be segmented to obtain the final license plate recognition system. In order to facilitate the testing and application of a single license plate recognition, a simple GUI interface was designed.

**Fig. 19.** The GUI interface after identifying

Figure above shows the GUI interface of the license plate recognition system. The function of the "Get Picture" button on the left is to read the picture. After pressing, the image is selected from the path, and the read image is conveniently displayed on the interface. After pressing the "recognition license plate" on the right side, wait a few seconds, the recognized license plate number will appear in the middle white rectangular box, which is quick and convenient and easy to operate. The recognition result is shown in Fig. 19.

Through the test, 213 license plates were identified, and a total of three cases showed recognition errors. The experimental screenshot is shown in Fig. 20.

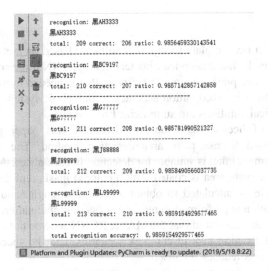

```
recognition: 黑AH3333
黑AH3333
total:  209 correct:  206 ratio: 0.9856459330143541
-------------------------------------------------
recognition: 黑BC9197
黑BC9197
total:  210 correct:  207 ratio: 0.9857142857142858
-------------------------------------------------
recognition: 黑G77777
黑G77777
total:  211 correct:  208 ratio: 0.985781990521327
-------------------------------------------------
recognition: 黑J88888
黑J99999
total:  212 correct:  209 ratio: 0.9858490566037735
-------------------------------------------------
recognition: 黑L99999
黑L99999
total:  213 correct:  210 ratio: 0.9859154929577465
-------------------------------------------------
total recognition accuracy:  0.9859154929577465
```

Platform and Plugin Updates: PyCharm is ready to update. (2019/5/18 8:22)

**Fig. 20.** Total recognition accuracy display

In Fig. 20, the first behavior recognition result of each small segment, the second behavior actual number of the license plate, and the third row are respectively the total number of license plates currently recognized, the total number of license plates currently correctly recognized, the total recognition accuracy rate, and the last single line. It shows that the total recognition accuracy of the calculated test set is 98.59%.

Table 2 compares the total recognition accuracy of the license plate recognition system of the two schemes, and compares the highest recognition rate of the disclosed conventional technical scheme with the scheme recognition rate of this paper. The new system's total recognition accuracy in this table is different from the character end-to-end recognition accuracy test in the previous chapter, including the MASK RCNN-based license plate location and character end-to-end recognition, which is the accuracy of the entire system. The calculation formula for the recognition accuracy used in this chapter is as follows:

$$\text{Total recognition accuracy} = \frac{\text{Identify the correct number of license plates for all characters}}{\text{Total number of license plates}} \times 100\%$$

(3)

**Table 2.** Comparison of total identification accuracy

|  | Total recognition accuracy |
|---|---|
| Traditional license plate recognition system | 93.74% |
| The license plate recognition system designed in this paper | 98.59% |

# 5   Conclusion

This paper designs a license plate recognition software based on convolutional neural network. Relying on the latest deep learning technology, CNN is the cornerstone, and it is applied to the license plate recognition software that is needed in modern smart city traffic. The comprehensive recognition accuracy of the whole system reaches 98.59%. The specific technical solutions are summarized as follows:

In the aspect of license plate location detection, this paper proposes a new MASK RCNN based license plate area detection method. The tensorflow-based framework trains a model that is suitable for detecting license plates. After the model is detected, the license plate area is coated with a colored mask, and then the minimum enveloping rectangle is calculated to obtain a correction angle, and finally the perspective transformation is performed to reduce the bending inclination. Compared with the traditional method, it is found that the MASK RCNN scheme has a large advantage in the license plate location task in complex scenarios, and the success rate of positioning in 213 test sets reaches 99.53%.

In the research of license plate character recognition, it is found that when the characters are stuck and the license plate is deformed, the segmentation will generate a great error, which will have a fatal effect on the subsequent character recognition. In this paper, the segmentation steps are rounded off, and the end-to-end identification scheme is adopted directly, and the complexity of the algorithm is greatly reduced. Firstly, CNN is used to extract features from finely positioned license plates, and then learn sequence features using two-way long-term and short-term memory networks. Finally, the output of the LSTM is processed using the CTC to obtain a corresponding string, and the identified license plate number is finally obtained. In the second phase of the program, the end-to-end recognition success rate of the character has been significantly improved, reaching 99.06%. In this article, the license plate in the picture showing the details of the vehicle is coded in the third last place, and the privacy of the individual is firmly protected.

# References

1. Zhu, K., Hao, Q., Li, S., Hu, C.: A review of license plate recognition. Mod. Inf. Technol. **2**(05), 4–6 (2018)
2. Dong, J., Zheng, B., Yang, Z.: License plate character recognition based on convolutional neural network. Comput. Appl. **37**(07), 2014–2018 (2017)
3. Azad, R., Azad, B., Shayegh, H.R.: Real-time and efficient method for accuracy enhancement of edge based license plate recognition system. arXiv preprint arXiv:1407.6498 (2014)
4. Yang, Z.: Research on improvement of key algorithms in license plate recognition. Guangxi Normal University (2016)
5. Chen, Y.: Research on license plate recognition system for intelligent parking lot. Zhejiang University of Technology (2015)
6. Xie, J.: Research and application of license plate recognition algorithm based on improved LM-BP neural network. Guangdong University of Technology (2016)

7. Liu, Y.: Design and implementation of license plate intelligent recognition system based on convolutional neural network. Suzhou University (2016)
8. Zhang, Y.: Research and implementation of key algorithms related to license plate recognition. Northern University for Nationalities (2017)
9. Dong, C.: Research on multiple license plate recognition technology based on video image. Chang'an University (2015)
10. Li, X., Ye, M., Li, T.: Review of target detection based on convolutional neural networks. Comput. Appl. Res. **34**(10), 2881–2886+2891 (2017)
11. Matsukawa, T., Suzuki, E.: Person re-identification using CNN features learned from combination of attributes. In: 2016 23rd International Conference on Pattern Recognition (ICPR), pp. 2428–2433. IEEE (2016)
12. Chen, C., Liu, M.-Y., Tuzel, O., Xiao, J.: R-CNN for small object detection. In: Lai, S.-H., Lepetit, V., Nishino, K., Sato, Y. (eds.) ACCV 2016. LNCS, vol. 10115, pp. 214–230. Springer, Cham (2017). https://doi.org/10.1007/978-3-319-54193-8_14
13. Girshick, R.: Fast R-CNN. In: Proceedings of the IEEE International Conference on Computer Vision, pp. 1440–1448 (2015)
14. Zhou, J., Zhao, Y.: Overview of application of convolutional neural networks in image classification and target detection. Comput. Eng. Appl. **53**(13), 34–41 (2017)
15. Zhou, J.: Research on image target detection method based on convolutional neural network. Chongqing University (2017)
16. Huang, T.: Research on vehicle type identification based on convolutional neural network. North China Electric Power University (2017)
17. Liu, Z.: Research and implementation of license plate recognition algorithm based on improved convolutional neural network. Zhejiang Sci-Tech University (2018)
18. Duan, M.: Research on image recognition method based on convolutional neural network. Zhengzhou University (2017)
19. Huachun, L.: Application of convolutional neural network in license plate recognition. Comput. Technol. Dev. **29**(04), 128–132 (2019)
20. Li, C.: Research on key technology of license plate recognition based on deep learning. University of Electronic Science and Technology (2018)
21. He, K., Gkioxari, G., Dollár, P., et al.: MASK R-CNN. In: Proceedings of the IEEE International Conference on Computer Vision, pp. 2961–2969 (2017)
22. Wu, Z.: Application of convolutional neural networks in image classification. University of Electronic Science and Technology (2015)
23. Wang, J., Bacic, B., Yan, W.Q.: An effective method for plate number recognition. Multimed. Tools Appl. **77**(2), 1679–1692 (2018)
24. Hua, K.: Design and implementation of license plate recognition system. Suzhou University (2015)
25. Zhu, W.: Algorithm Research and implementation of vehicle license plate recognition system. University of Electronic Science and Technology (2014)
26. Hanyang, L., Yongzhao, Z., Yuzhong, C.: Rapid detection and identification of motor vehicle driving license in complex scenes. Mini-Microcomput. Syst. **05**, 1076–1082 (2019)
27. Zhang, Z., Yang, W., Yuan, T., Li, D., Wang, X.: Traffic accident prediction based on LSTM neural network model [J/OL]. Computer Engineering and Applications, pp. 1–8, 21 May 2019
28. Kim, S., Hori, T., Watanabe, S.: Joint CTC-attention based end-to-end speech recognition using multi-task learning. In: 2017 IEEE International Conference on Acoustics, Speech and Signal Processing (ICASSP), pp. 4835–4839. IEEE (2017)

29. Hori, T., Watanabe, S., Zhang, Y., et al.: Advances in joint CTC-attention based end-to-end speech recognition with a deep CNN encoder and RNN-LM. arXiv preprint arXiv:1706.02737 (2017)

30. Shi, B., Bai, X., Yao, C.: An end-to-end trainable neural network for image-based sequence recognition and its application to scene text recognition. IEEE Trans. Pattern Anal. Mach. Intell. **39**(11), 2298–2304 (2016)

31. Li, H., Wang, P., You, M., et al.: Reading car license plates using deep neural networks. Image Vis. Comput. **72**, 14–23 (2018)

32. Wu, Z.: Application of convolutional neural networks in image classification. University of Electronic Science and Technology

# Author Index

Printed in the United States
By Bookmasters